STRUCTURE REPORTS

for 1986

Volume 53A

Structure Reports is prepared under the guidance of a Commission of the International Union of Crystallography. The members of the Commission sometime concerned with the preparation of this volume are listed below.

STRUCTURE REPORTS

for 1986

Volume 53A

METALS AND INORGANIC SECTIONS

General editor

G. Ferguson

Section editor

J. Trotter

Published for the

INTERNATIONAL UNION OF CRYSTALLOGRAPHY

by

KLUWER ACADEMIC PUBLISHERS

DORDRECHT / BOSTON / LONDON

First published in 1988

ISSN 0166–6983

ISBN 90–277–2708–2

Printed in the Netherlands

TABLE OF CONTENTS

INTRODUCTION

The present volume continues the aim of Structure Reports to present accounts of all crystallographic structure determinations, but with some changes from previous volumes. These stem from recent actions of the Executive Committee of the International Union of Crystallography, which decided, after much discussion and with regret, to discontinue the publication of the Organic volumes of Structure Reports after the 1985 Volume 52B is published. It was also decided to continue with the Metals and Inorganic volumes, but in a modified format, and to explore ways of collaborating in the production of these volumes with the Inorganic and Metals Crystallographic Data Bases. Since the latter collaboration has not yet been organized, the present volume is produced in the normal format, but with some modifications to achieve some economy in effort and cost.

The modifications in the present volume are mainly of a minor nature, and probably not immediately obvious to the reader. The principal changes are:
(i) The crystal system is not given specifically, but it is implied by the space group symbol.
(ii) The unit cell parameters are not named specifically (i.e., **a** = etc.), but the values are listed in the usual order: a, b, c (in Å), α, β, γ (in °) for triclinic; a, b, c, β for monoclinic etc. Possible ambiguities resulting from non-standard space group settings, rhombohedral/hexagonal settings etc. are given in full.
(iii) Cross-references to previous Structure Reports volumes are incorporated in the text (with volume and page numbers), rather than collected at the end of a report.
(iv) Figure legends are not given, but figures are situated usually exactly below the text of the report.
(v) In the Metals section, formulae have been used as report headings (rather than element names), and the reports are arranged alphabetically on that basis (mineral names are given where appropriate). In addition, the Metals reports are classified as binary alloys, borides, sulphides etc. (see pages V and 2).
(vi) Authors' names are not given. This should not inconvenience the reader greatly, but we apologize to the authors for the lack of acknowledgement that this omission causes. It was the change which made the greatest economies in effort, space, and cost.

These minor changes result in a 23% reduction in the size of this volume relative to the previous volume, in spite of a 5% increase in the number of reports.

University of Guelph
Guelph, Ontario, Canada

G. FERGUSON

15 December 1987

STRUCTURE REPORTS

SECTION I

METALS

Edited by

J. Trotter

(University of British Columbia)

ARRANGEMENT

The metals reports in this volume are arranged under the classifications: binary alloys, ternary alloys (within each of these classifications, the entries are sorted alphabetically by formula), hydrides, borides, carbides, silicides, pnictides (N, P, As), chalcogenides (S, Se, Te) (within each of these classifications, the entries are sorted alphabetically by formula on the elements other than that of the classification). See **47**A, VII for further details. To find particular substances the metals formula index may be used.

Al_3Ti

Acta Chem. Scand., **A40**, 157-159.

I4/mmm, 3.8537, 8.5839, Z = 2, powder data. Al(1) in 2(b); Al(2) in 4(d); Ti in 2(a). DO_{22}-type (**7**, 13, 100).

$AuCa_3$, Au_2Ca_5, Au_2Sr_3, Au_2Ba_3

Rev. Chim. Minér., **22**, 791-798.

$AuCa_3$, Pnma, 7.795, 10.000, 6.780, Z = 4. Fe_3C-type (**2**, 33; **29**, 34).

Au_2Ca_5, C2/c, 16.857, 6.809, 7.746, 97.26, Z = 4. Mn_5C_2-type (**26**, 76; **31**A, 31).

Au_2Sr_3, Au_2Ba_3, R$\bar{3}$, a = 9.604, 10.001, c = 18.36, 19.143 A, Z = 9. Er_3Ni_2-type (**40**A, 64).

Ca_3Ga_5

Z. anorg. Chem., **539**, 33-38.

Cmcm, 4.463, 10.799, 15.025, Z = 4. Ga(1) in 4(c): 0,y,1/4, y = 0.8367; Ga(2), Ga(3), Ca(1) in 8(f): 0,y,z, y = 0.2921, 0.1154, 0.5881, z = 0.5995, 0.4659, 0.8895; Ca(2) in 4(c): y = 0.1301. Ga framework in which Ga(1) has 2 Ga neighbours and 2 electron pairs, and Ga(2) and Ga(3) each have 4 Ga neighbours and 1 electron pair.

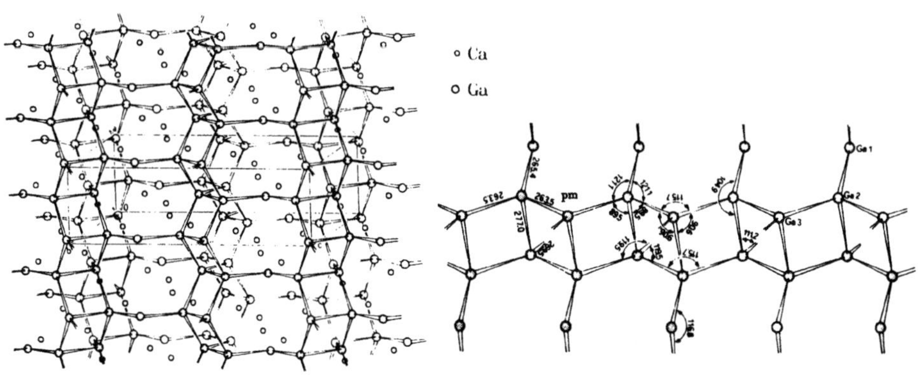

$Ca_{28}Ga_{11}$

Acta Cryst., C**42**, 394-396.

Imm2, 5.324, 61.445, 7.488, Z = 2, R = 0.051.

The $Ca_{28}Ga_{11}$ structure is characterized by three types of calcium polyhedra surrounding the gallium atoms: tricapped trigonal prisms, distorted cubes and distorted cubicosahedra (ten-vertex polyhedra similar to those found around Si in Co_2Si). The linkage of these polyhedra gives rise to columns or slabs which alternate along the greatest axis of the cell. No Ga–Ga contacts are present.

		x	*y*	*z*
Ca(1)	4(*d*)	0	0·3626	0·535
Ca(2)	4(*d*)	0	0·4564	0·126
Ca(3)	4(*d*)	0	0·1825	0·094
Ca(4)	4(*d*)	0	0·2759	0·528
Ca(5)	4(*d*)	0	0·0829	0·611
Ca(6)	4(*d*)	0	0·1467	0·716
Ca(7)	4(*d*)	0	0·4401	0·680
Ca(8)	4(*d*)	0	0·3832	0·986
Ca(9)	4(*d*)	0	0·2620	0·0*
Ca(10)	4(*d*)	0	0·0300	0·958
Ca(11)	4(*d*)	0	0·2058	0·676
Ca(12)	4(*d*)	0	0·0966	0·083
Ca(13)	4(*d*)	0	0·3240	0·921
Ca(14)	2(*b*)	0	0·5	0·746
Ca(15)	2(*a*)	0	0	0·550
Ga(1)	4(*d*)	0	0·1395 (1)	0·307
Ga(2)	4(*d*)	0	0·4092 (2)	0·338
Ga(3)	4(*d*)	0	0·2271 (1)	0·293
Ga(4)	4(*d*)	0	0·0440 (1)	0·358
Ga(5)	4(*d*)	0	0·3196 (1)	0·323
Ga(6)	2(*b*)	0	0·5	0·332

* Arbitrarily fixed.

Interatomic distances (Å) *in* $Ca_{28}Ga_{11}$ *involving the Ga atoms*

Ga(1)–	Ca(3)	3·09 (2)	Ga(2)–	Ca(8)	3·08 (2)
	– Ca(6)	3·09 (2)		– Ca(7)	3·19 (2)
	– Ca(12)	3·12 (2)		–2Ca(5)	3·20 (1)
	–2Ca(8)	3·29 (1)*		– Ca(1)	3·22 (2)
	–2Ca(1)	3·35 (1)*		–2Ca(12)	3·25 (1)
	–2Ca(13)	3·58 (1)*		– Ca(2)	3·31 (2)
Ga(3)–	Ca(9)	3·07 (1)	Ga(4)–	Ca(5)	3·05 (2)
	– Ca(3)	3·12 (2)		– Ca(15)	3·06 (1)
	–2Ca(9)	3·15 (1)		– Ca(10)	3·12 (2)
	– Ca(11)	3·15 (2)		–2Ca(7)	3·13 (1)
	–2Ca(4)	3·33 (1)		–2Ca(2)	3·33 (1)
	– Ca(4)	3·48 (2)		– Ca(12)	3·83 (2)
				–2Ca(14)	3·89 (1)
Ga(5)–	Ca(13)	3·02 (2)			
	– Ca(1)	3·08 (2)	Ga(6)–	2Ca(2)	3·09 (1)
	– Ca(4)	3·09 (2)		– Ca(14)	3·10 (2)
	–2Ca(11)	3·28 (1)*		–4Ca(10)	3·37 (1)*
	–2Ca(3)	3·35 (1)*		–2Ca(15)	3·40 (2)*
	–2Ca(6)	3·47 (1)*			

* Trigonal prism.

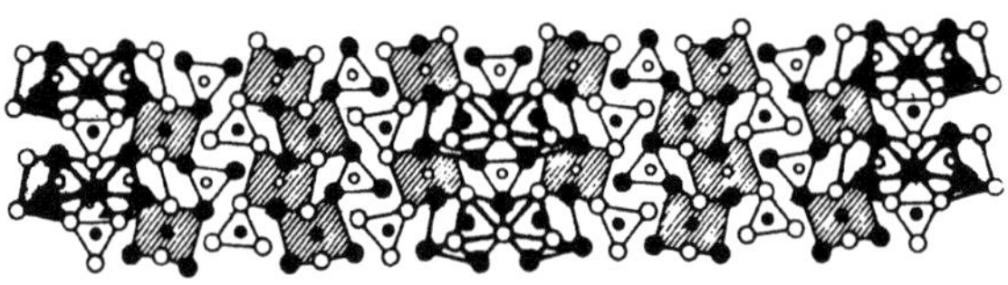

$CeCu_6$

Physica, B+C, **136**, 489-492; J. Phys. Soc. Japan, **55**, 454-456.

Above 230K, Pnma, a = 8.1009, 8.1088, b = 5.0978, 5.1004, c = 10.1548, 10.1621 A, at 250, 295K, Z = 4. Neutron powder data. Below 230K, $P2_1/c$, a = 5.084-5.095, b = 10.1279-10.1466, c = 8.0731-8.0931 A, β = 91.442-90.485°, at 10-200K, Z = 4, neutron powder data. Room-temperature structure as previously described (**24**, 99) Displacements in the low-temperature phase are mainly a shear.

$CoSb_2$

J. Solid State Chem., **63**, 23-30.

$P2_1/c$, a = 6.5051, 6.3833, 6.5410, 117.65, Z = 4, R = 0.054. Co: 0.2701,-0.0005,0.2817; Sb(1): 0.3466,0.3569,0.1675; Sb(2): 0.1489,0.6393,0.3671. Arsenopyrite-type structure (**26**, 35; **37**A, 67). Co-Sb = 2.508-2.603, Co-Co = 3.03, 3.73, Sb-Sb = 2.858 A.

Fe_6Sc_{29}

Dopov. Akad. Nauk Ukr.RSR, Ser. B: Geol., Khim. Biol. Nauki, No. 11, 28-31.

Pm3, 14.361, Z = 4. R = 0.067. Coordination numbers are Sc = 12-15, Fe= 11, 12.

Ga_2Gd_3

J. Less-Common Metals, **115**, 17-22.

I4/mcm, 11.666, 15.061, Z = 16, R = 0.072. New structure type, with 13-15 coordination for Gd, and 8-12 coordination for Ga atoms. Other Ga_2Ln_3 compounds are isostructural.

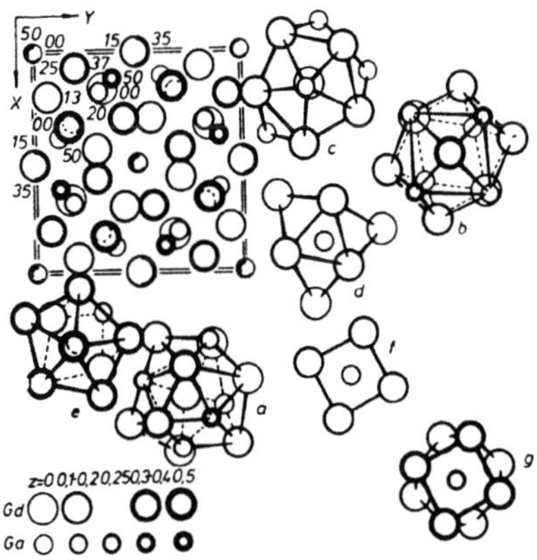

	x	y	z
Gd1	0,2029	0,0666	0,1356
Gd2	0,5	0	0,1510
Gd3	0,1688	−0,3312	0,5
Ga1	0,116	−0,384	0
Ga2	0,1777	−0,3223	0,2987
Ga3	0	0	0
Ga4	0	0	0,25

GaHf, Ga(V,Zr), $Ga_4V_2Zr_3$

Dopov. Akad. Nauk Ukr.RSR, Ser B: Geol., Khim. Biol. Nauki, No. 4, 43-47.

GaHf, Ga(V,Zr), Pbcm, a = 9.171, 9.250, b = 8.503, 8.577, c = 5.648, 5.638 A, Z = 4. Powder and single crystal data, R = 0.055, 0.075.

$Ga_4V_2Zr_3$, Pnma, 16.83, 5.474, 13.77, Z = 8. $Hf_3Cr_2Si_4$-type structure.

$GdMg_5$

Acta Cryst., C**42**, 138-141.

$F\bar{4}3m$, 22.344, Z ∿ 73, R = 0.048.

The final composition was determined to be $GdMg_{5.05}$. The structure, which is closely related to that of $Sm_{11}Cd_{45}$, consists of nested polyhedra units *A*, *B*, *C*, *D* with their centres at 0,0,0; $\frac{1}{4},\frac{1}{4},\frac{1}{4}$; $\frac{1}{2},\frac{1}{2},\frac{1}{2}$; and $\frac{3}{4},\frac{3}{4},\frac{3}{4}$, respectively. The units *A* and *D*, each containing 27 atoms, are typical of the body-centred cubic structure, while the units *B* and *C*, each containing 29 atoms, are found in the α-Mn structure. The *A* unit has two partially occupied Mg positions and another position filled two thirds by Gd and one third by Mg atoms. Cubes plus octahedra (CN 14) and Kasper polyhedra (CN 12, 14, 16) are the most common coordination polyhedra in this structure.

				x	y	z
A unit						
A C	4(*a*)		Gd	0	0	0
A IT	16(*e*)	70(5)%	Mg	0·0846	0·0846	0·0846
A OT	16(*e*)		Mg	0·9146	0·9146	0·9146
A OH	24(*f*)	86(4)%	Mg	0·1574	0	0
A CO	48(*h*)	{68(1)%	Gd	0·1741	0·1741	0·0111
		{32(1)%	Mg			
B unit						
B C	4(*c*)		Gd	$\frac{1}{4}$	$\frac{1}{4}$	$\frac{1}{4}$
B TT	48(*h*)		Mg	0·2959	0·2959	0·3902
B T	16(*e*)		Mg	0·1621	0·1621	0·1621
B CO	48(*h*)		Mg	0·4343	0·4343	0·2624
C unit						
C C	4(*b*)		Mg	$\frac{1}{2}$	$\frac{1}{2}$	$\frac{1}{2}$
C TT	48(*h*)		Mg	0·5457	0·5457	0·6416
C T	16(*e*)		Gd	0·4072	0·4072	0·4072
C CO	48(*h*)		Mg	0·6722	0·6722	0·5129
D unit						
D C	4(*d*)		Mg	$\frac{3}{4}$	$\frac{3}{4}$	$\frac{3}{4}$
D IT	16(*e*)		Mg	0·8300	0·8300	0·8300
D OT	16(*e*)		Gd	0·6598	0·6598	0·6598
D OH	24(*g*)		Mg	0·9066	$\frac{3}{4}$	$\frac{3}{4}$
D CO	48(*h*)		Mg	0·9151	0·9151	0·7664

Ir_3Zr_5

Acta Cryst., C**42**, 1101-1105.

$P6_122$, 7.9306, 17.010, Z = 6, R = 0.046. Deformation superstructure of the Mn_5Si_3 type (**4**, 24, 137, 246; **32**A, 102), with tripled c axis.

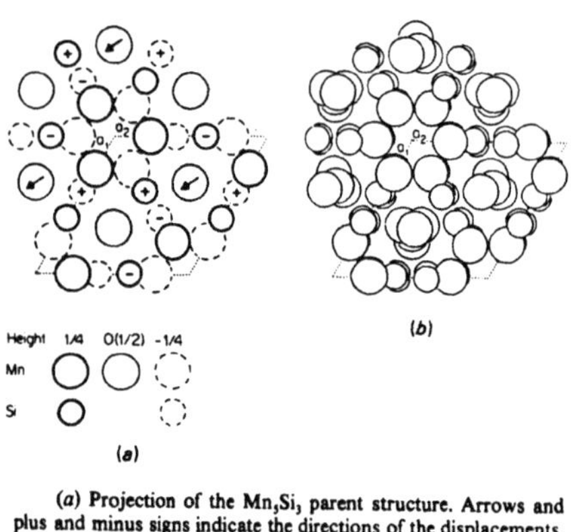

(*a*) Projection of the Mn_5Si_3 parent structure. Arrows and plus and minus signs indicate the directions of the displacements, observed in a layer of the Zr_5Ir_3-type superstructure between $z \simeq \frac{5}{6}$ and $z \simeq 1$, from the atom sites of the parent structure. (*b*) Projection along [001] of the real superstructure found with Zr_5Ir_3. Large circles indicate Zr atoms, smaller circles Ir atoms.

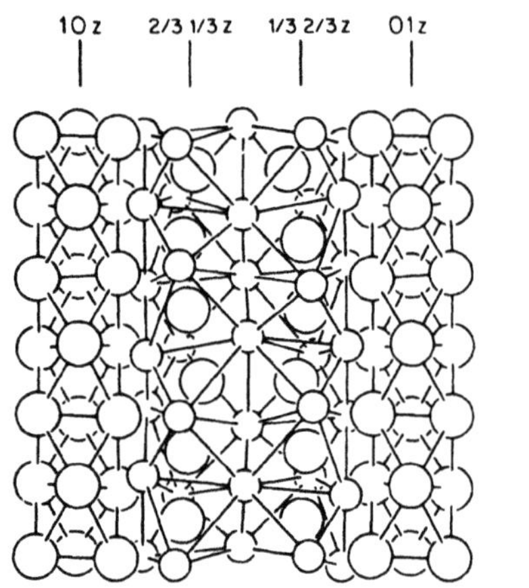

Partial projection of the Zr_5Ir_3 superstructure along [110] showing the columns of empty, face-sharing Zr octahedra along 10*z* and 01*z* and the columns of, equally face-sharing, Ir octahedra, centred by Zr(2*A*) and Zr(2*B*) atoms, along $\frac{2}{3}\frac{1}{3}z$ and $\frac{1}{3}\frac{2}{3}z$.

Zr_5Ir_3, superstructure
$P6_122$

		x	y	z
Zr(1*A*)	12(*c*)	0·2384	0·0062	0·3396
Zr(1*B*)	6(*a*)	0·2480	0	0
Zr(2*A*)	6(*b*)	0·3710	2*x*	$\frac{1}{4}$
Zr(2*B*)	6(*b*)	0·6557	2*x*	$\frac{1}{4}$
Ir(*A*)	12(*c*)	0·4137	0·0187	0·1444
Ir(*B*)	6(*a*)	0·6111	0	0

Ho_5Rh_3, Mn_5Si_3 type

$P6_122$		x	y	z	$P6_3/mcm$		x	y	z
Ho(1*A*)	12(*c*)	0·2457	0	$\frac{1}{4}$	Ho(1)	6(*g*)	0·2457	0	$\frac{1}{4}$
Ho(1*B*)	6(*a*)	0·2457	0	0					
Ho(2*A*)	6(*b*)	$\frac{1}{3}$	$\frac{2}{3}$	$\frac{1}{4}$	Ho(2)	4(*d*)	$\frac{1}{3}$	$\frac{2}{3}$	0
Ho(2*B*)	6(*b*)	$\frac{2}{3}$	$\frac{1}{3}$	$\frac{1}{4}$					
Rh(*A*)	12(*c*)	0·401	0	$\frac{1}{4}$	Rh	6(*g*)	0·599	0	$\frac{1}{4}$
Rh(*B*)	6(*a*)	0·599	0	0					

$Ni_{3+x}Sn_4$

Acta Chem. Scand., A**40**, 695-700.

C2/m, a = 12.199, 12.371, b = 4.0609, 4.0690, c = 5.2238, 5.2100 A, β = 105.17, 104.06°, for x = 0.08, 0.39, Z = 2, R = 0.026, 0.037. Ni(1) in 2(a): 0,0,0; Ni(2), Sn(1), Sn(2) in 4(i): x,0,z, x = 0.2126, 0.4263, 0.1748, z = 0.3389, 0.6927, 0.8179; x Ni(3) in 2(c): 0,0,1/2 (parameters for x = 0.39; slightly different values for x = 0.08). Closely related to CoGe and $Ni_{7+x}S_8$ types, with fragments of NiAs and $NiBi_3$-related types.

Ni_4Ti_3

J. Less-Common Metals, **125**, 157-166.

R$\bar{3}$, a = 6.704 A, α = 113.85°, Z = 2 (hexagonal cell, a = 11.24, c = 5.077 A, Z = 6), film data. Ti in 18(f): 5/7,1/7,0; Ni(1) in 18(f): 4/21,5/21,1/6; Ni(2) in 3(b): 0,0,1/2; Ni(3) in 3(a): 0,0,0.

Ni_5Zr, $(Ni,Cu)_5Zr$

J. Less-Common Metals, **124**, 21-27.

Ni_5Zr, F$\bar{4}$3m, 6.707, Z = 4, powder data. Zr in 4(a): 0,0,0; Ni(1) in 4(c): 1/4,1/4,1/4; Ni(2) in 16(e): x,x,x, x = 0.625. $AuBe_5$-type (**3**, 330; **22**, 48), as previously described (**21**, 158).

$(Ni,Cu)_5Zr$ has $AuBe_5$-type structure up to 30 atomic % Cu, and Cu_9Zr_2-type for phases richer in Cu.

PtU

J. Less-Common Metals, **120**, 113-122.

Cmcm, 3.703, 10.792, 4.391, Z = 4, neutron powder data. Atoms in 4(c): 0,y,1/4, y = 0.1380, 0.4111 for U, Pt. CrB-type structure (**12**, 30). Ferromagnetic structure at low temperature.

$Re_{25}Zr_{21}$

Acta Cryst., C**42**, 261-266.

R$\bar{3}$c, 25.847, 8.771, Z = 6, R = 0.068. Large segments of the hexagonal $MgZn_2$ (Laves) type structure. $Re_{25}Hf_{21}$ and $Mn_{25}Ti_{21}$ are isostructural.

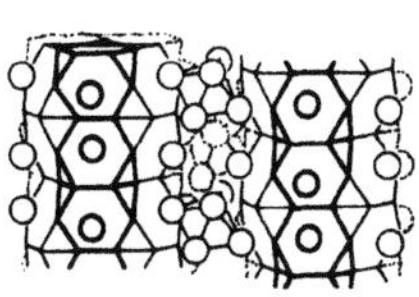

		x	y	z
Zr(1)	36(f)	0.0021	0.2306	0.0424
Re(1)	36(f)	0.0582	0.1778	0.2480
Zr(2)	36(f)	0.1161	0.1164	0.0659
Re(2)	36(f)	0.2342	0.1754	0.2477
Re(3)	36(f)	0.2350	0.1163	0.0000
Zr(3)	36(f)	0.3129	0.1084	0.2481
Re(4)	18(e)	0.0610	0	1/4
Re(5)	18(e)	0.4446	0	1/4
Zr(4)	18(e)	0.6943	0	1/4
Re(6)	6(b)	0	0	0

$AlCa_3Cd_{17}$

J. Less-Common Metals, **118**, 57-65.

R3, 22.134, 27.108 A, Z = 24, R = 0.061. Distorted variant of YCd_6 type (**37**A, 48), which is itself derived from the Ru_3Be_{17} type (**27**, 74). The structure contains Al_4 tetrahedra surrounded by a pentagonal dodecahedron of Cd, and a partial structure of Cd icosahedra. Coordination numbers vary from 9 to 16.

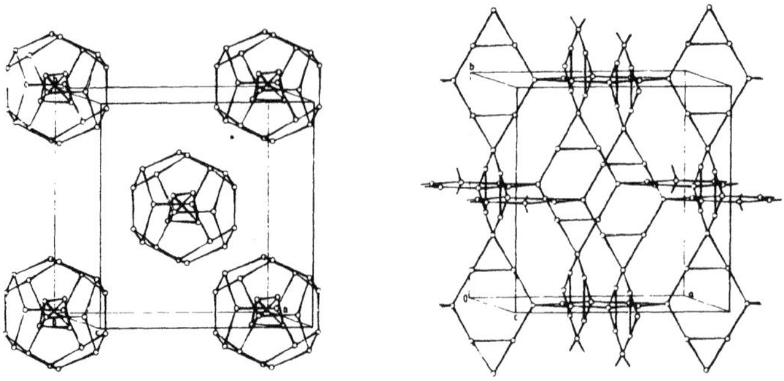

$Al_7Co_6Pr_7$

Kristallografija, **31**, 392-393 [Soviet Physics - Crystallography, **31**, 230-231].

P4/mbm, 13.784, 4.229, Z = 2, R = 0.089. A feature of the structure is a packing of columns of Co-centred trigonal prisms.

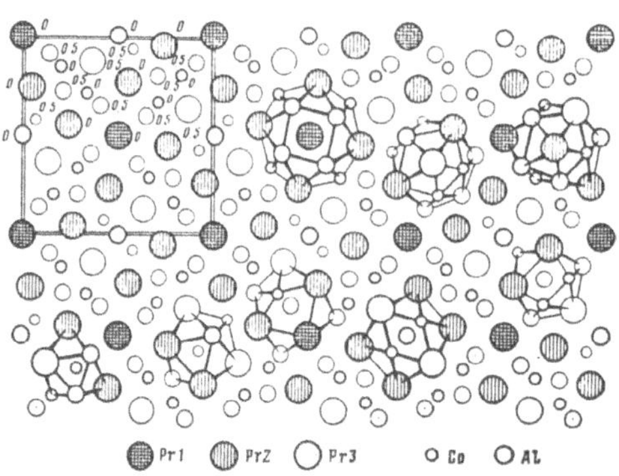

Projection of unit cell of structure of $Pr_7Co_6Al_7$ on XY plane and coordination polyhedra of atoms.

	x	y	z
Pr1	0	0	0
Pr2	0.2615	0.0467	0
Pr3	0.3684	0.8684	1/2
Co1	0.072	0.572	1/2
Co2	0.292	0.335	0
Al1	0.094	0.141	1/2
Al2	0.214	0.714	1/2
Al3	0	1/2	0

Interatomic Distances (Å) in Structure of $Pr_7Co_6Al_7$

Pr1–4Pr2	3.662(6)	Co2–Pr2	3.01(1)
–4Co2	3.65(1)	–Pr2	2.59(2)
–8Al1	3.15(3)	–2Pr3	3.10(1)
Pr2–Pr1	3.662(6)	–Co2	2.49(2)
–2Pr2	4.229	–2Al1	2.50(3)
–Pr2	3.739(9)	–2Al2	2.70(2)
–2Pr3	3.561(6)	Al1–2Pr1	3.15(3)
–2Co1	3.141(5)	–2Pr2	3.39(3)
–Co2	3.01(1)	–2Pr2	3.31(3)
–Co2	2.59(2)	–Pr3	3.17(5)
–2Al1	3.39(3)	–2Co2	2.50(3)
–2Al1	3.31(3)	–2Al1	3.30(6)
–2Al2	3.14(3)	Al2	2.84(4)
Al3	3.350(6)	Al2–4Pr2	3.14(3)
Pr3–4Pr2	3.561(6)	–Pr3	3.02(2)
–2Pr3	4.229	–Co1	2.76(3)
–2Co1	2.92(1)	–4Co2	2.70(2)
–4Co2	3.10(1)	–2Al1	2.84(4)
–2Al1	3.17(5)	Al3–4Pr2	3.350(6)
–Al2	3.02(2)	–4Pr3	3.324(4)
–2Al3	3.324(4)	–4Co1	2.536(6)
Co1–4Pr2	3.141(5)		
–2Pr3	2.92(1)		
–Al2	2.76(3)		
–2Al3	2.536(6)		

$CaCu_{0.1}Ga_{0.9}$, $SrAg_{0.9}Zn_{0.1}$

J. Less-Common Metals, **119**, 45-61.

$CaCu_{0.1}Ga_{0.9}$, Cmcm, 4.176, 11.436, 4.398, Z = 4, R = 0.026. Atoms in 4(c): 0,y,1/4, y = 0.1460, 0.4341 for Ca, Cu/Ga. CrB-type structure (**12**, 30). Other phases in this system have FeB, α-CaCu, β-CaCu, and TbNi-type structures.

$SrAg_{0.9}Zn_{0.1}$, $Pmn2_1$, 4.758, 24.86, 6.397, Z = 12, R = 0.069. Stacking variant of the FeB-type structure.

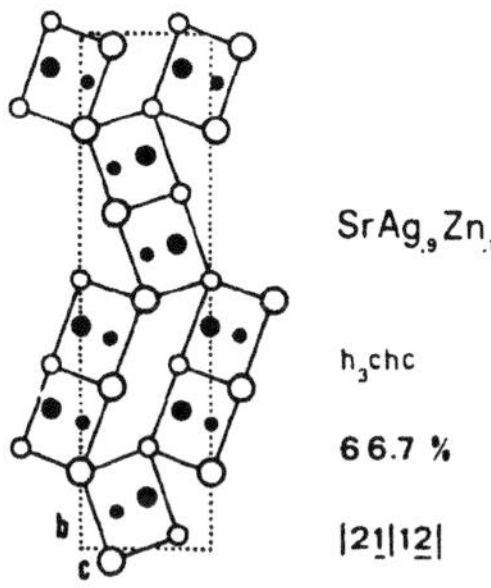

x = 0	y	z
Sr(1)	0.0234	0.750
Sr(2)	0.1901	0.518
Sr(3)	0.3542	0.756
Sr(4)	0.5235	0.995
Sr(5)	0.6923	0.750
Sr(6)	0.8575	0.514
Ag, Zn(1)	0.0687	0.263
Ag, Zn(2)	0.2405	0.005
Ag, Zn(3)	0.4084	0.243
Ag, Zn(4)	0.5692	0.504
Ag, Zn(5)	0.7370	0.259
Ag, Zn(6)	0.9067	0.007

$CeGa_{3.125}Pd_{0.875}$

J. Less-Common Metals, **118**, 335-342.

I4/mmm, 4.3092, 10.5154, Z = 2, R = 0.044. Ce in 2(a): 0,0,0; 3.55Ga + 0.45Pd in 4(d): 0,1/2,1/4; 2.49Ga + 1.51Pd in 4(e): 0,0,0.3791. $BaAl_4$-type (**3**, 45, 330).

$Co_{11}Ga_3La_6$

Kristallografija, **30**, 1077-1080 (1985) [Soviet Physics - Crystallography, **30**, 627-629 (1985)].

I4/mcm, 8.166, 23.132, Z = 4, R = 0.032. Stacking along **c** of fragments of the Ho_5Ga_3 and $U(Ni,Si)_{11}$ structures. Coordination numbers: La = 16, 17; Ga = 10; X = (0.5Ga + 0.5Co) = 12; Co = 12.

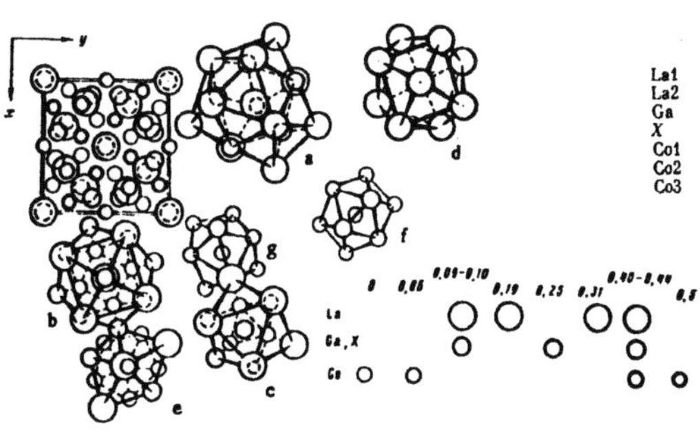

	x	y	z
La1	0.1691	0.6691	0.18595
La2	1/2	1/2	0.39777
Ga	0	0	1/4
X	0.3824	0.8824	0.0917
Co1	0.1767	0.6767	0.0557
Co2	0	1/2	0
Co3	0.0652	0.2103	0

$CoGa_5U$, Ga_5IrU

J. Less-Common Metals, **121**, 497-505.

P4/mmm, a = 4.2357, 4.317, c = 6.7278, 6.745 A, Z = 1, R = 0.027 and powder data. U in 1(a): 0,0,0; Co or Ir in 1(b): 0,0,1/2; Ga(1) in 1(c); 1/2,1/2,0; Ga(2) in 4(i): 0,1/2,z, z = 0.3082, 0.313. $HoCoGa_5$-type (45A,55).

Co_4GeNb_9

Dopov. Akad. Nauk Ukr.RSR, Ser. A, Fiz.-Mat. Tekh. Nauki, No. 2, 65-67.

I4/mcm, 13.278, 5.042. Two-dimensional hybrid of Al_2Cu and Cr_3Si types. Ge has square-antiprismatic, and Co icosahedral coordination.

$Co_5Ge_{2.2}Nd_6$

Kristallografija, **31**, 587-589 [Soviet Physics - Crystallography, **31**, 345-346].

$P\bar{6}m2$, 9.272, 4.188, Z = 1, R = 0.033.

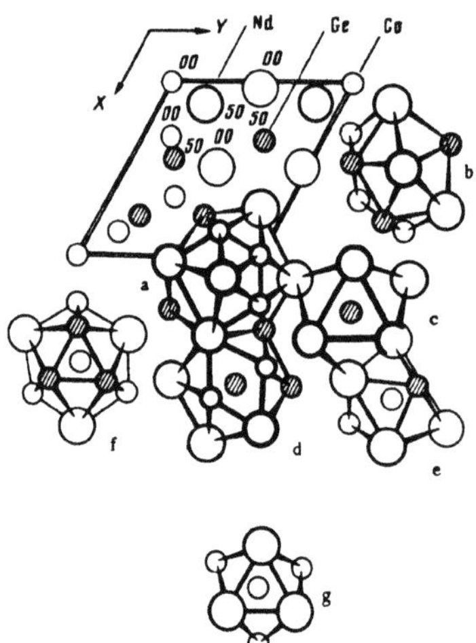

The neodymium has a 17-vertex figure $[Nd1Nd_8Co_6Ge_3]$ (Fig. 1a) and a 15-vertex figure $[Nd2Nd_6Co_3Ge_6]$ (Fig. 1b). For cobalt the coordination polyhedra are trigonal prisms $[Co1Nd_6CoGe_2]$ (Fig. 1e) and $[Co3Nd_6Co_3]$ (Fig. 1g) and an icosahedron $[Co2Nd_3Co_3Ge_6]$ (Fig. 1f). The germanium as coordination polyhedra has a trigonal prism $[Ge1Nd_9]$ (Fig. 1c) and a deformed cubooctahedron $[Ge2Nd_6Co_4Ge_2]$ (Fig. 1d).

Nd-2Co3	2.961(1)
-4Co1	3.100(2)
-Ge1	3.260(1)
-2Ge2	3.288(5)
-4Nd2	3.596(1)
-2Nd1	3.626(1)
-2Nd1	4.188(1)
Nd2-2Co1	2.939(2)
-Co2	2.964(1)
-2Ge1	3.177(1)
-4Ge2	3.313(3)
-4Nd1	3.596(1)
-2Nd2	4.188(1)
Co1-Co3	2.440(3)
-2Ge2	2.491(3)
-2Nd2	2.939(3)
-4Nd1	3.100(2)

Co2-6Ge2	2.613(3)
-3Co1	2.913(3)
-3Nd2	2.964(1)
Co3-3Co1	2.440(3)
-6Nd1	2.961(2)
Ge1-6Nd2	3.177(1)
-3Nd1	3.260(1)
Ge2-2Co1	2.491(3)
-2Co2	2.613(3)
-2Ge2	2.709(6)
-2Nd1	3.288(5)
-4Nd2	3.313(2)

	G, %	x	y	z
Nd1	100	0.1304	−0.1304	1/2
Nd2	100	0.4821	−0.4821	0
Co1	100	0.8481	−0.8481	0
Co2	100	2/3	1/3	0
Co3	100	0	0	0
Ge1	100	1/3	2/3	1/2
Ge2	40,0(1)	0.7641	−0.7641	1/2

$CoGe_2Tb_2$

Izv. Akad. Nauk SSSR, Met., No. 4, 214-216.

B2/m, 10.569, 10.209, 4.212, γ = 118.18°, Z = 4, R = 0.068. Isostructural with $CoSc_2Si_2$ (**45A**, 57; with different a axis), and related to Dy_3Ni_2 (**40A**, 62).

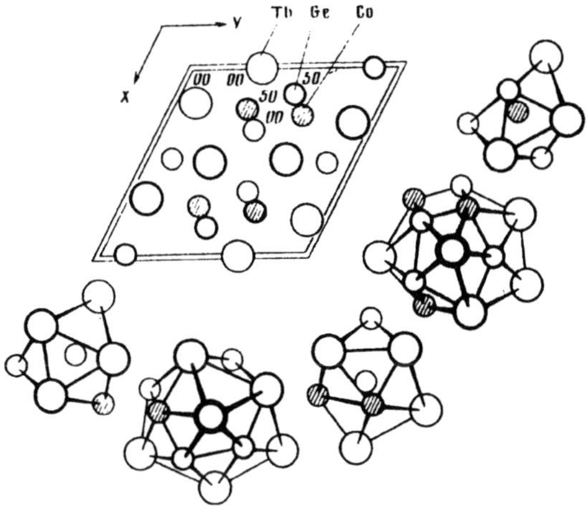

	x	y	z
Tb1	0,1870	0,1090	0
Tb2	0,0043	0,3308	0
Co	0,2250	0,3620	0
Ge1	0,4873	0,1235	0
Ge2	0,3484	0,4336	0

$Co_2Ge_4Tb_3$

Kristallografija, **31**, 394-396 [Soviet Physics - Crystallography, **31**, 231-232].

B2/m (c unique), 10.692, 8.067, 4.164, γ = 107.72°, Z = 2, R = 0.063. Stacking of trigonal prisms of Tb atoms, which can be derived from the AlB_2 type.

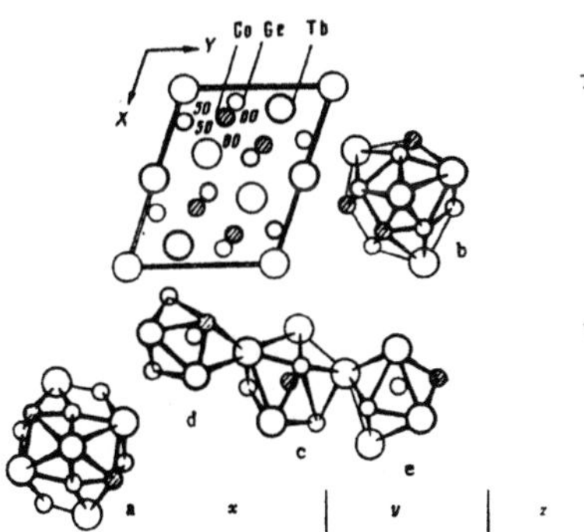

	x	y	z
Tb1	0.3771	0.3076	0
Tb2	0	0	0
Co	0.6904	0.3688	0
Ge1	0,0964	0.3998	0
Ge2	0.7105	0.0796	0

Tb1–Co	2.913(8)
–Ge2	2.975(6)
–2Ge2	2.982(4)
–2Co	3 027(6)
–2Ge1	3,054(5)
–Co	3.235(8)
–Ge1	3.305(6)
–Tb1	3.403(4)
–2Tb2	3.769(2)
–Tb2	4.039(3)
–2Tb1	4.164(2)
Tb2–4Ge2	2.989(4)
–2Ge1	3.072(6)
–2Ge2	3.347(6)
–4Co	3.694(6)
–4Tb1	3.769(2)
–2Tb1	4.039(3)
–2Tb2	4.164(2)

Co–2Ge1	2.358(4)
–Ge2	2.407(9)
–Ge1	2.468(9)
–Tb1	2.913(8)
–2Tb1	3.027(6)
–Tb1	3.235(8)
–2Tb2	3,694(6)
Ge1–2Co	2.358(4)
–Co	2.468(9)
–2Tb1	3.054(5)
–Tb2	3.072(6)
–2Tb1	3.095(5)
–Tb1	3.305(6)
Ge2–Co	2.407(9)
–2Ge2	2.713(5)
–Tb1	2.975(6)
–2Tb1	2.982(4)
–2Tb2	2.989(4)
–Tb2	3.347(6)

$Co_4Ge_{13}Y_3$

Izv. Akad. Nauk SSSR, Neorg. Mater., **22**, 1471-1474.

Pm3n, 8.769, Z = 2, R = 0.067. Co in 8(e): 1/4,1/4,1/4; Y in 6(d): 1/4,1/2,0; Ge(1) in 2(a): 0,0,0; 0.58 Ge(2) in 24(k): 0.327,0.161,0; 0.42 Ge(2') in 24(k); 0.273,0.142,0. Isostructural with $Pr_3Rh_4Sn_{13}$ (**46**A, 113), but with some Ge distributed in two sites 0.5 A apart.

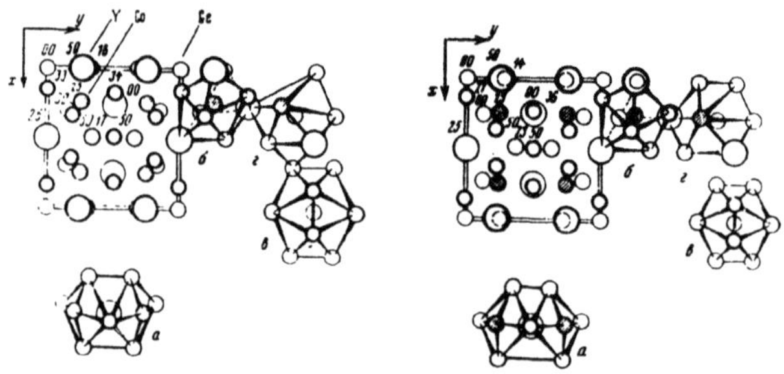

$Co_6Ho_7Sn_{23}$

Ukr. Fiz. Ž., **31**, 1537-1541.

P$\bar{3}$m1, 9.628, 9.846, Z = 1. R = 0.057. Coordination numbers are: Ho = 13, 18; Sn = 10-14, 16; Co = 9.

		x	y	z
Ho1	1 (a)	0	0	0
Ho2	6 (i)	0,4790	—0,4790	0,3064
Sn1	2 (d)	1/3	2/3	0,8195
Sn2	6 (i)	0,2188	—0,2188	0,5451
Sn3	6 (i)	0,1179	—0,1179	0,2764
Sn4	1 (b)	0	0	1/2
Sn5	6 (e)	0,3478	0	0
Sn6	2 (d)	1/3	2/3	0,1150
Co	6 (i)	0,1677	—0,1677	0,8015

$Cu_2Sb_3U_3$

J. Less-Common Metals, **125**, 239-245.

$P6_3$/mmc, 4.393, 2.3011, Z = 2, powder data. U(1) in 2(c): 1/3,2/3,1/4; U(2) in 4(f): 1/3,2/3,0.0786; Sb(1) in 4(f): 1/3,2/3,0.8408; Sb(2) in 2(a): 0,0,0; Cu in 4(e): 0,0,0.1853. Isostructural with $Ln_2Fe_4Si_9$ compounds (**45**A, 89), with U in the Ln and Fe sites, and Sb and Cu fully occupying the Si sites.

Ga_6Gd_2Mn

Dopov. Akad. Nauk Ukr.RSR, Ser. B: Geol., Khim. Biol. Nauki, No. 2, 42-44.

Fm3m, 8.757, Z = 4, R = 0.062. $MnGa_6$ octahedra, $GdGa_{12}$ cubooctahedra; Ga has 13 Ga, Gd, or Mn neighbours. Sm and Tb compounds are isostructural.

$Ga_{49.57}K_4Na_3$

J. Solid State Chem., **64**, 57-66.

R$\bar{3}$m, 16.399, 35.479, Z = 6, R = 0.079. Ga icosahedra and 27-vertex/28-atom polyhedral clusters linked to form a three-dimensional anionic lattice, with cavities containing alkali-metal cations.

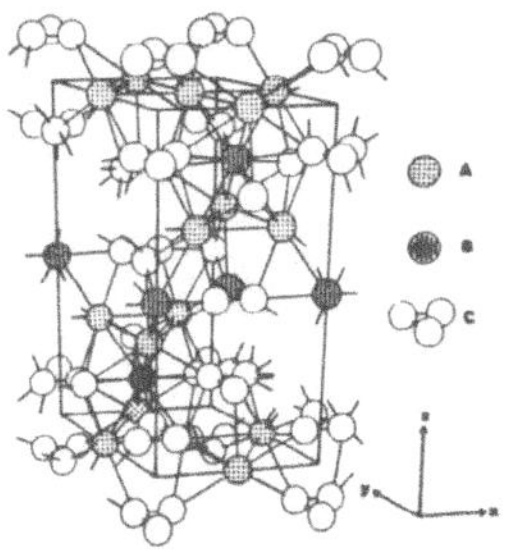

		Occupancy (%)	x	y	z
Ga(1)	36(*i*)	100	0.8406	0.3348	0.0102
Ga(2)	36(*i*)	100	0.3095	0.3449	0.2968
Ga(3)	36(*i*)	100	0.4580	0.3755	0.2553
Ga(4)	36(*i*)	100	0.4504	0.0875	0.4856
Ga(5)	18(*h*)	100	0.2786	−0.2786	0.2244
Ga(6)	18(*h*)	100	0.5547	−0.5547	0.9497
Ga(7)	18(*h*)	100	0.8737	−0.8737	0.7317
Ga(8)	18(*h*)	100	0.5787	−0.5787	0.8200
Ga(9)	18(*h*)	100	0.2355	−0.2355	0.4690
Ga(10)	18(*h*)	100	0.1765	−0.1765	0.5276
Ga(11)	18(*h*)	100	0.2479	−0.2479	0.7632
Ga(12)	18(*h*)	100	0.0553	−0.0553	0.8297
Ga(13)	18(*h*)	19.08	0.0531	−0.0531	0.0507
Ga(14)	6(*c*)	100	0.0	0.0	0.1113
K(1)	18(*h*)	100	0.1299	−0.1299	0.7574
K(2)	6(*c*)	100	0.0	0.0	0.2612
Na(1)	36(*i*)	100	0.2903	0.3705	0.2113
Na(2)	18(*h*)	100	0.2004	−0.2004	0.6758
Na(3)	18(*h*)	100	0.1112	−0.1112	0.6029
Na(4)	6(*c*)	100	0.0	0.0	0.3598

GaK_2Sb_2

J. Less-Common Metals, **119**, 291-296.

Pnma, 15.062, 10.373, 9.145, Z = 8, R = 0.059. $GaSb_2^{2-}$ polyanions, linked by six-coordinate K ions, K-Sb = 3.48-3.96A.

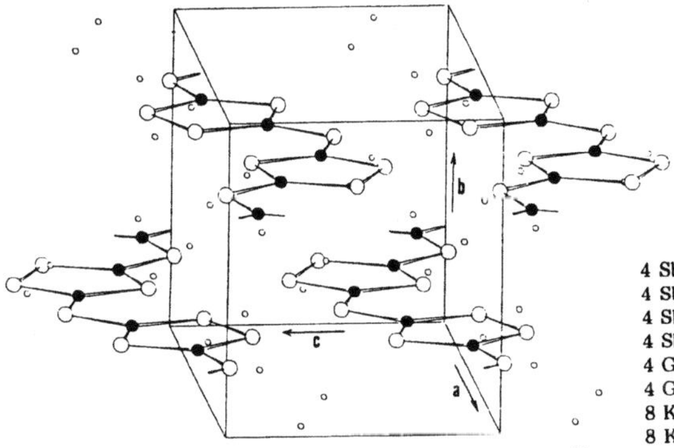

		x	y	z
4 Sb(1)	4c	0,5691	0,25	0,4969
4 Sb(2)	4c	0,3481	0,25	0,1509
4 Sb(3)	4c	0,1283	0,25	0,4938
4 Sb(4)	4c	0,2975	0,25	0,6325
4 Ga(1)	4c	0,4055	0,25	0,4154
4 Ga(2)	4c	0,6788	0,25	0,2763
8 K(1)	8d	0,7348	0,4684	0,5910
8 K(2)	8d	0,4850	0,0307	0,7745

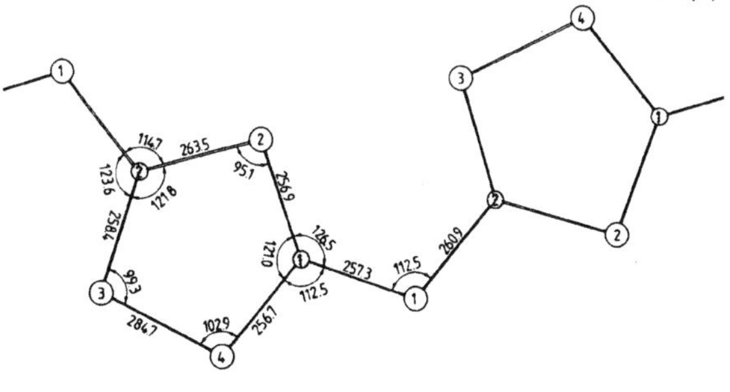

GaLaNi

Kristallografija, **31**, 181-183 [Soviet Physics - Crystallography, **31**, 106-108].

$P2_1/n$ (**c** unique), 6.679, 6.097, 6.341, γ = 106.35°, Z = 4, R = 0.043. Atoms in 4(e): Ga, 0.0753,0.3200,0.3648; La, 0.5850,0.3000,0.3543; Ni, 0.1801,0.0464,0.5891. EuGeNi-type structure (**50**A).

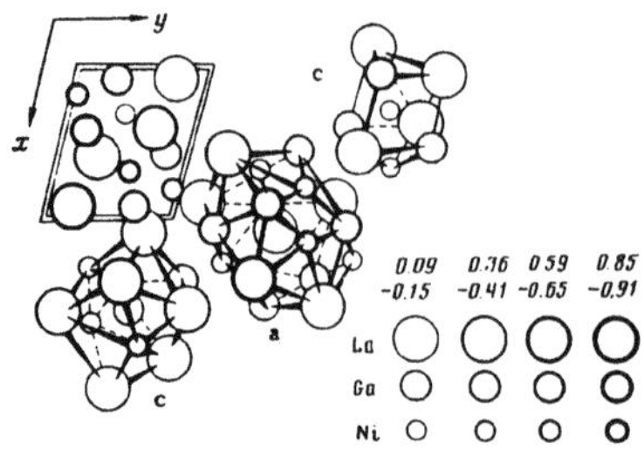

Projection of structure of LaGaNi on xy plane and coordination polyhedra of La (a), Ga (b), and Ni (c) atoms.

	x	y	z
La	0.5850	0.3000	0.3543
Ga	0.0753	0.3200	0.3648
Ni	0.1801	0.0464	0.5891

$Ga_3Na_2Sb_3$

Mater. Res. Bull., **21**, 331-336.

Pnma, 26.890, 4.283, 7.236, Z = 4, R = 0.055. Zintl phase with puckered sheets of Ga and Sb, and Na in channels.

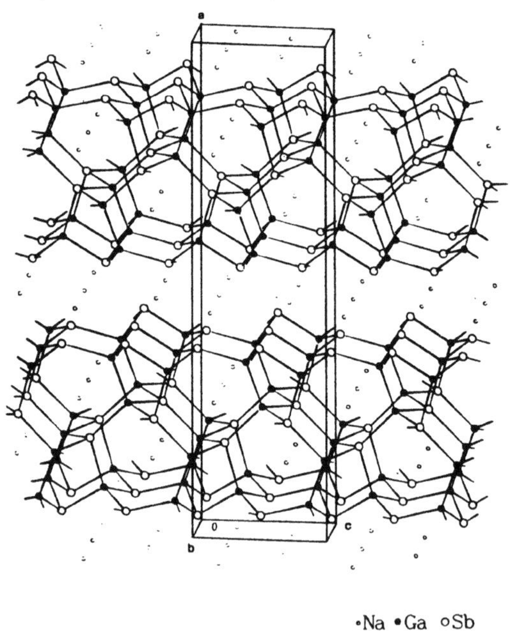

$Ga_{20}Ni_{11}U_4$

J. Nucl. Mater., **137**, 89-93.

C2/m, 20.734, 4.119, 15.338, 124.71, Z = 2, R = 0.056. Occupation variant of $Ga_{21}Ni_{10}Ho_4$-type (**46**A, 79), with Ni replacing Ga in the 2(d) sites at the centre of distorted octahedra. The Pd and Pt compounds are isostructural.

$Ge_5Ho_2Rh_3$

Mater. Res. Bull., **21**, 33-39.

C2/c, 11.353, 11.976, 5.813, 119.05, Z = 4, R = 0.034. $Lu_2Co_3Si_5$-type (**52**A, 37). Related compounds have $U_2Co_3Si_5$ (**43**A, 50), or $Lu_2Co_3Si_5$-type structures.

		x	y	z
Ho	8f	0,2695	0,3630	0,2664
Rh_1	8f	0,1001	0,1424	0,1172
Rh_2	4e	0	0,4987	1/4
Ge_1	8f	0,3313	0,0959	0,3319
Ge_2	4e	0	0,7204	1/4
Ge_3	4e	0	0,2886	1/4
Ge_4	4e	0	0,9929	1/4

$Ge_4La_3Ni_4$

Izv. Akad. Nauk SSSR, Neorg. Mater., **22**, 1573-1575.

Immm, 4.2234, 4.2295, 24.156, Z = 2, R = 0.064. $U_3Ni_4Si_4$-type (**45**A, 96).

$Ge_6La_{11}Ni_4$

Kristallografija, **31**, 312-316 [Soviet Physics - Crystallography, **31**, 184-186].

C2/m, 18.637, 4.384, 14.191, 106.13, Z = 2, R = 0.062. The structure is based on fragments of the AlB_2 and α-Fe types.

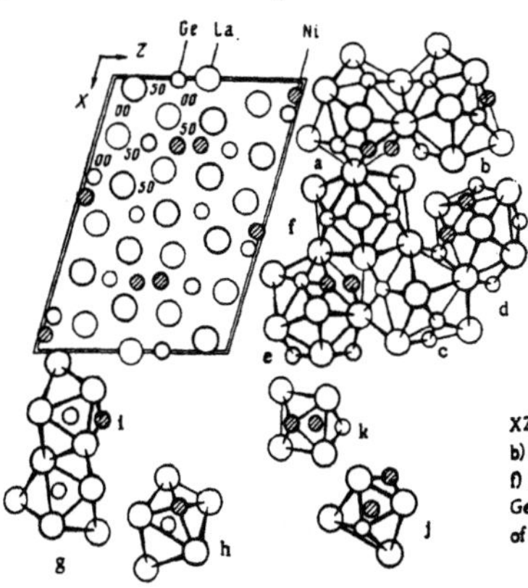

	x	y	z
La1	0	0	1/2
La2	0,1422	0	0,3494
La3	0.0957	0	0.7840
La4	0,2160	0	0,0999
La5	0,4515	0	0,8738
La6	0,3399	0	0.4065
Ni1	0.0594	0	0.9752
Ni2	0,2477	0	0.5503
Ge1	0.4864	0	0.6512
Ge2	0.3713	0	0.0490
Ge3	0,2614	0	0.7345

Projection of unit cell of structure of compound $La_{11}Ni_4Ge_6$ on XZ plane and coordination polyhedra of atoms. a) [$La2La_{10}Ni_3Ge_4$], b) [$La3La_{11}Ni_1Ge_5$], c) [$La4La_{11}Ge_6$], d) [$La5La_8Ni_4Ge_4$], e) [$La6La_{10}Ge_4$], f) [$La1La_{10}Ge_4$], g) [$Ge1La_9$], h) [$Ge2La_7Ni_2$], i) [$Ge3La_8Ni_1$], j) [$Ni1La_6Ni_1$-Ge_2], k) [$Ni2La_6Ni_2Ge_1$]. Figures mark Y coordinates of atoms in hundredths of unit cell parameter.

GeLiPr, GeLiYb, BiLiYb

Z. anorg. Chem., **535**, 7-12.

GeLiPr, GeLiYb, $P\bar{6}2m$, a = 7.316, 6.944, c = 4.373, 4.190 A, Z = 3, R = 0.029, 0.095. Ge(1) in 1(a): 0,0,0; Ge(2) in 2(d): 1/3,2/3,1/2; Li in 3(g): x,0,1/2, x = 0.235, 0.23; Pr or Yb in 3(f): x,0,0, x = 0.5804, 0.5732. Modified Fe_2P-type (**45**A, 86). GeLiTb and DyGeLi are isostructural.

BiLiYb [$P6_3mc$], 9.595, 7.635, [Z = 8], powder data, no atomic parameters determined. LiSnY-type (**45**A, 94).

Ge_2Ni_2Y

Dokl. Akad. Nauk SSSR, **291**, 1362-1365.

I4/mmm, 4.042, 9.776, Z = 2, R = 0.040. Ge in 4(e): 0,0,0.32705; Ni in 4(d): 0,1/2,1/4; Y in 2(a): 0,0,0. $ThCr_2Si_2$-type (**43**A, 99); electron-density studied.

$Ge_4Sc_2V_3$

Kristallografija, **31**, 868-873 [Soviet Physics - Crystallography, **31**, 513-516].

Pnma, 6.614, 12.774, 6.776, Z = 4, R = 0.021. Superstructure of Ge_4Sm_5 type. Coordination numbers are: Ge = 9, Sc = 18, V = 14 and 16.

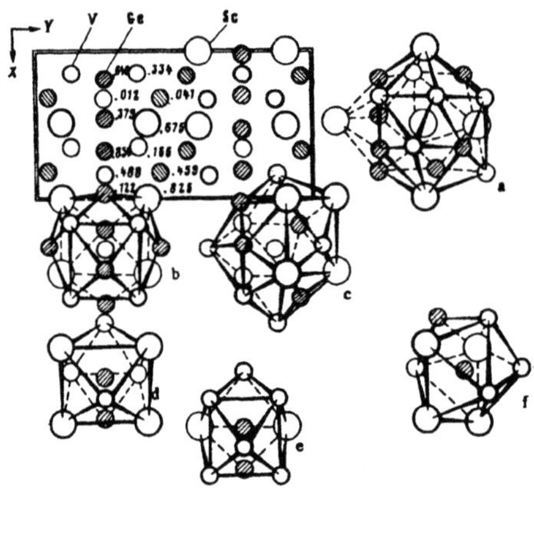

(× 10^4)

		x	y	z
Sc($R^{(3)}$)	8(d)	9911	930	8251
V1($R^{(1)}$)	4(c)	3351	1/4	123
V2($R^{(3)}$)	8(d)	1594	1270	3340
Ge1($X^{(2)}$)	4(c)	1912	1/4	6457
Ge2($X^{(2)}$)	4(c)	9635	1/4	1212
Ge3($X^{(3)}$)	8(d)	1782	9578	5411

$Ln_3Rh_4Sn_{13}$ (Ln = La-Gd, Yb, Ca, Sr, Th)

J. Solid State Chem., **63**, 358-368.

Pm3n, a $\sim$ 9.7A, Z = 2, R = 0.016-0.036. Sn(1) in 2(a): 0,0,0; Ln in 6(d): 1/4,1/2,0; Rh in 8(e): 1/4,1/4,1/4; Sn(2) in 24(k): 0,y,z, y = 0.2999-0.3085, z = 0.1527-0.1538. Structure as previously described for the Yb compound (**46**A, 115). Some of the materials have a distorted structure (phase I').

$LnRuSn_3$ (Ln = La, Ce, Pr, Nd)

J. Less-Common Metals, **123**, 89-94.

Pm3n, 9.773, 9.726, 9.708, 9.687, Z = 8, R = 0.059 for La compound. La(1) in 2(a): 0,0,0; La(2) in 6(d): 1/4,1/2,0; Ru in 8(e): 1/4,1/4,1/4; Sn in 24(k): 0,0.3075,0.1560. Isostructral with $Yb_3Rh_4Sn_{13} \equiv (Yb,Sn)RhSn_3$ (**46**A, 115). Strings of Sn atoms connected by Ru into a framework, with cage-like voids occupied by Ln.

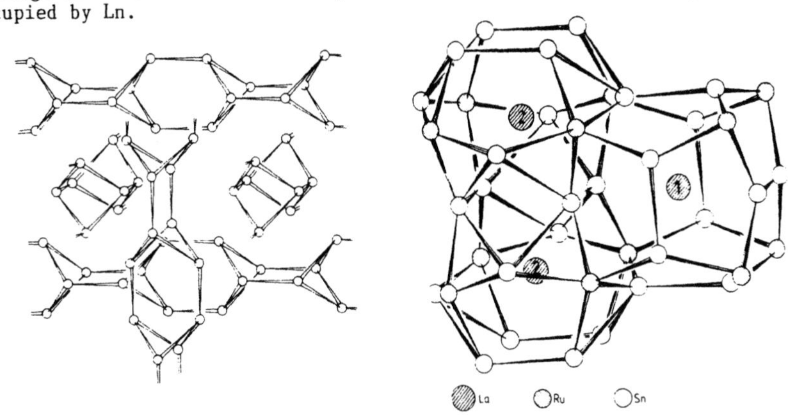

Lu_2NiSn_6

Dopov. Akad. Nauk Ukr.RSR, Ser B: Geol., Khim. Biol. Nauki, No. 12, 34-37 (1985).

Cmmm, 4.301, 22.78, 4.366, Z = 2, R = 0.086. Fragments of $ZrGa_2$ and CaF_2 types alternating along **b**.

MgPdSb, MgPtSb, LiMgPtSb

J. Less-Common Metals, **116**, 271-278.

$F\bar{4}3m$, 6.283, 6.263, 6.308, Z = 4, R = 0.042, 0.039, 0.039. Mg in 4(b): 1/2,1/2,1/2; Pd or Pt in 4(c): 1/4,1/4,1/4; Sb in 4(a): 0,0,0; Li in 4(d): 3/4,3/4,3/4. Filled zincblende-type structures.

Pd_2Sb_2Sr, $BaPd_2As_2$, $BaPd_2Sb_2$

Z. anorg. Chem., **536**, 7-14.

Pd_2Sb_2Sr (room-temperature), P4/nmm, 4.649, 10.625, Z = 2, R = 0.036. Sr in 2(c): 1/4,1/4,z, z = 0.7587; Pd(1) in 2(a): 3/4,1/4,0; Pd(2) in 2(c): z = 0.3714; Sb(1) in 2(b): 3/4,1/4,1/2; Sb(2) in 2(c): z = 0.1242. $CaBe_2Ge_2$-type (**43**A, 28).

Pd_2Sb_2Sr (high-temperature), $BaPd_2As_2$ (high-temperature), $BaPd_2Sb_2$, I4/mmm, a = 4.588, 4.486, 4.725, c = 10.737, 10.315, 10.772 A, Z = 2, R = 0.060, 0.027, 0.033. Sr or Ba in 2(a): 0,0,0; Pd in 4(d): 0,1/2,1/4; Sb, As or Sb in 4(e): 0,0,z, z = 0.3660, 0.3716, 0.3697. $ThCr_2Si_2$-type (**43**A, 99).

$PtSbSr$, $BaPtSb$, $PtSrAs$, $Pt_{0.70}SrAs_{0.90}$, $BaPt_{0.85}As_{0.90}$, $Pt_{0.75}SrP_{1.25}$, $BaPtP$, $BaPtAs$

Z. anorg. Chem., **535**, 110-122.

SrPtSb, BaPtSb, $P\bar{6}m2$, a = 4.504, 4.535, c = 4.507, 4.884, Z = 1, R = 0.089, 0.050. Sr or Ba in 1(a): 0,0,0; Pt in 1(d): 1/3,2/3,1/2; Sb in 1(f): 2/3,1/3,1/2. Ordered AlB_2-type (**3**, 28, 311; **4**, 121; **20**, 5; **46A**, 123).

SrPtAs, $P6_3/mmc$, 4.246, 8.967, Z = 2, R = 0.072. Sr in 2(a): 0,0,0; Pt in 2(c); 1/3,2/3,1/4; As in 2(d): 2/3,1/3,1/4. Modified Ni_2In-type (**9**, 91), which can be considered as an ordered AlB_2 superstructure.

$SrPt_{0.70}As_{0.90}$, $BaPt_{0.85}As_{0.90}$, $P\bar{6}m2$, a = 4.262, 4.325, c = 4.428, 4.763, Z = 1, R = 0.072. Sr or Ba in 1(a): 0,0,0; 0.70 or 0.85 Pt in 1(d): 1/3,2/3,1/2; 0.90 As in 1(f): 2/3,1/3,1/2. Ordered AlB_2-type.

$SrPt_{0.75}P_{1.25}$, P6/mmm, 4.087, 4.432, Z = 1, R = 0.067. Sr in 1(a): 0,0,0; Pt/P in 2(d): 1/3,2/3,1/2. AlB_2-type, with Pt/P disorder.

BaPtP, BaPtAs, $P2_13$, 6.533, 6.716, Z = 4, R = 0.081, 0.080. Atoms in 4(a): x,x,x, x = 0.1373, 0.8487, 0.416 for Ba, Pt, P; x = 0.1136, 0.4040, 0.8293 for Ba, Pt, As. $SrSi_2$-type derivative (**30A**, 97; **31A**, 63; **38A**, 145; **44A**, 98), with ordered occupation of the Si sites by Pt and P (or As).

λ_1-$Fe_2ScD_{2.5}$

Ž. Neorg. Khim., **31**, 2500-2503 [Russ. J. Inorg. Chem. **31**, 1443-1446].

$P6_3/mmc$, 5.223, 8.500, Z = 4, neutron powder data. Sc in 4(f): 1/3,2/3,0.0593; Fe(1) in 2(a): 0,0,0; Fe(2) in 6(h): x,2x,1/4, x = 0.8362; 2.9 and 2.4 D in 6(h) sites, x = 0.182, 0.478; 2.0 D in 12(k): x,2x,z, x = 0.565, z = 0.130; 2.8 D in 24(ℓ): 0.273,0.311,0.974. $MgZn_2$-type C14 structure (**1**, 180; **46A**, 96), with D in four cavities.

$Fe_{1.5}ScSi_{0.5}D_2$

Latv. PSR Zinat. Akad. Vestis, Fiz. Teh. Zinat. Ser., No. 1, 95-99.

$P6_3/mmc$, 5.113, 8.324, Z = 4, neutron radiation, R = 0.10. D atom occupancies determined.

POTASSIUM TETRAHYDRIDOPLATINATE(II)
K_2PtD_4

J. Less-Common Metals, **116**, 9-15.

295K, Fm3m, 8.025, Z = 4, neutron powder data. K in 8(c): 1/4,1/4,1/4; Pt in 4(a): 0,0,0; 2/3 D in 24(e): 0.202,0,0. K_2PtCl_6-type, with 2/3 occupancy of D site.

15K, $P4_2/mnm$, 5.582, 8.085, Z = 2, neutron powder data. K in 4(d): 0,1/2,1/4; Pt in 2(a): 0,0,0; D in 4(e): 0,0,0.195; D(2) in 4(f): 0.201,0.201,0.

β-$LaNi_5D_7$

J. Phys. F: Met. Phys., **16**, 675-685.

$P6_3mc$, 5.388, 8.559, Z = 2, neutron powder data. Doubled **c** axis relative to $LaNi_5$. D in tetrahedral and octahedral sites.

Mg_2NiH_4, Mg_2NiD_4

J. Less-Common Metals, **115**, 65-78; Inorg. Chem., **25**, 3590-3593.

C2/c, a = 14.363, 14.343, b = 6.405, 6.4038, c = 6.486, 6.4830 A, β = 113.62, 113.52°, Z = 8. X-ray and neutron powder data. D positions differ from previous studies (51A, 41, 57), and surround Ni tetrahedrally, Ni-D = 1.54 A.

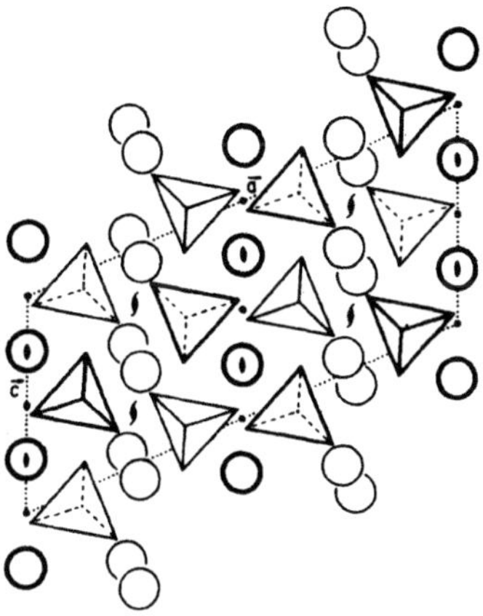

	x	y	z
Ni in (8f)	0.1194	0.2308	0.0832
Mg(1) in (8f)	0.2652	0.4827	0.0754
Mg(2) in (4e)	0	0.0144	0.25
Mg(3) in (4e)	0	0.5130	0.25
D(1) in (8f)	0.2113	0.2995	0.3037
D(2) in (8f)	0.1360	0.3163	0.8811
D(3) in (8f)	0.0105	0.2868	0.0537
D(4) in (8f)	0.1306	0.9950	0.0815

$ThZr_2D_x$ (x = 3.6, 4.8, 6.0, 6.3)

J. Less-Common Metals, **121**, 455-460.

Fd3m, 9.042, 9.112, 9.151, 9.154, Z = 8, neutron powder data. $MgCu_2$-type, with D in tetrahedral 32(e) and 96(g) sites.

$Ti_5Si_3D_{0.9}$

Nippon Kinzoku Gakkaishi, **50**, 769-777; Trans. Jpn. Inst. Met., **27**, 639-647.

$P6_3/mcm$, neutron powder data. D in 2(b), surrounded octahedrally by 6 Ti.

β_1-V_2H (TETRAGONAL and MONOCLINIC), β-V_2D

Acta Cryst., **B42**, 529-533.

Tetragonal, $I4_1/amd$, 6.0012, 6.6188, Z = 8, R = 0.049. V in 16(h): 0,0.49964,0.26712; H in 8(c): 0,0,0. Monoclinic, Cm, 4.4566, 3.0022, 4.4760, 95.609, Z = 2, R = 0.080. Atoms in 2(a): x,0,z, x = 0.2832, 0.75, 0.0, z = 0.2171, 0.75, 0.5 for V(1), V(2), H (H parameters tentative in both structures). Both structures contain a distorted body-centred arrangement of V atoms.

Z. Phys. Chem. (Munich), **145**, 27-42 (1985).

β_1-V_2H, $I4_1/amd$, 6.035, 6.872, Z = 8, neutron radiation, R = 0.13. V in 16(h): 0.498,0,0.247; H distributed in 8(c): 0,0,0, 16(f): 0.119,0,0, 16(h): 0,0.121,0, with occupancies 0.79, 0.08, 0.08 respectively.

β-V_2D, Cm (or C2 or C2/m), reanalysis of above data.

$Ca_2Os_3B_5$, $Eu_2Os_3B_5$

Z. anorg. Chem., **533**, 30-36.

C2, a = 13.265, 13.260, b = 4.013, 4.071, c = 7.971, 8.001 A, β = 139.5, 138.4, Z = 2, R = 0.049, 0.047, and neutron powder data for the Ca compound. Puckered Os-B layers connected by B atoms.

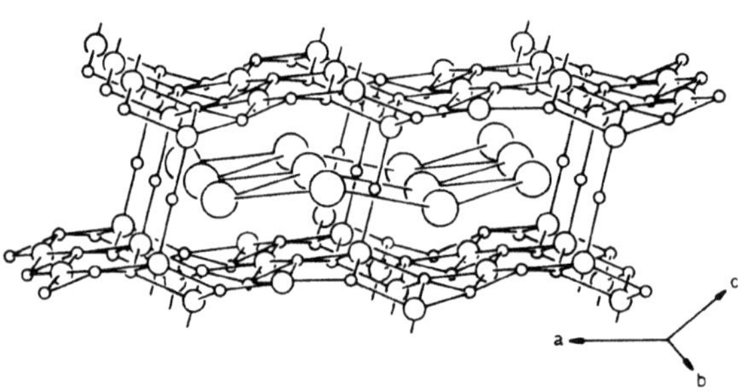

		x	y	z
Os(1)	2a	0	0	0
		0	0	0
Os(2)	4c	0.6342	0.2352	0.8659
		0.6346	0.2338	0.8641
Ca	4c	0.8411	0.2130	0.4424
Eu		0.8355	0.2172	0.4389
B(1)	2b	0.5	0.205	0.5
		0.5	0.233	0.5
B(2)	4c	0.469	0,992	0.058
		0.472	0.002	0,065
B(3)	4c	0.145	0,205	0,982
		0.145	0.213	0.977

$CeRh_3B$

J. Less-Common Metals, **120**, L7-L10.

Pm3m, 4.1883, Z = 1, R = 0.020. Ce in 1(a): 0,0,0; Rh in 3(c): 1/2,1/2,0; B in 1(b): 1/2,1/2,1/2. Anti-perovskite structure.

$Co_4Nb_3B_7$

Kristallografija, **31**, 184-186 [Soviet Physics - Crystallography, **31**, 108-110].

Cmcm, 3.202, 18.432, 8.917, Z = 4, R = 0.055. Corrugated sheets of trigonal prisms.

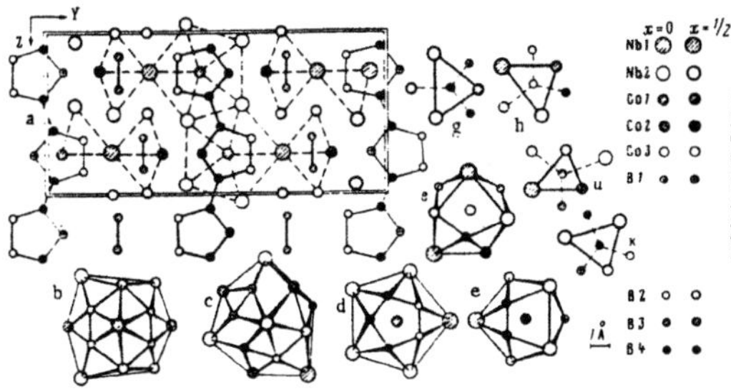

		x	y	z
Nb1	4(c)	0	0.3023	0.25
Nb2	8(f)	0	0.4193	0.5551
Co1	4(c)	0	0.4582	0.25
Co2	4(c)	0	0.1552	0.25
Co3	8(f)	0	0.1987	0.5053
B1	4(c)	0	0.041	0.25
B2	8(f)	0	0.109	0.645
B3	8(f)	0	0.289	0.648
B4	8(f)	0	0.015	0.591

$Co_{2.85}Tb_{4.15}B_6$

J. Less-Common Metals, **117**, 297-302.

$R\bar{3}m$, 5.400, 23.20 A, Z = 3, powder data. 0.4 Tb(1) + 2.6 Co(1) in 3(a): 0,0,0; Tb(2), Tb(3), Co in 6(c): 0,0,z, z = 0.2497, 0.4147, 0.112; 0 in 18(g): 0.333,0,1/2. $Co_2Pr_5B_6$-type (**50**A).

$ErRh_4B_4$

Acta Cryst., C**42**, 1469-1472.

$I4_1/acd$, 7.461, 14.804, Z = 8, R = 0.051. Er in 8(b): 0,1/4,1/8; Rh in 32(g): 0.11733,0.09915,0.93841; B in 32(g): 0.8334,0.1092,0.9584. $LuRu_4B_4$-type structure (**43**A,35), with rotation of Rh_4 and B_4 tetrahedra by 7° and 5°, respectively, relative to the $P4_2/nmc$ form (**51**A, 23).

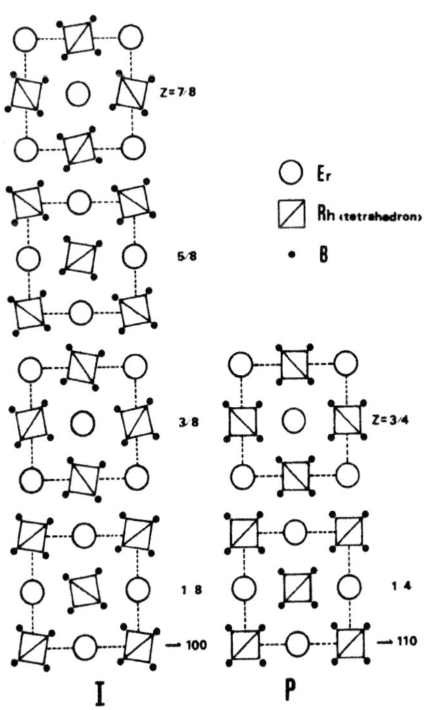

	I	P	
Er–8 Er^{i}	5·255 (1)	5·255 (2)	
Er–4 Er^{ii}	5·276 (1)	5·292 (4)	
Er–4 Rh	3·109 (1)	2·930 (1)	
Er–4 Rh^{iii}	2·904 (1)	3·149 (2)	
Er–4 Rh^{iv}	3·209 (1)		
Er–4 B^{v}	2·97 (1)	3·02 (1)	
Er–4 B^{vi}	3·20 (1)		
Er–4 B^{vii}	2·95 (1)	3·09 (1)	
Rh–1 Rh^{viii}	2·851 (1)	2·634 (2)	Intra-Rh tetrahedron
Rh–2 Rh^{ix}	2·755 (1)	2·840 (1)	
Rh–1 Rh^{iv}	2·692 (1)	3·064 (2)	Inter-Rh tetrahedron
Rh–1 Rh^{x}	2·929 (1)		
Rh–1 Rh^{xi}	3·084 (1)	2·658 (2)	
Rh–1 B^{xii}	2·14 (1)	2·18 (1)	Intra-$(RhB)_4$ unit
Rh–1 B^{vii}	2·23 (1)		
Rh–1 B^{xiii}	2·18 (1)	2·24 (1)	
Rh–1 B^{xiv}	2·21 (1)	2·20 (1)	Inter-$(RhB)_4$ unit
Rh–1 B^{xv}	2·26 (1)		
B–1 B^{xvi}	3·25 (1)	3·46 (1)	Intra-B tetrahedron
B–2 B^{xvii}	3·38 (1)	3·33 (2)	
B–1 B^{xviii}	1·75 (1)	1·83 (2)	Inter-B tetrahedron
B–1 B^{xix}	3·20 (1)	3·12 (1)	
B–1 B^{xx}	3·22 (1)		

$FeNd_2B_3$

J. Less-Common Metals, **115**, 357-366.

R$\bar{3}$m, 5.463, 24.281 A, powder data. Trial Nd and Fe positions given.

$Fe_4Nd_{1.11}B_4$, $Fe_4Gd_{1.14}B_4$

J. Less-Common Metals, **115**, L7-L10 and **123**, 109-116.

Pccn, a = b = 7.117, 7.051, c = 35.07, 27.33 A, Z = 18, 14, R = 0.081, 0.070. The structures contain two substructures with different periodicities along **c** (52A, 19), with a modulation of Fe_4 tetrahedra which allows an increase in the shortest Fe...Ln distances.

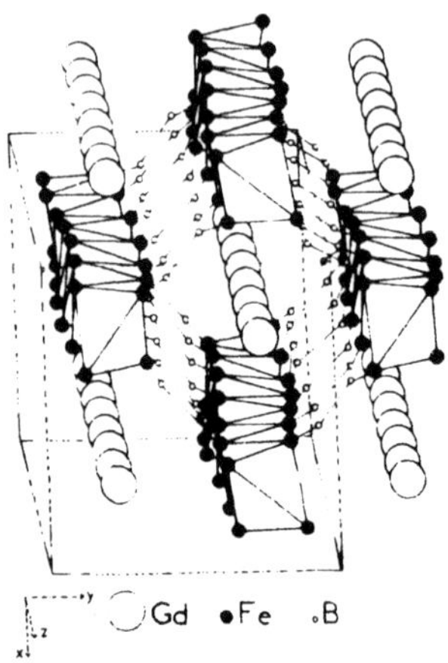

$IrMg_2B_{2-x}$

Z. anorg. Chem., **543**, 89-96.

C2/m, 8.580, 4.846, 6.836, 115.72, Z = 4, R = 0.046. Distorted Mg_4Ir_2 prisms, with partially disordered B atoms on Ir_2 edges or in pairs inside the prisms.

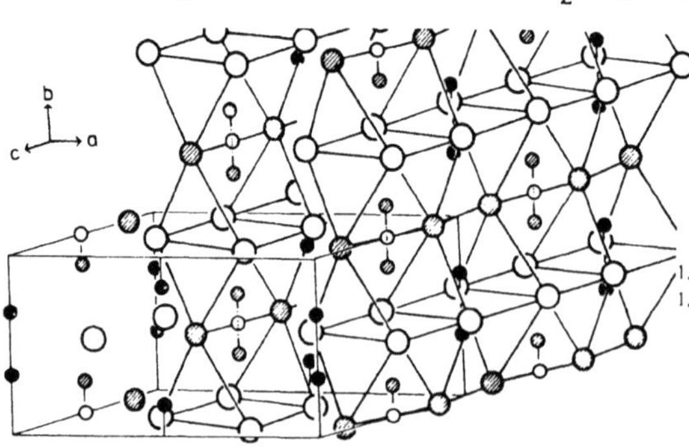

			x	y	z
4 Ir(1)	4i	1	0,01369	0,0	0,2172
4 Mg(1)	4i	1	0,3096	0,0	0,1636
4 Mg(2)	4i	1	0,3066	0,0	0,6362
4 B(1)	4g	1	0,0	0,326	0,0
1,08 B(2)	4h	0,27	0,0	0,170	0,5
1,52 B(3)	2c	0,76	0,0	0,0	0,5

LaB_6

J. Less-Common Metals, **117**, 73-81.

Pm3m, 4.157, Z = 1, R = 0.011-0.014 for three crystals. 1 La in 1(a): 0,0,0; 5.8 B in 6(f): x,1/2,1/2, x = 0.1996. Structure as previously described (**23**, 60), with possible B vacancies.

LaB_6, CeB_6, SmB_6

Kristallografija, **31**, 803-805 [Soviet Physics - Crystallography, **31**, 475-477; in this translation, pages 475 and 476 are interchanged].

Pm3m, 4.1570, 4.1408, 4.1347, Z = 1, R not given. Ln in 1(a): 0,0,0; B in 6(f): x,1/2,1/2, x = 0.1993, 0.1985, 0.1986 (some possible defects). CaB_6-type (**2**, 37, 308; **3**, 334; **18**, 62).

MoB_2

J. Less-Common Metals, **123**, 277-283.

R$\bar{3}$m, 3.0152, 20.971 A, Z = 6, R = 0.026. Mo, B(1), B(2) in 6(c): 0,0,z, z = 0.07572, 0.3320, 0.1821. Mo layers stacked AABBCC with trigonal prisms and tetrahedra centre by boron, and empty octahedra; this results in planar (H) and puckered (K') boron layers.

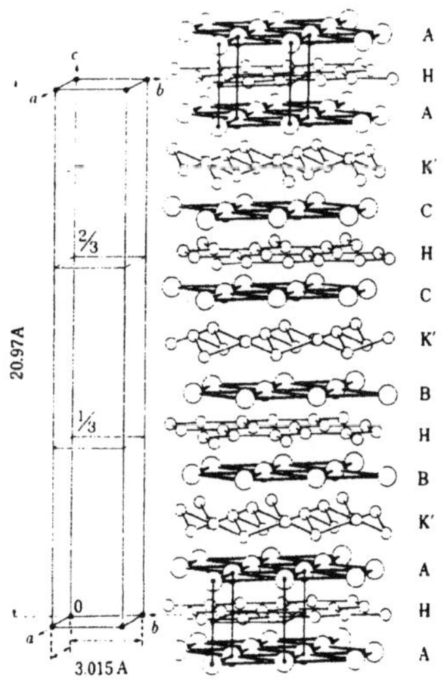

$Pr_2Re_3B_6$

Dopov. Akad. Nauk Ukr.RSR, Ser. B: Geol., Khim. Biol. Nauki, No. 8, 42-44.

C2/c, 6.779, 10.623, 7.569, 90.84, Z = 4, R = 0.099. Zigzag chains of B with trigonal prismatic coordinations; Pr and Re have coordination numbers 20 and 14, respectively.

ReY_3B_7, FeY_3B_7

J. Less-Common Metals, **117**, 29-35.

Cmcm, a = 3.537, 3.423, b = 15.859, 15.658, c = 9.396, 9.295 A, Z = 4, R = 0.16, powder data. Er_3CrB_7-type structure, as previously described (**42**A, 58).

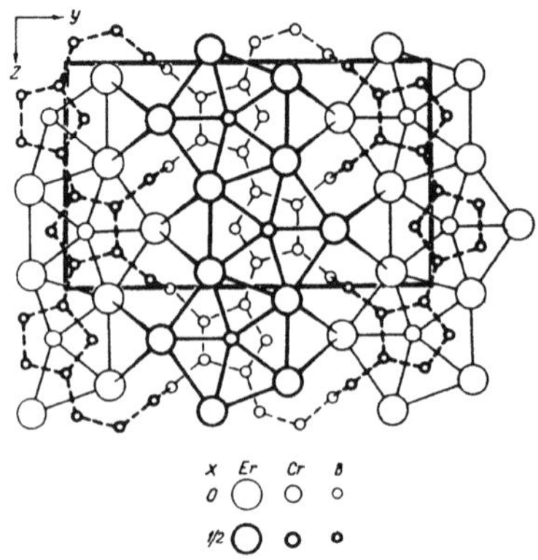

		Y_3ReB_7		Y_3FeB_7	
		y	*z*	*y*	*z*
Y(1)	8(f)	0.1056	0.0597	0.1061	0.0552
Y(2)	4(c)	0.7568	0.2500	0.7580	0.2500
M	4(c)	0.9471	0.2500	0.9468	0.2500
B(1)	8(f)	0.274	0.023	0.275	0.024
B(2)	8(f)	0.367	0.352	0.366	0.353
B(3)	8(f)	0.481	0.408	0.480	0.406
B(4)	4(c)	0.544	0.250	0.543	0.250

x = 0 for all the positions.

SiB_6

J. Solid State Chem., **63**, 31-45.

Pnnm, 14.397, 18.318, 9.911, Z = 43, R = 0.061. Boron framework of 18 M_{12} icosahedra, 4 M_{15} icosihexahedra, and 16 M_1 single-atom positions, not all filled.

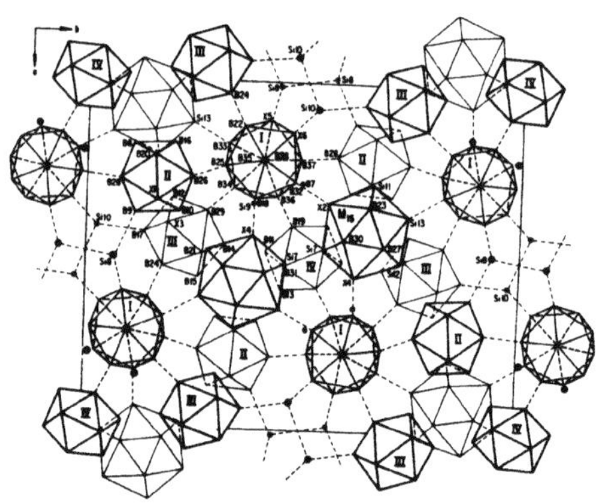

VB_{65}, VB_{165}

J. Solid State Chem., **63**, 424-430.

$R\bar{3}m$, a = 10.972, 10.949, c = 23.908, 23.840 A, Z ∿ 315 B atoms, R = 0.054, 0.049. β-Rhombohedral boron structure (**27**, 394; **35**A, 127; **38**A, 37; **42**A, 34), with V in the same sites as the metal atoms in CrB_{41} and FeB_{49} (**35**A, 34; **42**A, 57).

		x	y	z
B(1)	36(i)	1762	1761	1768
B(2)	36(i)	3191	2960	1288
B(3)	36(i)	2611	2169	4196
B(4)	36(i)	2357	2516	3469
B(5)	18(h)	0547	1095	−0558
B(6)	18(h)	0864	1728	0132
B(7)	18(h)	1105	2211	−1132
B(8)	18(h)	1700	3401	0279
B(9)	18(h)	1295	2590	−2340
B(10)	18(h)	1022	2044	−3018
B(11)	18(h)	0565	1130	3268
B(12)	18(h)	0898	1796	3986
B(13)[a]	18(h)	0564	1128	−4457
B(14)	6(c)	0	0	3856
B(15)	3(b)	0	0	5000
B(16)	18(h)	Vacant		
V(1)[b]	6(c)	0	0	1342
V(2)[c]	18(h)	1990	3980	1738

Note. All sites, except for B(16) and the three listed ones, are fully occupied.

[a] Occupancy 74.6%.

[b] Occupancy 64.0%.

[c] Occupancy 5.3%.

	x	y	z
B(1)	1746	1752	1769
B(2)	3190	2962	1291
B(3)	2612	2171	4196
B(4)	2353	2515	3469
B(5)	0545	1089	−0560
B(6)	0866	1731	0132
B(7)	1101	2202	−1136
B(8)	1702	3404	0277
B(9)	1292	2584	−2337
B(10)	1021	2042	−3016
B(11)	0565	1131	3268
B(12)	0898	1797	3989
B(13)[b]	0575	1151	−4462
B(14)	0	0	3855
B(15)	0	0	5000
B(16)[c]	0539	1079	1182
V(1)[d]	0	0	1343
V(2)	Vacant		

Note. All sites, except for V(2) and the three listed ones, are fully occupied.

[b] Occupancy 76.3%.

[c] Occupancy 14.7%.

[d] Occupancy 31.6%.

Al_3ScC_3

Kristallografija, **31**, 791-792 [Soviet Physics - Crystallography, **31**, 467-468].

$P6_3/mmc$, 3.355, 16.776, Z = 2, R = 0.032. Sc in 2(a): 0,0,0; Al(1) in 4(f): 1/3,2/3,0.1293; Al(2) in 2(d): 1/3,2/3,3/4; C(1) in 4(f): 1/3,2/3,0.5876; C(2) in 2(c): 1/3,2/3,1/4. 8-Layer close-packing of Sc and Al, with C in all octahedral holes and in trigonal bipyramids between three layers of Al. The Y and Ln compounds are isostructural.

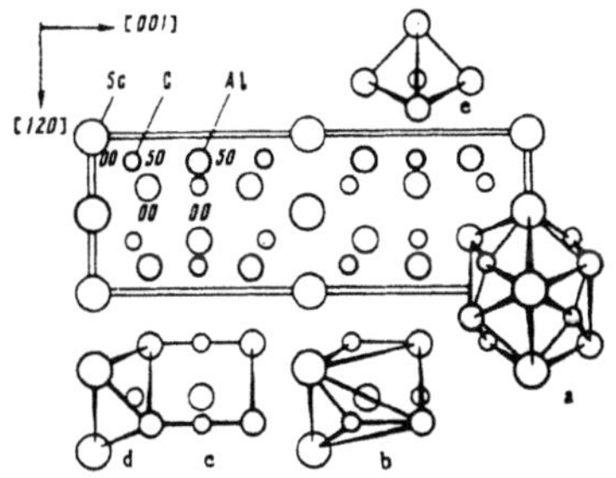

Projection of structure of $ScAl_3C_3$ on (100) plane and coordination polyhedra of atoms. a) Sc; b) Al1; c) Al2; d) C1; e) C2.

$Ce_xNd_{1-x}C_2$ (x = 0, 0.25, 0.50, 0.75, 1)

J. Solid State Chem., **62**, 172-176.

I4/mmm, a = 3.827-3.881, c = 6.407-6.487, Z = 2, neutron powder data. (Ce,Nd) in 2(a): 0,0,0; C in 4(e): 0,0,z, z = 0.3984-0.4010. Structures as previously described (**1**, 741, 783; **31A**, 32; **32A**, 118), with acetylenic C-C groups, C-C = 1.29(1) A.

$CoDyC_2$, $DyNiC_2$

J. Less-Common Metals, **116**, 147-157.

Amm2, a = 3.544, 3.571, b = 4.515, 4.508, c = 6.036, 6.041 A, Z = 2, R = 0.012, 0.023. Dy in 2(a): 0,0,0; Co or Ni in 2(b): 1/2,0,z, z = 0.6152, 0.6116; C in 4(e): 1/2,y,z, y = 0.1513, 0.1523, z = 0.3057, 0.3016. $CeNiC_2$-type (**45**A, 50). Other $CoLnC_2$ compounds have $CeCoC_2$ structure type (**46**A, 51).

$CoSc_3C_4$

Kristallografija, **31**, 788-790 [Soviet Physics - Crystallography, **31**, 466-467].

Immm, 3.3933, 4.3727, 11.989, Z = 2, R = 0.021. Co in 2(b): 0,1/2,1/2; Sc(1) in 2(d): 1/2,0,1/2; Sc(2) in 4(i): 0,0,0.18802, C in 8(ℓ): 0,0.1673,0.3753. The structure contains C-C pairs (1.46 A) surrounded by a Sc_6Co_2 tetragonal antiprism plus an additional Sc atom. The Fe and Ni compounds are isostructural.

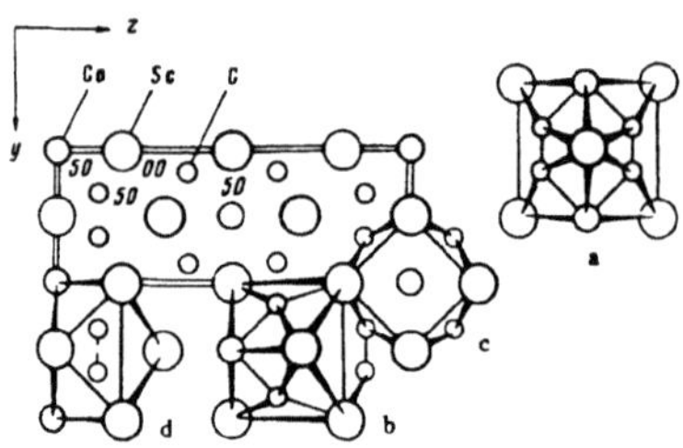

Projection of structure of Sc_3CoC_4 on yz plane and coordination polyhedra of atoms. a) Sc1; b) Sc2; c) Co; d) C_2.

$CoUC_2$

Mater. Res. Bull., **21**, 209-216.

P4/nmm, 3.4969, 7.313, Z = 2, R = 0.016. Co in 1(a): 3/4,1/4,0; U, C(1), C(2) in 2(c): 1/4,1/4,z, z = 0.32082, 0.8416, 0.6396. New structure which can be considered as an anti-PbFCl-type, with a pair of C atoms on the Pb site. C-C = 1.48, U-9C = 2.33-2.74, U-4Co = 2.93, U-8U = 3.50, 3.60, Co-4C = 2.10, Co-4Co = 2.47, Co-4U = 2.93 A.

CoYC

Z. Naturforsch., **41**B, 946-950.

$P4_2/mmc$, 3.6500, 6.8636, Z = 2, R = 0.048. Co in 2(b): 1/2,1/2,0; Y in 2(e): 0,0,1/4; C in 2(c): 0,1/2,0. CsCl-type arrangement of Co and Y, with insertion of C on octahedral sites formed by 2 Co and 4 Y giving a superstructure with doubled **c** axis. CoLnC compounds (Ln = Gd, Tb, Dy, Ho, Er, Tm, Lu) are isostructural.

$Cr_2Ho_2C_3$

Z. Metallk., **77**, 788-793.

C2/m, 10.470, 3.3650, 5.5403, 106.31, Z = 2, R = 0.027. Ho in 4(i): 0.39299,0,0.18347; Cr in 4(i): 0.15544,0,0.3853; C(1) in 2(c): 0,0,1/2; C(2) in 4(i): 0.7161,0,0.263. Related to $UMoC_2$-type.

$La_2Ni_5C_3$

Kristallografija, **31**, 73-75 [Soviet Physics - Crystallography, **31**, 39-41].

P4/mbm, 8.331, 4.0283, Z = 2, R = 0.081. La in 4(g): x,1/2+x,0, x = 0.3390; Ni(1) in 8(j): 0.2125,0.0752,1/2; Ni(2) in 2(a): 0,0,0; C(1) in 2(b): 0,0,1/2; C(2) in 4(h): x,1/2+x,1/2, x = 0.060. Derivative of ThB_4 type (**13**, 36; **17**, 65).

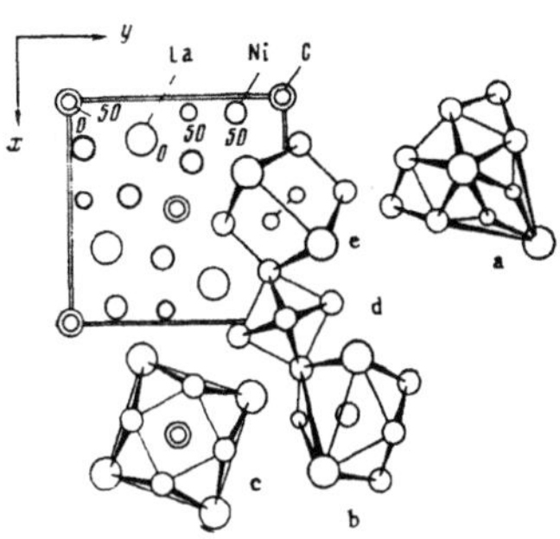

$Mn_{17}Pr_2C_{3-x}$ (x = 1.23)

Inorg. Chem., **25**, 279-282.

$R\bar{3}m$, a = 8.8714, c = 12.783 A, Z = 3, R = 0.023. Th_2Zn_{17}-type structure (**20**, 196), with C in Mn_4Pr_2 octahedral voids. La, Ce, Nd, and Sm compounds are isostructural.

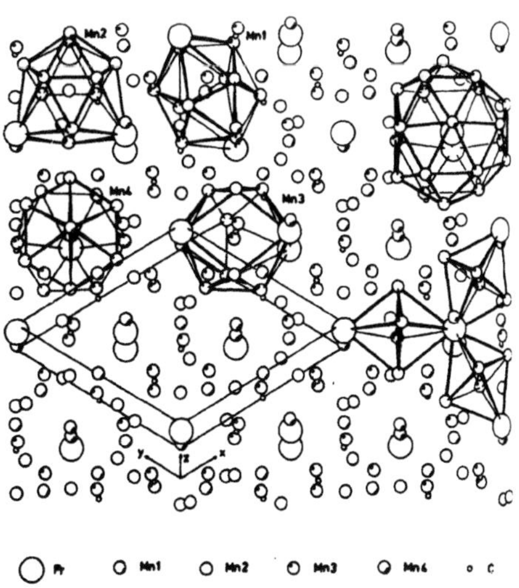

Interatomic Distances (Å) in the Structure of $Pr_2Mn_{17}C_{3-x}$

Pr	3 C	2.566	Mn(3)	1 C	1.935
	3 Mn(3)	3.186		2 Mn (1)	2.528
	1 Mn(4)	3.197		2 Mn(2)	2.659
	6 Mn(2)	3.220		2 Mn(3)	2.672
	3 Mn(3)	3.241		1 Mn(4)	2.715
	3 Mn(3)	3.421		2 Mn(2)	2.744
	3 Mn(1)	3.433		1 Pr	3.186
	1 Pr	3.950		1 Pr	3.241
				1 Pr	3.421
Mn(1)	4 Mn(2)	2.501		(2 C	3.492)
	4 Mn(3)	2.528			
	2 Mn(4)	2.718	Mn(4)	1 Mn(4)	2.438
	(4 C	3.331)		3 Mn(3)	2.715
	2 Pr	3.433		3 Mn(1)	2.718
				6 Mn(2)	2.773
Mn(2)	1 C	1.946		1 Pr	3.197
	2 Mn(2)	2.490			
	2 Mn(1)	2.501	C	2 Mn(3)	1.935
	2 Mn(3)	2.659		2 Mn(2)	1.946
	2 Mn(3)	2.744		2 Pr	2.566
	2 Mn(4)	2.773			
	2 Pr	3.220			

		occup	x	y	z
Pr	6c	1	0	0	0.34548
Mn(1)	9d	1	1/2	0	1/2
Mn(2)	18f	1	0.2807	0	0
Mn(3)	18h	1	0.50536	0.49464	0.15126
Mn(4)	6c	1	0	0	0.0954
C	9e	0.59	1/2	0	0

NiU_2C_3

Z. Kristallogr., **175**, 203-210.

Pnma, 6.4145, 3.5457, 13.2315, Z = 4, R = 0.029. New structure type, with one C in octahedra of 4U and 2Ni, and two C forming a C_2 pair.

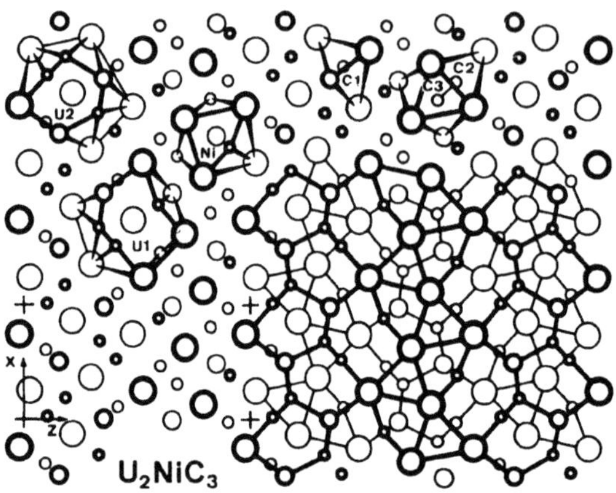

	Occupancy		x	y	z
U(1)	1.004	4c	0.24856	1/4	0.02163
U(2)	0.994	4c	0.37043	1/4	0.29143
Ni	0.978	4c	0.4918	1/4	0.8471
C(1)	1.10	4c	0.104	1/4	0.411
C(2)	0.95	4c	0.475	1/4	0.599
C(3)	0.93	4c	0.319	1/4	0.676

U(1):	1 C(2)	2.372 Å	U(2):	1 C(1)	2.329	C(1):	2 Ni	2.058
	1 C(1)	2.448		2 C(2)	2.496		1 U(2)	2.329
	2 C(1)	2.486		2 C(3)	2.637		1 U(1)	2.448
	2 C(2)	2.500		2 Ni	2.699		2 U(1)	2.486
	2 C(3)	2.739		2 C(3)	2.702			
	1 Ni	2.787		2 Ni	3.014	C(2):	1 C(3)	1.428
	2 Ni	2.989		2 U(2)	3.390		1 U(1)	2.372
	1 U(2)	3.464		1 U(1)	3.464		2 U(2)	2.496
	2 U(1)	3.546		2 U(2)	3.546		2 U(1)	2.500
	2 U(2)	3.606		2 U(1)	3.606			
	1 U(2)	3.654		1 U(1)	3.654	C(3):	1 C(2)	1.428
	2 U(1)	3.693					1 Ni	2.121
	2 U(1)	3.725	Ni:	2 C(1)	2.058		1 Ni	2.521
				1 C(3)	2.121		2 U(2)	2.637
				1 C(3)	2.521		2 U(2)	2.702
				2 U(2)	2.699		2 U(1)	2.739
				1 U(1)	2.787			
				2 U(1)	2.989			
				2 U(2)	3.014			

Ta_2C, W_2C

J. Less-Common Metals, **120**, 239-245.

Ta_2C, $P\bar{3}m1$, 3.1037, 4.9394, Z = 1, neutron powder data. Ta in 2(d): 1/3,2/3,0.2550; C in 1(a): 0,0,0. Anti-CdI_2-type (**3**, 309; **30**A, 40). Ta-C = 2.190 A.

W_2C, $P\bar{3}1m$, 5.1809, 4.7216, Z = 3, neutron powder data. W in 6(k): 0.3270,0,0.2553; 0.63 C(1) in 2(d): 1/3,2/3,1/2; 0.82 C(2) in 1(a): 0,0,0; 0.22 C(3) in 2(c): 1/3,2/3,0. ε-Fe_2N-type (**21**, 140; **44**A, 46), but carbon deficient ($W_2C_{0.84}$) and C distributed in three sites. W-C = 2.079-2.120 A.

$Ba_8Ga_{16}Si_{30}$, $Ba_8Ga_{16}Ge_{30}$, $Ba_8Al_{16}Ge_{30}$ (I, II, III), $Ba_8Ga_{16}Sn_{30}$ (IV)

J. Less-Common Metals, **118**, 43-55.

I, II, III, Pm3n, 10.530, 10.767, 10.835, Z = 1, R = 0.053, 0.079, 0.052. Ba(1) in 2(a): 0,0,0; Ba(2) in 6(d): 1/4,1/2,0; (Ga,Si), (Ga,Ge), Al(1) in 6(c): 1/4,0,1/2; (Ga,Si),(Ga,Ge), (Al,Ge)(2) in 16(i): x,x,x, x = 0.1850, 0.1842, 0.1844; (Ga,Si), (Ga,Ge), (Al,Ge)(3) in 24(k): 0,y,z, y = 0.3063, 0.3082, 0.3079, z = 0.1192, 0.1181, 0.1168. Na_8Si_{46}-type clathrate structure (**30**A, 96; **37**A, 133).

IV, $I\bar{4}3m$, 11.595, Z = 1, R = 0.073. Ba in 8(c): x,x,x, x = 0.1846; (Ga,Sn)(1) in 12(d): 1/4,1/2,0; Sn(2) in 2(a): 0,0,0; (Ga,Sn)(3) in 24(g): x,x,z, x = 0.4159, z = 0.1491; (Ga,Sn)(4) in 8(c): x,x,x, x = 0.3656. New clathrate-type structure, with tetrahedral coordinations for (Ga,Sn) and Sn.

CaPtSi

J. Solid State Chem., **62**, 133-137.

$P2_13$, 6.320, Z = 4, R = 0.050. Atoms in 4(a): x,x,x, x = 0.1276, 0.4189, 0.8345 for Ca, Pt, Si. LaIrSi-type (**49**A), Pt-Si = 2.30 A (very short).

$CeCo_{0.83}Si_{2.17}$, $LuCo_{0.64}Si_2$

J. Less-Common Metals, **125**, 147-156.

Cmcm, a = 4.143, 3.928, b = 16.505, 15.888, c = 4.076, 3.857 A, Z = 4, R = 0.026, 0.038. Si(1), Si(2), (0.83Co + 0.17Si) or 0.64 Co, and Ce or Lu in 4(c): 0,y,1/4, y = 0.0401, 0.7499, 0.1805, 0.3928; 0.0442, 0.7500, 0.1833, 0.3946. Non-stoichiometric $CeNiSi_2$- or $BaCuSn_2$-type (**34**A, 63; **35**A, 96; **40**A, 33), with different types of defects.

$CePd_2Si_2$

J. Less-Common Metals, **117**, 375-383.

I4/mmm, 4.2367, 9.8880, Z = 2, powder data. Ce in 2(a): 0,0,0; Pd in 4(d): 0,1/2,1/4; Si in 4(e): 0,0,0.3713. $ThCr_2Si_2$-type (**43**A, 99), as previously described (**42**A, 62). The Ru, Rh, Os, and low-temperature Ir compounds also have this structure; the Pt compound has a primitive tetragonal structure, and the high-temperature Ir compound has the $CaBe_2Ge_2$-type structure (**43**A, 28).

$Co_4Lu_5Si_4$

Acta Cryst., C**42**, 945-949.

$P2_1/c$, 12.245, 7.756, 7.684, 98.87, Z = 2, R = 0.048.

The structure can be considered as an intergrowth of three different types of slabs. One slab is built up of close-packed Si-centered trigonal rare-earth prisms arranged as in the AlB_2 type. The second slab corresponds to a segment of a ternary $BaAl_4$-derivative structure, constructed of Lu_4Si_4 square antiprisms, but with only one of every two of the available prism centers occupied by Co atoms. The third slab is made up of rare-earth-centered Si cuboctahedra, connected by vertices, and with extra Co atoms positioned in between the polyhedra.

		x	y	z
Co(1)	4(e)	0·0000	0·1261	0·1275
Si(1)	4(e)	0·0006	0·3335	0·3351
Si(2)	4(e)	0·1792	0·2256	0·0512
Si(3)	4(e)	0·1791	0·7751	0·0382
Si(4)	4(e)	0·1795	0·0060	0·2686
Si(5)	4(e)	0·1798	0·5068	0·3191
Co(2)	4(e)	0·2506	0·5004	0·0611
Lu(1)	4(e)	0·36551	0·25154	0·34236
Si(6)	4(e)	0·4393	0·0004	0·1105
Si(7)	4(e)	0·4410	0·4988	0·1098
Lu(2)	4(e)	0·63427	0·25089	0·15988
Lu(3)	2(c)	0	0	½

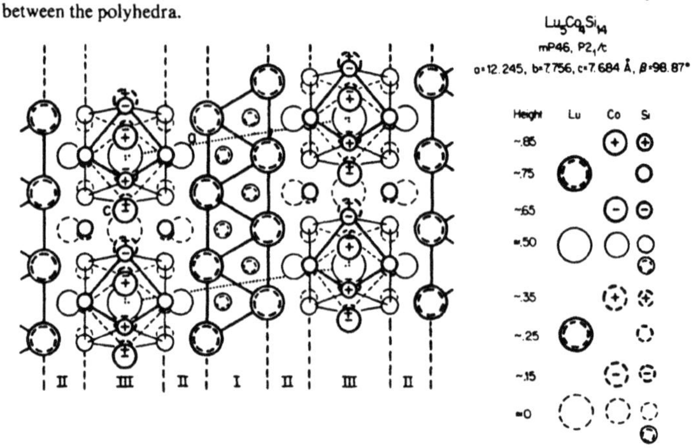

$Eu_2Ni_8Si_3$

Dopov. Akad. Nauk Ukr.RSR, Ser. A: Fiz.-Mat. Tekh. Nauki, No. 12, 63-65.

$P4_2/nmc$, a = 9.980, 7.485, R = 0.034.

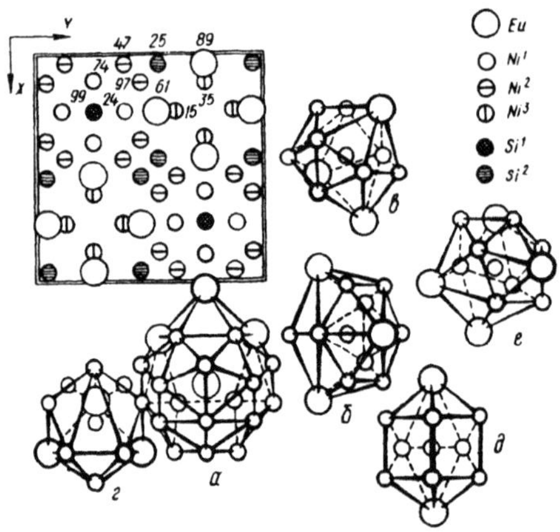

	x	y	z
Eu	1/4	0,95570	0,6099
Ni^1	1/4	0,1121	0,2432
Ni^2	0,1234	0,0423	0,9679
Ni^3	1/4	0,8766	0,1469
Si^1	1/4	1 4	0,9852
Si^2	0,0415	00,9585	1 4

$Ga_2Mo_{2.85}Si_{4.15}$, $Ga_2Ge_{4.2}Mo_{2.8}$

J. Less-Common Metals, **119**, 91-97.

$P6_222$, a = 4.706, 4.839, c = 6.543, 6.686 A, Z = 1. Powder data. (2.85Mo + 0.15Ga) or (2.8Mo + 0.2Ga) in 3(d): 1/2,0,1/2; (1.85Ga + 4.15Si) or (1.8Ga + 4.2Ge) in 6(j): x,2x,1/2, x = 0.1631, 0.1670. $CrSi_2$-type structures (**3**, 35, 628).

La_8MgSi_6

J. Solid State Chem., **62**, 199-206.

$P2_1/m$, 12.701, 4.347, 10.507, 107.58, Z = 2, R = 0.021. Isolated Si atoms and planar, five-membered, quasi-aromatic Si_5 rings which form $LiSi_5$ sandwich stacks. Si-Si = 2.35-2.39 A.

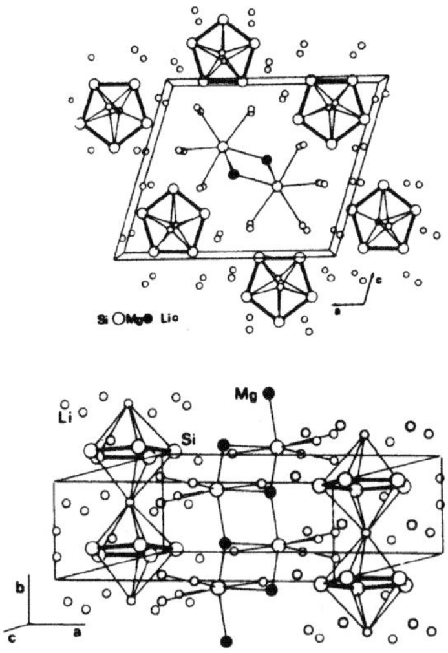

SOF	Atom	x	y	z
1	Si1	0.66091	1/4	0.00979
1	Si2	0.65567	1/4	0.23190
1	Si3	0.84075	1/4	0.37706
1	Si4	0.95736	1/4	0.23917
1	Si5	0.84860	1/4	1.01471
1	Si6	0.34404	1/4	0.38244
1	Li1	0.4424	1/4	0.9008
1	Li2	0.2938	1/4	0.6167
1	Li3	0.0275	1/4	0.6005
1	Li4	0.0017	1/4	0.8899
1	Li5	0.2113	1/4	0.8448
1	Li6	0.4230	1/4	0.1688
1	Li7	0.1693	1/4	0.1540
0.5	Li8	0.8521	0.616	0.6011
0.422	Mg11	0.5669	1/4	0.4591
0.06	Mg12	0.548	1/4	0.525
0.04	Mg13	0.512	1/4	0.601
0.04	Mg14	0.490	0.100	0.634
0.07	Mg15	0.643	1/4	0.751
0.13	Mg16	0.7028	1/4	0.7669
0.11	Mg17	0.7950	1/4	0.609
0.03	Mg18	0.686	1/4	0.509

$LaNi_2Si_2$

Kexue Tongbao, **31**, 455-459.

I4/mmm, 4.111, 9.690, Z = 2, R = 0.10. $BaAl_4$-type structure (**3**, 45, 330); Si in 4(e): x = 0.34-0.40.

$Sc_6Ni_{18}Si_{11}$

Kristallografija, **31**, 796-799 [Soviet Physics - Crystallography, **31**, 470-472].

Immm, 17.949, 12.223, 8.029, Z = 4, R = 0.062. Sc atoms have 15- and 17-coordinations, Ni 11-, 12- and 13-, and Si 10- and 12-coordinations.

($\times 10^4$)	x	y	z
Sc1	1288	1/2	0
Sc2	3096	1/2	0
Sc3	0	1831	2034
Sc4	2845	1835	0
Ni1	0	3810	0
Ni2	4311	0	0
Ni3	7901	0	8351
Ni4	9239	0	7305
Ni5	1137	1004	0
Ni6	4214	3288	0
Ni7	2179	3453	1697
Ni8	4155	1476	2219
Si1	0	0	0
Si2	1/2	1/2	0
Si3	1/2	0	7321
Si4	0	6753	1/2
Si5	3320	0	1779
Si6	1168	2840	0
Si7	1566	1688	2587

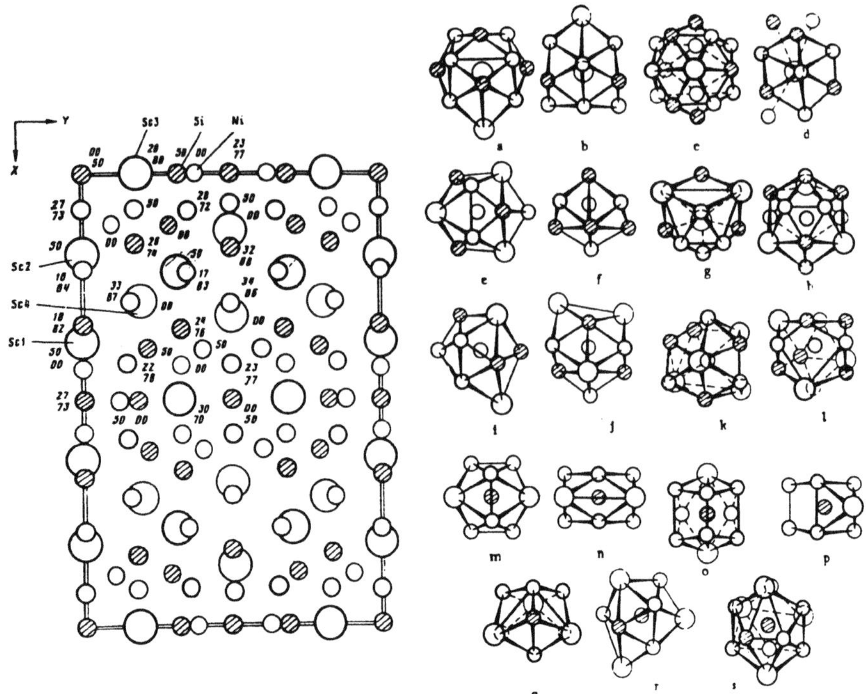

Coordination polyhedra of atoms in structure. a) Sc1; b) Sc2; c) Sc3; d) Sc4; e) Ni1; f) Ni2; g) Ni3; h) Ni4; i) Ni5; j) Ni6; k) Ni7; l) Ni8; m) Si1; n) Si2; o) Si3; p) Si4; q) Si5; r) Si6; s) Si7.

$Sc_2V_3Si_4$

Kristallografija, **31**, 793-795 [Soviet Physics - Crystallography, **31**, 469-470].

$P4_12_12$, 6.644, 12.270, Z = 4, R = 0.026. Isostructural with the Re (**44**A, 95) and Cr compounds. The structure contains Si-Si pairs, with each Si cooordinated in addition to 8 Sc or V atoms. Si-Si = 2.42, Si-Sc = 2.67-2.87, Si-V = 2.43-2.64 A.

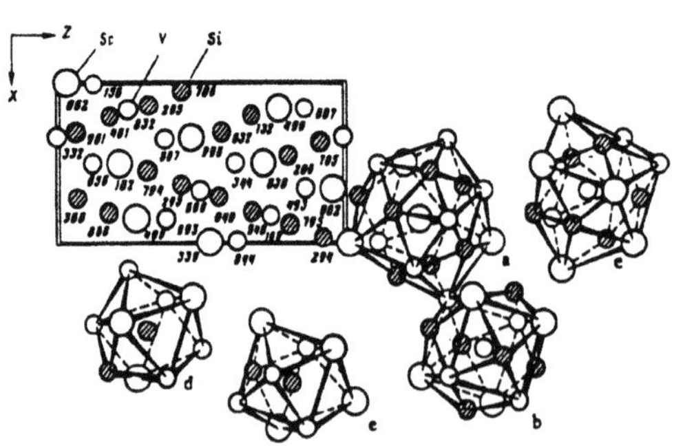

x 10^4

	x	y	z
Sc	19	6617	306
V1	3318	3318	0
V2	68	1557	1267
Si1	2940	9605	643
Si2	7053	3679	705

$AlZr_3N$

Z. Kristallogr., **175**, 211-215.

Cmcm, 3.3690, 11.498, 8.9825, Z = 4, R = 0.042. Al in 4(c): 0,y,1/4, y = 0.7474; Zr(1) in 8(f): 0,y,z, y = 0.3719, z = 0.0420; Zr(2) in 4(c): y = 0.0446; N in 4(a): 0,0,0. Filled Re_3B-type structure (**32**A, 24; **37**A, 55). The Hf compound is isostructural.

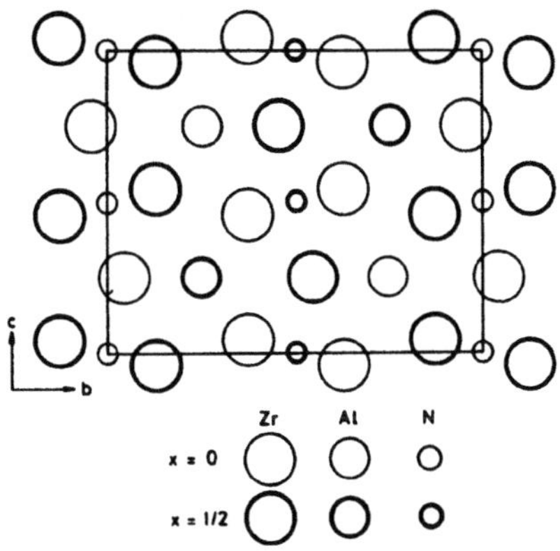

BORON NITRIDE (CUBIC)
BN

J. Less-Common Metals, **117**, 61-71.

F$\bar{4}$3m, 3.6153, Z = 4, R = 0.0054. B in 4(a): 0,0,0; N in 4(c): 1/4,1/4,1/4. Zincblende structure as previously described (**21**, 194; **28**, 43; **40**A, 103). Charge density studies indicate a transfer of 0.46 electron from B to N.

Li_3BN_2

J. Solid State Chem., **65**, 6-12.

$P2_1/c$, 5.1502, 7.0824, 6.7908, 112.956, Z = 4, R = 0.023. Two layers alternate parallel to (100), one with Li and B, the other with N. Each Li is coordinated to 4 N in a distorted tetrahedron, and B is coordinated linearly to 2 N. Li-N = 1.982-2.509, B-N = 1.336, 1.339(1) A.

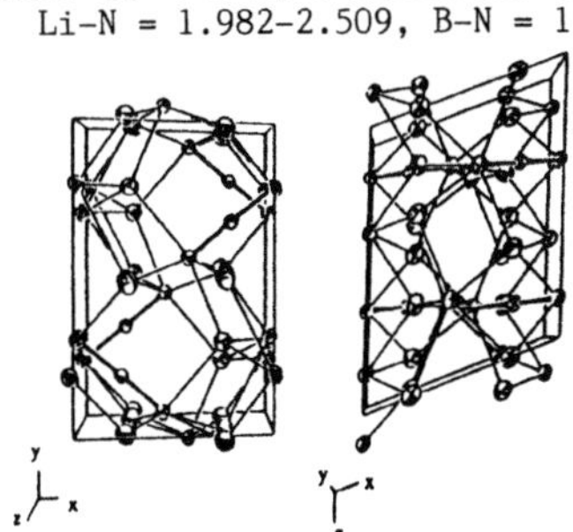

FRACTIONAL ATOMIC COORDINATES ($\times 10^5$)

	x	y	z
Li(1)	24995	48518	49811
Li(2)	25228	1212	37502
Li(3)	74360	20714	31402
B	21509	31982	17648
N(1)	43403	43735	21904
N(2)	99439	20472	13461

η-Ti_3N_{2-x}

J. Less-Common Metals, **125**, 127-134.

$R\bar{3}m$, a = 7.4236 A, α = 23.16°, Z = 1 (hexagonal cell, a = 2.9809, c = 21.664 A, Z = 3), powder data. 6Ti(1) in 6(c): 0,0,z, z = 0.2222; 3Ti(2) in 3(a): 0,0,0; 2.6N(1) in 6(c): z = 0.3889; 1.3N(2) in 3(b): 0,0,1/2. Isostructural with ε-Hf_3N_{2-x} and Ta_2VC_{2-x} (35A, 43, 68), with a nine-layer $(hhc)_3$ metal close packing, and N in octahedral sites. Ti-N = 2.10, Ti-Ti = 2.98, N-N = 2.41 A.

ζ-Ti_4N_{3-x}

J. Less-Common Metals, **120**, 153-159.

$R\bar{3}m$, a = 9.8070 A, α = 17.47°, Z = 1 (hexagonal cell, a = 2.9795, c = 28.965 A, Z = 3), powder data. Ti(1) in 6(c): 0,0,z, z = 0.1255; Ti(2) in 6(c): z = 0.2910; 0.583 N(1) in 6(c): z = 0.4150; 0.583 N(2) in 3(a): 0,0,0; 0.583 N(3) in 3(b): 0,0,1/2. Close packed Ti arrangement with $(hhcc)_3$ stacking, and N probably randomly distributed in three sites, as for Hf_4N_3 (35A, 69).

$Ba_3Sn_2P_4$

Z. anorg. Chem., **532**, 73-80.

$P2_1$, 7.886, 19.278, 7.869, 112.8, Z = 4, R = 0.092. Tetrahedral P_3Sn-SnP_3 groups share edges and corners to form infinite chains [description in original paper is in terms of $(Sn_2)P_6$ distorted octahedra]; Ba ions have 6-coordinations, Ba-P = 3.14-3.47 A.

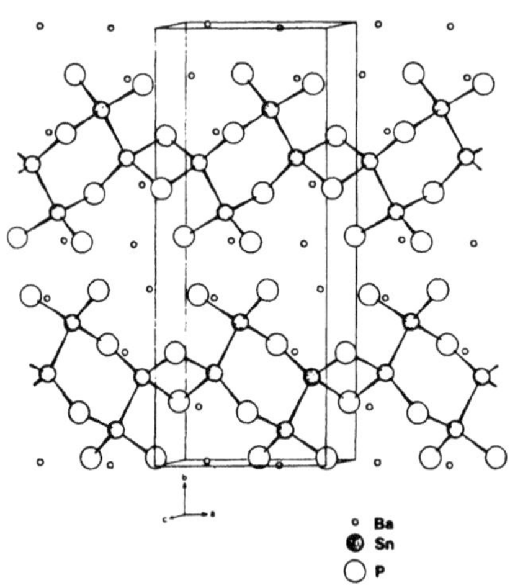

	x	y	z
Ba1	0,2990	0,3769	0,6906
Ba2	0,2975	0,5000	0,1749
Ba3	0,7645	0,2530	0,6964
Ba4	0,2258	0,1302	0,7846
Ba5	0.8206	0.3892	0.1933
Ba6	0.1849	0.9998	0.2971
Sn1	0.6483	0.0725	0.3390
Sn2	0,2340	0,1980	0,3059
Sn3	0.3535	0,3129	0,1638
Sn4	0,7837	0,1865	0.2033
P1	0,0197	0,2456	0.4616
P2	0,9652	0,1314	0,0361
P3	0.5310	0,2604	0,9795
P4	0,4612	0,1175	0,5330
P5	0,0634	0,3664	0,9470
P6	0.5513	0.5013	0,9395
P7	0.0680	0,5088	0.4438
P8	0,5692	0,3800	0.4355

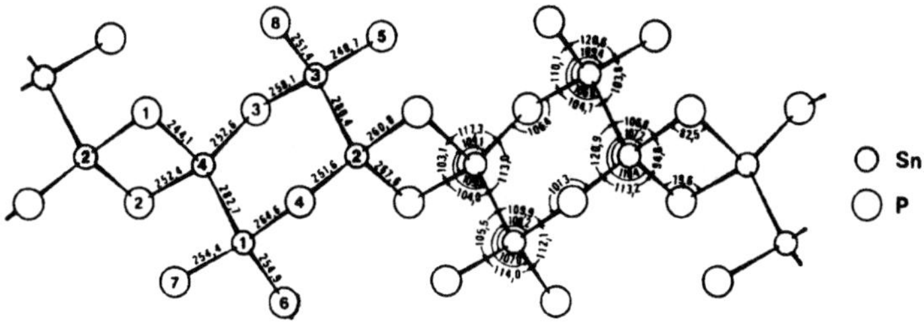

$CaPtP$, $EuPtP$, $CaPtAs$, $EuPtAs$, $CaPtSb$, $EuPtSb$, $EuPt_{0.65}P_{1.35}$, $EuPt_{0.55}As_{1.45}$

Z. anorg. Chem., **543**, 49-62.

$CaPtP$, $Pmc2_1$, 4.344, 8.311, 6.519, Z = 4, R = 0.035. Ca(1) in 2(a): 0,y,z, y = 0.5860, z = 0.5719; Ca(2) in 2(a): y = 0.8722, z = 0.1362; Pt(1), Pt(2), P(1), P(2) in 2(b): 1/2,y,z, y = 0.1698, 0.3420, 0.3836, 0.0733, z = 0, 0.3778, 0.7562, 0.3321. AlB_2-related structure with deformed PtP hexagonal nets.

$EuPtP$, $P6_3/mmc$, 4.086, 8.630, Z = 2, R = 0.052. Eu in 2(a): 0,0,0; Pt in 2(c): 1/3,2/3,1/4; P in 2(d): 2/3,1/3,1/4. Eu-Pt, P = 3.197, Pt-P = 2.359(1) A. Ni_2In-type derivative.

$CaPtAs$, $EuPtAs$, $I4_1md$, a = 4.183, 4.230, c = 43.711, 14.878 A, Z = 12, 4, R = 0.052, 0.046. Atoms in 4(a): 0,0,z; for CaPtAs, z = 0, 0.1665, 0.8335, 0.3044, 0.4707, 0.6379, 0.3608, 0.5265, 0.6937 for Ca(1,2,3), Pt(1,2,3), As(1,2,3); for EuPtAs, z = 0, 0.4149, 0.5822 for Eu, Pt, As. AlB_2-related and modified $ThSi_2$-type structures.

$CaPtSb$, $EuPtSb$, Pnma, a = 7.317, 7.581, b = 4.574, 4.642, c = 7.869, 7.893 A, Z = 4, R = 0.048, 0.040. Atoms in 4(c): x,1/4,z; for CaPtSb, x = 0.0064, 0.3035, 0.1879, z = 0.6963, 0.4154, 0.0908 for Ca, Pt, Sb; for EuPtSb, x = 0.0083, 0.3066, 0.1837, z = 0.6947, 0.4140, 0.0893 for Eu, Pt, Sb. Distorted Ni_2In-type, with puckered PtSb layers.

$EuPt_{0.65}P_{1.35}$, $EuPt_{0.55}As_{1.45}$, P6/mmm, a = 4.065, 4.181, c = 4.290, 4.319 A, Z = 1. Eu in 1(a): 0,0,0; Pt/P or Pt/As in 1(d): 1/3,2/3,1/2. AlB_2-type (**3**, 28).

$CaRu_2P_2$, $EuRu_2P_2$, $CaPt_2P_{1.43}$, $EuPt_2P_{1.62}$

Z. Naturforsch., **40**B, 38-43.

Ru compounds, I4/mmm, a = 4.046, 4.029, c = 9.771, 10.772, Z = 2, R = 0.018, 0.029. Ca or Eu in 2(a): 0,0,0; Ru in 4(d): 0,1/2,1/4; P in 4(e): 0,0,z, z = 0.3682, 0.3467. $ThCr_2Si_2$-type (**43**A, 99).

Pt compounds, I4/mmm, a = 4.099, 4.143, c = 18.951, 19.120 A, Z = 4, R = 0.043, 0.071. Ca or Eu in 4(e): 0,0,z, z = 0.3708, 0.3704; Pt(1) in 4(e): z = 0.1949, 0.1953; Pt(2) in 4(c): 1/2,0,0; 0.43 or 0.62 P(1) in 4(d): 1/2,0,1/4; P(2) in 4(e): z = 0.0746, 0.0742. Stacking variation of the $CaBe_2Ge_2$-type (**43**A, 28).

$(Ce,La)_{12}Rh_{30}P_{21}$

J. Less-Common Metals, **120**, 247-254.

$P6_3/m$, 17.475, 3.948, Z = 1, R = 0.049. New structure type, with LnP_6 trigonal prisms, and Rh with square pyramidal, tetrahedral, and triangular coordinations to P. There is some P and Rh disorder. Ln-P = 3.04-3.12, Rh-P = 2.28-2.72(1) A.

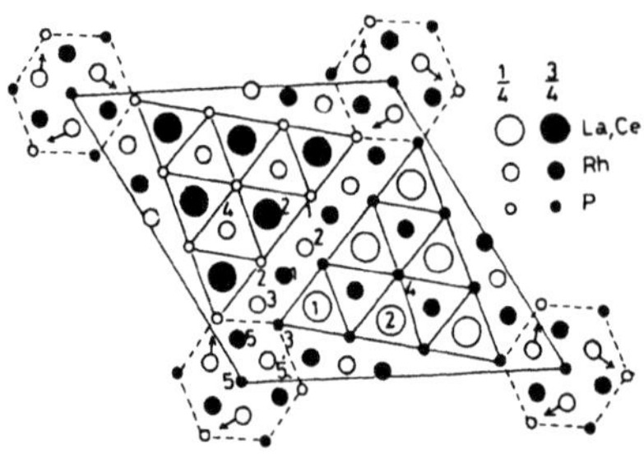

		x	y	z
Ln(1)	6h	0.43745	0.81864	3/4
Ln(2)	6h	0.23394	0.87407	3/4
Rh(1)	6h	0.6492	0.9585	3/4
Rh(2)	6h	0.9897	0.4377	3/4
Rh(3)	6h	0.2528	0.0703	3/4
Rh(4)	6h	0.7745	0.4927	3/4
[a]Rh(5)	6h	0.0846	0.9333	3/4
[a]Rh'(5)	6h	0.9322	0.0552	1/4
P(1)	6h	0.4643	0.3904	3/4
P(2)	6h	0.7187	0.5920	3/4
P(3)	6h	0.2107	0.1854	3/4
P(4)	2c	2/3	1/3	3/4
[a]P(5)	2a	0	0	3/4

[a]Occupancy factor 0.5

$Co_4La_{0.2}P_{12}$, $Ce_{0.25}Co_4P_{12}$

J. Solid State Chem., **65**, 1-5.

Im3, 7.745, 7.738, Z = 2, R = 0.014 for Ce compound. 0.4La or 0.5Ce in 2(a): 0,0,0; 8Co in 8(c): 1/4,1/4,1/4; 24P in 24(g): 0,y,z, y = 0.3503, 0.35060, z = 0.1455, 0.14571. Isostructural with $LaFe_4P_{12}$ (**43**A, 71), with partial occupancy of Ln sites.

$Co_{0.2}Mn_{0.8}P$

Acta Chem. Scand., A**40**, 634-636.

Pnma, 5.239, 3.163, 5.885, Z = 4, R = 0.049. Atoms in 4(c): x,1/4,z, x = 0.0037, z = 0.1979 for Mn,Co, x = 0.1887, z = 0.5690 for P. MnP-type (**3**, 17, 264; **27**, 319; **30**A, 98; **51**A, 69).

$Co_{19}Zr_5P_{12}$

J. Less-Common Metals, **120**, 197-201.

$P\bar{6}2m$, a = 11.885, 3.596, Z = 1, R = 0.033. Isostructural with $Hf_2Co_4P_3$ (**33**A, 67), except for 0.5 Co(5) in position 2(e), rather than 1 Co in 1(a), and with Co in the 3(f) Hf sites.

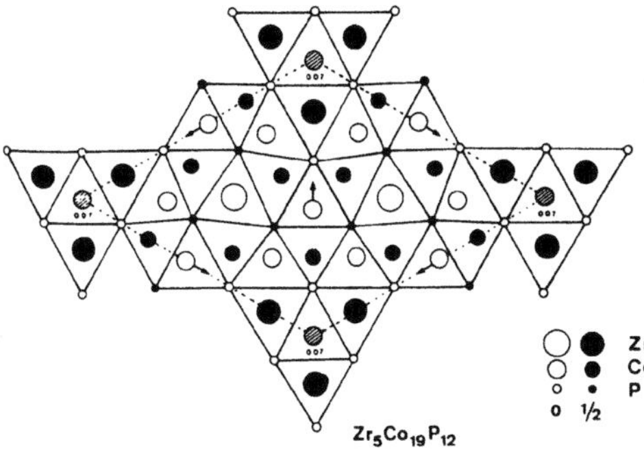

	X	Y	Z
Zr1	0.0000	0.1849	0.5000
Zr2	0.3330	0.6660	0.0000
Co1	0.0000	0.5507	0.0000
Co2	0.0000	0.7166	0.5000
Co3	0.1981	0.3760	0.0000
Co4	0.1281	0.4823	0.5000
Co5	0.0000	0.0000	0.074
P1	0.0000	0.8232	0.0000
P2	0.3165	0.4827	0.5000
P3	0.0000	0.3612	0.0000

$Ho_2Ni_{12}P_7$, $Ho_6Ni_{20}P_{13}$

J. Less-Common Metals, **118**, 191-200.

$Ho_2Ni_{12}P_7$, $P\bar{6}$, 9.063, 3.673, Z = 1, R = 0.030. $Zr_2Fe_{12}P_7$-type (**33**A, 103; **51**A, 71).

$Ho_6Ni_{20}P_{13}$, $P6_3/m$, 12.677, 3.730, Z = 1, R = 0.023. Essentially $Zr_6Ni_{20}P_{13}$ type (**51**A, 71), with disorder of one P and two Ni sites.

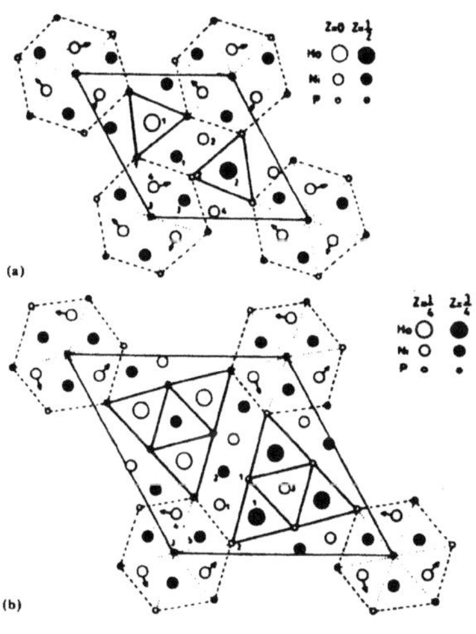

		x	y	z
(a) $Ho_2Ni_{12}P_7$				
Ho(1)	1c	1/3	2/3	0
Ho(2)	1f	2/3	1/3	1/2
Ni(1)	3k	0.0595	0.6257	1/2
Ni(2)	3j	0.4324	0.0495	0
Ni(3)	3k	0.8457	0.7234	1/2
Ni(4)	3j	0.1214	0.2156	0
P(1)	3k	0.2930	0.8854	1/2
P(2)	3j	0.4056	0.2932	0
P(3)	1b	0	0	1/2
(b) $Ho_6Ni_{20}P_{13}$				
Ho(1)	6h	0.52969	0.81133	1/4
Ni(1)	6h	0.75403	0.07386	1/4
Ni(2)	6h	0.41554	0.97903	1/4
Ni(3)	2d	2/3	1/3	1/4
Ni(4)[a]	6h	0.0923	0.8950	1/4
Ni(5)[a]	6h	0.9267	0.0817	3/4
P(1)	6h	0.6139	0.1386	1/4
P(2)	6h	0.9503	0.2326	1/4
P(3)[a]	2a	0	0	3/4

[a] Occupancy factor, 0.5.

$In_2Sr_3P_4$, $Ca_3In_2As_4$

Z. Naturforsch., **41B**, 1416-1419.

Pnnm, a = 16.323, 16.212, b = 6.828, 6.595, c = 4.289, 4.306, Z = 2, R = 0.076, 0.072. InX_4 tetrahedra (X = P, As) share edges and corners to form strings of alternating 4- and 8-membered rings. In-4P = 2.56-2.65, In-4As = 2.62-2.69, Sr-6P = 3.00-3.67, Ca-6As = 2.92-3.36 A.

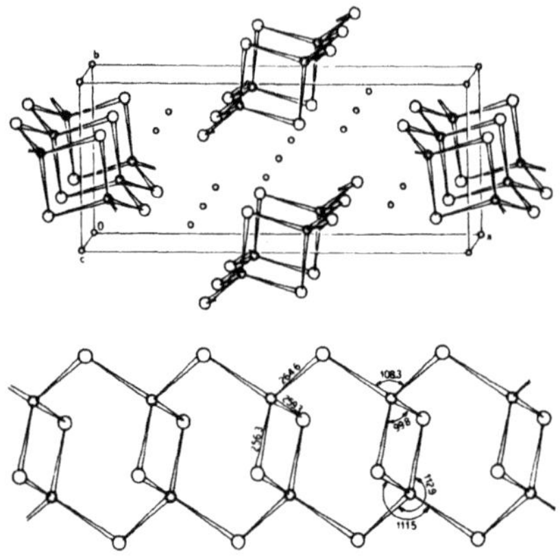

		x	y	z
$Sr_3In_2P_4$				
4 Sr1	4g	0.1986	0.7146	0
2 Sr2	2a	0	0	0
4 In1	4g	0.5850	0.1335	0
4 P1	4g	0.4325	0.2395	0
4 P2	4g	0.1654	0.2454	0
$Ca_3In_2As_4$				
4 Ca1	4g	0.1961	0.7244	0
2 Ca2	2a	0	0	0
4 In1	4g	0.5879	0.1381	0
4 As1	4g	0.4294	0.2567	0
4 As2	4g	0.1721	0.2302	0

Li_3P_7

Z. anorg. Chem., **539**, 95-109.

$P2_12_12_1$, 9.742, 10.535, 7.596, Z = 4, R = 0.023. P_7^{3-} cage anions linked by Li^+ ions. P-P = 2.14-2.27, Li-P = 2.50-2.98 A.

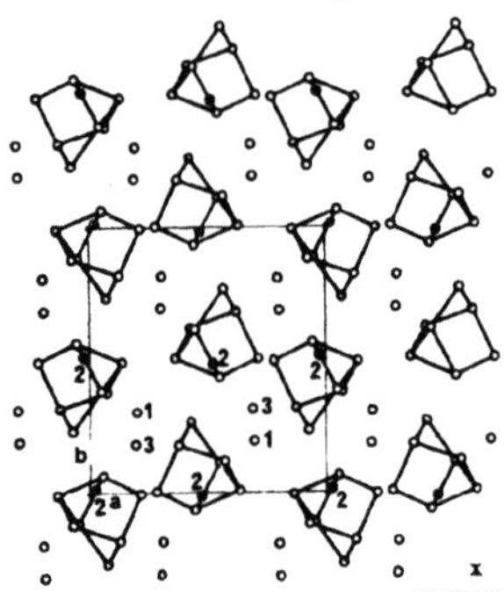

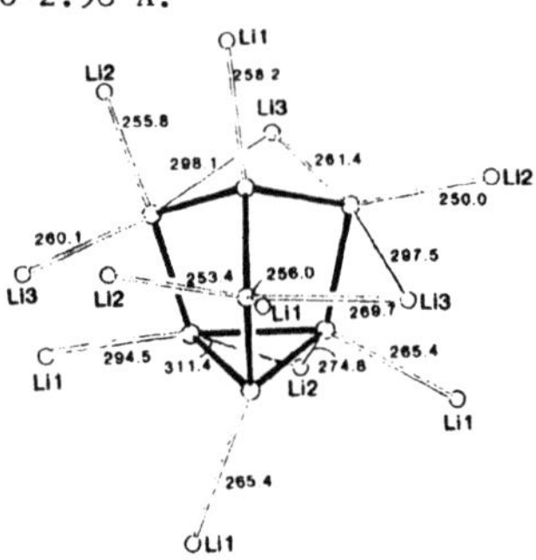

	x	y	z
P21	−0,0712	0,2338	0,5324
P22	−0,2174	0,4960	0,3708
P23	0,0732	0,3811	0,1827
P30	−0,1251	0,3194	0,2803
P31	0,0499	0,3866	0,6362
P32	−0,0537	0,5648	0,5335
P33	0,1373	0,4926	0,4054
Li1	−0,3007	0,1942	0,094
Li2	−0,0253	0,5150	−0,064
Li3	−0,3068	0,3160	0,651

MnNaX (X = P, As, Sb, Bi), KMnAs, LiMnAs

Z. anorg. Chem., **539**, 175-182.

P4/nmm, a = 4.086, 4.206, 4.478, 4.571, 4.391, 4.263, c = 6.884, 7.077, 7.555, 7.743, 7.791, 6.179 A, Z = 2, neutron powder data. Na, K in 2(c): 0,1/2,z, z = 0.637, 0.640, 0.651, 0.65, 0.64; Li in 2(b): 0,0,1/2; Mn in 2(a): 0,0,0; X in 2(c): z = 0.205, 0.213, 0.220, 0.23, 0.18, 0.758. Structure as previously described (**48**A, 73). The structures were also determined at 13-643K, and an antiferromagnetic structure with doubled c axis found for the K and Li compounds.

Mn_2TiP_{12}

Z. anorg. Chem., **540/541**, 234-242.

C2/c, 16.058, 5.794, 10.636, 115.23, Z = 4, R = 0.021. Ti has square-antiprismatic coordination to 8 P, and Mn octahedral coordination to 6 P, with a Mn-Mn bond; P atoms have tetrahedral coordinations to metal and P.

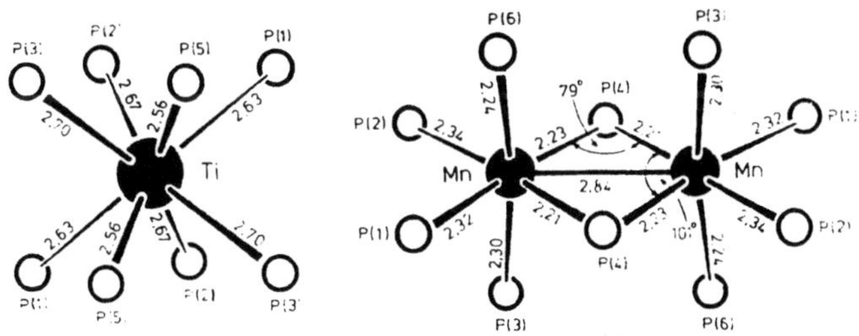

		x	y	z
Ti	4e	0	0,30297	1/4
Mn	8f	0,18221	0,19021	0,04657
P(1)	8f	0,15581	0,17586	0,24430
P(2)	8f	0,46792	0,45113	0,07280
P(3)	8f	0,38319	0,05220	0,49347
P(4)	8f	0,32250	0,32712	0,16590
P(5)	8f	0,43761	0,14830	0,33357
P(6)	8f	0,27358	0,31946	0,42491

$Pd_{15}P_2$, $Pd_{15}P_2D_{0.46}$

Z. Phys. Chem. (Munich), **145**, 43-49 (1985).

$R\bar{3}$, a = 7.1067, 7.1166, c = 17.0867, 17.1281 A, Z = 3, neutron powder data. Pd(1) in 18(f): 0.3990,0.2836,0.9636; Pd(2) in 18(f): 0.2570,0.2422,0.7964; Pd(3) in 6(c): 0,0,0.0795; Pd(4) in 3(b): 0,0,1/2; P in 6(c): 0,0,0.2866: 0.076 D in 18(f): 0.171,0.080,0.016. Structure of $Pd_{15}P_2$ as previously described (**43**A, 89), with D in Pd_4 tetrahedral interstices.

PtYbP, PtYAs, NdPtSb, PtSbY

Z. Kristallogr., **176**, 125-134.

PtYbP, $P\bar{6}m2$, 4.077, 3.777, Z = 1, R = 0.050. Pt in 1(d): 1/3,2/3,1/2; Yb in 1(a): 0,0,0; P in 1(f): 2/3,1/3,1/2. AlB_2-type (**3**, 28). Yb-12 Pt/P = 3.018, Pt-3 P = 2.354(1) A. Other LnPtP compounds are isostructural.

PtYAs, $P6_3/mmc$, 4.250, 15.165, Z = 4, R = 0.099. Pt in 4(f): 1/3,2/3,0.1365; Y(1) in 2(a): 0,0,0; Yb(2) in 2(b): 0,0,1/4; As in 4(f): 1/3,2/3,0.6204. AlB_2-derivative structure. Y-6 Pt 3.210, 2.997(3), Y-6 As = 3.059, 3.144(7), Pt-3 As = 2.466(1) A. Other LnPtAs compounds are isostructural.

NdPtSb, $P6_3mc$, 4.534, 7.866, Z = 2, R = 0.042. Nd in 2(a): 0,0,0; Pt in 2(b): 1/3,2/3,0.2863; Sb in 2(b): 1/3,2/3,0.7365. Modified $CaIn_2$-type (**29**, 3). Nd-6 Pt = 3.111, 3.453(4), Nd-6 Sb = 3.211, 3.339(5), Pt-3 Sb = 2.647(1) A. The Pr compound is isostructural.

PtSbY, $F\bar{4}3m$, 6.538, Z = 4, R = 0.023. Pt in 4(c): 1/4,1/4,1/4; Sb in 4(b): 1/2,1/2,1/2; Y in 4(a): 0,0,0. MgAgAs-type structure (**8**, 28). Y-4 Pt and 6 Sb = 2.831, 3.269, Pt- 4 Y and 4 Sb = 2.831, 2.831, Sb-6 Y and 4 Pt = 3.269, 2.831(1) A.

$Sn_2Sr_5P_6$, SnSrP

J. Less-Common Metals, **116**, 251-258.

$Sn_2Sr_5P_6$, Pbam, 13.842, 12.161, 4.118, Z = 2, R = 0.070. $Ca_5Sn_2As_6$-type structure (**52**A, 13), with chains of edge-sharing SnP_4 distorted tetrahedra.

SnSrP, P4/nmm, 4.345, 8.942, Z = 2, R = 0.049. NbCrN-type structure (**37**A, 66; **38**A, 72) with puckered nets of Sn_4 rings, separated by P and Sr.

$Sr_5Sn_2P_6$

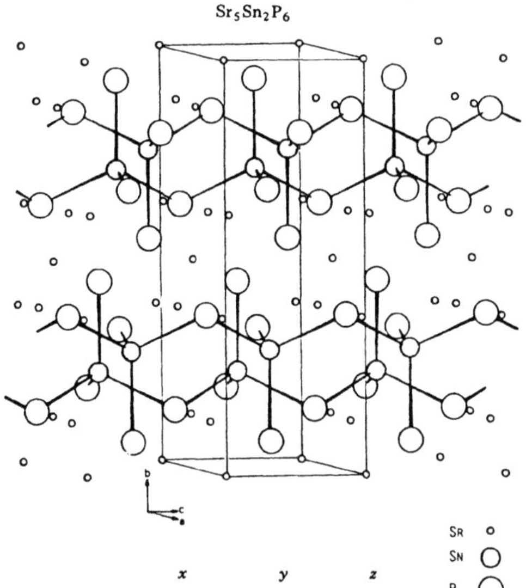

		x	y	z
4 Sr1	4g	0,2600	0,3811	0,0
2 Sr2	2a	0,0	0,0	0,0
4 Sr3	4g	0,4297	0,1266	0,0
4 Sn1	4h	0,1035	0,2158	0,5
4 P1	4h	0,9248	0,1707	0,5
4 P2	4h	0,1061	0,4206	0,5
4 P3	4g	0,1931	0,1358	0,0

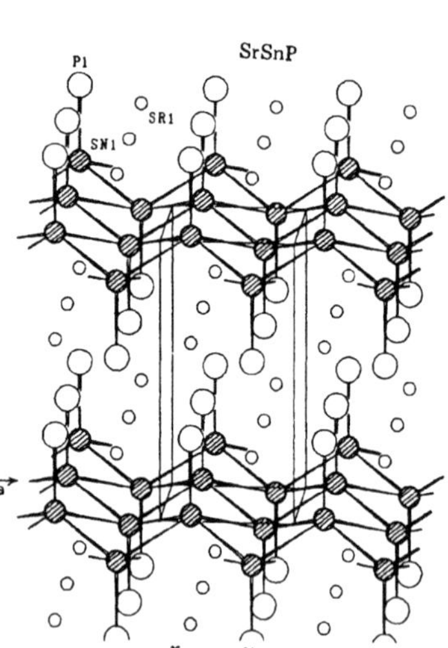

	x	y	z
2 Sr1	0,25	0,25	0,3329
2 Sn1	0,25	0,25	0,9474
2 P	0,25	0,25	0,6722

Tc_2P_3

J. Solid State Chem., **63**, 377-385.

$P\bar{1}$, 6.266, 6.325, 7.683, 95.79, 101.76, 104.34, Z = 4. R = 0.035 (twinned crystal). Tc_2As_3-type structure (**52**A, 14), which is a superstructure of the Mo_2As_3-type (**30**A, 19; **31**A, 22; this volume, p. 42)

	x	y	z
Tc(1)	0.92338	0.13096	0.63174
Tc(2)	0.46119	0.13780	0.63769
Tc(3)	0.61995	0.58315	0.85986
Tc(4)	0.09992	0.58510	0.85520
P(1)	0.5373	0.1828	0.9622
P(2)	0.0328	0.1771	0.9563
P(3)	0.6297	0.8494	0.6785
P(4)	0.1378	0.8643	0.6815
P(5)	0.7650	0.4375	0.6189
P(6)	0.2644	0.4231	0.6364

ThP_7

J. Less-Common Metals, **116**, 259-270.

$P2_12_12_1$, 10.218, 10.401, 5.671, Z = 4, R = 0.025. Three-dimensional P network, with some P disorder (some superstructure reflexions),and 10-coordinate Th in channels along **c**.

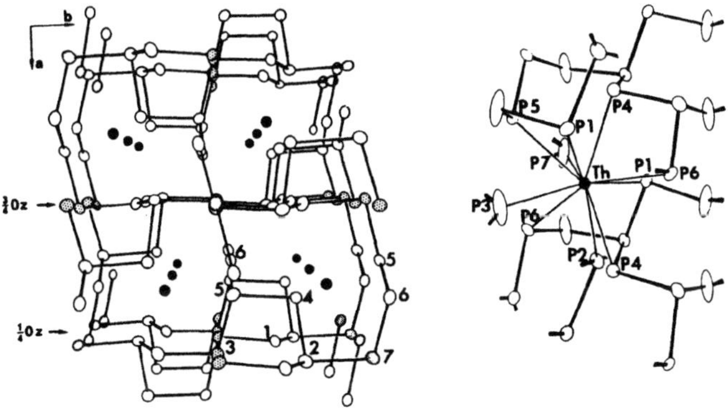

	x	y	z
Th	0.51106	0.16883	0.8651
P(1)	0.7538	0.7946	0.6804
P(2)	0.7310	0.2871	0.5483
P(3)	0.7589	0.0020	0.2752
P(3A)	0.7616	0.0006	0.2340
P(3B)	0.7562	0.0045	0.3150
P(4)	0.9410	0.2607	0.6366
P(5)	0.5504	0.9487	0.2079
P(6)	0.4482	0.9483	0.5553
P(7)	0.2402	0.9978	0.4883
P(7A)	0.2386	0.9941	0.5240
P(7B)	0.2418	0.0016	0.4510

Cu_4NaAs_2, Cu_4NaSb_2, Cu_4KAs_2

Z. anorg. Chem., **535**, 143-147.

$R\bar{3}m$, a = 4.100, 4.262, 4.120, c = 23.893, 25.413, 25.956 A, Z = 3, R = 0.061, 0.094, 0.072. Na or K in 3(a): 0,0,0; Cu(1) in 6(c): 0,0,z, z = 0.1429, 0.1411, 0.1444; Cu(2) in 6(c): z = 0.4384, 0.4488, 0.4443; As or Sb in 6(c): z = 0.2476, 0.2470, 0.2406. $CaCu_4P_2$-type (**46**A, 50).

Mo_2As_3

J. Solid State Chem., **63**, 377-385.

C2/m, 13.361, 3.2337, 9.6385, 124.57, Z = 4, R = 0.031. Atoms in 4(i): x,0,z. Structure as previously described (**30**A, 19; **31**A, 22).

	x	*z*
Mo(1)	0.24801	0.37127
Mo(2)	0.35103	0.13539
As(1)	0.12674	0.04751
As(2)	0.41512	0.68540
As(3)	0.10383	0.62754

Rb_2SiAs_2

Z. Naturforsch., **41**B, 731-735.

Ibam, 7.266, 13.771, 6.389, Z = 4, R = 0.047. Rb in 8(j): 0.1620,0.1454,0; As in 8(j): 0.3388,-0.0959,0; Si in 4(b): 1/2,0,1/4. K_2SiP_2-type (**51**A, 77), with $[SiAs_2{}^{2-}]$ chains of edge-sharing tetrahedra, linked by six-coordinate Rb^+ ions. Si-As = 2.381(1) Rb-As = 3.510-3.701(2) A, As-Si-As = 95.7-121.1, Si-As-Si = 84.3°.

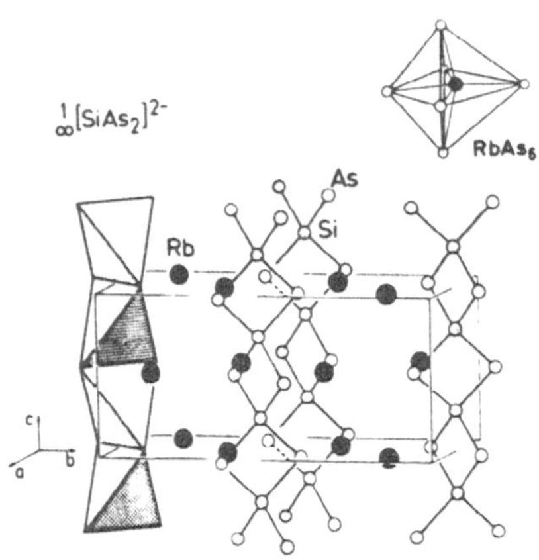

$Ag_{0.5}V_{0.5}PS_3$

Mater. Res. Bull., **21**, 917-928.

P2/a, 5.921, 10.684, 6.755, 106.62, Z = 4, R = 0.051. Isostructural with the Cr analogue (**50**A).

$Ba_{1.14}K_{0.72}VS_4$

J. Solid State Chem., **61**, 332-337.

Pna2_1, 9.158, 12.144, 6.729, Z = 4, R = 0.031. Structure related to that of β-K_2SO_4, with VO_4 tetrahedra linked by 9-coordinate Ba and 11-coordinate K/Ba ions. V-S = 2.109-2.164(2) A.

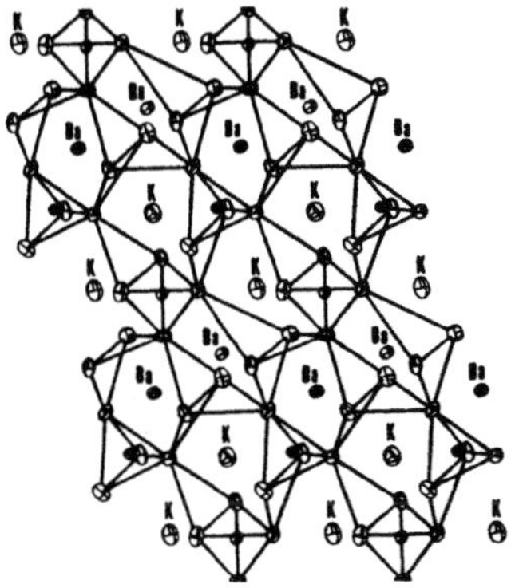

	x	y	z
Ba	0,02618	0,33022	¼
K-Ba(1)	0,3585	0,5954	0,2560
V	0,76319	0,57831	0,2548
S(1)	−0,0016	0,5950	0,2468
S(2)	0,1971	0,0921	0,2458
S(3)	0,3291	0,3439	0,0053
S(4)	0,1758	0,8431	−0,0014

$BaMo_6S_8$

Physica B+C, **136**, 485-488.

295K, R$\bar{3}$, a = 9.2974, c = 11.7734 A, Z = 3. Neutron powder data. Ba in 3(a): 0,0,0; Mo in 18(f): 0.0125,0.1709,0.4064; S(1) in 18(f): 0.3207,0.2937, 0.4160; S(2) in 6(c): 0,0,0.2490. 19K, P$\bar{1}$, 6.6976, 6.6545, 6.5685, 87.929, 88.978, 88.887, Z = 1, neutron powder data. See original paper for atomic parameters. Chevrel phases, with distorted Mo_6 octahedra and a shear of the axial S atoms in the low-temperature (below 175K) phase.

$BaPdS_2$

J. Less-Common Metals, **119**, 159-165.

Cmcm, 6.783, 10.634, 5.627, Z = 4, R = 0.042. Pd in 4(a): 0,0,0; Ba in 4(c): 0,0.34915,1/4; S in 8(g): 0.2367,0.0958,1/4. Isostructural with $BaNiO_2$ (**37**A, 263) with chains of edge-sharing PdS_4 square planes, linked by 8-coordinate Ba ions.

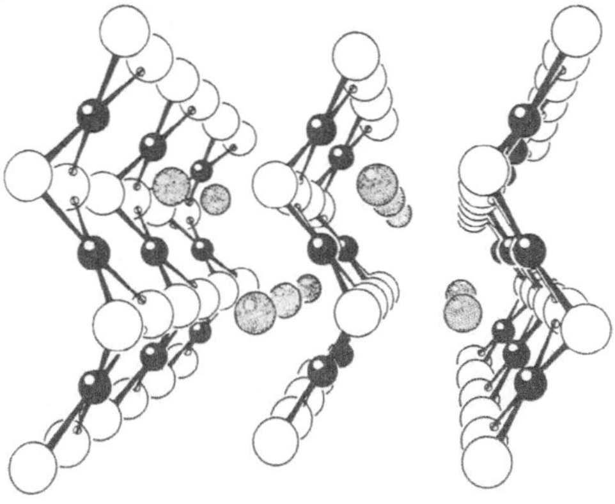

Pd-Pd	2.814(1) 2×
Pd-S	2.365(1) 4×
Pd-Ba	3.970(1) 2×
Pd-Ba	4.007(1) 4×
Ba-S	3.136(1) 2×
	3.173(1) 2×
	3.384(1) 4×
S-S around Pd	3.211(3) 2×
	3.474(1) 2×
Ba-Ba	4.267(1) 2×
	4.885(1) 4×

$Bi_2Eu_{1.1}S_4$

Acta Cryst., C42, 259-261.

$P6_3/m$, 24.820, 4.080, Z = 12, R = 0.053. Eu atoms have 8-fold bicapped trigonal-prismatic coordinations and Bi atoms octahedral coordinations; large channels contain a partially-occupied (0.5) Eu site.

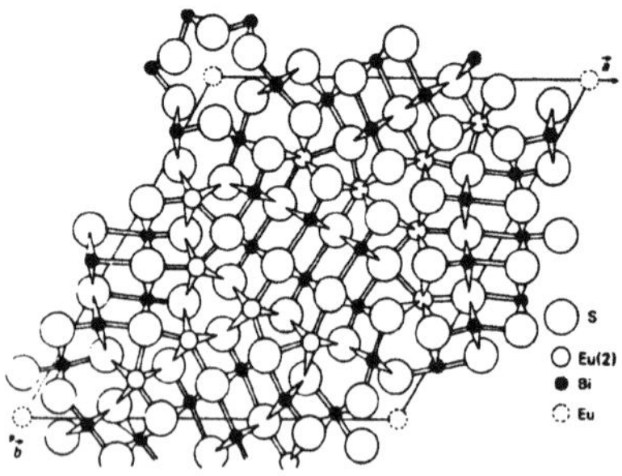

	x	y	z
Bi(1)	0,3328	0,0586	1/4
Bi(2)	0,1569	0,1776	1/4
Bi(3)	0,0534	0,4625	1/4
Bi(4)	0,3467	0,4954	1/4
Eu(1)	0,7679	0,1232	1/4
Eu(2)	0,6765	0,2337	1/4
S(1)	0,5481	0,0993	1/4
S(2)	0,4511	0,1858	1/4
S(3)	0,3679	0,3652	1/4
S(4)	0,1523	0,4562	1/4
S(5)	0,3430	0,5952	1/4
S(6)	0,1344	0,0590	1/4
S(7)	0,0358	0,2553	1/4
S(8)	0,1885	0,3082	1/4
Eu	0	0	1/4

(Å)

Bi(1)–S(7)	2,611 (14)	Bi(2)–S(6)	2,689 (14) ×2
Bi(1)–S(8)	2,784 (9) ×2	Bi(2)–S(6)	2,707 (15)
Bi(1)–S(3)	2,890 (9) ×2	Bi(2)–S(8)	2,930 (14)
Bi(1)–S(2)	3,052 (14)	Bi(2)–S(7)	2,944 (11) ×2
Bi(3)–S(4)	2,537 (12)	Bi(4)–S(5)	2,525 (12)
Bi(3)–S(1)	2,793 (9) ×2	Bi(4)–S(2)	2,689 (8) ×2
Bi(3)–S(3)	2,881 (9) ×2	Bi(4)–S(1)	2,996 (9) ×2
Bi(3)–S(1)	3,630 (15)	Bi(4)–S(3)	3,523 (14)
Eu(1)–S(4)	3,025 (10) ×2	Eu(2)–S(5)	3,061 (9) ×2
Eu(1)–S(7)	3,104 (11) ×2	Eu(2)–S(4)	3,076 (10) ×2
Eu(1)–S(2)	3,116 (13)	Eu(2)–S(2)	3,101 (15)
Eu(1)–S(3)	3,158 (15)	Eu(2)–S(5)	3,139 (10) ×2
Eu(1)–S(8)	3,170 (10) ×2	Eu(2)–S(1)	3,264 (15)
	Eu–S(6)	2,896 (31) ×3	
	Eu–S(6)	3,543 (26) ×6	

$Bi_4In_2Pb_4S_{13}$

Acta Cryst., C42, 1089-1091.

Pcma, 21.344, 4.002, 26.494, Z = 4, R = 0.054.

The structure consists of In–S tetrahedra and octahedra, mono- and bicapped trigonal prisms of Pb–S, and Bi–S octahedra; all polyhedra are distorted and form chains parallel to [010]. The site-occupancy factors >1 of the octahedrally coordinated In atoms indicate a partial substitution of In^{3+} by Bi^{3+}.

atomic coordinates (×10^4)

	x	y	z
Pb(1)	359	5000	1586
Pb(2)	1724	5000	2788
Pb(3)	1830	5000	4324
Pb(4)	3754	5000	2501
In(1)	0	0	0
In(2)	2255	0	1624
In(3)	5000	0	0
Bi(1)	232	0	3431
Bi(2)	1547	5000	249
Bi(3)	3573	0	3903
Bi(4)	3834	5000	848
S(1)	91	5000	4292
S(2)	262	5000	2785
S(3)	833	0	725
S(4)	1211	0	2025
S(5)	1467	0	3578
S(6)	2452	5000	1100
S(7)	2729	0	2452
S(8)	2761	0	4740
S(9)	2929	5000	3520
S(10)	3819	0	87
S(11)	3951	0	1574
S(12)	4348	5000	4502
S(13)	4381	0	3134

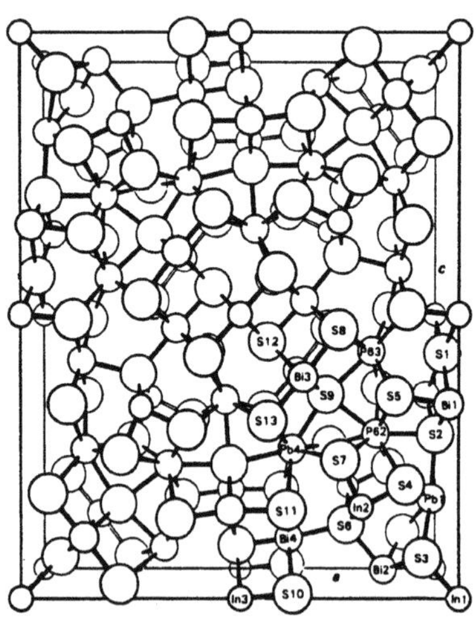

Pb(1)–S(3)	3·197 (12)	2×	
S(4)	2·944 (11)	2×	
S(13)	2·985 (11)	2×	
S(2)	3·184 (14)		
Pb(2)–S(5)	2·947 (10)	2×	
S(4)	3·049 (12)	2×	
S(7)	3·066 (10)	2×	
S(2)	3·120 (14)		
S(9)	3·235 (18)		
Pb(3)–S(5)	2·917 (10)	2×	
S(8)	3·028 (12)	2×	
S(10)	3·165 (11)	2×	
S(9)	3·166 (15)		
Pb(4)–S(7)	2·968 (10)	2×	
S(11)	3·196 (13)	2×	
S(13)	2·934 (11)	2×	
S(9)	3·225 (15)		
S(2)	3·306 (14)		
In(1)–S(3)	2·618 (15)	2×	
S(12)	2·771 (11)	4×	
In(2)–S(7)	2·417 (15)		
S(4)	2·469 (15)		
S(6)	2·472 (9)	2×	
In(3)–S(10)	2·531 (14)	2×	
S(1)	2·750 (11)	4×	

Bi(1)–S(2)	2·634 (9)	2×	(Å)
S(1)	3·050 (12)	2×	
S(5)	2·666 (14)		
S(11)	2·732 (16)		
Bi(2)–S(3)	2·814 (11)	2×	
S(8)	2·829 (11)	2×	
S(12)	2·752 (16)		
S(6)	2·967 (14)		
Bi(3)–S(9)	2·631 (10)	2×	
S(12)	3·043 (12)	2×	
S(13)	2·670 (15)		
S(8)	2·812 (16)		
Bi(4)–S(11)	2·787 (11)	2×	
S(10)	2·841 (10)	2×	
S(1)	2·707 (15)		
S(6)	3·025 (13)		

$Bi_7In_3Pb_4S_{18}$

Acta Cryst., C**42**, 249-251.

$P2_1/m$, 21.021, 4.014, 18.898, 97.07, Z = 2, R = 0.042. InS_6 octahedra (with In partially substituted by Bi), Pb-S mono- and bicapped trigonal prisms, BiS_6 octahedra, and BiS_5 pyramids with two additional S below the basal plane.

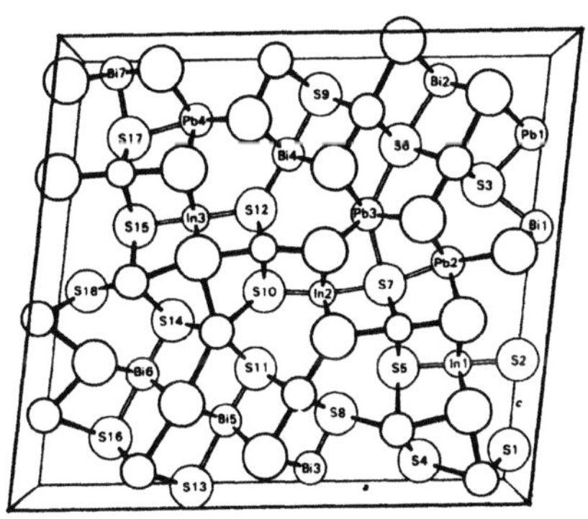

Bond distances **(Å)**

Pb(1)–	S(3)	3·021 (12)	
	S(17)	3·370 (11)	
	S(16)	3·352 (9)	2×
	S(1)	3·127 (8)	2×
	S(2)	2·811 (8)	2×
Pb(2)–	S(7)	2·926 (10)	
	S(14)	2·995 (7)	2×
	S(15)	3·143 (10)	2×
	S(18)	3·112 (8)	2×
Pb(3)–	S(6)	3·297 (13)	
	S(7)	3·136 (10)	
	S(10)	3·228 (9)	2×
	S(11)	3·087 (8)	2×
	S(14)	2·936 (10)	2×
Pb(4)–	S(17)	3·036 (11)	
	S(13)	3·464 (13)	
	S(4)	2·906 (7)	2×
	S(5)	3·090 (8)	2×
	S(8)	3·183 (7)	2×
In(1)–	S(2)	2·620 (12)	
	S(5)	2·609 (11)	
	S(15)	2·853 (8)	2×
	S(17)	2·643 (9)	2×
In(2)–	S(7)	2·771 (10)	
	S(10)	2·583 (12)	
	S(10)	2·718 (7)	2×
	S(12)	2·661 (8)	2×
In(3)–	S(12)	2·707 (13)	
	S(15)	2·668 (12)	
	S(5)	2·743 (7)	2×
	S(7)	2·687 (7)	2×

Bi(1)–	S(3)	2·958 (10)	
	S(15)	3·398 (12)	
	S(18)	2·817 -10)	
	S(18)	2·753 (7)	2×
	S(2)	3·011 (9)	2×
Bi(2)–	S(6)	3·287 (13)	
	S(4)	2·625 (10)	
	S(13)	3·018 (8)	2×
	S(16)	2·737 (7)	2×
Bi(3)–	S(8)	2·633 (10)	
	S(9)	3·099 (12)	
	S(9)	2·951 (8)	2×
	S(13)	2·786 (8)	2×
Bi(4)–	S(9)	2·919 (12)	
	S(12)	2·774 (13)	
	S(8)	2·829 (6)	2×
	S(11)	2·892 (7)	2×
Bi(5)–	S(11)	2·638 (11)	
	S(13)	3·227 (13)	
	S(6)	2·795 (7)	2×
	S(9)	2·940 (7)	2×
Bi(6)–	S(14)	2·654 (11)	
	S(16)	3·121 (12)	
	S(3)	2·672 (7)	2×
	S(6)	2·994 (8)	2×
Bi(7)–	S(17)	2·838 (11)	
	S(1)	2·840 (9)	
	S(16)	3·509 (12)	
	S(1)	2·735 (7)	2×
	S(4)	3·015 (7)	2×

CdY_2S_4

Acta Cryst., B**42**, 364-371.

Fd3m, 11.216, Z = 8, R = 0.045. 0.98Cd in 8(a): 1/8,1/8,1/8; (0.01Cd + 0.99Y) in 16(d): 0,0,1/2; (0.01Cd + 0.01Y) in 16(c): 0,0,0; S in 32(e): x,x,x, x = 0.2564. Domains of a disordered NaCl structure (2%) within ordered domains of spinel structure (98%).

$CuEu_2S_3$

Acta Cryst., C**42**, 390-391.

Pnam, 10.347, 12.863, 3.954, Z = 4, R = 0.057. Chains along **c** of corner-sharing CuS_4 tetrahedra, linked by $Eu(III)S_6$ octahedra and $Eu(II)S_7$ polyhedra.

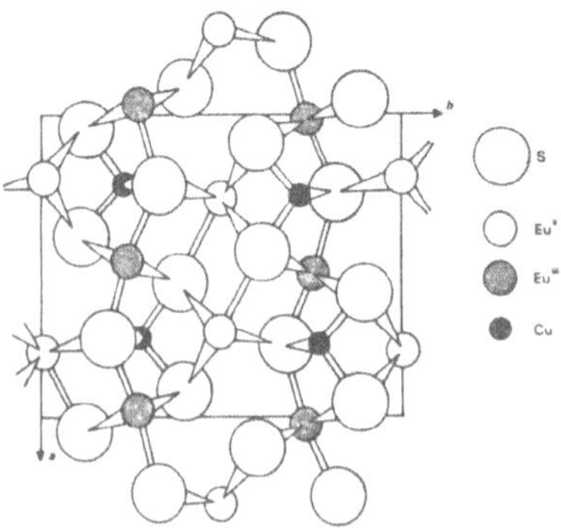

	x	y	z
Eu(1)	0,0171	0,7393	1/4
Eu(2)	0,7807	0,0019	1/4
Cu	0,2327	0,2228	1/4
S(1)	0,0492	0,1168	1/4
S(2)	0,4046	0,1046	1/4
S(3)	0,2628	0,8281	1/4

Eu(1)–S(2)	2,751 (4) ×2	Eu(2)–S(3)	2,982 (4) ×2 (Å)
Eu(1)–S(3)	2,771 (5)	Eu(2)–S(1)	3,056 (4) ×2
Eu(1)–S(3)	2,787 (5)	Eu(2)–S(2)	3,076 (4) ×2
Eu(1)–S(1)	2,793 (4) ×2	Eu(2)–S(1)	3,147 (5)
Cu–S(1)	2,337 (6)		
Cu–S(2)	2,341 (6)		
Cu–S(3)	2,397 (4) ×2		

$EuMo_6S_8$

J. Solid State Chem., **63**, 62-69.

$R\bar{3}$, a = 6.5531 A, α = 88.931°, Z = 1 (hexagonal cell, 9.1806, 11.5601 A, Z = 3), R = 0.024. Eu in 1(a): 0,0,0; Mo in 6(f): 0.22863,0.41817,0.56278; S(1) in 6(f): 0.3882,0.1251,0.7436; S(2) in 2(c): x,x,x, x = 0.2447. Chevrel-type phase (**39**A, 85; **42**A, 67).

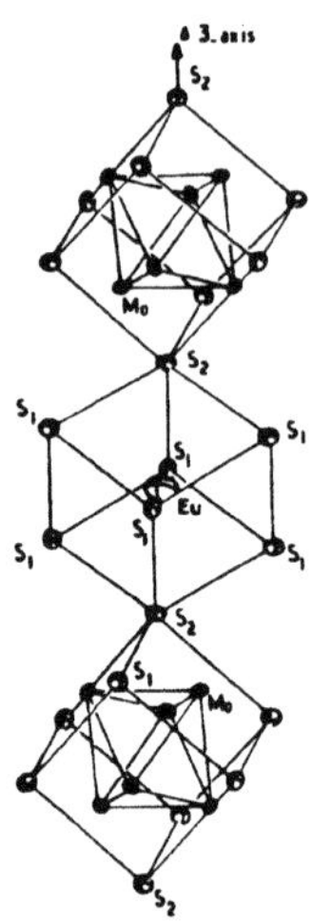

Shortest Interatomic Distances (in Å) in $EuMo_6S_8$

Mo_6S_8 cube		
	$(Mo{-}Mo)_\Delta$ intra	2 × 2.665(1)
	$(Mo_\Delta{-}Mo_\Delta)$ intra	2 × 2.716(1)
	(Mo–Mo) inter	3.278(1)
	(Mo–S(1)) intra	2.448(1)
		2.460(1)
		2.508(1)
	(Mo–S(1)) inter	2.583(1)
	Mo–S(2)	2.391(1)
EuS_8 cube		
	Eu–S(1)	6 × 3.104(1)
	Eu–S(2)	2 × 2.830(1)

$FeNa_5S_4$

Z. anorg. Chem., **532**, 23-30.

Pbca, 11.947, 7.091, 21.54, Z = 8, R = 0.030. Na_5GaO_4-type structure (**41**A, 223), with FeS_4 tetrahedra, linked by (4+1)- and (4+2)-coordinate Na. Fe-S = 2.277-2.293(1), Na-S = 2.702-3.921(2) A.

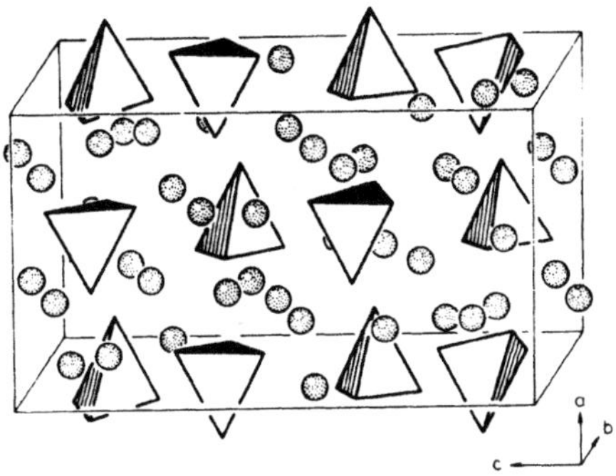

	x	y	z
Na(1)	0,1615	0,1057	0,82994
Na(2)	0,0589	0,0611	0,43284
Na(3)	0,4091	0,1240	0,75034
Na(4)	0,3538	0,0918	0,49883
Na(5)	0,2866	0,0497	0,13330
Fe	0,04030	0,16794	0,12542
S(1)	0,35532	0,2539	0,37527
S(2)	0,13211	0,2599	0,21337
S(3)	0,13011	0,2068	0,54180
S(4)	0,44541	0,1519	0,62070

$Fe_2Ta_9S_6$, $Ni_2Ta_9S_6$

J. Less-Common Metals, **115**, 177-189.

$P\bar{6}2m$, a = 10.310, 10.134, c = 3.2977, 3.3704 A, Z = 1. Ta tetrakaidecahedra sharing faces along **c** and vertices in the hexagonal plane, with additional bridging via S atoms. Fe (Ni) occupies the centre of a tricapped trigonal prism of Ta atoms.

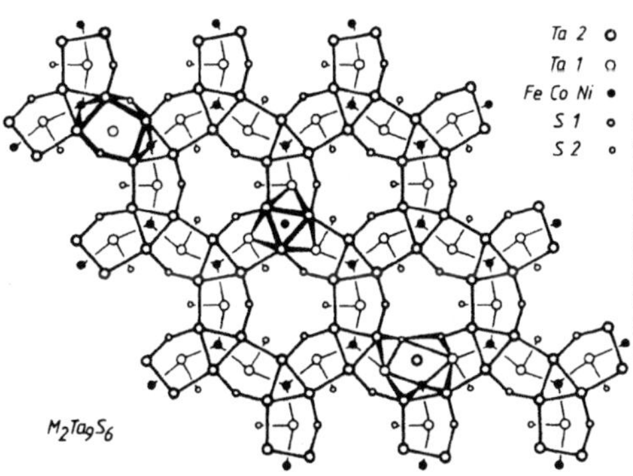

		x	y	z
Ta(1)	(Fe)	0.4580	0	0
Ta(1)	(Ni)	0.4534		
Ta(2)	(Fe)	0.2006	0.7387	$\frac{1}{2}$
Ta(2)	(Ni)	0.19683	0.73653	
Fe(1)		$\frac{1}{3}$	$\frac{2}{3}$	0
Ni(1)				
S(1)	(Fe)	0.2753	0	$\frac{1}{2}$
S(1)	(Ni)	0.2684		
S(2)	(Fe)	0.7058	0	0
S(2)	(Ni)	0.7047		

$Fe_2Ta_9S_6$, $Co_2Ta_9S_6$

J. Less-Common Metals, **124**, 125-134.

$P\bar{6}2m$, a = 10.266, 10.244, c = 6.583, 6.574 A, Z = 2. R = 0.045, 0.023. The structures are now described in a cell with a doubled **c** axis resulting from displacement of Fe and Co atoms from the centres of the trigonal prisms of Ta. The previous description of the Ni compound is retained (preceding report).

Ga_2NiS_4

Z. anorg. Chem., **533**, 118-124.

$P\bar{3}m1$, 3.6249, 11.9956, Z = 1, R = 0.051. Ni in 1(b): 0,0,1/2; Ga, S(1), S(2) in 2(d): 1/3,2/3,z, z = 0.2145, 0.8679, 0.3985. $FeGa_2S_4$-type structure (**46**A, 79), with hexagonal close-packed S, and Ni and Ga in one-quarter of the octahedral and tetrahedral holes, respectively. Ni-S = 2.421, Ga-S = 2.207, 2.315(1) A.

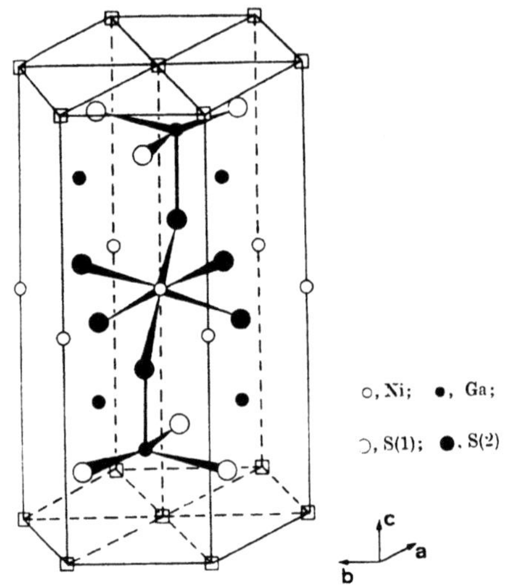

$GeMn_2S_4$

J. Solid State Chem., **65**, 280-282.

Pnma, 12.776, 7.441, 6.033, Z = 4, R = 0.038. Olivine-type structure (**34**A, 91). Ge-4S = 2.18-2.23, Mn-6S = 2.54-2.63 A.

	x	*y*	*z*
Mn(1)	**0**	**0**	**0**
Mn(2)	**0,22893**	**1/4**	**0,5069**
Ge	**0,40999**	**1/4**	**0,08628**
S(1)	**0,4069**	**1/4**	**0,7247**
S(2)	**0,56824**	**1/4**	**0,2417**
S(3)	**0,33203**	**0,0179**	**0,2502**

$HgKSbS_3$

Mater Res. Bull., **21**, 493-501.

A2/a, 17.084, 8.647, 8.951, 105.78, Z = 8, R = 0.082 for twinned crystal. SbS_3 trigonal pyramids linked by Hg ions into (100) sheets, which are weakly connected by K ions. Sb-3S = 2.38-2.45, Hg-4S = 2.44, 2.45 (angle 159°), 2.77, 2.79 A.

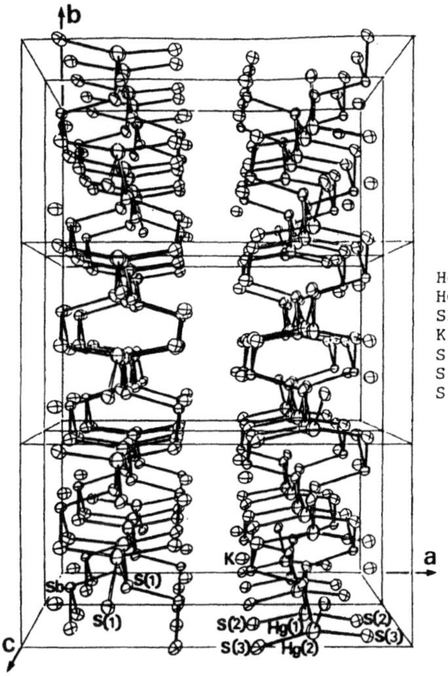

			x	y	z
Hg(1)	4 e 2		0.25	0.4402	0
Hg(2)	4 e 2		0.25	0.4217	0.5
Sb	8 f 1		0.4044	0.1613	0.3254
K	8 f 1		0.0805	0.1790	0.1532
S(1)	8 f 1		0.2661	0.2429	0.2535
S(2)	8 f 1		0.3947	0.4898	0.0375
S(3)	8 f 1		0.1051	0.4725	0.3828

$La_6NiSi_2S_{14}$, $La_6CoSi_2S_{14}$

Yingyong Huaxue, **2**, 42-46 (1985).

$P6_3$, a = 10.293, 10.290, c = 5.744, 5.763, Z = 1, R = 0.036, 0.096. La has bicapped trigonal prismatic coordination, mean La-S = 2.990, 2.992 A, Ni and Co have octahedral, and Si tetrahedral coordination.

Mo_2S_3

Ž. Neorg. Khim., **31**, 1102-1106 [Russ. J. Inorg. Chem., **31**, 627-630].

$P2_1/m$, 6.087, 3.214, 8.627, 102.43, Z = 2, R = 0.028. Atoms in 2(e): x,1/4,z, x = 0.3097, 0.1092, 0.5103, 0.9718, 0.7281, z = 0.0092, 0.6307, 0.8035, 0.1565, 0.5135 for Mo(1,2), S(1,2,3). Structure as previously described (**26**, 200; **35**A, 79), with chains of Mo atoms, each Mo octahedrally coordinated to S. Mo-Mo = 2.86-3.23, Mo-S = 2.32-2.64 A.

Mo_2VS_4, MoV_2S_4

J. Solid State Chem., **63**, 369-376.

I2/m, a = 5.949, 5.902, b = 3.236, 3.262, c = 11.670, 11.550, β = 92.91, 92.65°, Z = 2, powder data. Cr_3S_4-type structure (**21**, 97).

		x	y	z
V	2(*a*)	0.0	0.0	0.0
Mo	4(*i*)	0.9465	0.0	0.2561
S_I	4(*i*)	0.3401	0.0	0.3675
S_{II}	4(*i*)	0.3431	0.0	0.8938

		x	y	z
V_I	2(*a*)	0.0	0.0	0.0
V_{II} } Mo }	4(*i*)	0.9475	0.0	0.2572
S_I	4(*i*)	0.3413	0.0	0.3665
S_{II}	4(*i*)	0.3418	0.0	0.8930

$Ni_3Ta_2S_8$, $Pt_3Ta_2S_8$, $Pd_{0.89}Ta_2S_5$, $Co_2Nb_4PdSe_{12}$

J. Solid State Chem., **64**, 261-269.

$Ni_3Ta_2S_8$, $Pt_3Ta_2S_8$, Pbam, a = 14.092, 15.003, b = 10.078, 10.687, c = 3.314, 3.541 Å, Z = 2 [not 4], R = 0.049, 0.067. Isostructural with $Nb_2Pd_3Se_8$ (**51**A, 74; **52**A, 72), with Ni in square-planar and square-pyramidal sites.

$Pd_{0.89}Ta_2S_5$, C2/m, 12.204, 3.269, 15.183, 104.48, Z = 4, R = 0.041. Isostructural with $Nb_2Pd_{0.71}Se_5$ (**52**A, 72), with Pd(2) site 78.6% occupied.

$Co_2Nb_4PdSe_{12}$, C2/m, 12.871, 3.402, 19.436, 110.87°, Z = 2, R = 0.051. Isostructural with the Ta compound (**52**A, 72).

		x	y	z
		$Ta_2Ni_3S_8$		
Ta	4g	0.119005	0.203894	½
Ni(1)	2a	0	0	0
Ni(2)	4g	0.22461	0.37118	0
S(1)	4g	0.99122	0.22059	0
S(2)	4g	0.15540	0.02894	0
S(3)	4g	0.28684	0.24675	½
S(4)	4h	0.12952	0.44275	½
		$Ta_2Pt_3Se_8$		
Ta	4g	0.11973(11)	0.21332	½
Pt(1)	2a	0	0	0
Pt(2)	4g	0.21778	0.37395	0
Se(1)	4g	0.99162	0.22842	0
Se(2)	4g	0.15818	0.04147	0
Se(3)	4g	0.28908	0.24854	½
Se(4)	4g	0.11804	0.45328	½
		$Ta_2Pd_{0.89}S_5$		
Ta(1)	4i	0.075093	½	0.182425
Ta(2)	4i	0.151169	0	0.376921
Pd(1)	2a	0	0	0
Pd(2)[a]	2c	0	0	½
S(1)	4i	0.35171	0	0.49455
S(2)	4i	0.25134	½	0.30386
S(3)	4i	0.17395	0	0.10883
S(4)	4i	0.42399	½	0.12598
S(5)	4i	0.49872	0	0.31617
		$Co_2Nb_4PdSe_{12}$		
Pd	2a	0	0	0
Nb(1)	4i	0.52803	0	0.149074
Nb(2)	4i	0.66431	0	0.406732
Se(1)	4i	0.65590	½	0.116682
Se(2)	4i	0.48991	½	0.609313
Se(3)	4i	0.82711	½	0.453107
Se(4)	4i	0.41639	0	0.232661
Se(5)	4i	0.37884	½	0.072614
Se(6)	4i	0.68584	0	0.280668
Co	4i	0.55212	½	0.28070

[a] Occupancy = 0.786

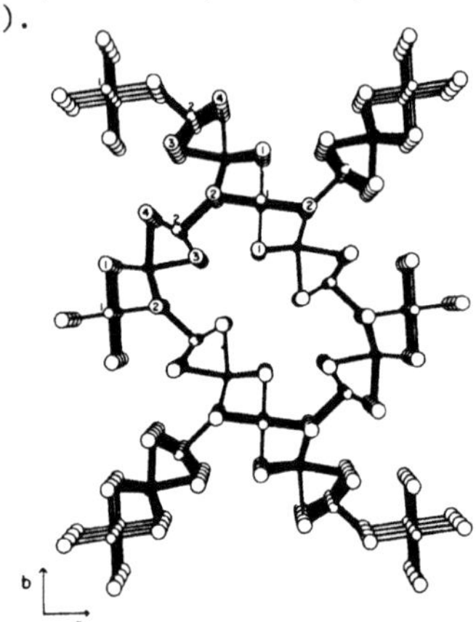

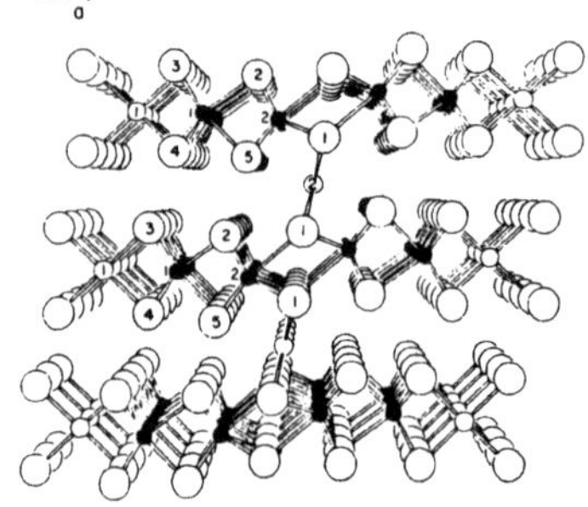

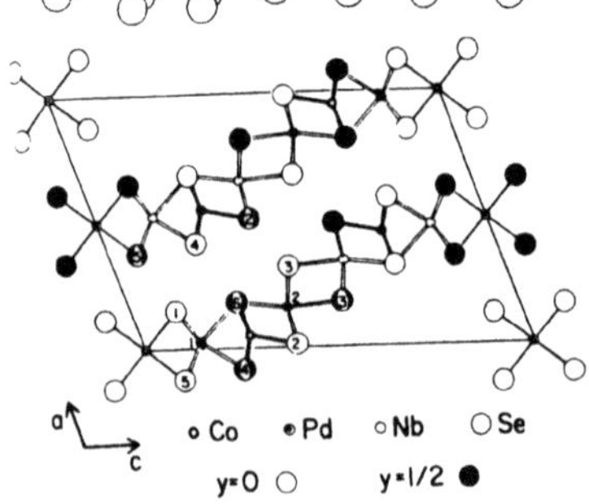

PdU_2S_4

J. Less-Common Metals, **115**, 253-259.

Fdd2, 10.725, 13.148, 8.381, Z = 8, R = 0.036. U-7S = 2.75-2.83, Pd-4S = 2.38, 2.60, Pd-4U = 3.10, 3.43 A.

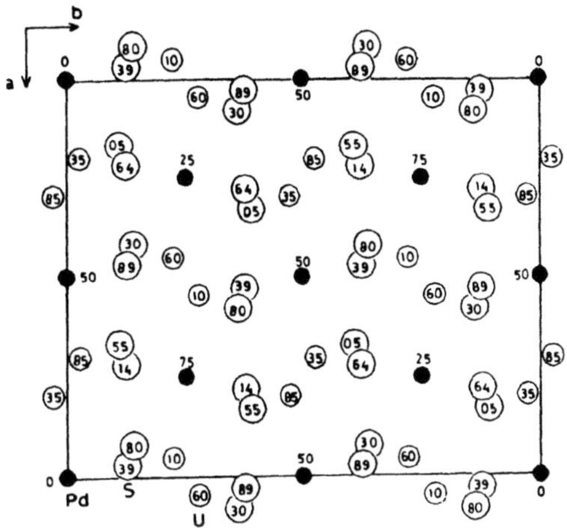

		x	*y*	*z*
U	16(b)	0,4528	0,2223	0,6050
Pd	8(a)	0	0	0
S1	16(b)	0,1698	0,6104	0,5569
S2	16(b)	0,2793	0,3759	0,6469

$Pd_3U_{0.92}S_4$

Inorg. Chim. Acta, **117**, 183-185.

Pm3n, 6.639, Z = 2, R = 0.044. U in 2(a): 0,0,0; Pd in 6(d): 1/4,1/2,0; S in 8(e): 1/4,1/4,1/4. Pt-bronze type structure (**13**, 273), as for $Ln_xPd_3S_4$ (**50**A). U-8S = 2.875, U-12Pd = 3.711, Pd-4S = 2.347, Pd-2Pd = 3.319, Pd-4U = 3.711 A.

Sm_xS_4 (x = 2.71, 2.88, 3.05)

J. Less-Common Metals, **116**, 1-8.

$I\bar{4}3d$, a = 8.445, 8.501, 8.548 A, Z = 4. Sm in 12(a): 3/8,0,1/4; S in 16(c): x,x,x, x = +0.07160, -0.07180, ±0.07237. Th_3P_4-type (**7**, 15, 112), as previously described (**44**A, 96).

Tl_4SiS_4, Tl_4SiSe_4

Acta Cryst., C**42**, 528-534.

Tl_4SiS_4, Cc, 12.518, 11.241, 7.567, 112.80, Z = 4, R = 0.066. Tl_4SiSe_4, C2/c, 11.664, 7.277, 24.903, 99.93, Z = 8, R = 0.064. Tl_4SiS_4 is isostructural with Tl_4GeS_4 (**44**A, 66), with tetrahedral SiS_4^{4-} anions linked by Tl^+ cations which have irregular 6-coordination.

$Tl_4[SiSe_4]$ is built up from slabs parallel to (001) which consist of almost planar 3^6 nets of Tl^+ cations enclosed between two parallel layers of $[SiSe_4]^{4-}$ tetrahedra. Further Tl^+ cations are located in voids between the $[SiSe_4]^{4-}$ tetrahedra of each layer and between the slabs. The Tl atoms in three general and two special positions are coordinated to essentially six, five, and four Se atoms. The lone pairs of 6*s* electrons of univalent thallium exhibit strong stereochemical effects with two of the Tl atoms. The pseudohexagonal properties of the structure are described by means of a group–subgroup relation to a hypothetical hexagonal aristotype.

	x	*y*	*z*
$Tl_4[SiS_4]$			
Tl(1)	0·4863	0·1113	0·5064
Tl(2)	0·1200	0·8751	0·4132
Tl(3)	0·1935	0·2096	0·4378
Tl(4)	0·2834	0·4999	0·4133
Si	0·4106	0·7887	0·4781
S(1)	0·3283	0·2784	0·1539
S(2)	0·0781	0·6474	0·1620
S(3)	0·3140	0·9332	0·3029
S(4)	0·4300	0·6516	0·3012
$Tl_4[SiSe_4]$			
Tl(1)	0·1911	0·4621	0·0008
Tl(2)	0·3887	0·2757	0·6793
Tl(3)	0·3110	0·3705	0·8350
Tl(4)	0	0	0
Tl(5)	0	0·1674	0·75
Si	0·0655	0·3766	0·6243
Se(1)	0·3060	0·2754	0·4169
Se(2)	0·0175	0·3528	0·0772
Se(3)	0·4014	0·2969	0·1194
Se(4)	0·3521	0·0442	0·2905

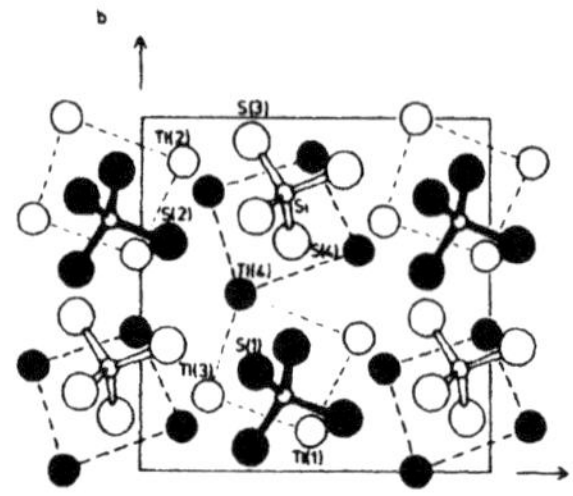

Projection of the structure of $Tl_4[SiS_4]$ down [001]. The Tl and S atoms of the upper pseudotetragonal layer are depicted by filled circles, those of the lower layer by open circles. The broken lines indicate the 'squares' of Tl atoms.

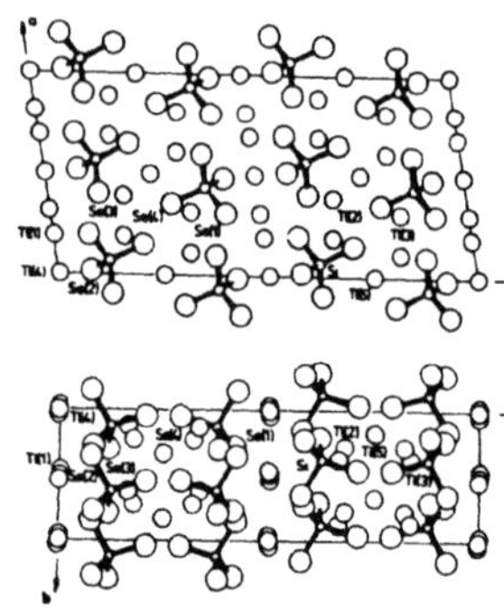

Projection of the structure of $Tl_4[SiSe_4]$ down [010] (top) and down [100] (bottom).

$Tl_{0.82}V_5S_{4.36}Se_{3.64}$

Acta Cryst., C**42**, 6-9.

C2/m, 17.841, 3.383, 8.741, 104.88, Z = 2, R = 0.037. Framework of face- and edge-sharing VX_6 octahedra, with Tl in large rectangular channels.

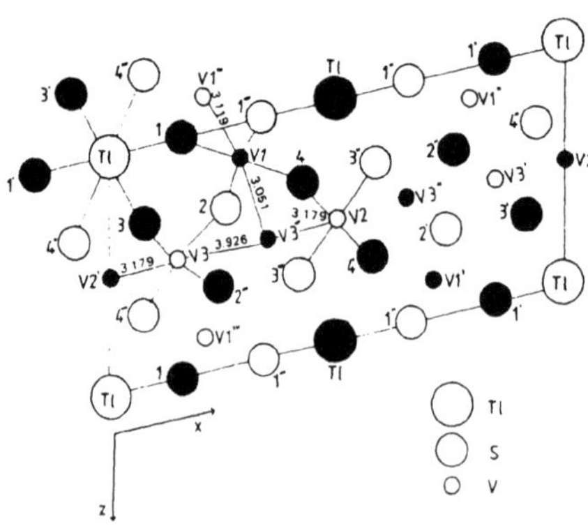

	x	*y*	*z*	s.o.f
Tl	0·0000	0·0000	0·0000	0·2048
V(1)	0·2910	0·5000	0·1477	0·5000
V(2)	0·5000	0·0000	0·5000	0·2500
V(3)	0·1498	0·0000	0·4923	0·5000
Se(3)	0·0830	0·5000	0·3111	0·2794
S(3)	0·0830	0·5000	0·3111	0·2206
Se(2)	0·2576	0·0000	0·3396	0·2590
S(2)	0·2576	0·0000	0·3396	0·2410
Se(1)	0·1584	0·5000	−0·0001	0·2488
S(1)	0·1584	0·5000	−0·0001	0·2512
Se(4)	0·4262	0·5000	0·3257	0·1233
S(4)	0·4262	0·5000	0·3257	0·3767

θ-Yb_2S_3

Ž. Neorg. Khim., **31**, 2210-2214 [Russ. J. Inorg. Chem., **31**, 1273-1276].

Ia3, 12.449, Z = 16, R = 0.034. Yb(1) in 8(b): 1/4,1/4,1/4; Yb(2) in 24(d): 1/4,0.45211,0; S in 48(e): 0.38346,0.39270,0.16116. Mn_2O_3-type structure (**2**, 38; **27**, 497). Yb-6S = 2.672-2.708(3) A (distorted octahedra).

$ZnPS_3$

Mater. Res. Bull., **21**, 195-200 and 643-644.

C2/m, 5.9717, 10.3424, 6.7565, 107.139, Z = 4, R = 0.040. Zn in 4(g): 0,0.3324,0; P in 4(i): 0.0563,0,0.1695; S(1) in 4(i): 0.7530,0,0.2473; S(2) in 8(j): 0.2466,0.1643,0.2493. Isostructural with other MPS_3 compounds (**35**A, 183; **39**A, 76; **52**A, 80, 81). P-P = 2.188, P-3S = 2.027, 2.030, Zn-6S = 2.562-2.575 A.

COVELLITE
CuS

Z. Kristallogr., **173**, 119-128 (1985).

$P6_3/mmc$, 3.796-3.729, 16.382-16.11, at 1-33kbar, Z = 6, R = 0.022-0.060. Structure as previously described (**1**, 771; **2**, 10, 229; **18**, 380; **33**A, 79; **41**A, 60; **42**A, 78), with shorter Cu-S and longer S-S bonds as pressure is increased.

Pressure		1 bar	12 kbar	18 kbar	21 kbar	33 kbar
Cu(1)	x	1/3	1/3	1/3	1/3	1/3
	y	2/3	2/3	2/3	2/3	2/3
	z	3/4	3/4	3/4	3/4	3/4
Cu(2)	x	1/3	1/3	1/3	1/3	1/3
	y	2/3	2/3	2/3	2/3	2/3
	z	0.10724	0.1107	0.1111	0.1197	0.123
S(1)	x	1/3	1/3	1/3	1/3	1/3
	y	2/3	2/3	2/3	2/3	2/3
	z	1/4	1/4	1/4	1/4	1/4
S(2)	x	0	0	0	0	0
	y	0	0	0	0	0
	z	0.0637	0.066	0.069	0.069	0.076

Pressure	1 bar	12 kbar	18 kbar	21 kbar	33 kbar
Triangle					
Cu(1)−S(1)×3	2.1915(4)	2.175(2)	2.171(1)	2.169(1)	2.153(1)
Area (Å^2)	6.239(2)	6.145(7)	6.123(3)	6.111(3)	6.022(7)
Tetrahedron					
Cu(2)−S(1)	2.339(2)	2.27(3)	2.25(1)	2.11(1)	2.05(4)
Cu(2)−S(2)×3	2.305(1)	2.29(2)	2.28(1)	2.32(1)	2.28(3)
Average	2.314	2.29	2.27	2.27	2.22
Volume (Å^3)	6.35(2)	6.1(3)	6.0(1)	6.0(1)	5.6(3)
S(2)−S(2)	2.086(7)	2.16(13)	2.22(8)	2.22(9)	2.44(20)
S(1)−Cu(2)−S(2)	108.04(8)	108(1)	107.6(9)	110.9(8)	109(2)
S(2)−Cu(2)−S(2)	110.87(8)	110(1)	111.2(9)	108.0(9)	109(2)

FREIBERGITE
$(Cu,Fe,Zn,Ag)_{12}Sb_4S_{13}$

TETRAHEDRITE
$(Cu,Fe,Zn)_{12}Sb_4S_{13}$

Miner. Mag., **50**, 717-721.

$I\bar{4}3m$, a = 10.610, 10.364, Z = 2, R = 0.053, 0.032. Cu(1) in 12(d): 1/4,1/2,0; Cu(2) in 12(e): x,0,0, x = 0.2167, 0.2165; Sb in 8(c): x,x,x, x = 0.2685, 0.2697; S(1) in 24(g): x,x,z, x = 0.1213, 0.1159, z = 0.3683, 0.3615; S(2) in 2(a): 0,0,0. Structure as previously described (**29**, 15; **37**A, 4; **45**A, 14), with Ag in freibergite in the 3-coordinate Cu(2) site.

PADERAITE
$Cu_{5.9}Ag_{1.3}Pb_{1.6}Bi_{11.2}S_{22}$

Canad. Miner., **24**, 513-521.

$P2_1/m$, 28.44, 3.90, 17.55, 106.0, Z = 2, R = 0.17. Cuprobismutite-related structure (**41**A, 32; **42**A, 50), with two types of layer.

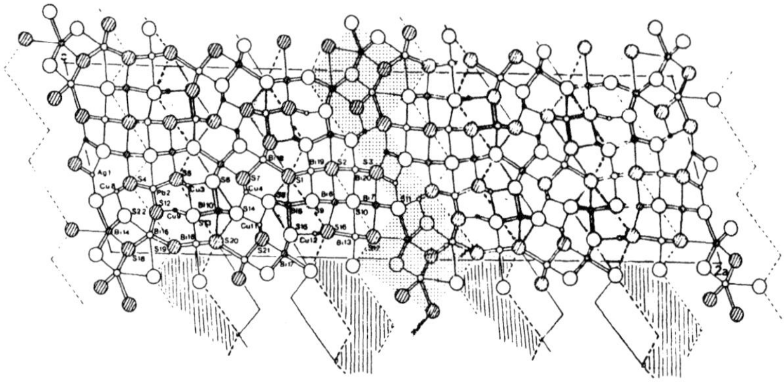

As_2Se_3

Z. Kristallogr., **173**, 185-191 (1985).

$P2_1/n$, 12.0774, 9.9037, 4.2835, 90.458, Z = 4, R = 0.038. Structure as previously described (**30**A, 22; **39**A, 15), with $AsSe_3$ trigonal pyramids which share Se to form undulating layers; four more-distant Se result in 7-coordination for As. As-Se = 2.390-2.452, 3.288-4.143(2) A.

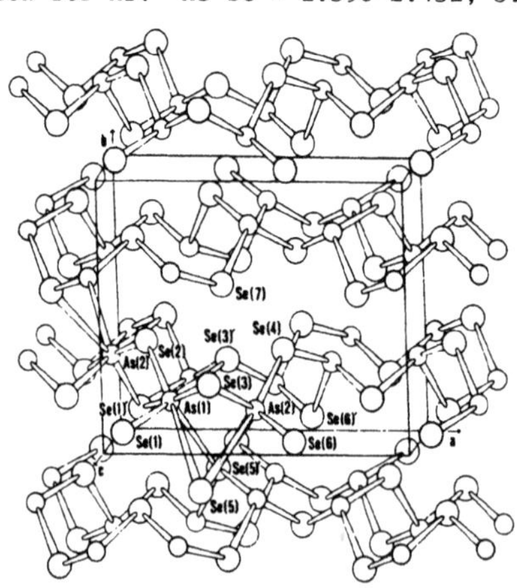

	x	y	z
As(1)	0.2362	0.1967	0.1113
As(2)	0.0108	0.3253	0.3810
Se(1)	0.0977	0.1182	0.4726
Se(2)	0.1534	0.4104	0.0257
Se(3)	0.3785	0.2981	0.4436

$CsSb_2Se_4$

Z. anorg. Chem., **536**, 114-118.

P$\bar{1}$, 7.051, 9.806, 6.523, 93.58, 105.44, 80.50, Z = 2, R = 0.041. Polymeric $(Sb_2Se_4^-)_n$ anions and 10-coordinate Cs^+ ions. Sb-Se = 2.56-2.91, 3.18-3.46, Se-Se = 2.38, Cs-Se = 3.47-3.97 Å.

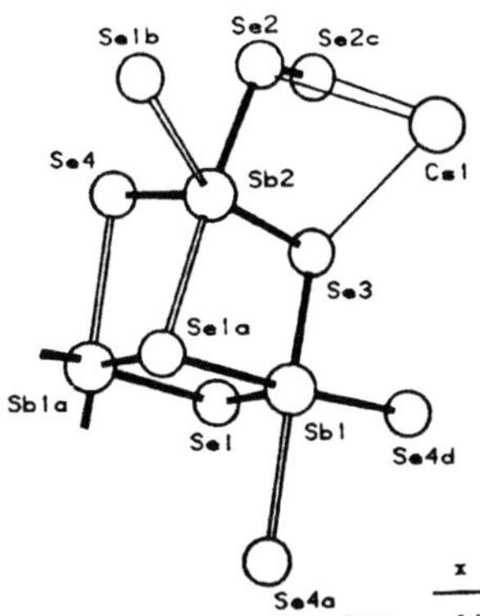

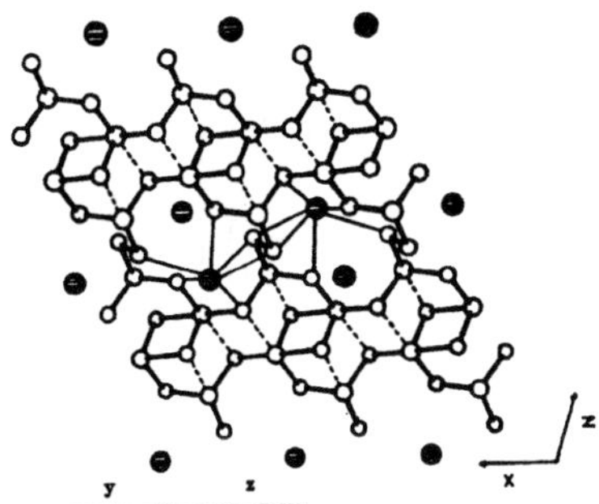

	x	y	z
Cs(1)	0,1753	0,0604	1,3298
Sb(1)	−0,1832	0,4600	0,6617
Sb(2)	0,4042	0,3321	0,9347
Se(1)	−0,1525	0,3920	0,2861
Se(2)	0,5897	0,0841	1,1044
Se(3)	0,0698	0,2388	0,8283
Se(4)	0,5175	0,2903	0,5893

$InLiSe_2$

Z. anorg. Chem., **543**, 161-168.

Pna2_1, 7.183, 8.398, 6.781, Z = 4, R = 0.032. In: 0.0787, 0.1261,0; Se(1): 0.0801,0.1283,-0.3800; Se(2): 0.4158,0.1221,0.1278; Li: 0.089,0.631,-0.005. β-$NaFeO_2$-type (**18**, 422; **28**, 133). In-4Se = 2.555-2.577(3), Li-4Se = 2.49-2.62(4) Å.

In_2ZnSe_4

Z. Kristallogr., **173**, 199-203.

I$\bar{4}$2m, 5.7095, 11.449, Z = 2, R = 0.021. M(1) in 2(a): 0,0,0; M(2) in 4(d): 0,1/2,1/4; Se in 8(i): 0.2302,0.2302,0.3841; vacancy in 2(b): 0,0,1/2. Isostructural with $MgIn_2Te_4$ and $MnIn_2Te_4$ (**41**A, 79). The structure is a defect-zincblende structure with vacancy ordering, and random In/Zn distribution. M-4Se = 2.541, 2.551(1) Å.

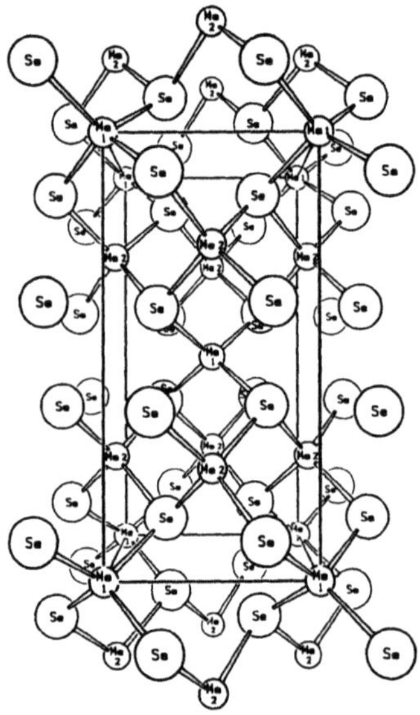

$MoSe_2$, MoS_2

Z. anorg. Chem., **540/541**, 15-17.

$MoSe_2$, $P6_3/mmc$, 3.289, 12.927, Z = 2, R = 0.12. Mo in 2(c): 1/3,2/3,1/4; Se in 4(f): 1/3,2/3,z, z = 0.6210. C7-type (**1**, 164, 214). Mo-6Se (trigonal prism) = 2.527 A.

MoS_2, $P6_3/mmc$, 3.1602, 12.294, Z = 2. C7 structure, as for $MoSe_2$, z(S) = 0.621. In previous reports (**1**, 164, 214) there is some confusion in origin choice, and value of the **z** parameter [z'$\sim$ 1.25-z$\sim$ z].

$NiTa_2Se_7$, $PtTa_2Se_7$

Inorg. Chem., **25**, 4355-4358.

C2/m, a = 13.827, 13.954, b = 3.482, 3.530, c = 18.577, 18.694 A, β = 108.80, 109.28°, at 123K, Z = 4, R = 0.043, 0.043. Layers of $TaSe_6$ octahedra, $TaSe_8$ bicapped trigonal prisms, and MSe_6 (M = Ni or Pt) distorted octahedra, with one Se-Se pair. Ta-Se = 2.473-2.745, Ni-Se = 2.394-2.467, 2.729, Pt-Se = 2.490-2.589, 2.703, Se-Se = 2.549, 2.597(3) A.

	Ta_2NiSe_7		
	x	*y*	*z*
Ta(1)	0.151 270	0	0.399 142
Ta(2)	0.228 901	0	0.089 761
Ni	0.689 72	0	0.209 30
Se(1)	0.366 27	0	0.012 96
Se(2)	0.562 24	0	0.294 83
Se(3)	0.581 03	1/2	0.137 03
Se(4)	0.846 69	0	0.173 53
Se(5)	0.019 39	1/2	0.418 33
Se(6)	0.198 98	0	−0.446 65
Se(7)	0.259 23	0	0.299 70

	Ta_2PtSe_7		
	x	*y*	*z*
Ta(1)	0.150 919	0	0.398 874
Ta(2)	0.231 065	0	0.091 873
Pt	0.688 086	0	0.207 920
Se(1)	0.360 37	0	0.011 77
Se(2)	0.565 67	0	0.293 87
Se(3)	0.574 46	1/2	0.127 19
Se(4)	0.851 87	0	0.171 62
Se(5)	0.023 16	1/2	0.419 09
Se(6)	0.200 37	0	−0.447 35
Se(7)	0.264 58	0	0.303 78

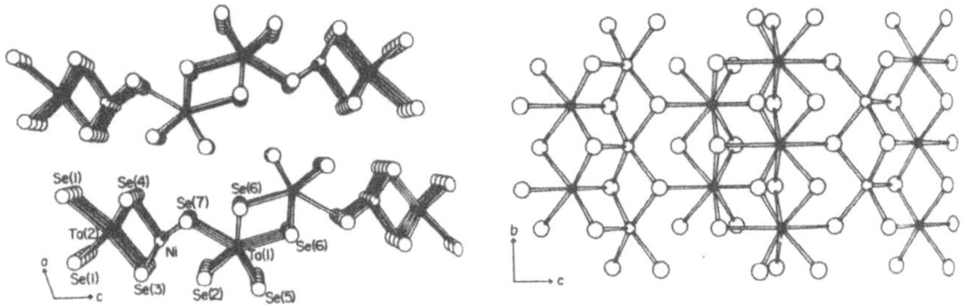

$Tl_4P_2Se_6$

Z. Naturforsch., **40**A, 1248-1252 (1985).

$P2_1/c$, 12.216, 9.038, 12.300, 98.9, Z = 4, R = 0.084. Two independent Se_3P-PSe_3^{4-} anions with staggered conformations (approximate D_{3d} symmetry) lying on symmetry centres, linked by 6- and 8-coordinate Tl^+ ions. Mean P-P = 2.26, P-Se = 2.18 A; Tl-Se = 3.14-3.98 A.

$AgTl_3Te_2$

Z. Naturforsch., **41**B, 941-945.

$P2_1/c$, 11.020, 7.427, 9.888, 116.62, Z = 4, R = 0.083. (100) $[AgTe_2]$ layers of edge-sharing $AgTe_4$ tetrahedra, linked by 5- or (3+2)-coordinate Tl ions. Ag-Te = 2.83-3.04, Tl-Te = 3.25-4.09 A.

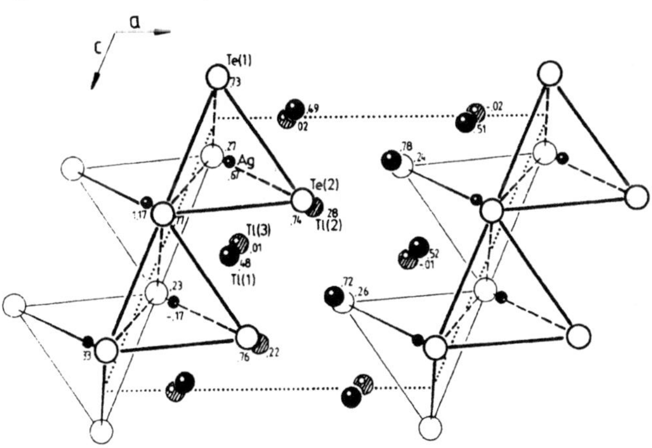

	x	y	z
Tl(1)	0.2163	0.0156	0.0107
Tl(2)	0.4121	0.2842	0.3316
Tl(3)	0.2356	0.0088	0.4674
Ag	0.9070	0.1690	0.3259
Te(1)	0.0380	0.2345	0.6458
Te(2)	0.6242	0.2646	0.6888

As_2Te_3 (metastable β-form)

Mater. Res. Bull., **21**, 1509-1514.

R3m, 4.06, 29.59 A, Z = 3, powder data. Te(1) in 3(a): 0,0,0; Te(2) in 3(a): 2/3,1/3,0.1178; Te(3) in 3(a): 0,0,0.2139; As(1) in 3(a): 1/3,2/3,0.0702; As(2) in 3(a): 1/3,2/3,0.2616. Sheets with all atoms in distorted octahedral coordinations; Te-As = 2.74-3.16, Te-Te = 3.69 A.

$InLiTe_2$

Z. anorg. Chem., **532**, 150-156.

I$\bar{4}$2d, 6.398, 12.460, Z = 4, R = 0.02. In in 4(b): 0,0,1/2; Li in 4(a): 0,0,0; Te in 8(d): 0.2441,1/4,1/8. Chalcopyrite-type structure (**1**, 279; **2**, 48). In-4Te = 2.769, Li-Te = 2.725(1) A.

MnPdTe

J. Less-Common Metals, **123**, 169-173.

[F$\bar{4}$3m], 6.2605, Z = 4, neutron powder data. Mn in 4(c): 1/4,1/4,1/4; Pd in 4(a) 0,0,0; Te in 4(d): 3/4,3/4,3/4. $C1_b$-type structure (**1**, 148; **8**, 28).

$Mn_3Si_2Te_6$

J. Solid State Chem., **63**, 349-352.

P$\bar{3}$1c, 7.029, 14.255, Z = 2, R = 0.041. Two layers along **c** of condensed Te_6 octahedra; one layer is fully occupied by Mn ions and Si_2 pairs, and the other partially occupied by Mn. Mn-6Te = 2.907-2.951, Si-3Te = 2.501, Si-Si = 2.324 A.

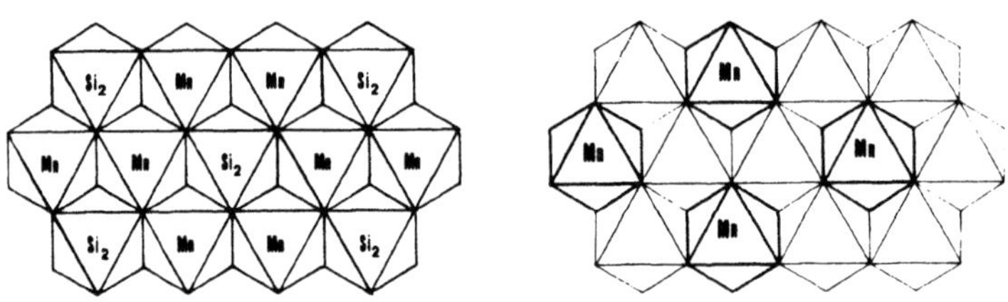

		x	y	z
Mn(1)	4f	1/3	2/3	0.00068
Mn(2)	2c	1/3	2/3	1/4
Si	4e	0	0	0.08153
Te	12i	0.65869	0.00361	0.12788

$NaTe_3$

Z. anorg. Chem., **542**, 144-152.

$P\bar{3}c1$, 9.033, 21.930, Z = 12, R = 0.015. Te_6^{2-} chains (Te-Te = 2.77-2.98 A) linked into infinite chains (Te...Te = 3.16 A), with chains linked by Na ions, Na-6Te = 3.07-3.31 A.

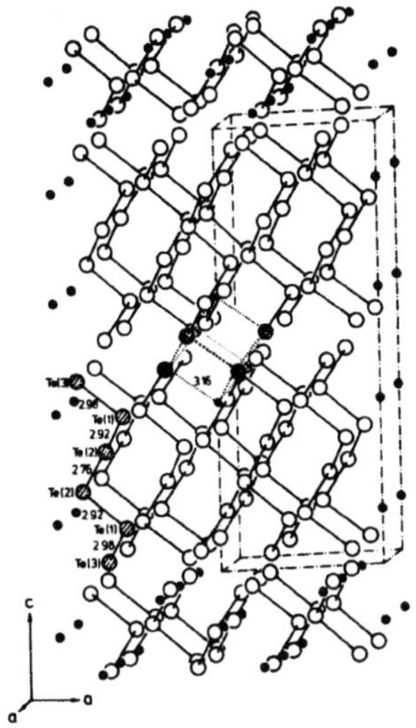

	x	y	z
Te(1)	0,4747	−0,0073	0,1233
Te(2)	0,8045	0,6706	0,2088
Te(3)	0,6675	0,8390	0,5411
Na(1)	1/3	2/3	0,7007
Na(2)	1/3	2/3	0,4538
Na(3)	0	0	0,6256

$NbTe_4$

Acta Cryst., B**42**, 43-50.

Subcell, P4/mcc, 6.499, 6.837, Z = 2, R = 0.065. Nb in 2(a): 0,0,1/4; Te in 8(m): 0.23599,0.09193,0. Structure essentially as previously described (**29**, 77). Incommensurately modulated superstructure, 2**a** x 2**b** x ∿3**c**.

Re_6Te_{15}

Mater. Res. Bull., **21**, 1335-1342.

Pcab, 12.958, 13.034, 14.240, Z = 4, R = 0.092. Structure as previously described (**50**A), with Re_6Te_8 clusters linked by additional Te atoms.

$ZrTe_5$, $HfTe_5$

Solid State Comm., **60**, 91-93.

Cmcm, a = 3.9875, 3.9797; 3.9713, 3.9640, b = 14.530, 14.470; 14.499, 14.443, c = 13.724, 13.676; 13.729, 13.684 A, for Zr compound at 293, 10K; Hf compound at 293, 10K, Z = 4. Neutron powder data. For $HfTe_5$ at 293: Hf in 4(c): 0,y,1/4, y = 0.3155; Te(1) in 4(c): y = 0.6575; Te(2) in 8(f): 0,y,z, y, z = 0.9340, 0.1487; Te(3) in 8(f): y,z=0.2092, 0.4350; similar values for others. $HfTe_5$-type structures (**39**A, 69).

TABLE I

Some structural information has also been given for the following materials (listed with 1986 references).

Compound	Structure	Reference
$(Fe,Cr,Ni)_2(Mo,Ti,Al)$, C-phase	B2/m, related to C14-type. Electron microscope study	Acta Cryst., B42, 21
$Au_{79}Mn_{21}$	Electron micrographs	Ibid., B42, 39
Kostovite, $(Cu,Au)Te_4$	Electron microscope study of modulated structure	Ibid., B42, 121
LiH LiD	NaCl-type. Thermal parameters determined at three temperatures	Ibid., B42, 131
Ni-based alloys	Electron microscope study of F, K, and J phases	Ibid., B42, 152 and 159
Cu_2ZnGeS_4	Electron diffraction study of tetragonal (Cu_2FeSnS_4-type) and orthorhombic polymorphs	Ibid., B42, 236
GaAs	Study of valence charge density	Ibid., B42, 443
ZnS	14 New polytypes	Ibid., C42, 518
SnS_2	26 New polytypes	Ibid., C42, 653
(M,Ni)P, M = Ti-Co	MnP	Acta Chem. Scand., A40 8
(V,Ti)As	MnP, NiAs, and TiP types	Ibid., A40, 17
Co_3Sn_2 Ni_3Sn_2	Ni_3Sn_2-type, with NiAs-Ni_2In type at higher temperatures	Ibid., A40, 23
(Mn,Fe)P	MnP	Ibid., A40, 227
MM'Si, M = Nb, Ta M' = Ru, Re	TiFeSi	Bull. Mater. Sci., 7, 215
ZrRuSi (Zr,M)RuSi, M = Nb,Ta	ZrRuSi	
Ullmannite, NiSbS	Pseudo-$P2_13$, probably twinned $Pca2_1$	Canad. Miner., 24, 27

$BaVS_3$	X-ray single crystal and neutron powder data give similar results, x(S) = 0.1654, y(S) = 0.3308 (48A, 28)	Chem. Scripta, 26A, 91
Palladium phosphide hydrides	Neutron study of H positions	Ibid., 26A, 99
ZnP_4	$P4_12_12$, 5.0154, 14.540, Z = 4	Cryst., Res. Technol., 21, K93
$CeLiGe_2$	$CaLiGe_2$	Dopov. Akad. Nauk Ukr.RSR,Ser.A: Fiz.-Mat.Tekh. Nauki, No. 7, 76
$TℓGaS_2$ $TℓGaSe_2$ $TℓInS_2$	Powder diffraction study of materials with excess Tℓ	Izv. Akad. Nauk SSSR, Neorg. Mater., 22, 521
CuTiGe	Pnma, 6.275, 3.8184, 7.278, R = 0.16	Ibid., 22, 689
$HoFe_6Sn_6$	YCo_6Ge_6	Ibid., 22, 690
$La_2Ni_5B_4$	$Nd_2Ni_5B_4$	Ibid., 22, 1920
TaN_x, x = 0.94-1.04	Hexagonal ε-phase, (44A, 85)	Izv. Akad. Nauk UzSSR, Ser. Fiz.-Mat. Nauk, No. 5, 53
$Mn_2Ga_{0.5}As_{0.5}$ Fe_3GaAs $Co_2Ga_{0.5}As_{0.5}$	Ni_2In	J. Appl. Cryst., 19, 80
$Mo_{3.25}Ir$	A15	Ibid., 19, 411
$(Ag,Ga)_4Pr$	Al_4Ba	J. Less-Common Metals 115, 367
Fe_5Nd	$CaCu_5$	Ibid., 115, L11
γ-Sn	A possible Ammm form is reported	Ibid., 116, 307
Nd	A3' form transforms to f.c.c. at 3.8 GPa, with another transition at 18 GPa	Ibid., 116, 313
$CoYb_3$ $NiYb_3$ Ni_2Yb_7	Fe_3C Al_3Ni Gd_2Co_7	Ibid., 118, 103
CuFeY CuFeGd $(Cu,Fe)_2Yb$ Cu_2Yb	C15 C15 C15 C14	Ibid., 118, 173

$Rh_{50}Al_{50-x}Cu_x$	CsCl	J. Less-Common Metals, 119, 75
(Nb,Si)M_2 (Nb,Al)M_2 M = transition metal	C14 and C15	Ibid., 119, 297
LnCoX, X = Si, Ge, Sn	TiNiSi	Ibid., 119, 319
$LnInAu_2$	B2, Heusler-type, and tetragonal phases	Ibid., 120, 101
Cf	Triclinic structure proposed for a high pressure phase	Ibid., 120, 345
UP USb	NaCl NaCl and CsCl	Ibid., 121, 157
UCu_2P_2 UCu_2As_2 UNi_2P_2	$CaAl_2Si_2$ $CaAl_2Si_2$ $ThCr_2Si_2$	Ibid., 121, 193
UNTe UNSe ThNTe ThNSe	PbFCl	Ibid., 121, 233
NpAs	NaCl, with incommensurate tetragonal phase and commensurate cubic phase at low temperatures	Ibid., 121, 325
$PuRu_3B_2$ $PuOs_3B_2$ $PuRh_3B_2$ $PuIr_3B_2$ $PuRu_4B_4$ $PuOs_4B_4$ $PuIr_4B_4$	URu_3B_2 URu_3B_2 $ErIr_3B_2$ $ErIr_3B_2$ $LuRu_4B_4$ YOs_4B_4 $NdCo_4B_4$	Ibid., 121, 431
UC	NaCl, with orthorhombic phase at high pressure	Ibid., 121, 445
$NpAl_2$	C15	Ibid., 121, 461
$TiS_{1.748}$	12R structure (41A, 110)	Ibid., 123, 101
$Th_{14}Au_{51}$ $ThAu_2$ Th_3Au_4 ThAu Th_3Au_2 Th_2Au	$Gd_{14}Ag_{51}$ AlB_2 Pu_3Pd_4 CrB Cubic (?) $CuAl_2$	Ibid., 124, 245
Gd(Al,Ga)$_2$	$MgCu_2$ and AlB_2 phases	Ibid., 124, L13

Gd_5Sb_3 Gd_4Sb_3 α-GdSb $GdSb_2$	Mn_5Si_3 anti-Th_3P_4 NaCl -	J. Less-Common Metals, 125, 1
$ZrCr_2$ $Zr(Cr,Fe)_2$	C14 polytypes	Ibid., 125, 33
$HgTh_2$	$CuAl_2$-type, I4/mcm, 7.696, 5.902 A, x(Th) = 0.158	Ibid., 125, L5
Hg_2Th	$CaIn_2$, $P6_3/mmc$, 4.822, 7.438 A, x(Hg) = 0.460	
Hg_3Th	Ni_3Sn, $P6_3/mmc$, 6.716, 4.902, x(Hg) = 0.833	
$(Y,Zr)Co_{2.9}$	$PuNi_3$ and $MgCu_2$	J. Phys. Soc. Japan, 55, 4435
$MnAs_{0.9}P_{0.1}$	MnP and NiAs phases	J. Solid State Chem., 64, 123
$(Co,Rh)S_2$ $(Co,Ru)S_2$ $(Rh,Ru)S_2$	Pyrite	Ibid., 64, 365
$CaRh_2B_2$	Description of relationship with the $BaAl_4$ type	Kristallografija, 31, 400 [Soviet Physics - Crystallography, 31, 234]
AB_5X_8 AB_6X_8 A = Tl, In, K Rb B = Ti, V, Cr X = S, Se	TlV_5S_8 TlV_6S_8	Mater. Res. Bull., 21, 69
Mg_2NiH_4	H probably in 24(e) (compare 46A, 72; 51A, 57; this vol., p. 19)	Ibid., 21, 415
$CoGaInS_4$	$P\bar{3}m1$, 3.759, 12.184 A. α-$FeGa_2S_4$-type	Ibid., 21, 709
$Mo_6Fe_{11}C_5$ (ζ-phase)	Correction of misprint in a Table heading (already corrected in 52A, 65)	Mh. Chem., 117, 55
V(P,As)	NiAs, MnP, and hexagonal phases	Ibid., 117, 773
Tc_4Si_7 Mo_9Ge_{16}	Structures proposed	Ibid., 117, 921

α-Mn β-Mn γ-Mn δ-Mn	A12 A13 A1 A2	Mh. Chem., 117, 1349
Li_3N	Neutron data. Results in accord with previous X-ray data	Mineral. J., 13, 28
Li	Below 80K, hexagonal 9R phase, ABCBCACAB stacking; neutron powder data	Physica B+C, 136, 481
$La_6MSi_2S_{14}$ M = Ni, Co	High-pressure studies	Ibid., 139-140, 337
$RuIn_3$	$CoGa_3$	Powder Diffr., 1, 20
$Ag_{0.4}Ga_{1.6}Y$	$CaIn_2$	Ibid., 1, 328
Gd_2Te_3	Sb_2S_3	Sci. Tech. Aerosp. Rep., 25, Abs. No, N87-14169; Chemical Abstracts, 106, 224943
$CeRu_2Si_2$	EXAFS study	Solid State Comm., 59, 241
$Mn_{1.8}Co_{0.2}Sb$	Mn_2Sb (4, 9), z = 0.307-0.290 and -0.271 to -0.275	Tohoku Gakuin Daigaku Kogakubu Kenkyu Hokoku, 21, 35
Cu_2Sb $TlAsPd_5$	Similarities between these structure types are reviewed	Z. Kristallogr., 175, 187
IrPTe	CoSbS	Z. Naturforsch., 40B, 130
$DyFe_3$	$PuNi_3$	J. Less-Common Metals, 118, 269
Co_5As_2	Pd_5Sb_2	Ibid., 118, 327

PAPERS REFERRED TO LATER YEARS

Preliminary notes:

α-B $B_{12}P_2$ $B_{12}As_2$ B_4C	AIP Conf. Proc., 140, 70 and 109
UCr_4C_4	Z. Kristallogr., 174, 11
$Ln_2Mn_{17}C_{3-x}$	Ibid., 174, 19

NaTl	Z. Kristallogr., 174, 26
UCr_6P_4	Ibid., 174, 27
BaD_2	Ibid., 174, 28
GeTe	Ibid., 174, 31
YCoC $UCoC_2$ U_2NiC_3	Ibid., 174, 63 [This volume, p. 26, 28]
Na_3As_7	Ibid., 174, 82
$Er_8Rh_5C_{12}$	Ibid., 174, 85
$Al_2Ru_3B_2$ $Al_3Ru_4B_2$	Ibid., 174, 109
Al_7Te_{10}	Ibid., 174, 157
$Fe_3B_{0.94}$	Ibid., 174, 165
NaMnX (X = P, As, Sb, Bi) LiMnAs KMnAs	Ibid., 174, 186 [This volume, p. 39]
MPtX (M = Ca, Eu; X = P, As, Sb)	Ibid., 174, 200 [This volume, p. 35]

The following reports were prepared too late for inclusion in the main text.

IrU

J. Less-Common Metals, **125**, 167-174.

$B2_1$, 7.271, 10.578, 8.511, 90.16, Z = 16, R = 0.082. Distortion variant of the CrB type, similar to the structure of BiPd (**45**A, 39). U-Ir = 2.80-3.28, U-U = 3.41-3.84, Ir-Ir = 2.89-3.41(2) A.

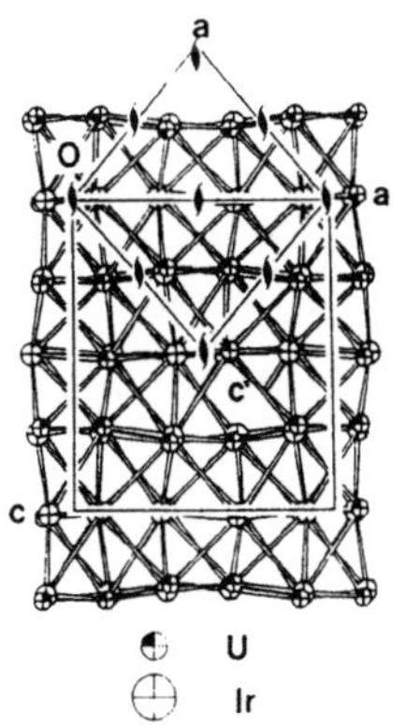

	x	y	z
U1	0.115	0.0	0.992
U2	0.636	0.996	0.017
U3	0.1254	0.7181	0.2738
U4	0.6216	0.720	0.224
Ir1	0.0945	0.2643	0.9990
Ir2	0.6582	0.262	0.0027
Ir3	0.139	0.454	0.2545
Ir4	0.608	0.4536	0.2412

$Ba_7Ga_4Sb_9$

Z. anorg. Chem., **534**, 137-142.

Pmmn, 18.024, 10.861, 7.100, Z = 2, R = 0.067. Edge- and corner-sharing $GaSb_4$ tetrahedra, linked by long Sb-Sb bonds into sheets which are connected by Ba ions.

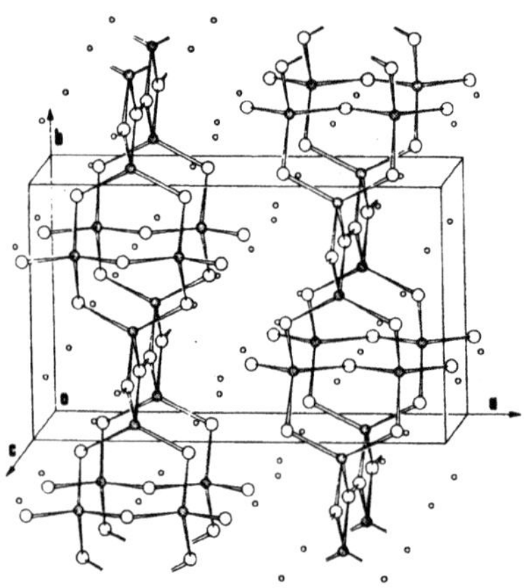

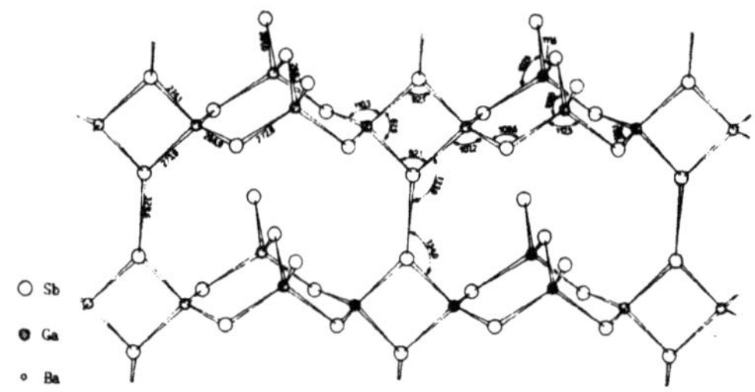

		x	y	z
4 Ba1	4f	0,0374	0,25	0,0406
8 Ba2	8g	0,1258	0,4796	0,5438
2 Ba3	2b	0,25	0,75	0,9207
4 Ga1	4f	0,8772	0,25	0,7420
4 Ga2	4e	0,25	0,0684	0,0587
8 Sb1	8g	0,1175	0,9589	0,0449
4 Sb2	4f	0,5074	0,25	0,5197
2 Sb3	2a	0,25	0,25	0,7910
2 Sb4	2a	0,25	0,25	0,3268
2 Sb5	2b	0,25	0,75	0,4541

Nb_6C_5

Kristallografija, **30**, 1112-1115 (1985) [Soviet Physics - Crystallography, **30**, 648-650 (1985)].

C2/m, 5.460, 9.457, 5.460, 109.47, Z = 2, neutron radiation, R = 0.082. Nb(1): 0.245,0,0.755; Nb(2): 0.264,0.660,0.750; C(1): 0,1/2,1/2; C(2): 0,0.320,0; C(3): 0,0.184,1/2; vacancy: 0,0,0. Structure derived from those of higher-symmetry phases (**33**A, 57; **35**A, 110).

STRUCTURE REPORTS

SECTION II

INORGANIC COMPOUNDS

Edited by

J. Trotter

(University of British Columbia)

ARRANGEMENT

To find particular inorganic compounds the subject index should be used. The general arrangement in the text is: elements, boron hydrides, carbonyls, phosphorus-nitrogen and sulphur-nitrogen compounds, halides, cyanides, oxides, double oxides, hydroxides, sulphides, borates, carbonates, nitrates, phosphates, arsenates, sulphates, perchlorates, iodates, silicates, silicate minerals. Only complete structure analyses are described; incomplete structural data are given in a Table, electron diffraction studies in the gas phase and compounds which have been described only in preliminary communications are tabulated.

NITROGEN
ε-N_2

J. Chem. Phys., **84**, 2837-2845.

$R\bar{3}c$, a = 5.928 A, α = 85.14° (hexagonal cell, a = 8.020, c = 11.104 A), at 110K and 7.8GPa, Z = 8 (24 in hexagonal cell), powder data. N(1) in 12(c): 0,0,0.0495; N(2) in 36(f): 0.2731,0.2127,0.2828 (hexagonal cell).

CYCLOUNDECASULPHUR
S_{11}

CYCLOTRIDECASULPHUR
S_{13}

Z. Naturforsch., **41**B, 958-970.

S_{11}, $Pca2_1$, 14.933, 8.321, 18.086, at 163K, Z = 8, R = 0.032. 11-membered ring with approximate C_2 symmetry, S-S = 2.032-2.110(3) A.

S_{13}, $P2_1/c$, 12.95, 12.36, 17.61, 110.41, at 173K, Z = 8, R = 0.055. 13-membered ring with approximate C_2 symmetry, S-S = 1.978-2.113(4) A.

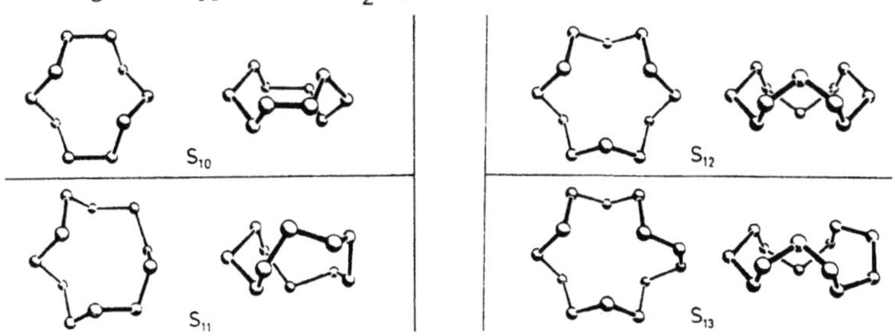

1,2'-BIS(PENTABORANE)
$[B_5H_8]_2$

Inorg. Chem., **25**, 4618-4622.

$Pna2_1$, 11.020, 5.870, 14.988, Z = 4, R = 0.049. Two pentaborane clusters linked by a B-B bond.

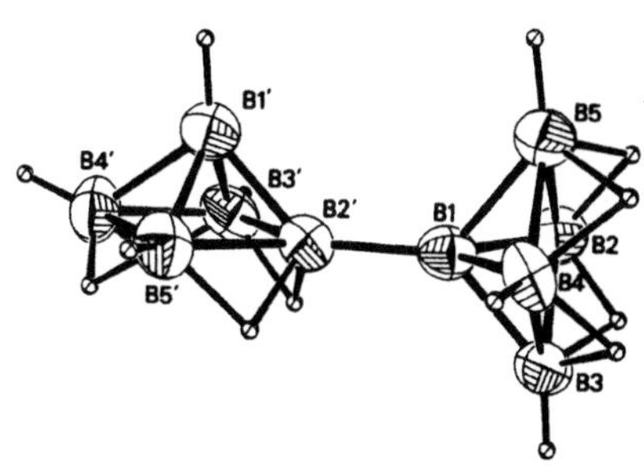

	x	y	z
B1	0.7072	0.114	0.775
B2	0.6959	-0.129	0.8338
B3	0.6846	0.144	0.8858
B4	0.8249	0.229	0.8350
B5	0.8352	-0.049	0.7821
B1′	0.6571	0.178	0.5754
B2′	0.6354	0.206	0.6848
B3′	0.5201	0.094	0.6121
B4′	0.5365	0.327	0.5371
B5′	0.6513	0.450	0.6098
H23	0.700	-0.047	0.911
H45	0.896	0.049	0.858
H34′	0.468	0.23	0.586
H34	0.794	0.181	0.918
H25	0.823	-0.191	0.841
H23′	0.517	0.163	0.699
HB2	0.639	-0.277	0.827
HB3	0.634	0.215	0.927
HB4	0.866	0.394	0.825
HB5	0.896	-0.117	0.732
HB1′	0.720	0.107	0.533
HB3′	0.461	-0.052	0.603
HB5′	0.718	0.600	0.605
HB4′	0.510	0.333	0.473
H25′	0.626	0.439	0.696
H45′	0.554	0.50	0.578

Interatomic Distances (Å)

B1-B2	1.683 (10)	B1′-B2′	1.665 (11)
B1-B3	1.686 (7)	B1′-B3′	1.680 (10)
B1-B4	1.713 (9)	B1′-B4′	1.693 (11)
B1-B5	1.708 (10)	B1′-B5′	1.680 (11)
B2-B3	1.787 (11)	B2′-B3′	1.798 (10)
B2-B5	1.783 (11)	B2′-B5′	1.827 (11)
B3-B4	1.794 (10)	B3′-B4′	1.782 (11)
B4-B5	1.815 (11)	B4′-B5′	1.818 (11)
		B1′-B2	1.660 (8)
B2-HB2	1.08 (5)	B1′-HB1′	1.03 (5)
B2-H23	1.26 (5)	B2′-H23′	1.35 (5)
B2-H25	1.45 (5)	B2′-H25′	1.38 (5)
B3-HB3	0.93 (5)	B3′-HB3′	1.08 (5)
B3-H23	1.19 (5)	B3′-H23′	1.36 (5)
B3-H34	1.32 (5)	B3′-H34′	1.07 (6)
B4-HB4	1.09 (5)	B4′-HB4′	1.01 (5)
B4-H34	1.32 (5)	B4′-H34′	1.20 (5)
B4-H45	1.36 (5)	B4′-H45′	1.18 (6)
B5-HB5	1.08 (5)	B5′-HB5′	1.15 (5)
B5-H25	1.22 (5)	B5′-H25′	1.32 (5)
B5-H45	1.44 (5)	B5′-H45′	1.21 (5)

TUNGSTEN CARBONYL SELENIDE HEXAFLUOROARSENATE
$[W_2(CO)_{10}Se_4][AsF_6]_2$

Inorg. Chem., **25**, 2057-2061.

$P2_1/c$, 8.688, 7.036, 21.767, 95.30, Z = 2, R = 0.040. Centrosymmetric cations with weak Se-Se bonds and pentagonal-pyramidal coordination at W, and octahedral anions.

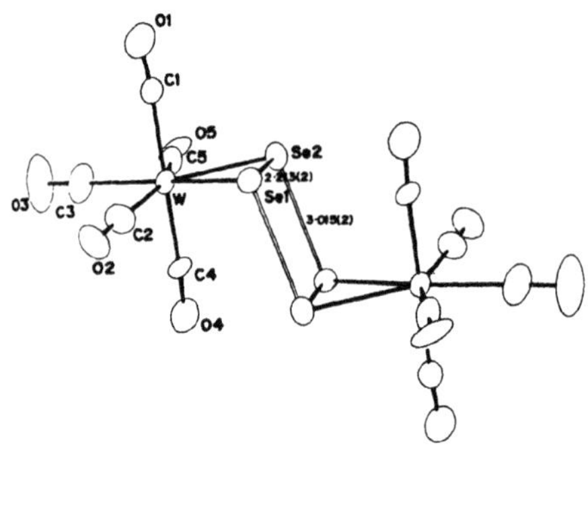

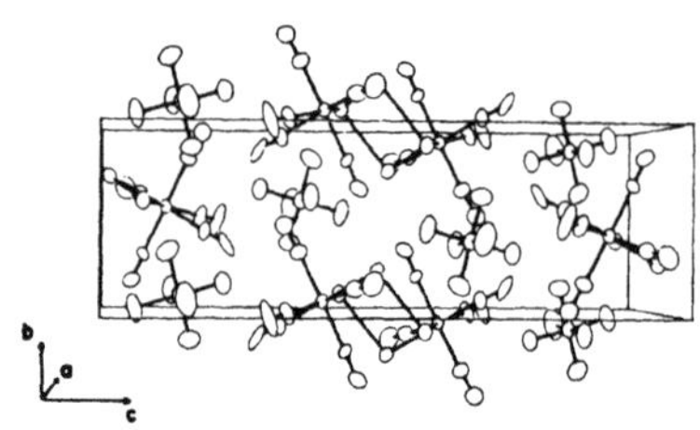

RHENIUM GALLIUM CARBONYL
$Re_4Ga_2(CO)_{18}$

Acta Cryst., C**42**, 665-667.

$I4_1/a$, 13.894, 28.836, Z = 8, R = 0.058. Ga_2Re_2 rhombus with an Re-Re bond (3.139(2) A), with two attached $Re(CO)_5$ groups in a trans configuration. Re-Ga = 2.569-2.599(4) A.

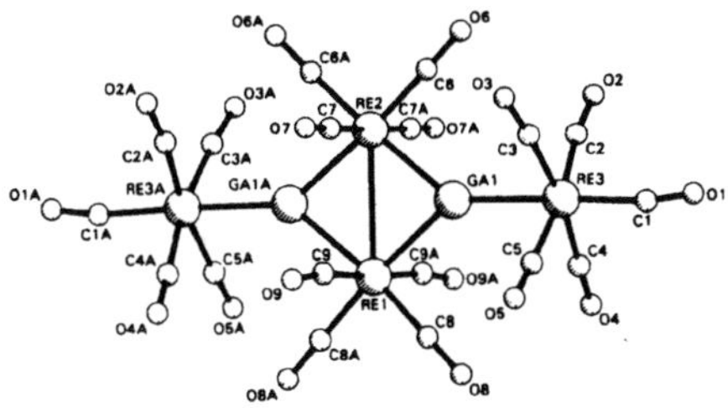

	x	y	z
Re(1)	0	0·25	0·2232
Re(2)	0	0·25	0·3322
Re(3)	−0·1042	−0·0669	0·2769
Ga	0·0531	0·3892	0·2779
C(1)	0·134	0·705	0·271
C(2)	0·007	0·576	0·324
C(3)	0·196	0·534	0·325
C(4)	0·009	0·578	0·227
C(5)	0·203	0·534	0·229
C(6)	0·030	0·348	0·376
C(7)	0·139	0·204	0·332
C(8)	0·042	0·341	0·177
C(9)	0·124	0·190	0·226
O(1)	0·157	0·783	0·276
O(2)	−0·047	0·593	0·355
O(3)	0·257	0·504	0·353
O(4)	−0·052	0·588	0·201
O(5)	0·260	0·513	0·203
O(6)	−0·048	0·096	0·405
O(7)	0·215	0·179	0·333
O(8)	−0·076	0·100	0·150
O(9)	0·202	0·153	0·220

RHENIUM MANGANESE DECACARBONYL
$ReMn(CO)_{10}$

Inorg. Chem., **25**, 3706-3707.

I2/a, 14.390, 7.112, 14.736, 105.54, Z = 4, R = 0.033. Isostructural with the Mn and Re analogues (**48**A, 116), with Mn/Re disorder.

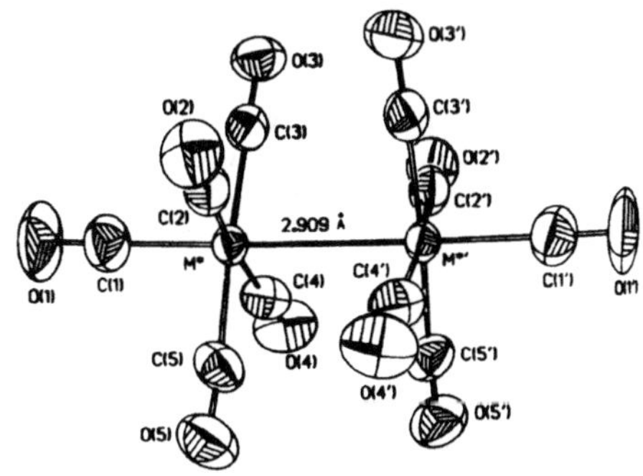

Atomic Coordinates (×10⁴) for $MnRe(CO)_{10}$

	x	y	z
Mn/Re	1557.9	2320.9	9319.4
O(1)	−340	2454	7808
O(2)	775	3994	10877
O(3)	2090	6262	8777
O(4)	2602	468	8009
O(5)	1092	−1485	10025
C(1)	347	2397	8387
C(2)	1062	3407	10289
C(3)	1885	4792	8954
C(4)	2212	1160	8497
C(5)	1299	−90	9763

OSMIUM CARBONYL SULPHIDES
$Os_4(CO)_{12}S$ (I), $Os_7(CO)_9S$ (II)

Organometallics, **2**, 990-996 (1983).

I, C2/c, 29.110, 8.491, 16.610, 104.98, Z = 8, R = 0.031. II, $P2_1$, 10.107, 11.569, 12.791, 90.38, Z = 2, R = 0.036. Molecular structures are shown below.

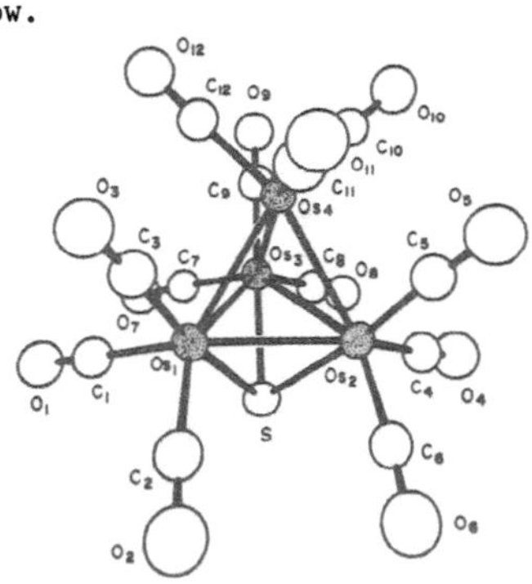

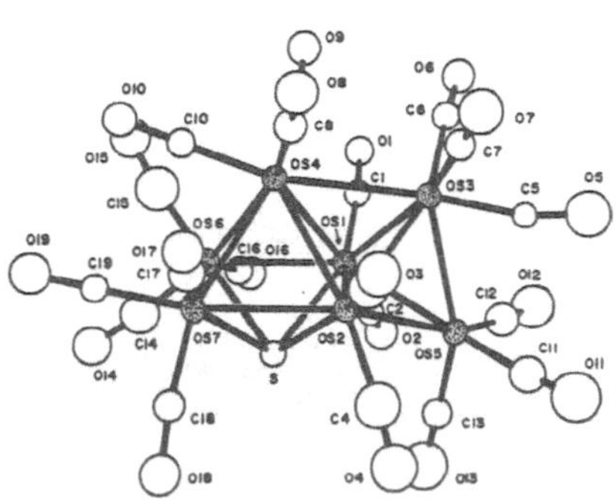

OSMIUM CARBONYL HYDRIDE SULPHIDE

$Os_4(CO)_{12}(\mu\text{-}H)_2\ (\mu_3\text{-}S)$

Inorg. Chem., **25**, 2534-2538.

$P2_1/m$, 7.320, 17.097, 8.155, 104.79, Z = 2, R = 0.051. Butterfly tetrahedral metal cluster, with S bridging one open triangular face; H atoms not located, but they probably bridge the Os(1)-Os(3) and Os(1')-Os(3) bonds (2.970(1) A; other Os-Os = 2.815, 2.858(1) A). The structure of $Os_4(CO)_{12}(NHMe_2)(\mu_3\text{-}S)$ is also described.

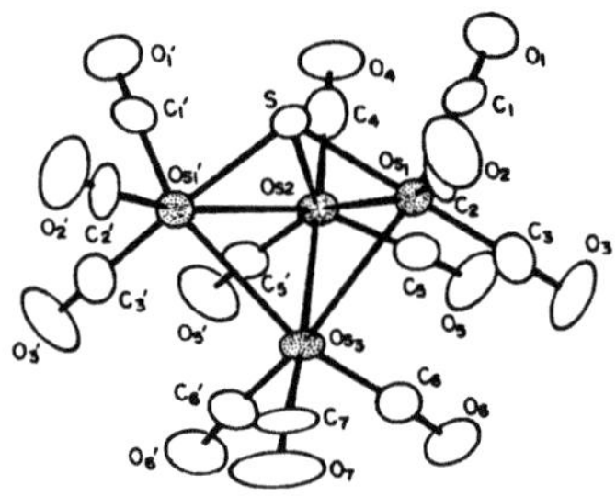

OSMIUM CARBONYL HYDRIDE SULPHIDES

$H_2Os_6(CO)_{17}S_2$ (I, two isomers), $H_2Os_6(CO)_{18}S_2$ (II)

Organometallics, **2**, 996-1005 (1983).

I, isomer 1 (red), C2/c, 20.364, 9.764, 32.544, 114.01, Z = 8, R = 0.037. I, isomer 2 (green), $P2_1/n$, 11.124, 14.090, 18.694, 92.00, Z = 4, R = 0.054. II, $P\bar{1}$, 9.022, 11.986, 14.370, 89.94, 78.23, 83.39, Z = 2, R = 0.044. Molecular structures are shown below.

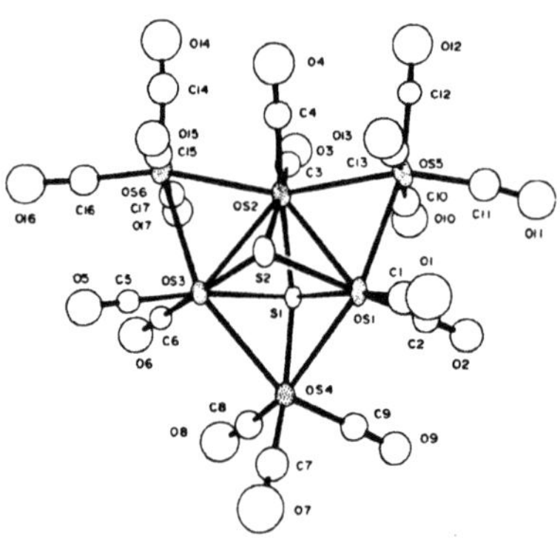

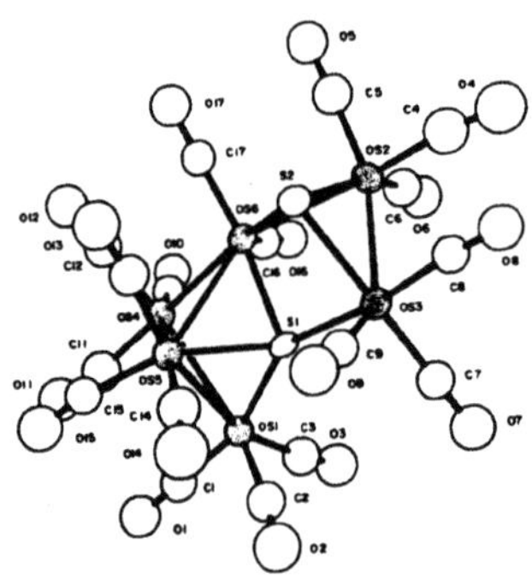

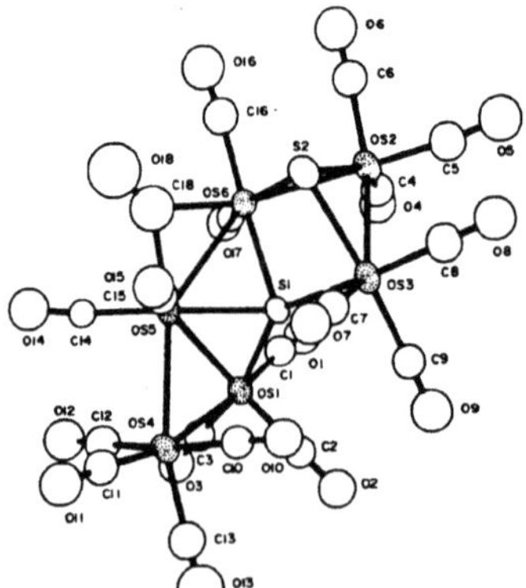

OSMIUM CARBONYL HYDROXIDE HYDRIDE SULPHIDE
$Os_6(CO)_{18}(\mu\text{-}OH)(\mu_3\text{-}S)(\mu\text{-}H)$

Inorg. Chem., **25**, 1122-1127.

$P\bar{1}$, 12.127, 14.009, 9.155, 103.24, 98.18, 90.00, Z = 2, R = 0.035. Folded chain of six Os atoms, Os-Os = 2.795-2.962(1). The structure of $Os_3(CO)_8(NMe_3)(\mu\text{-}OH)(\mu_3\text{-}S)(\mu\text{-}H)$ is also described.

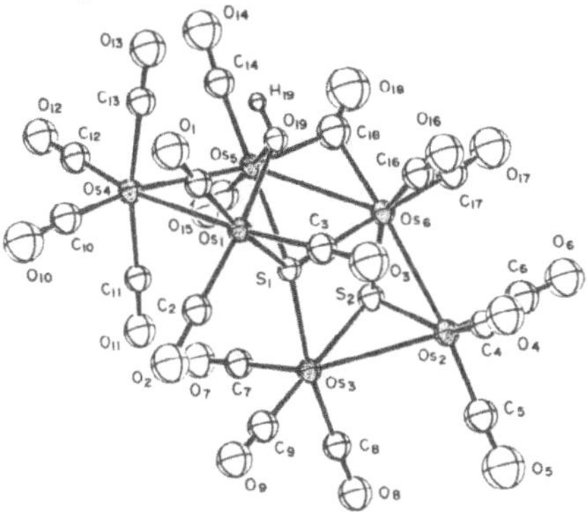

RUTHENIUM RHODIUM HYDRIDE CARBONYL
$Ru_2Rh_2H_2(CO)_{12}$

J. Chem. Soc., Dalton, 681-686 (1986).

$P\bar{1}$, 9.841, 9.883, 10.033, 94.70, 90.84, 98.13, Z = 2, R = 0.027. Tetrahedral metal cluster, with disorder of 1 Ru + 1 Rh. The Co analogue is described in **50A**.

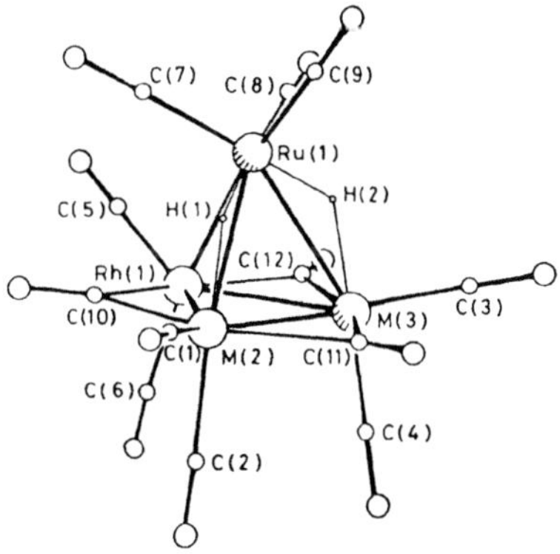

RUBIDIUM IODIDE AZIDE
Rb_2IN_3

CAESIUM IODIDE AZIDE
Cs_2IN_3

J. Less-Common Metals, **124**, 61-72.

Rb, P4/mbm (or possibly $P\bar{4}b2$), 6.252, 8.081, Z = 2, R = 0.021. Rb in 4(e): 0,0,0.26287; I in 2(d): 0,1/2,0; N(1) in 2(c): 0,1/2,1/2; N(2) in 4(h): 0.135,0.635,1/2. Cs, P4/mmm, 4.602, 8.550, Z = 1, R = 0.072. Cs in 2(g): 0,0,0.252; I in 1(c): 1/2,1/2,0; N(1) in 1(d): 1/2,1/2,1/2; 0.5 N(2) in 4(o): 0.20,1/2,1/2. The Rb structure contains layers of azide anions, and of Rb^+ and I^- ions. The Cs compound is essentially isostructural, but with disordered azide positions giving a smaller unit cell (a/√2).

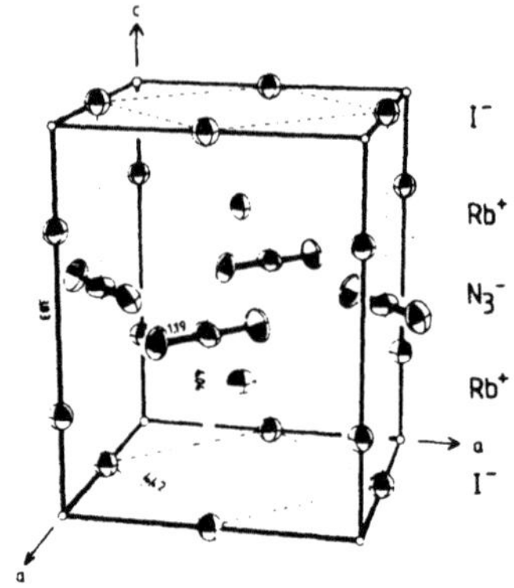

MAGNESIUM AZIDE HEXAHYDRATE
$Mg(N_3)_2.6H_2O$

Z. Naturforsch., **41** B, 935-937.

Fmmm, 6.443, 11.344, 13.335, Z = 4, R = 0.037. $Mg(H_2O)_6^{2+}$ octahedra and linear azide anions. Mg-O = 2.024 (x 4), 2.083 (x 2), N-N = 1.170(2) A.

ZINC AZIDE HYDRATE
$Zn(N_3)_2.2.5H_2O$

Z. anorg. Chem., **533**, 191-196.

P2/n, 9.962, 5.948, 10.184, 90.16, Z = 4, R = 0.043. Previous study in Z. anorg. Chem., **413**, 94 (1975). Zn is coordinated octahedrally by four azide groups and two water molecules, the octahedra sharing edges to form chains along **b**. The water molecules are arranged in chain-like clusters, and two of the ten water molecules in the cell are not coordinated to Zn. The azide groups are asymmetric and nearly linear. Zn-N = 2.11-2.14, Zn-O = 2.20, 2.23, N-N = 1.20, 1.16 A.

CAESIUM TETRAAMIDE AZIDE
$Cs_5(NH_2)_4N_3$

J. Less-Common Metals, **118**, 261-268.

P4/mnc, 8.918, 8.807, Z = 2, R = 0.036. Amide and linear azide anions, linked by 6- and 10-coordinate Cs cations. N-N = 1.20(3), Cs-N = 3.08-3.99(2) A.

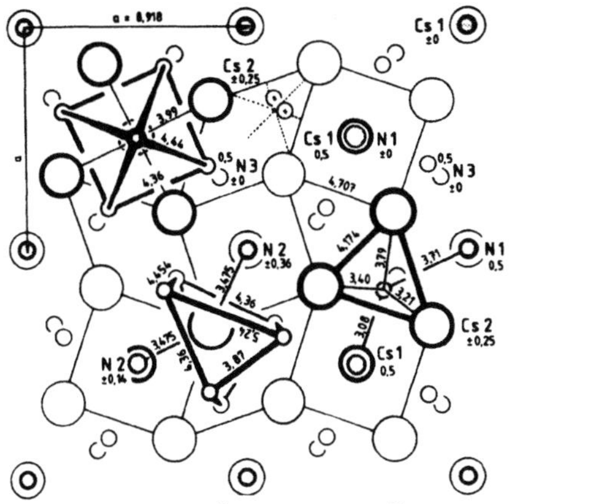

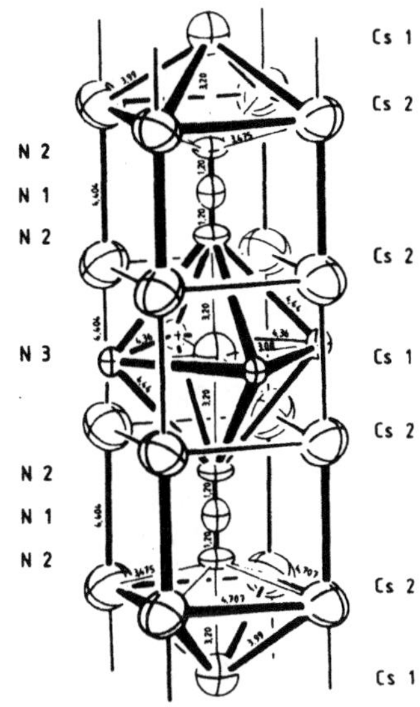

	x	y	z
2 Cs(1)	0	0	0
8 Cs(2)	0,3345(3)	$\frac{1}{2}$ + x	$\frac{1}{4}$
2 N(1) } Azidion	0	0	$\frac{1}{2}$
4 N(2) } Azidion	0	0	0,363
8 N(3) Amidion	0,322(3)	0,124(2)	0

POTASSIUM YTTRIUM AMIDE
$KY(NH_2)_4$

RUBIDIUM YTTRIUM AMIDE
$RbY(NH_2)_4$

J. Less-Common Metals, **119**, 99-113.

K, C2/c, 7.199, 16.692, 5.989, 120.26, Z = 4, R = 0.024. Rb, P$\bar{1}$, 6.05, 6.770, 8.760, 104.47, 98.33, 110.9, Z = 2, R = 0.050. The K compound contains cubic close packed amide anions, with K and Y in chains of edge-sharing anion octahedra. In the Rb salt, Rb^+ and NH_2^- ions form a distorted cubic close packing, and Y again has octahedral coordination.

	x	y	z
KY(NH₂)₄			
Y	0	0,06508	1/4
K	0	0,30492	1/4
N(1)	0,2135	0,1629	0,2088
N(2)	0,2881	0,4555	0,2936
H(11)	0,332	0,147	0,218
H(12)	0,169	0,190	0,072
H(21)	0,290	0,417	0,168
H(22)	0,152	0,476	0,220
RbY(NH₂)₄			
Y	0,1838	0,3982	0,3722
Rb	0,2193	0,8331	0,1001
N(1)	0,534	0,749	0,536
N(2)	0,949	0,618	0,375
N(3)	0,308	0,419	0,136
N(4)	0,162	0,936	0,775

CAESIUM LANTHANON AMIDES
$Cs_3Ln_2(NH_2)_9$ (M = Gd, Yb, Y)

J. Less-Common Metals, **118**, 31-41.

Gd, $P6_3/mmc$, 7.153, 17.930, Z = 2, R = 0.089. Yb and Y, $R\bar{3}c$, a = 12.141, 12.220, c = 18.083, 18.321, Z = 6, R = 0.059, 0.030. In the Gd compound, Cs^+ and NH_2^- form a distorted close packing with sequence ABACBC; Gd occupies hexagonal face-sharing double octahedra, $[Gd_2(NH_2)_9]^{3-}$. The Yb and Y compounds are hexagonal perovskites.

Gd

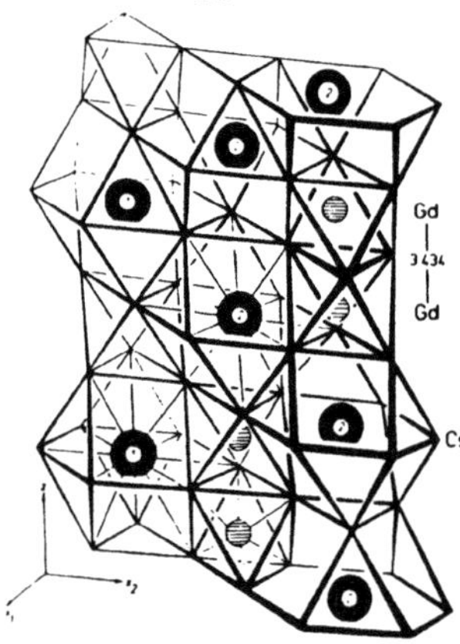

Yb, Y

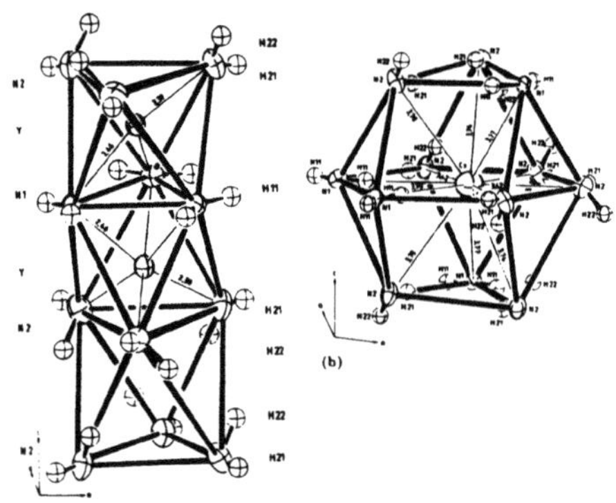

	x	*y*	*z*
$Cs_3Gd_2(NH_2)_9$ $P6_3/mmc$			
Gd 4f	1/3	2/3	0,1542
Cs(1) 2b	0	0	1/4
Cs(2) 4f	1/3	2/3	0,5834
N(1) 6h	0,4797	2*x*	1/4
N(2) 12k	0,1707	2*x*	0,086
$Cs_3Yb_2(NH_2)_9$ $R\bar{3}c$			
Yb 12c	0	0	0,1559
Cs 18e	0,3555	0	1/4
N(1) 18e	0,8586	0	1/4
N(2) 36f	0,1648	0,1554	0,5889
$Cs_3Y_2(NH_2)_9$ $R\bar{3}c$			
Y 12c	0	0	0,15555
Cs 18e	0,35585	0	1/4
N(1) 18e	0,8575	0	1/4
N(2) 36f	0,1661	0,1549	0,5899
H(11) 36f	0,87	0,087	0,247
H(21) 36f	0,22	0,21	0,595
H(22) 36f	0,08	0,20	0,459

CYCLOTHIAZENO-VANADIUM DICHLORIDE
β-$VCl_2(N_3S_2)$

Z. Naturforsch., **41**B, 1420-1425.

$P\bar{1}$, 5.700, 7.799, 7.353, 90.85, 108.26, 98.77, Z = 2, R = 0.034. Isostructural with the bromide (**52**A, 110), with chains of V atoms linked alternately by chloro and nitrogen bridges. The chains are rotated by 11° compared with the α-form (**50**A).

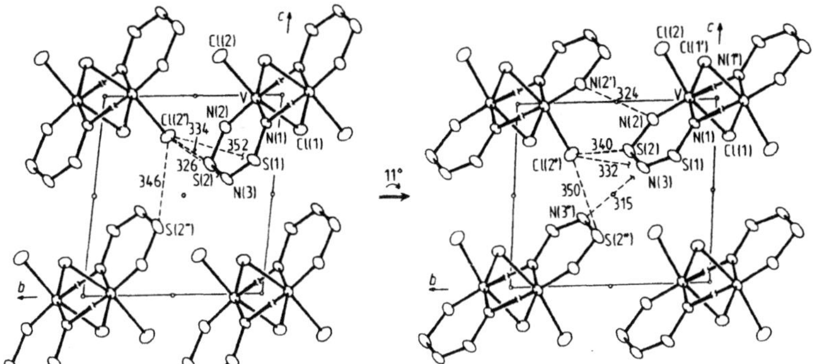

α-VCl$_2$(N$_3$S$_2$)

β-VCl$_2$(N$_3$S$_2$)

	x	y	z
V	.2531	.1364	.0118
Cl(1)	.3336	−.0677	−.2055
Cl(2)	.2635	.2826	.2845
S(1)	−.2290	.2000	−.3081
S(2)	.2307	.4281	−.2737
N(1)	−.0793	.1062	−.1286
N(2)	.3461	.3143	−.1016
N(3)	−.0617	.3549	−.3673

SODIUM CYCLODINITRIDOIMIDODISULPHOPHOSPHATE DIHYDRATE
$Na_3[HN_3S_2PO_6].2H_2O$

Koord. Khim., **12**, 840–845.

P2$_1$/n, 7.832, 9.954, 13.281, 91.06, Z = 4, R = 0.043. Nitridoimide tautomer, with some P/S disorder.

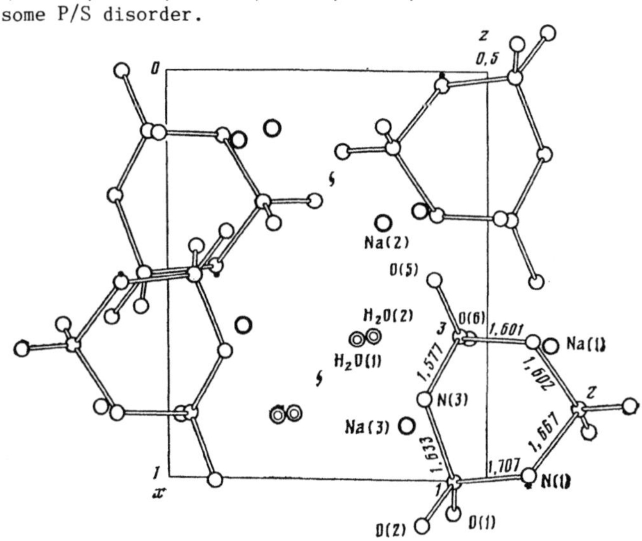

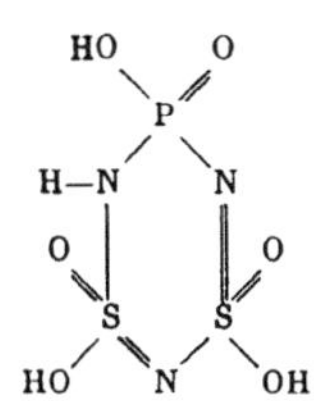

	x	y	z
1	0,4985	0,28403	0,04125
2	0,8461	0,23087	0,03307
S	0,6758	0,27124	−0,13836
Na(1)	0,6676	0,4337	−0,3948
Na(2)	0,8674	0,5441	−0,1572
Na(3)	0,6290	−0,0324	0,1225
O(1)	0,4322	0,1441	0,0395
O(2)	0,3933	0,3874	0,0896
O(3)	0,6955	0,3626	−0,2333
O(4)	0,6344	0,1305	−0,1725
O(5)	1,0026	0,2780	0,0795
O(6)	0,8408	0,0839	0,0332
O_{w1}	0,6540	0,0262	0,3024
O_{w2}	0,8516	−0,1818	0,1802
N(1)	0,5151	0,3317	−0,0813
N(2)	0,8461	0,2851	−0,0804
N(3)	0,6907	0,2937	0,0904
H(1)(N(1))	0,780	0,471	0,155
$H(1)_{w1}$	0,192	0,585	−0,174
$H(2)_{w1}$	−0,019	0,924	0,573

HYDROGENTETRASULPHURTETRANITRIDE AQUOPENTACHLOROSTANNATE(IV)
$S_4N_4H^+[SnCl_5(H_2O)]^-$ (I)

AMINOTRISULPHURDINITRIDE HEXACHLOROSTANNATE(IV)
$(S_3N_2NH_2^+)_2[SnCl_6]^{2-}$ (II)

Z. Naturforsch., **41**B, 581-586.

I, $P2_1/c$, 9.084, 23.758, 6.587, 101.86, Z = 4, R = 0.044. II, $P2_1/n$, 9.882, 12.873, 6.907, 93.77, Z = 2, R = 0.029. Cyclic cations and octahedral anions. S-N = 1.51-1.65, S-S = 2.17, Sn-Cl = 2.38-2.44, Sn-O = 2.19 A.

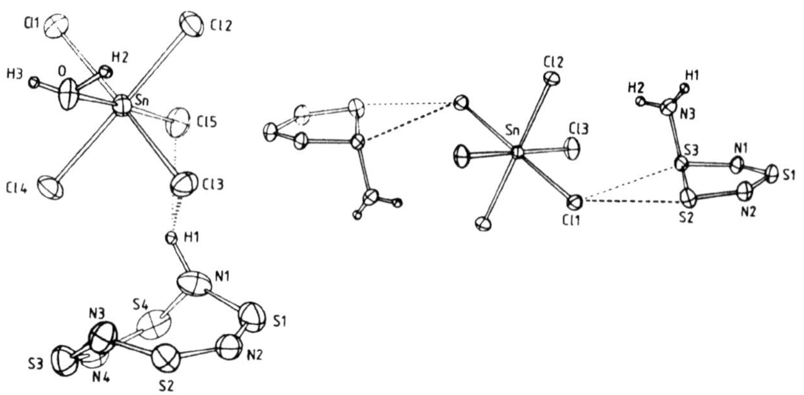

HYDROGENTETRASULPHURTETRANITRIDE TETRACHLOROFERRRATE(III)
$S_4N_4H[FeCl_4]$

Z. Naturforsch., **41**B, 1061-1066.

$P2_1/c$, 8.038, 10.039, 15.077, 104.52, Z = 4, R = 0.035. Protonated 8-membered S_4N_4 rings and tetrahedral $FeCl_4^-$ anions, with disorder of N(3), N(4) and S(4).

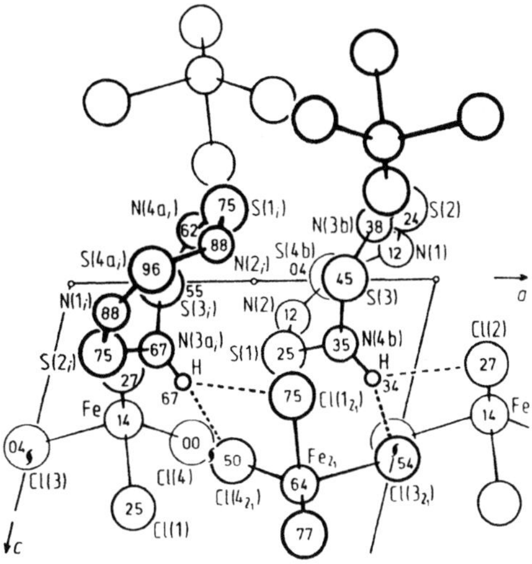

	x	y	z
Anion			
Fe	0.2335	0.1398	0.2442
Cl(1)	0.3184	0.2464	0.3748
Cl(2)	0.1922	0.2733	0.1255
Cl(3)	−0.0122	0.0413	0.2380
Cl(4)	0.4344	−0.0037	0.2360
Cation			
S(1)	0.6313	0.2526	0.1073
S(2)	0.8710	0.2440	−0.1045
S(3)	0.7476	0.4501	−0.0028
S(4a)	0.7906	0.0393	0.0141
S(4b)	0.7174	0.0423	−0.0033
N(1)	0.8706	0.1176	−0.0436
N(2)	0.6341	0.1187	0.0537
N(3a)	0.7242	0.3349	−0.0969
N(3b)	0.7954	0.3799	−0.0801
N(4a)	0.7021	0.3837	0.0749
N(4b)	0.7857	0.3457	0.0946

PENTASULPHUR PENTANITRIDE CHLOROOCTAARSENATE - TETRASULPHUR TETRANITRIDE
$(S_5N_5)_4[As_8Cl_{28}].2S_4N_4$

Z. anorg. Chem., **537**, 145-153.

$P\bar{4}2_1c$, 15.966, 15.201, Z = 2, R = 0.059. Positionally-disordered planar cyclic $S_5N_5^+$ cations, S_4N_4 molecules (also disordered), and $As_8Cl_{28}^{4-}$ anions which contain a cubane-like $As_4Cl_{16}^{4-}$ unit to which four $AsCl_3$ molecules are attached by chloro bridges. As-Cl = 2.13-2.30, 2.86-3.05, 3.14-3.28 A.

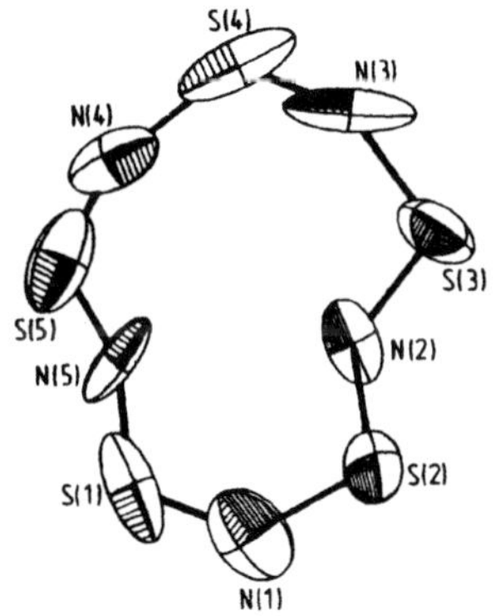

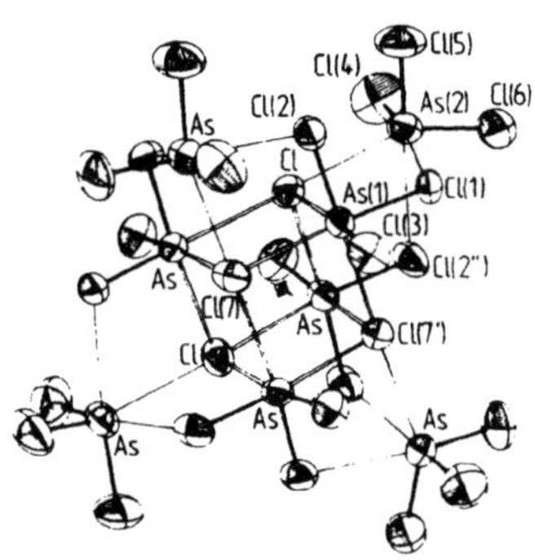

CYCLOPENTAAZATHIENIUM TETRACHLOROAURATE(III)
$[S_5N_5]AuCl_4$

Acta Cryst., C**42**, 1263-1264.

$P2_1/m$, 5.340, 11.997, 10.160, 99.51, Z = 2, R = 0.044. Azulene-like cations and square-planar anions.

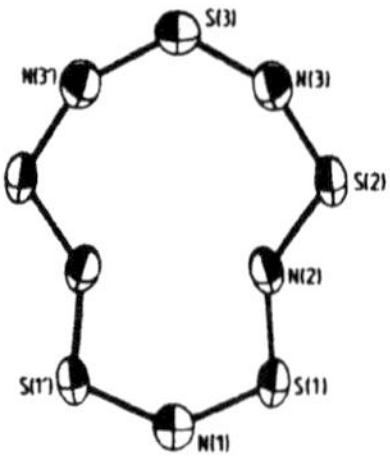

	x	y	z
Au	0,1549	0,25	0,29616
Cl(1)	0,3840	0,1155	0,4197
Cl(2)	−0,0728	0,1157	0,1700
N(1)	1,007	0,25	0,611
S(1)	0,9270	0,1293	0,6495
N(2)	0,741	0,1397	0,7530
S(2)	0,5851	0,0629	0,8308
N(3)	0,425	0,1341	0,9151
S(3)	0,3371	0,25	0,9637

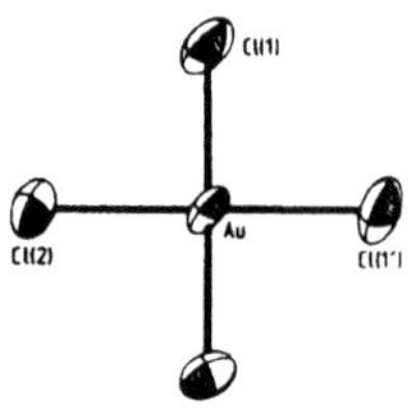

	Å		°
Au–Cl(1)	2,274 (3)	Cl(1)–Au–Cl(2)	89,8 (1)
Au–Cl(2)	2,281 (3)	Cl(1)–Au–Cl(1')	90,4 (1)
		Cl(2)–Au–Cl(2')	89,9 (1)
N(1)–S(1)	1,577 (6)	S(1)–N(1)–S(1')	133,3 (9)
S(1)–N(2)	1,566 (9)	N(1)–S(1)–N(2)	108,7 (6)
N(2)–S(2)	1,544 (9)	S(1)–N(2)–S(2)	138,7 (6)
S(2)–N(3)	1,562 (9)	N(2)–S(2)–N(3)	110,2 (5)
N(3)–S(3)	1,572 (9)	S(2)–N(3)–S(3)	150,9 (7)
		N(3)–S(3)–N(3')	124,4 (5)

POTASSIUM HYDROGEN FLUORIDES
KF.2.5HF, KF.3HF

J. Amer. Chem. Soc., **108**, 6634-6636.

KF.2.5HF, $P2_1/a$, 8.315, 8.384, 10.031, 101.63, at 173K, Z = 8, R = 0.035. KF.3HF, $R\bar{3}c$, a = 7.458, c = 11.818 A, at 173K, Z = 6, R = 0.011. The compounds can be formulated as $K_2[H_2F_3][H_3F_4]$ and $K[H_3F_4]$ (with two isomeric $H_3F_4^-$ anions).

	x	y	z
		KF·2.5HF	
K(1)	0.000 15	0.249 30	0.500 50
K(2)	0.637 11	0.273 82	0.038 69
F(1)	0.215 0	0.871 2	0.353 0
F(2)	0.155 5	0.754 4	0.135 5
F(3)	0.210 7	1.147 7	0.337 8
F(4)	0.351 4	0.501 9	0.351 1
F(5)	0.077 6	0.502 5	0.352 9
F(6)	−0.078 2	0.451 9	0.123 3
F(7)	0.390 7	0.449 9	0.124 4
H(2)	0.179	0.803	0.197
H(3)	0.215	1.034	0.345
H(45)	0.209	0.499	0.360
H(6)	−0.040	0.474	0.210
H(7)	0.387	0.475	0.192
		KF·3HF	
K	0.0	0.0	0.0
F(1)	0.0	0.0	0.25
F(2)	0.321 95	0.0	0.25
H	0.218	0.0	0.25

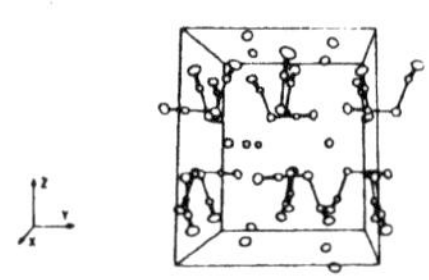

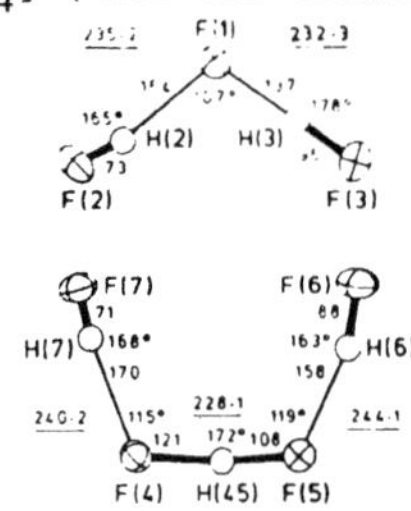

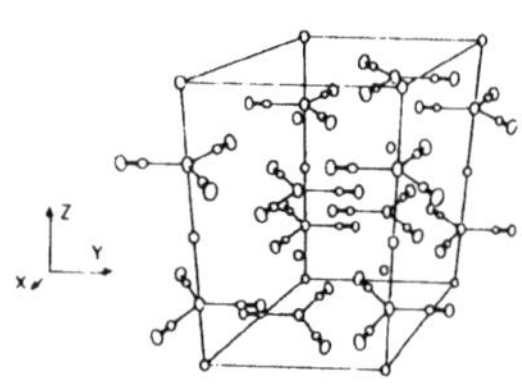

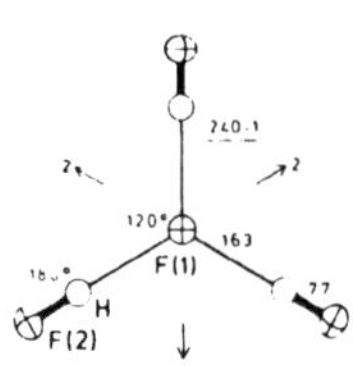

IRON(III) FLUORIDE
FeF_3

Mater. Res. Bull., **21**, 971-978.

Fd3m, 10.325, Z = 16, powder data. Fe in 16(c): 0,0,0; F in 48(f): 0.3104,1/8,1/8. Pyrochlore-type structure. Other forms have been described (**2**, 34, 291; **21**, 206; **50A**; **52A**, 117).

IRON FLUORIDE HYDRATE
$Fe_2F_5.2H_2O$

J. Solid State Chem., **62**, 274-277.

Imma, 7.477, 10.862, 6.652, Z = 4, R = 0.029. Inverse weberite structure, as previously described (**43**A, 122).

		x	y	z
Fe^{2+}	4*a*	0	0	0
Fe^{3+}	4*c*	¼	¼	¼
F1	16*j*	0.2024	0.1252	0.0544
F2	4*e*	0	¼	0.3343
O	8*h*	½	0.5690	0.2034

LITHIUM TETRAFLUOROBORATE MONOHYDRATE
$LiBF_4.H_2O$

Kristallografija, **31**, 1076-1080 [Soviet Physics - Crystallography, **31**, 637-639].

Pbnn, 5.126, 10.250, 13.262, Z = 8, R = 0.034. BF_4^- tetrahedra, linked by $LiF_4(H_2O)_2$ octahedra. B-F = 1.386-1.389(2), Li-F/O = 2.025-2.144(4) A.

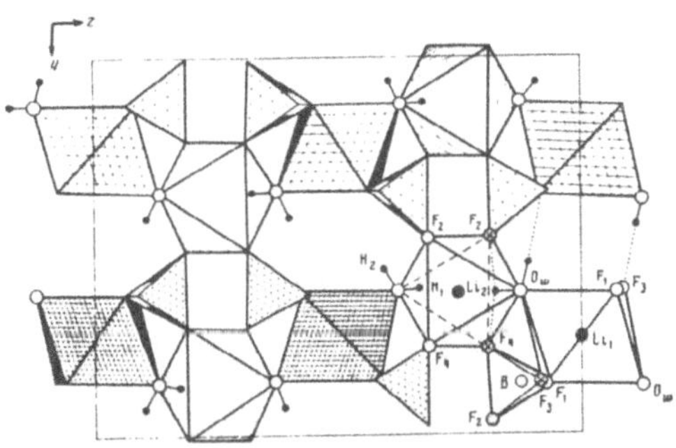

	x	y	z
Li_1	0.5285	0.2500	0,5000
Li_2	0.7500	0.3690	0.2500
B	0,0104	0.1297	0,3744
F_1	0.7954	0.12856	0,43783
F_2	0.0098	0.01672	0.31695
F_3	0.2391	0.13365	0.43050
F_4	−0.0009	0.23893	0.31287
O_w	0.5054	0.12197	0.62420
H_1	0.363	0.125	0.639
H_2	0.521	0.058	0.606

BARIUM HEXAFLUOROSILICATE
$BaSiF_6$

Acta Chem. Scand., A**40**, 631-633.

$R\bar{3}m$, a = 7.198, c = 7.020 A, at 180K, Z = 3, R = 0.013. Ba in 3(a): 0,0,0; Si in 3(b): 0,0,1/2; F in 18(h): 0.2191,0.1095,0.3588. Structure as previously described (**9**, 196), with a distorted CsCl-type packing of Ba^{2+} and octahedral SiF_6^{2-} ions. Si-F = 1.688(2), Ba-F = 2.796, 2.865(2) A.

HYDRAZINIUM(2+) DIFLUORIDE HEXAFLUOROGERMANATE
$(N_2H_6)_2F_2GeF_6$ (I)

HYDRAZINIUM(2+) PENTAFLUOROANTIMONATE(III)
$N_2H_6SbF_5$ (II)

Rev. Chim. Minér., **23**, 668-675.

I, Pccn, 5.408, 13.721, 10.832, Z = 4, R = 0.048. Isostructural with the Ti compound (**46**A, 160), with hydrazinium, fluoride, and hexafluorogermanate ions. II, Pnma, 5.649, 12.880, 7.039, Z = 4, R = 0.062. Hydrazinium and SbF_5^{2-} ions.

AMMONIUM TIN FLUORIDE
$(NH_4)_3Sn_3F_{11}$

Rev. Chim. Minér., **23**, 313-317.

$P2_12_12$ or Pmmn, 6.472, 12.656, 8.028, Z = 2, R = 0.044. NH_4^+, $Sn_2F_5^-$, and octahedral (disordered) SnF_6^{2-} ions.

OXONIUM HEXAFLUOROARSENATE(V)
H_3OAsF_6

Inorg. Chem., **25**, 3095-3097.

$P\bar{1}$, 4.868, 6.471, 7.525, 86.52, 85.23, 81.09, at 208K, Z = 2, R = 0.032. Trigonal pyramidal cations and octahedral anions linked by hydrogen bonds. As-F = 1.70-1.74, O-H...F = 2.62-2.67 A. The room-temperature form is cubic.

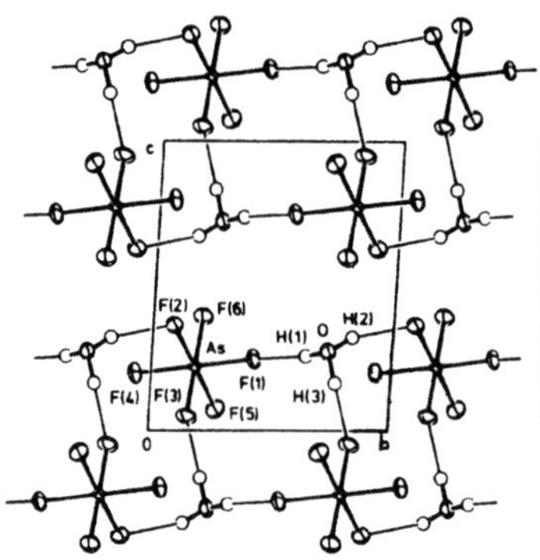

	x	y	z
As	−0.44870	0.23414	0.24376
F(1)	−0.3253	0.4677	0.2628
F(2)	−0.1888	0.1084	0.3753
F(3)	−0.2127	0.1839	0.0592
F(4)	−0.5562	−0.0020	0.2248
F(5)	−0.6943	0.3580	0.1084
F(6)	−0.6687	0.2841	0.4290
O	0.0370	0.7287	0.2712
H(1)	−0.075	0.652	0.261
H(2)	−0.023	0.833	0.324
H(3)	0.112	0.756	0.158

PENTAIODINIUM HEXAFLUOROARSENATE(V)
$I_5^+AsF_6^-$

Inorg. Chem., **25**, 422-426.

C2/c, 10.529, 18.568, 8.320, 128.51, Z = 4, R = 0.066. Planar centrosymmetric cations and octahedral anions, with I...F interactions (2.98-3.27(1) A).

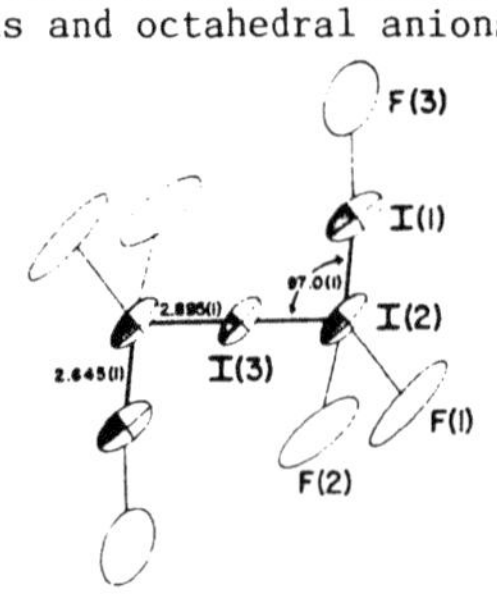

	x	y	z
I(1)	0.73630	0.02873	0.56838
I(2)	0.58140	0.14310	0.55769
I(3)	0.75000	0.25000	0.50000
As(1)	0.00000	0.15594	0.25000
F(1)	0.1943	0.1566	0.3353
F(2)	0.0483	0.2161	0.4317
F(3)	0.0494	0.0883	0.4104

SULPHUR-SELENIUM FLUOROANTIMONATE
$(S_3Se)_2(Sb_4F_{17})(SbF_6)_3$

Inorg. Chem., **25**, 2053-2057.

$B2_1/c$ [no reason given for unconventional space group], 15.267, 13.440, 16.437, β = 91.59°, Z = 4, R = 0.063. Two independent square-planar $(S_3Se)^{2+}$ cations on crystallographic centres of symmetry, with S/Se disorder, with S/Se...F contacts to two independent SbF_6^- anions and a novel $F_5Sb(III)$-F-$Sb(V)F_2$-F-$Sb(V)F_2$-F-$Sb(III)F_5^-$ anion. S/Se-S/Se = 2.109-2.165(3) Å.

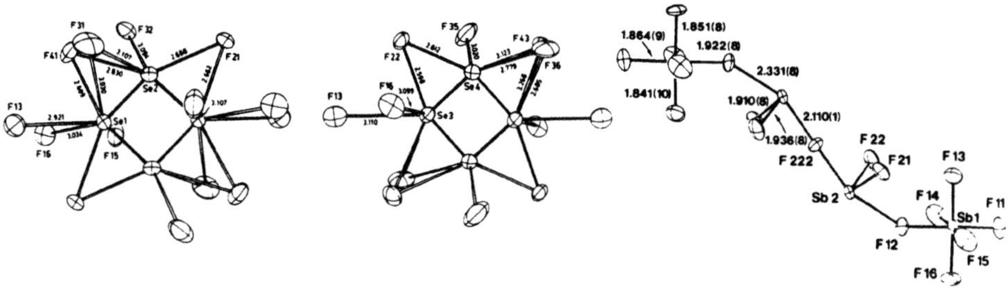

MERCURY HEXAFLUOROANTIMONATE and TANTALATE
$Hg_{2.866}SbF_6$, $Hg_{2.858}TaF_6$

Acta Cryst., B**42**, 209-213.

Monoclinic, pseudo-$I4_1/amd$, a = 7.655, 7.634, c = 12.558, 12.610 A, for Sb, Ta compounds at 173, 150K, respectively, Z = 4. R = 0.057, 0.058 for Sb compound at 293, 173K, R = 0.039 for Ta compound at 150K. The room-temperature structures are as previously described (**40**A, 138; **44**A, 143; **49**A; **51**A, 133). Below 190K both compounds transform to a phase with long-range order of the Hg chains.

AMMONIUM PENTAFLUOROTELLURATE(IV) HYDROCHLORIDE HYDRATE
$NH_4TeF_5.0.25HCl.0.25H_2O$

Mater. Res. Bull., **21**, 1503-1508.

$P\bar{4}2_1c$, 13.741, 6.0199, Z = 8, R = 0.045. Square-pyramidal TeF_5^- (lone-pair completing an octahedron) and NH_4^+ ions, with tunnels containing ordered HCl and H_2O molecules.

BIS(FLUOROSULPHURYL)AMIDOXENON(II) HEXADECAFLUOROTRIANTIMONATE(V)
$XeN(SO_2F)_2^+Sb_3F_{16}^-$

Inorg. Chem., **25**, 563-571.

$P2_1/c$, 10.320, 10.642, 18.167, 108.94, Z = 4, R = 0.061. Discrete fluorine-bridged cation-anion pairs. Xe-N = 2.02(1), Xe...F = 2.46(1) A, N-Xe-F = 178°. See also **49**A.

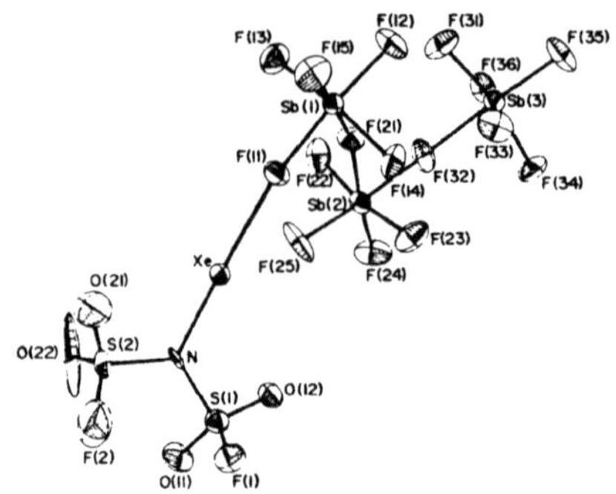

SODIUM HEXAFLUOROTITANATE(IV)
SODIUM HEXAFLUOROSILICATE
Na_2TiF_6, Na_2SiF_6

Z. Kristallogr., **175**, 269-276.

Na_2TiF_6, C2 [compare **19**, 325], 15.50, 8.93, 5.08, $\beta = 90^o$, Z = 6, R = 0.036 (for twinned crystal, which gives pseudo-P321 symmetry).

Na_2SiF_6, data of **29**, 264 reinterpreted in terms of a twinned model in C2 (rather than P321), R = 0.029.

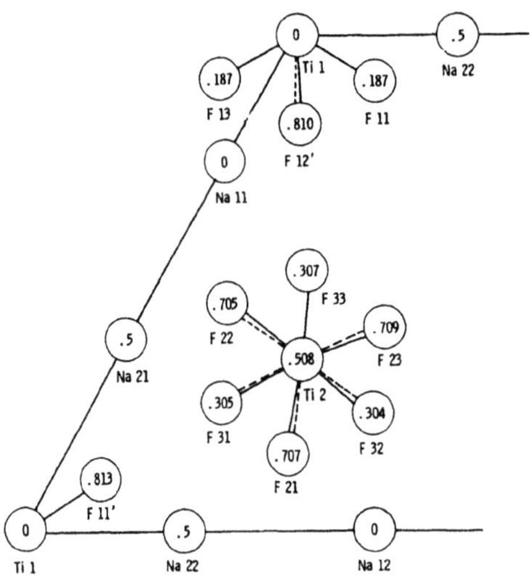

Na_2TiF_6, deviations from trigonal symmetry (dotted lines)

Na_2TiF_6 (hexagonal system)

	x	y	z
Ti1	0	0	0
Ti2	0.3337	0.6670	0.4920
Na11	0.6254	0.0031	−0.0005
Na12	0	0.6261	0
Na21	0.2841	0.0017	0.5009
Na22	0	0.2851	0.5000
F11	0.0961	0.1830	0.1867
F12	−0.1808	−0.0824	0.1898
F13	0.0842	−0.0958	0.1870
F21	0.1534	0.5535	0.2927
F22	0.4483	0.6026	0.2951
F23	0.4004	0.8492	0.2911
F31	0.2567	0.4850	0.6947
F32	0.2268	0.7429	0.6963
F33	0.5176	0.7773	0.6932

Na_2SiF_6 (hexagonal system)

	x	y	z
Si1	0	0	0
Si2	0.3341	0.6668	0.4944
Na11	0.6190	−0.0030	0.0015
Na12	0	0.6191	0
Na21	0.2826	−0.0008	0.4972
Na22	0	0.2865	0.5000
F11	0.0977	0.1794	0.1884
F12	−0.1817	−0.0870	0.1928
F13	0.0868	−0.0923	0.1902
F21	0.1540	0.5541	0.3013
F22	0.4428	0.5957	0.3057
F23	0.4037	0.8438	0.2956
F31	0.2627	0.4896	0.6889
F32	0.2266	0.7368	0.6900
F33	0.5130	0.7714	0.6868

POTASSIUM PENTAFLUOROZIRCONATE(IV) MONOHYDRATE
$KZrF_5.H_2O$

Z. Kristallogr., **175**, 177-185.

$P2_1/c$, 7.628, 12.081, 11.100, 99.62, Z = 8, R = 0.024. $(H_2O)F_4Zr-F_3-ZrF_4(H_2O)$ units connected into chains along **a** by single F bridges, with hydrogen bonding within the chains, and chains linked into (010) layers by further hydrogen bonds; layers are connected via K ions (irregular coordinations). Zr-F = 2.00-2.33, Zr-O = 2.30, K-F = 2.61-3.43, K-O = 2.90, O-H...F = 2.62-3.09 A.

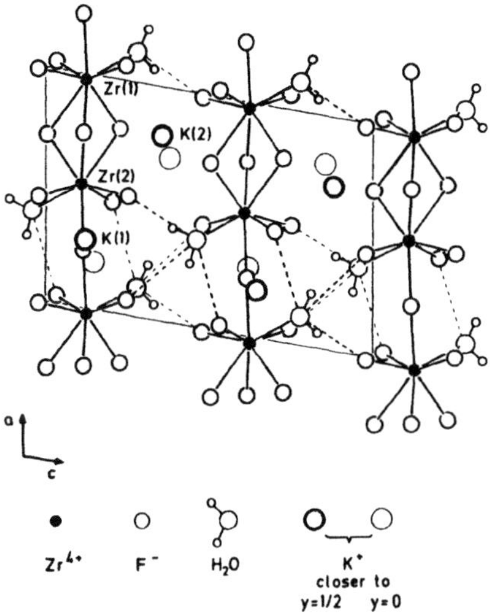

	x	y	z
Zr(1)	0.97715	0.26426	0.12399
Zr(2)	0.52846	0.25585	0.10764
K(1)	0.2946	0.5258	0.1203
K(2)	0.7959	0.4606	0.3542
F(1)	0.4744	0.1213	0.1996
F(2)	0.2522	0.2750	0.1142
F(3)	0.2081	0.7024	0.2736
F(4)	0.5118	0.3550	0.2504
F(5)	0.7153	0.2022	0.0039
F(6)	0.0122	0.6393	0.0223
F(7)	0.0643	0.3820	0.2457
F(8)	0.0297	0.1274	0.0317
F(9)	0.2494	0.8746	0.3841
F(10)	0.4438	0.3685	−0.0215
O(1)	0.1279	0.1566	0.2786
O(2)	0.3789	0.1432	−0.0420
H(1)	0.217	0.157	0.307
H(2)	0.071	0.140	0.325
H(3)	0.282	0.146	−0.055
H(4)	0.407	0.136	−0.107

POTASSIUM HEXAFLUOROHAFNATE(IV)
K_2HfF_6

Z. Kristallogr., **175**, 159-164.

C2/c, 6.592, 11.485, 6.910, 90.56, Z = 4, R = 0.021. Chains of edge-sharing HfF_8 polyhedra, linked by 8-coordinate K. Hf-F = 2.04-2.21, K-F = 2.69-2.83 A.

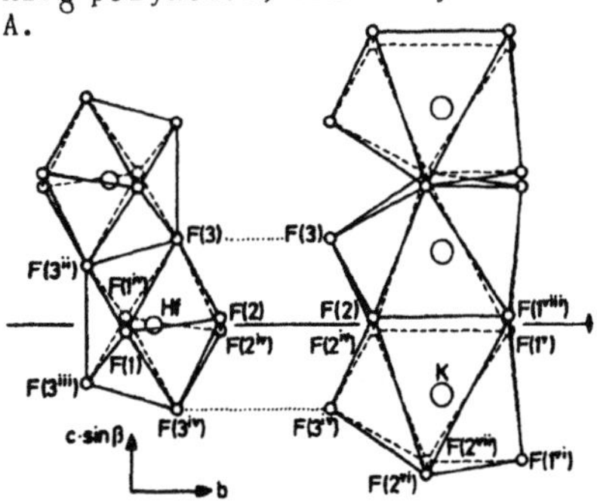

	x	y	z
K	0.0169	0.3431	0.0018
Hf	0.0	0.04831	0.25
F(1)	0.2887	−0.0133	0.2234
F(2)	0.1898	0.1924	0.2728
F(3)	−0.0297	0.0983	0.5455

LITHIUM and SODIUM HEXAFLUORONIOBATE(IV)
Li_2NbF_6, Na_2NbF_6

J. Solid State Chem., **62**, 212-219.

Li, P$\bar{3}$1m, 4.9634, 4.6172, Z = 1, powder data. Li in 2(d): 1/3,2/3,1/2; Nb in 1(a): 0,0,0; F in 6(k): 0.329,0,0.259. Li_2ZrF_6-type (**24**, 277; **39**A, 154). Nb-6F = 2.02, Li-6F = 2.00(1) A.

Na, $P4_2/mnm$, 5.042, 10.2767, Z = 2, powder data. Na in 4(e): 0,0,0.336; Nb in 2(a): 0,0,0: F(1) in 4(f): 0.285,0.285,0; F(2) in 8(j): 0.309,0.309,0.359. Trirutile structure. Na_2SnF_6, previously described as monoclinic (**31**A, 260), also has this structure.

POTASSIUM TETRAFLUOROCHROMATE(III)
$KCrF_4$

POTASSIUM DICHLOROHEXAFLUOROTRICHROMATE(II)
$K_2Cr_3Cl_2F_6$

J. Chem. Soc., Dalton, 2623-2627 (1986).

$KCrF_4$, Pnma, 15.76, 7.43, 18.38, Z = 24, R = 0.041. Approximately planar Cr_3F_9 units joined by fluorine bridges into columns, which are linked by K^+ ions. Cr atoms have octahedral coordinations, Cr-F = 1.84-1.96 A.

$K_2Cr_3Cl_2F_6$, $P2_1/c$, 6.726, 11.157, 7.391, 125.31, Z = 2, R = 0.041. Two independent distorted trans-octahedral $Cr(II)F_4Cl_2$ groups sharing F corners and F...Cl edges to form (100) sheets, which are linked by Cl bridges, with K ions in holes. Cr-F = 1.96-1.98, Cr-Cl = 2.83-2.93 A.

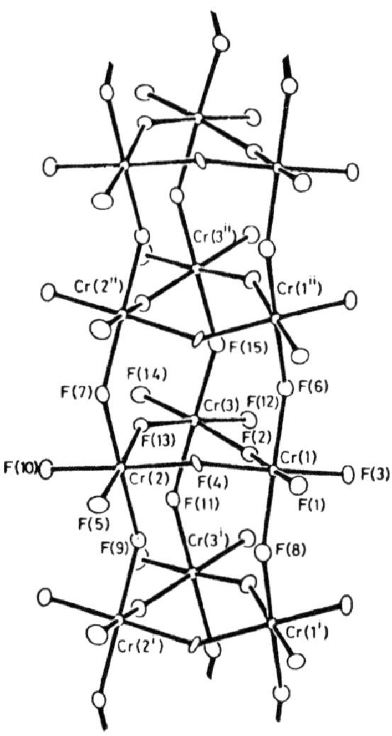

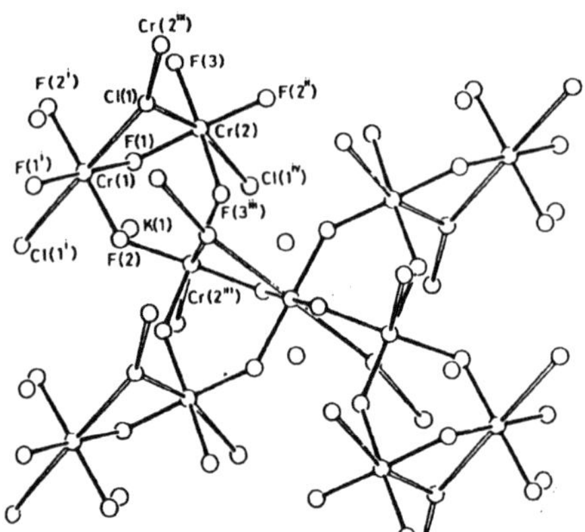

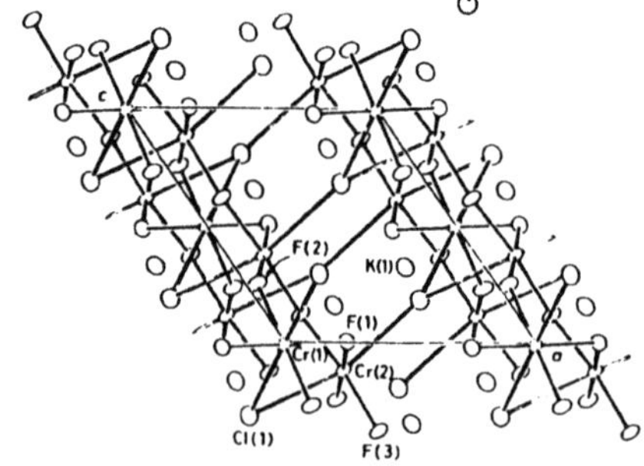

ELPASOLITE STRUCTURES
K_2NaCrF_6, Rb_2NaFeF_6, Rb_2KFeF_6

Rev. Chim. Minér., **23**, 508-519.

Fm3m, a = 8.275, 8.462, 8.869 A, Z = 4, R = 0.021, 0.013, 0.019. Elpasolite structures (**2**, 498; **40**A, 137). F was refined in 24(e) site, but may be disordered in a 96(j) site. M-F = 1.908, 1.939, 1.941 A.

SODIUM PENTAFLUOROMANGANATE(III)
Na_2MnF_5

Acta Cryst., C**42**, 644-647.

$P2_1/c$, 7.719, 5.236, 10.862, 108.99, Z = 4, R = 0.027. Chains of trans-corner-sharing MnF_6 octahedra, linked by 7-coordinate Na ions.

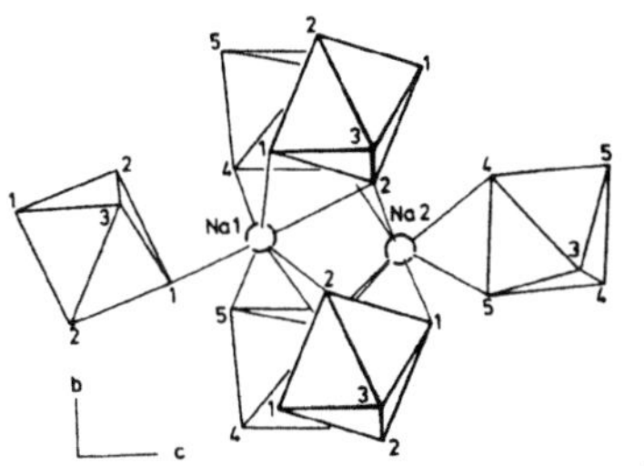

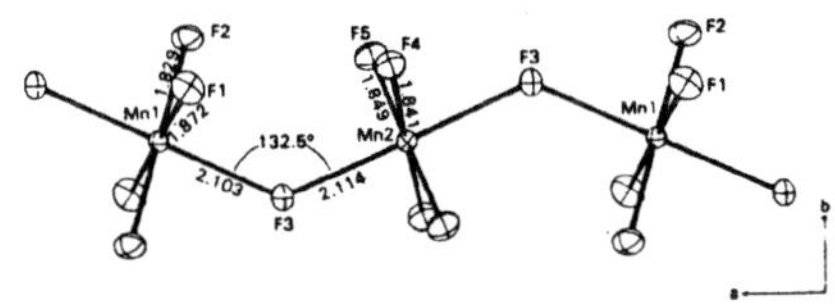

	x	y	z
Mn(1)	0	0	0
Mn(2)	½	0	0
Na(1)	0,1523	−0,0075	0,3656
Na(2)	0,3441	0,4441	0,1799
F(1)	0,0459	0,1523	0,1634
F(2)	−0,1082	0,2971	−0,0757
F(3)	0,2507	0,1621	0,0032
F(4)	0,6182	0,2179	0,1341
F(5)	0,4917	0,2451	−0,1243

Mn(1)–F(1)	2×	1,872 (1)	Na(1)–F(1)	2,242 (1)
Mn(1)–F(2)	2×	1,829 (1)	Na(1)–F(1′)	2,302 (1)
Mn(1)–F(3)	2×	2,103 (1)	Na(1)–F(2′)	2,549 (1)
			Na(1)–F(2″)	2,622 (1)
Mn(2)–F(3)	2×	2,114 (1)	Na(1)–F(3′)	2,311 (1)
Mn(2)–F(4)	2×	1,841 (1)	Na(1)–F(4′)	2,281 (1)
Mn(2)–F(5)	2×	1,849 (1)	Na(1)–F(5′)	2,929 (1)
Mittelwerte:			Na(2)–F(1)	2,719 (1)
$Mn–F_{axial}$		2,109	Na(2)–F(2‴)	2,256 (1)
$Mn–F_{äquatorial}$		1,848	Na(2)–F(3)	2,342 (1)
$Mn–F_{gesamt}$		1,935	Na(2)–F(4)	2,610 (1)
			Na(2)–F(4″)	2,416 (1)
Na–F		2,438	Na(2)–F(5′)	2,286 (1)
			Na(2)–F(5″)	2,262 (1)
F(1)–Mn(1)–F(2)		89,0 (<1)	F(3)–Mn(2)–F(4)	87,6 (<1)
F(1)–Mn(1)–F(3)		85,6 (<1)	F(3)–Mn(2)–F(5)	85,3 (<1)
F(2)–Mn(1)–F(3)		87,0 (<1)	F(4)–Mn(2)–F(5)	87,8 (<1)

BARIUM TETRAFLUOROMANGANATE(II)
$BaMnF_4$

Mater. Res. Bull., **21**, 843-851.

Incommensurate modulated structure at 100K, with displacement of atoms from room-temperature positions (**35**A, 164), involving rotation of the MnF_6 octahedra about **a**, and a correlated translation of Ba ions in the (001) plane.

POTASSIUM TETRAFLUOROFERRATE(III)
$KFeF_4$

Acta Cryst., B**42**, 258-262 (1986) and B**43**, 111 (1987).

Phase II, 415K, Amma, 7.68, 3.92, 12.39, Z = 4, R = 0.032. Phase III, 295K, Pmcn, 7.64, 7.81, 12.33, Z = 8, R = 0.044. Both phases contain sheets of corner-sharing FeF_6 octahedra, linked by K ions. In phase II, the octahedra are tilted by 12.4° along **b** (relative to phase I, Ammm, a' = a/2, above 563K), and phase III exhibits an additional 6.8° tilt along **c**.

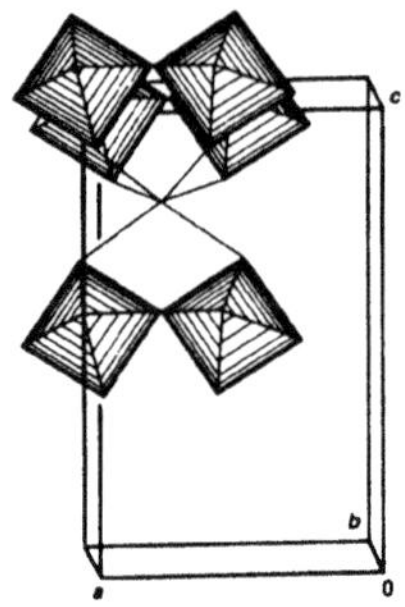

Phase II

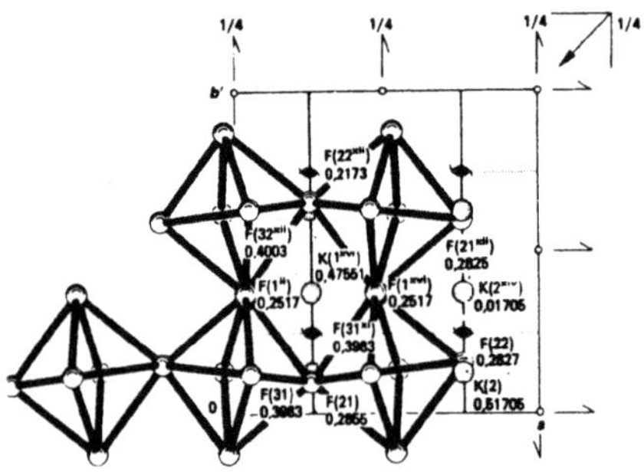

Phase III

	x	y	z
Phase II			
K	¼	0	0,72852
Fe	0	0	0
F(1)	0	0	½
F(2)	¼	0	0,9659
F(3)	0,0463	0	0,1488
Phase III			
K(1)	¼	0,1210	0,97551
K(2)	¾	0,1201	0,51705
Fe	0,00035	0,1255	0,24894
F(1)	0,0313	0,1227	0,7517
F(21)	¼	0,0959	0,2145
F(22)	¾	0,1540	0,2827
F(31)	0,0476	0,1202	0,3983
F(32)	0,9543	0,1308	0,0997

RUBIDIUM TETRAFLUOROFERRATE(III)
$RbFeF_4$

Phys. Status Solidi, A, **97**, 387-395.

Phase II, above 416K, P4/mbm, 5.383, 6.410, at 440K, Z = 2, R = 0.086. Rb in 2(c): 0,1/2,1/2; Fe in 2(a): 0,0,0; F(1) in 4(g): 0.300,0.800,0; F(2) in 4(e): 0,0,0.279. Phase IV, below 380K, $P2_12_12$, 7.620, 7.647, 6.273, at 300K, Z = 4, R = 0.126. Rb(1) in 2(b): 0,1/2,0.549; Rb(2) in 2(a): 0,0,0.512; Fe in 4(c): 0.282,1/4,0; F(1) in 4(c): 0.262,0,0.064; F(2) in 4(c): 0,0.251,-0.083; F(3) in 4(c): 0.234,0.304,0.293; F(4) in 4(c): 0.309,0.224,0.773. Pseudo-$TlAlF_4$ type structures, with tilted FeF_6 octahedra.

RUBIDIUM LITHIUM HEXAFLUOROFERRATE(III)
CAESIUM SODIUM HEXAFLUOROFERRATE(III)
Rb_2LiFeF_6, Cs_2NaFeF_6

Z. anorg. Chem., **539**, 87-94.

Fm3m, 8.244, 8.739, Z = 4, R = 0.027, 0.022. Metastable high pressure phases. Rb or Cs in 8(c): 1/4,1/4,1/4; 1/6 Li in 24(e): 0,464,0,0, or Na in 4(b): 1/2,0,0; Fe in 4(a): 0,0,0; F in 24(e): 0.2360,0,0 or 0.2231,0,0. Elpasolite structures (**2**, 498; **40**A, 137). In comparison with the normal pressure rhombohedral phases (**41**A, 163; **42**A, 179; **46**A, 163), the Rb-F distances (12-coordinate) are decreased by 2-3% and the other M-F distances (octahedral) are increased by 1-4%.

CAESIUM TETRAFLUOROFERRATE(III) (PHASE II)
$CsFeF_4$

Phys. Status Solidi, A, **95**, 413-421.

P4/nmm, 7.7874, 6.5402, Z = 4, R = 0.081. Cs(1) in 2(c): 1/4,1/4,0.5416; Cs(2) in 2(b): 3/4,1/4,1/2; Fe in 4(d): 0,0,0; F(1) in 8(i): 1/4,0,-0.0329; F(2) in 8(j): 0.0202,0.0202,0.2860. Pseudo-$TlAlF_4$ type structure, with tilt of FeF_6 octahedra and displacement of Cs, relative to phase I (P4/mmm, above 475K) and phase III ($P2_12_12$, below 250K).

CAESIUM ENNEAFLUORODIFERRATE(III)
$Cs_3Fe_2F_9$

J. Solid State Chem., **63**, 446-451.

$P6_3$/mmc, 6.347, 14.805, Z = 2, R = 0.026. The structure is quite similar to that previously described (**37**A, 187), but in a higher symmetry space group, Fe...Fe = 2.916 A.

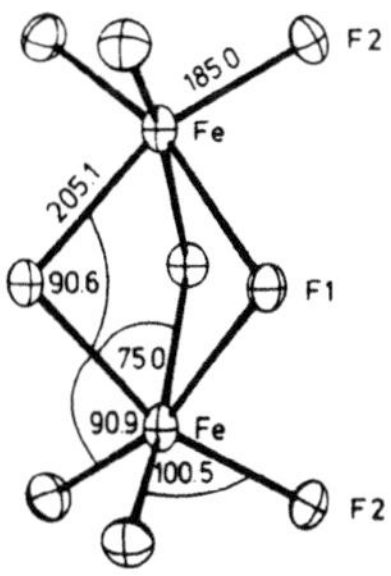

		x	y	z
Cs1	(2d)	1/3	2/3	3/4
Cs2	(4f)	1/3	2/3	0.43271
Fe	(4e)	0	0	0.15153
F1	(6h)	0.1312	2x	1/4
F2	(12k)	0.1494	2x	0.5940

CAESIUM TETRAFLUOROCUPRATE(II)
Cs_2CuF_4

Z. anorg. Chem., **540/541**, 135-141.

I4/mmm, 4.4026, 14.032, Z = 2, R = 0.051. Cs in 4(e): 0,0,0.3544; Cu in 2(a): 0,0,0; F(1) in 4(e): 0,0,0.1399; F(2) in 4(c): 1/2,0,0, or disordered in 8(i): 0.4388,0,0, with half-occupancy. K_2NiF_4-type (**17**, 332; **19**, 323), but probably with split F(2) site. Cu-6 F = 1.93, 1.97, 2.47, Cs-9 F = 3.01-3.12 A.

CAESIUM POTASSIUM HEXAFLUOROCUPRATE(III)
Cs_2KCuF_6

Z. anorg. Chem., **532**, 17-22.

Fm3m, 8.894, Z = 4, R = 0.067. Cs in 8(c): 1/4,1/4,1/4; K in 4(b): 1/2,1/2,1/2; Cu in 4(a): 0,0,0; F in 24(e): x,0,0, x = 0.2060. Cu-F = 1.83 A (very short).

BARIUM COPPER(II) FLUORIDES

α-$Ba_2Cu_5F_{14}$

Rev. Chim. Minér., **23**, 352-361.

C2/c, 18.170, 6.652, 10.328, 117.10, Z = 4, R = 0.058. (100) layers built from pairs of edge-sharing Jahn-Teller distorted CuO_6 octahedra which share corners with neighbouring octahedra, and corner-share along **a**. Ba ions have (8+6)-coordination.

$Ba_6Cu_{11}F_{34}$

J. Solid State Chem., **62**, 164-171.

$P\bar{1}$, 7.490, 10.031, 10.271, 82.98, 73.88, 70.42, Z = 1, R = 0.054. Three-dimensional framework of units of three edge-sharing CuF_6 distorted octahedra connected by complex groups of octahedra; Ba ions have 11- and 12-coordinations. Cu-F = 1.88-2.03, 2.16-2.56(1) A.

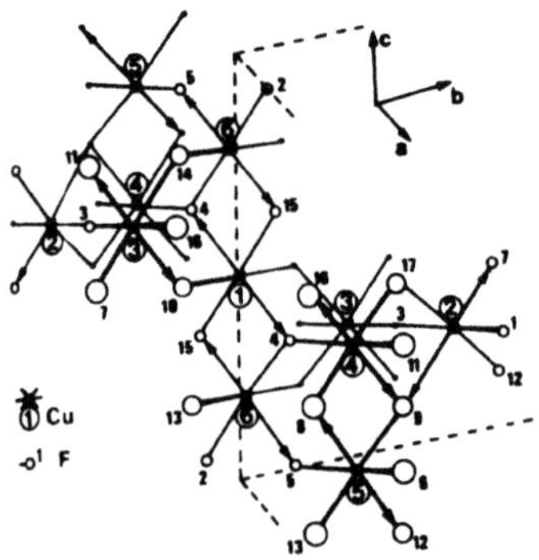

BARIUM COPPER(II) VANADIUM(III) FLUORIDE

$Ba_2CuV_2F_{12}$

J. Solid State Chem., **62**, 158-163.

$P\bar{1}$, 5.365, 6.950, 7.433, 65.05, 70.26, 73.19, Z = 1, R = 0.033. Layers of edge-sharing pairs of MF_6 octahedra (M = Cu+V) linked by single corner-sharing with single VF_6 octahedra; layers are linked by 12-coordinate Ba ions.

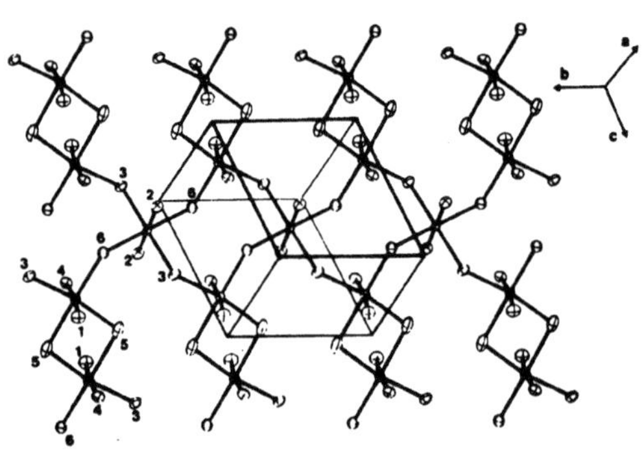

	x	y	z
Ba	0.7075	0.7968	0.7641
V	$\frac{1}{2}$	$\frac{1}{2}$	$\frac{1}{2}$
Cu/V	0.0522	0.7753	0.1679
F 1	0.1730	0.6542	0.9626
F 2	0.4649	0.7813	0.4946
F 3	0.1236	0.4870	0.6513
F 4	0.0224	0.0950	0.6419
F 5	0.2517	0.0363	0.9542
F 6	0.6039	0.3894	0.7616

INTERATOMIC DISTANCES (Å) AND BOND ANGLES (°) IN $Ba_2CuV_2F_{12}$

Barium polyhedron		
Ba–F 4: 2.716(3)	Ba–F 4: 2.745(3)	
Ba–F 2: 2.783(3)	Ba–F 3: 2.793(3)	
Ba–F 2: 2.833(3)	Ba–F 5: 2.836(4)	
Ba–F 5: 2.878(4)	Ba–F 1: 2.951(3)	
Ba–F 1: 2.989(3)	Ba–F 1: 3.037(3)	
Ba–F 6: 3.048(3)	Ba–F 6: 3.212(3)	
⟨Ba–F⟩ 2.90(14)		
Single V octahedron		Angle at V
V–$F2_t$: 2 × 1.894(2)	F2–F6: 2 × 2.679(3)	87.6(1)
V–$F3_b$: 2 × 1.959(3)	F2–F3: 2 × 2.690(3)	88.5(1)
V–$F6_b$: 2 × 1.976(3)	F2–F3: 2 × 2.759(4)	91.4(1)
	F3–F6: 2 × 2.779(4)	89.8(1)
⟨V – F⟩ 1.943(35)	F3–F6: 2 × 2.787(4)	90.1(1)
	F2–F6: 2 × 2.794(3)	92.4(1)
Bioctahedral CuV unit		Angle at *M*
M–$F4_t$: 1.859(3)	F1–F6: 2.579(4)	83.7(1)
M–$F1_t$: 1.886(3)	F1–F3: 2.678(4)	86.4(1)
M–$F6_b$: 1.978(3)	F5–F5: 2.696(6)	81.9(2)
M–$F5_b$: 1.990(4)	F4–F5: 2.737(7)	93.8(3)
M–$F3_b$: 2.022(3)	F1–F5: 2.745(4)	87.6(2)
M–$F5_b$: 2.122(4)	F4–F6: 2.747(4)	91.4(1)
	F1–F5: 2.779(5)	90.1(2)
⟨*M*–F⟩ 1.976(87)	F4–F5: 2.812(6)	86.6(2)
	F3–F6: 2.826(5)	89.9(1)
F1–*M*–F4: 172.4(3)°	F5–F6: 2.885(9)	89.3(2)
F3–*M*–F5: 174.0(5)°	F3–F4: 2.960(8)	99.3(2)
F5–*M*–F6: 169.5(5)°	F3–F5: 3.034(9)	98.3(3)
M–F5–*M*: 98.1(3)°		
Intermetallic distances		
V–*M*: 3.590(1)		
M–*M*: 3.107(1)		

BARIUM ZINC FLUORIDE
$Ba_2Zn_7F_{18}$

Rev. Chim. Minér., **23**, 497-507.

$P\bar{1}$, 7.032, 7.292, 7.505, 94.24, 98.82, 116.39, Z = 1, R = 0.066. Three-dimensional framework with groups of 3 and 4 edge-sharing ZnF_6 octahedra linked by corner-sharing. Ba ions have 12-coordination.

STRONTIUM LANTHANUM FLUORIDE
$Sr_{0.69}La_{0.31}F_{2.31}$

Kristallografija, **31**, 248-251 [Soviet Physics - Crystallography, **31**, 145-147].

Fm3m, 5.849, Z = 4, neutron radiation, R = 0.013. Sr,La in 4(a): 0,0,0; 7.5 F(1) in 8(c): 1/4,1/4,1/4; 1.7 F(2) in 32(f): x,x,x, x = 0.419. Fluorite-type structure, with additional interstitial F.

BARIUM PRASEODYMIUM FLUORIDE
$Ba_{0.73}Pr_{0.27}F_{2.27}$

Kristallografija, **31**, 661-665 [Soviet Physics - Crystallography, **31**, 390-392].

Fm3m, a = 6.076, 6.118 A, at 293, 573, Z = 4, neutron radiation, R = 0.020. 4 Ba/Pr in 4(a): 0,0,0; 6.18 F(1) in 8(c): 1/4,1/4,1/4; 2.30 F(2) in 48(i): 1/2,x,x, x = 0.362; 0.60 F(3) in 32(f): x,x,x, x = 0.410. Fluorite structure, with additional F sites, as previously described (**51**A, 138).

SODIUM LITHIUM YTTERBIUM FLUORIDE
$NaLiYb_2F_8$

J. Solid State Chem., **65**, 205-214.

C2/c, 10.3516, 8.2069, 6.9674, 90, Z = 4, R = 0.021. Yb_2F_{14} groups of two edge-sharing YbF_8 polyhedra, with further edge sharing to form a three-dimensional framework, with cavities containing 7-coordinate Na and 6-coordinate Li. Yb-F = 2.18-2.40, Na-F = 2.17-2.50, Li-F = 1.88-2.23 A.

		x	*y*	*z*
Yb	**8(*f*)**	**0,68682**	**0,23711**	**0,25000**
Na	**4(*e*)**	**0**	**0,6547**	**¾**
F(1)	**4(*e*)**	**½**	**0,4115**	**¼**
F(2)	**8(*f*)**	**0,7188**	**0,0261**	**0,7488**
F(3)	**8(*f*)**	**0,8616**	**0,2034**	**0,4487**
F(4)	**4(*e*)**	**½**	**0,1101**	**¼**
F(5)	**8(*f*)**	**0,3621**	**0,7035**	**0,0516**
Li	**4(*a*)**	**0**	**0**	**0**

CALCIUM THORIUM FLUORIDE
$Ca_{1-x}Th_xF_{2+2x}$

J. Solid State Chem., **61**, 359-368.

Fm3m, 5.521-5.594, for x = 0.075-0.18, Z = 4, neutron powder data. Fluorite-related structure, with excess anions in x,x,x, x = 0.41 and 0.3.

CALCIUM CHLORIDE HEXAHYDRATE
STRONTIUM CHLORIDE HEXAHYDRATE
$CaCl_2.6H_2O$, $SrCl_2.6H_2O$

Acta Cryst., C**42**, 141-143.

P321, a = 7.8759, 7.9596, c = 3.9545, 4.1243 A, Z = 1, neutron data, R = 0.029, 0.036. Structures as previously described (**2**, 499; **8**, 133; **43**A, 131; **51**A, 140).

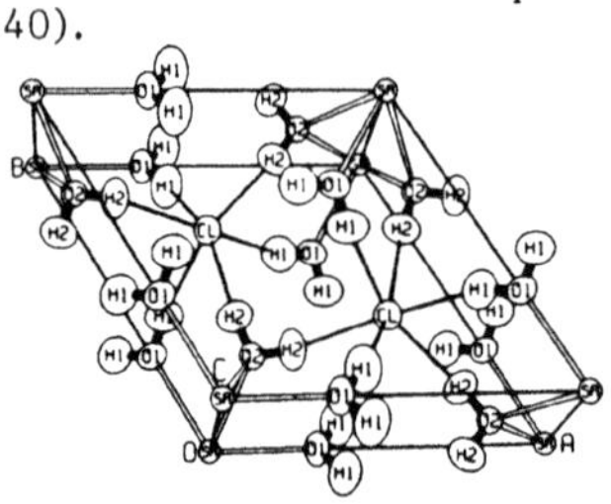

	CaCl₂.6H₂O	SrCl₂.6H₂O
(*a*) Distances (Å)		
M–O(1)	2·453 (1)	2·570 (2)
M–O(2)	2·594 (1)	2·715 (1)
O(1)–H(1)	0·953 (2)	0·947 (3)
O(1)–H(1) corr.	0·986 (2)	0·981 (3)
O(2)–H(2)	0·966 (2)	0·966 (2)
O(2)–H(2) corr.	0·995 (2)	0·995 (2)
H(1)···Cl	2·266 (2)	2·286 (3)
H(2)···Cl	2·279 (2)	2·265 (2)
O(1)···Cl	3·194 (1)	3·202 (1)
O(2)···Cl	3·218 (1)	3·208 (1)
(*b*) Angles (°)		
H(1)–O(1)–H(1)	107·2 (3)	107·0 (3)
H(2)–O(2)–H(2)	104·1 (3)	104·0 (3)
O(1)–H(1)···Cl	164·4 (2)	162·6 (2)
O(2)–H(2)···Cl	163·8 (2)	165·0 (2)
H(1)···Cl···H(1)	89·7 (1)	88·4 (1)
H(2)···Cl···H(2)	117·53 (3)	117·23 (4)

	CaCl₂.6H₂O			SrCl₂.6H₂O		
	x	*y*	*z*	*x*	*y*	*z*
Cation	0	0	0	0	0	0
Cl	3333	6667	4249	3333	6667	4190
O(1)	3114	0	0	3229	0	0
O(2)	−2132	0	5000	−2219	0	5000
H(1)	4326	988	−926	4419	964	−899
H(2)	−2330	1113	4835	−2415	1103	4887

SILICON TETRACHLORIDE (PHASE II)
$SiCl_4$

Kristallografija, **31**, 171-172 [Soviet Physics - Crystallography, **31**, 99-100].

$P2_1/c$, 9.608, 6.356, 9.672, 102.91, at 163K, Z = 4, R = 0.023. Tetrahedral molecules.

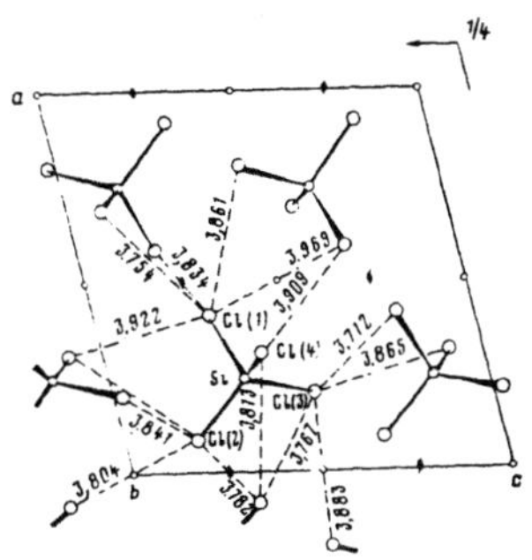

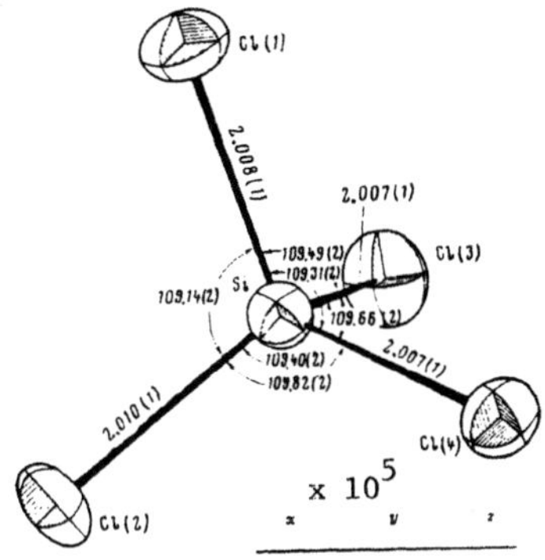

x 10^5

	x	y	z
Si	24878	7217	35721
Cl(1)	41170	−7751	30100
Cl(2)	7822	6825	19322
Cl(3)	19991	−7464	52382
Cl(4)	30579	37055	40956

PENTAAMMINECHLOROMETAL(III) CHLORIDES
$[M(NH_3)_5Cl]Cl_2$ (M = Cr, Ru, Os, Ir)

Inorg. Chem., **25**, 4553-4558.

Pnma, a = 13.441, 13.418, 13.393, 13.393, b = 10.504, 10.501, 10.520, 10.483, c = 6.755, 6.777, 6.774, 6.768 A, Z = 4, R = 0.023, 0.020, 0.021, 0.027. Isostructural with the Co and Rh compounds (**3**, 125, 492; **28**, 90; **33**A, 208; **39**A, 165; **44**A, 157), with octahedral cations and Cl^- anions linked by N-H...Cl hydrogen bonds. M-Cl = 2.327-2.371, M-N = 2.051-2.115 A.

Positional Parameters (×10⁴) for $[Cr(NH_3)_5Cl]Cl_2$

	x	y	z
Cr(1)	1026	2500	1787
Cl(1)	−279	2500	−466
Cl(2)	3525	10	1632
N(1)	2199	2500	3768
N(2)	2058	2500	−501
N(3)	1027	527	1781
N(4)	41	2500	4141

Positional Parameters (×10⁴) for $[Os(NH_3)_5Cl]Cl_2$

	x	y	z
Os(1)	1027	2500	1794
Cl(1)	−305	2500	−502
Cl(2)	3515	13	1618
N(1)	2218	2500	3817
N(2)	2061	2500	−581
N(3)	1014	491	1790
N(4)	2	2500	4157

Positional Parameters (×10⁴) for $[Ru(NH_3)_5Cl]Cl_2$

	x	y	z
Ru(1)	1029	2500	1802
Cl(1)	−288	2500	−470
Cl(2)	3518	12	1633
N(1)	2210	2500	3824
N(2)	2063	2500	−522
N(3)	1027	495	1798
N(4)	19	2500	4185

Positional Parameters (×10⁴) for $[Ir(NH_3)_5Cl]Cl_2$

	x	y	z
Ir(1)	1036	2500	1804
Cl(1)	−302	2500	−485
Cl(2)	3514	18	1602
N(1)	2207	2500	3796
N(2)	2044	2500	532
N(3)	1020	498	1779
N(4)	14	2500	4151

SAMARIUM(II) CHLORIDE
$SmCl_2$

J. Less-Common Metals, **116**, 187-197.

Pbnm, 9.011, 7.563, 4.508, Z = 4, R = 0.056. Sm, Cl(1), Cl(2) in 4(c): x,y,1/4, x = 0.11607, 0.4288,-0.1680, y = 0.24978, 0.1439, 0.0245. $PbCl_2$-type structure (**2**, 16). Other $LnCl_2$ compounds have $PbCl_2$, $SrBr_2$, and SrI_2 type structures.

BISMUTH TETRACHLOROALUMINATE
$Bi_5(AlCl_4)_3$

J. Less-Common Metals, **116**, 159-168.

$R\bar{3}c$, 11.860, 30.100 A, Z = 6, R = 0.052. Bi_5^{3+} cations and $AlCl_4^-$ anions. Bi-Bi = 3.01, 3.32, Al-Cl = 2.13, 2.14 A.

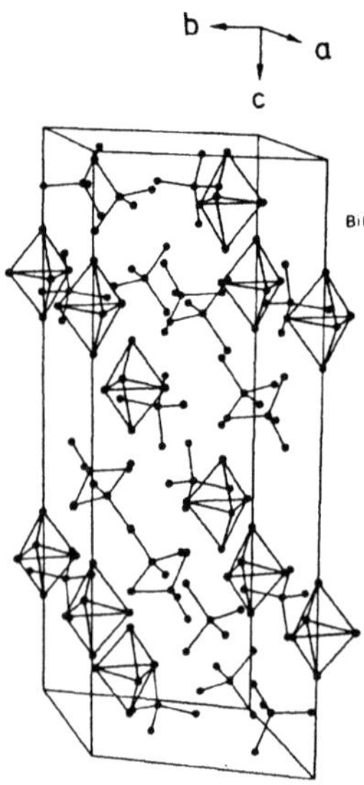

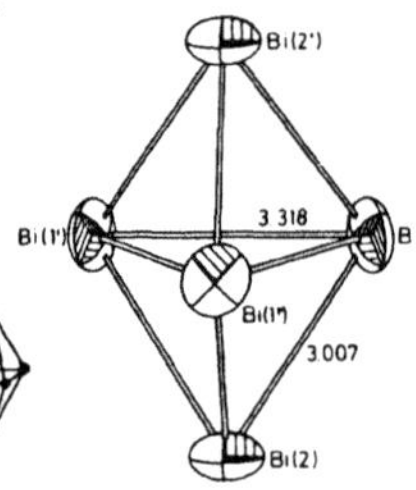

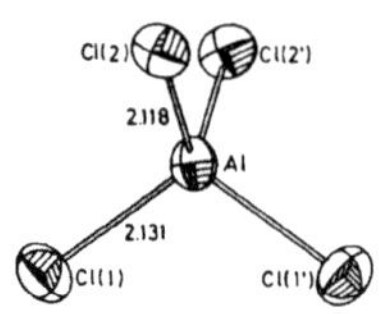

		Point symmetry	x	y	z
Bi(1)	18e	2	0.16155	0.0	0.25
Bi(2)	12c	3	0.0	0.0	0.32700
Al	18e	2	0.5231	0.0	0.25
Cl(1)	36f	1	0.4212	0.0067	0.3074
Cl(2)	36f	1	0.7105	0.1659	0.2426

TITANIUM(II) TETRABROMOALUMINATE
$Ti(AlBr_4)_2$

Z. anorg. Chem., **543**, 63-65.

I2/c, 13.769, 8.066, 12.441, 92.00, Z = 4, R = 0.079, positional parameters not given. Isostructural with $Co(AlCl_4)_2$ (**27**, 459). Al-4Br = 2.19-2.34(5), Ti-6Br = 2.69-2.72(3) A.

cis-TETRACHLOROBIS(CYCLOOCTASULPHURMONOXIDE)TIN(IV)
$SnCl_4(S_8O)_2$

Z. Naturforsch., **41**B, 951-957.

$P2_1/n$, 8.212, 13.319, 20.717, 91.82, Z = 4, R = 0.038. The molecule has approximately C_2 symmetry, with cis-octahedral coordination at Sn. Sn-Cl = 2.33-2.39, Sn-O = 2.19, 2.24, S-O = 1.52, S-S = 2.18 (S bonded to O), 2.00-2.07 A.

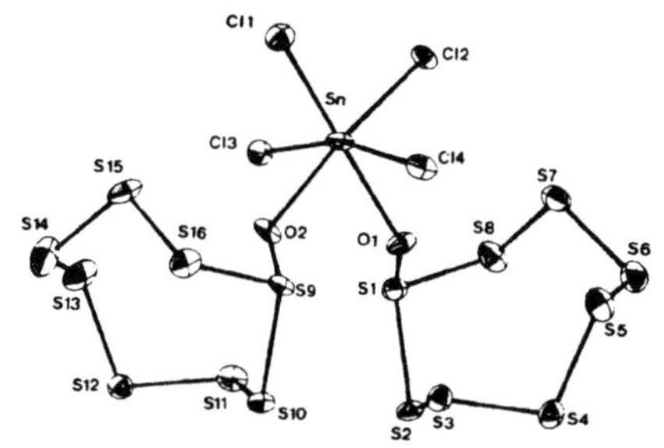

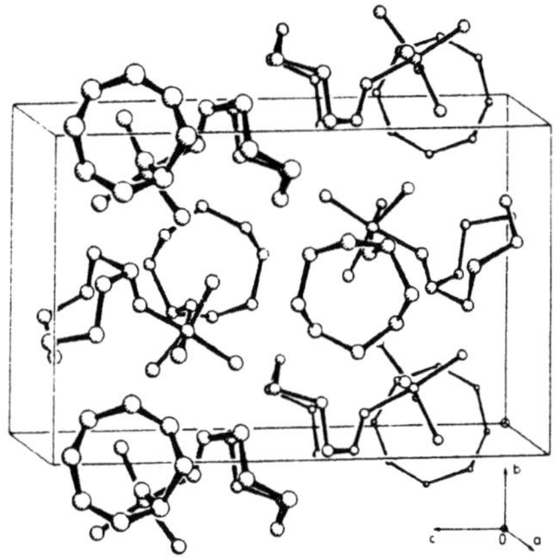

CAESIUM HEXACHLOROBISMUTHATE(III) (MONOCLINIC)
Cs_3BiCl_6

CAESIUM POTASSIUM HEXACHLOROBISMUTHATE(III)
CsK_2BiCl_6

J. Solid State Chem., **65**, 13-26.

C2/c, a = 27.017, 25.653, b = 8.252, 7.799, c = 13.121, 12.874 A, β = 99.70, 99.24°, Z = 8, R = 0.069, 0.044 (superstructure reflexions for the Cs_3 compound neglected). Two $BiCl_6^{3-}$ octahedra, linked by 11-coordinate Cs and 8-coordinate Cs or K. Bi-Cl = 2.66-2.73(1) A. An orthorhombic form of the Cs_3 compound is also found, isostructural with the Rb_3BiBr_6 (**44**A, 163).

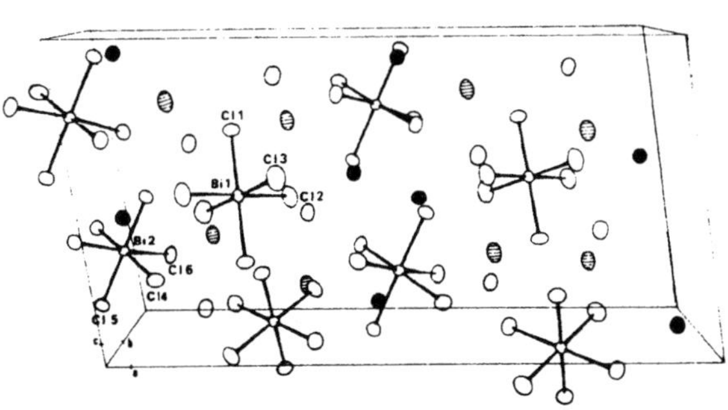

Atomic parameters $\times 10^4$

		x	y	z
Cs_3BiCl_6	Bi(1)	¼	¼	½
	Bi(2)	0	7835	¼
	Cs(1)	496	7444	9317
	Cs(2)	1614	8133	2982
	Cs(3)	3462	1863	8546
	Cl(1)	2492	3890	6871
	Cl(2)	3235	4548	4639
	Cl(3)	3269	549	5813
	Cl(4)	572	5536	1815
	Cl(5)	−583	7739	641
	Cl(6)	576	10263	1928
CsK_2BiCl_6	Bi(1)	¼	¼	½
	Bi(2)	0	7786	¼
	Cs	512	7363	9307
	K(1)	1622	8129	3004
	K(2)	3432	1906	8756
	Cl(1)	2503	3906	6939
	Cl(2)	3275	4665	4708
	Cl(3)	3267	274	5771
	Cl(4)	643	5405	1845
	Cl(5)	−589	7653	600
	Cl(6)	642	10310	1991

TRIHALOGENOCHALCOGEN SALTS
SCl_3SbCl_6 (I)
$SeCl_3SbCl_6$ (II)
$SBr_{1.5}Cl_{1.8}SbCl_6$ (III)
$TeCl_3AlCl_4$ (triclinic) (IV)
$TeCl_3AsF_6$ (V)
$TeCl_3SbF_6$ (VI)
$(TeF_3)_2SO_4$ (VII)

Inorg. Chem., **25**, 777-788.

I, C2/m, 12.385, 7.774, 13.959, 108.06, Z = 4, R = 0.038. II, III, $R\bar{3}m$, a = 8.840, 8.871 A, α = 74.89, 74.11°, Z = 2, R = 0.062, 0.031. IV, $P\bar{1}$, 6.554, 16.691, 8.391, 92.79, 97.31, 96.11, Z = 4, R = 0.030. V, $P2_1/n$, 8.827, 10.009, 10.592, 108.27, Z = 4, R = 0.038. VI, Pnma, 17.031, 8.460, 6.398, Z = 4, R = 0.029, VII, $P2_12_12_1$, 8.758, 8.983, 9.946, Z = 4, R = 0.053. All compounds contain trigonal pyramidal MX_3^+ cations, with significant cation-anion secondary interactions of varying strengths and geometries.

V

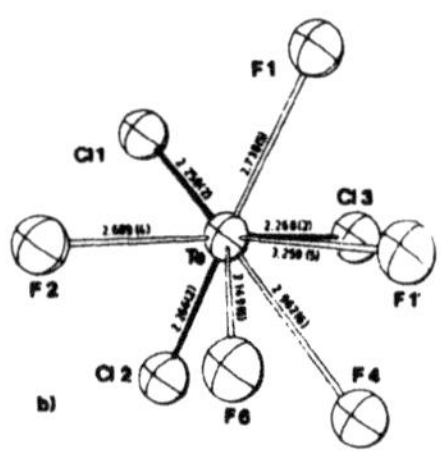

VI

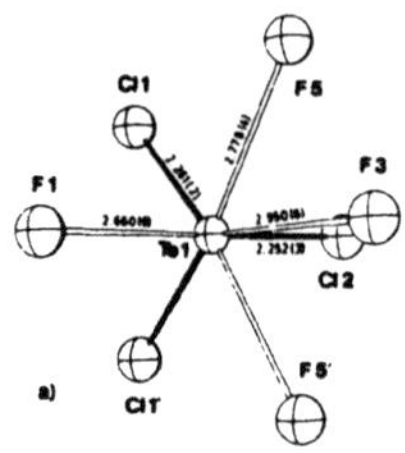

VII

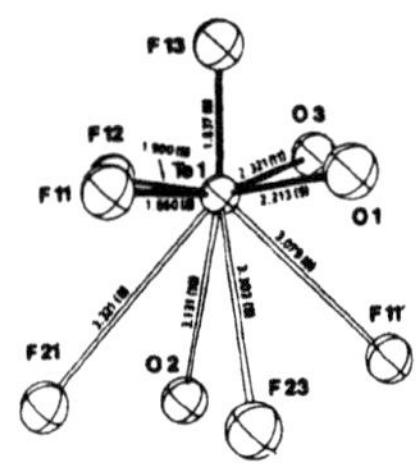

AMMONIUM HEXACHLOROSELENATE(IV)
$(NH_4)_2SeCl_6$

Acta Cryst., C**42**, 113-115.

Fm3m, 9.9616, Z = 4, R = 0.056. Antifluorite structure (**1**, 429), x(Cl) = 0.24111, disordered H sites, Se-Cl = 2.411(1) A (corrected for libration).

CAESIUM ENNEACHLORODIVANADATE(III)
$Cs_3V_2Cl_9$

RUBIDIUM ENNEABROMODIVANADATE(III)
$Rb_3V_2Br_9$

Inorg. Chem., **25**, 2930-2935.

$P6_3/mmc$, a = 7.240, 7.361, c = 17.95, 18.53 A, Z = 2, neutron powder data. Cs or Rb(1) in 2(b): 0,0,1/4; Cs or Rb(2) in 4(f): 1/3,2/3,z, z = 0.0763, 0.0715; V in 4(f): z taken as 0.8380, 0.8388; Cl or Br(1) in 6(h): x,2x,1/4, x = -0.4833, -0.4904; Cl or Br(2) in 12(k): x = x,2x,z, x = -0.1765, -0.1681, z = 0.0919, 0.0884. Isostructural with $Cs_3Cr_2X_9$ (**21**, 277; **37**A, 206).

MOLYBDENUM CHLORIDES
$PbMo_6Cl_{14}$, $CdMo_6Cl_{14}$, $Cu_2Mo_6Cl_{14}$, $Rb_2Mo_6Cl_{14}$, $Cs_2Mo_6Cl_{14}$, $AgMo_6Cl_{13}$

Mater Res. Bull., **21**, 1239-1245.

Pb, Cd, Cu, Pn3, 12.95, 12.67, 12.77, Z = 4, no details of analyses, no positional parameters given. Isostructural with the Hg compound (**37**A, 201), with $(Mo_6Cl_8)Cl_6$ clusters; in the Cu compound, an octahedral site is occupied by 2 Cu with trigonal planar coordinations.

Rb, C2/c, 18.90, 14.80, 9.20, 115, no details. Again $(Mo_6Cl_8)Cl_6$ units.

Cs, P31c or P$\bar{3}$1c, 9.77, 14.22, no details.

Ag, P$\bar{1}$, 8.83, 9.08, 13.90, 95.75, 102.21, 114.56, no details. $(Mo_6Cl_8)Cl_6$ units share two bridging Cl to form zigzag chains; Ag has tetrahedral coordination.

AMMONIUM DIAQUODECACHLOROTRIRHENATE DIHYDRATE
$NH_4[Re_3Cl_{10}(OH_2)_2].2H_2O$

Z. anorg. Chem., **534**, 115-120.

$P4_12_12$, 11.576, 16.145, Z = 4, R = 0.099. Isolated trinuclear anions which contain strong Cl...H-O-H...Cl hydrogen bonds. Ammonium sites (occupancy = 0.5) have 7 Cl and 2 H_2O neighbours.

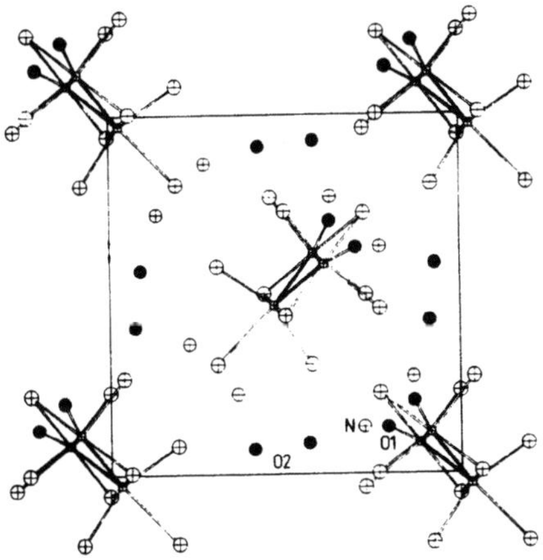

		x	y	z
N	8b	0,3685	0,7741	0,4363
Re1	4a	0,0285	0,0285	0
Re2	8b	0,8839	0,9139	0,0745
Cl1	8b	0,2281	0,4626	0,6778
Cl2	8b	0,0813	0,8082	0,4714
Cl3	8b	0,9950	0,0563	0,1449
Cl4	8b	0,7683	0,9971	0,8729
Cl5	4a	0,7726	0,7726	0
Cl6	4a	0,1901	0,1901	0
O1	8b	0,7875	0,8640	0,1910
O2	8b	0,9316	0,4186	0,8114

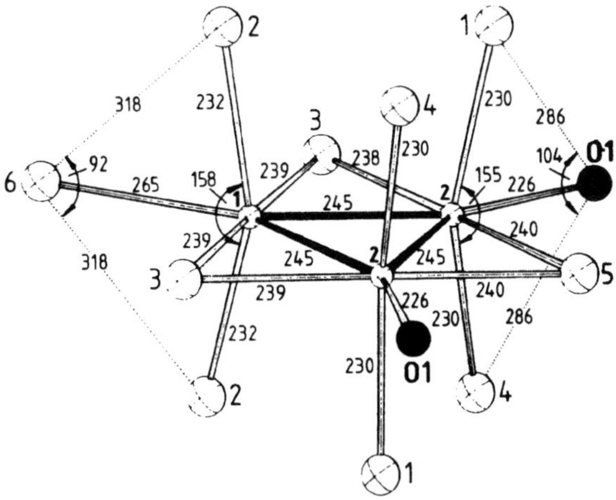

RUBIDIUM and CAESIUM RHENIUM CHLORIDES
$M_3[Re_3Cl_{12}]$ (M = Rb, Cs)

J. Less-Common Metals, **119**, 31-44.

Rb, Cmcm, 10.4951, 14.8932, 12.1293, Z = 4, R = 0.069. Cs, C2cm, 14.0872, 14.0498, 10.6667, Z = 4, R = 0.072. The Cs structure is as previously described (**28**, 87), both structures containing trinuclear anions.

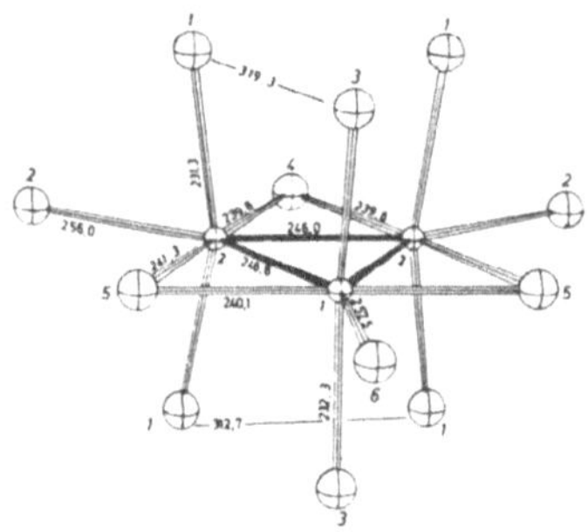

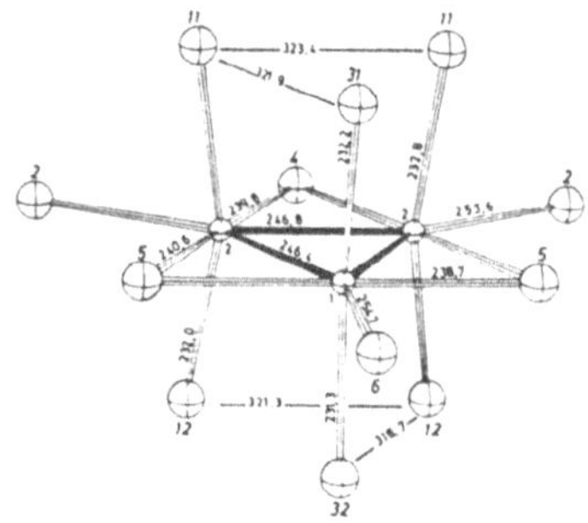

		x	*y*	*z*
Rb1	4*c*	0.5	0.2470	0.75
Rb2	8*e*	0.7506	0.5	0
Re1	4*c*	0.5	0.1465	0.25
Re2	8*f*	0.5	0.2902	0.3514
Cl1	16*h*	0.7171	0.3046	0.1211
Cl2	8*f*	0.5	0.6254	0.0353
Cl3	8*g*	0.2175	0.3823	0.75
Cl4	4*c*	0.5	0.4284	0.25
Cl5	8*f*	0.5	0.1485	0.4479
Cl6	4*c*	0	0.5264	0.75

		x	*y*	*z*
Cs1	4*a*	0	0	0
Cs2	4*b*	0.6184	0.1790	0.25
Cs3	4*b*	0.8939	0.4089	0.25
Re1	4*b*	0.2770	0.3596	0.25
Re2	8*c*	0.2299	0.2153	0.1343
Cl11	8*c*	0.3795	0.1504	0.0984
Cl12	8*c*	0.0714	0.2516	0.0994
Cl2	8*c*	0.7004	0.3683	0.0728
Cl31	4*b*	0.4405	0.3388	0.25
Cl32	4*b*	0.1325	0.4377	0.25
Cl4	4*b*	0.1843	0.0763	0.25
Cl5	8*c*	0.2775	0.3578	0.0262
Cl6	4*b*	0.8336	0.0318	0.25

AMMONIUM HEXACHLORORHODATE(III) MONOHYDRATE
$(NH_4)_3RhCl_6.H_2O$

Z. Naturforsch., **41**B, 1-3.

Pnma, 12.193, 7.007, 14.167, Z = 4, R = 0.037. Octahedral anions and ammonium cations, with water molecules hydrogen bonded to the cations. Rh-Cl = 2.332-2.361(3) A.

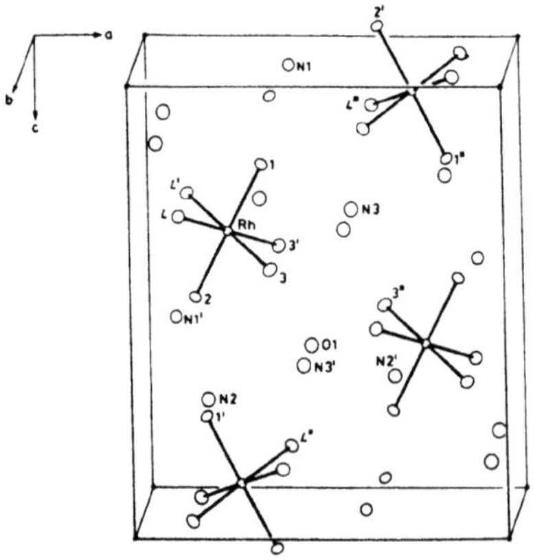

	x	*y*	*z*
Rh	0.22487	0.25	0.40538
Cl1	0.3141	0.25	0.2575
Cl2	0.1350	0.25	0.5507
Cl3	0.3443	0.4846	0.4632
Cl4	0.1047	0.4833	0.3481
O1	0.4400	0.25	0.6566
N1	0.3904	0.25	0.0375
N2	0.166	0.25	0.7764
N3	0.5566	0.25	0.3560

RUBIDIUM TRICHLOROCUPRATE(II)
$RbCuCl_3$

Z. Kristallogr., **176**, 135-139.

Pbcn (pseudo $P6_3/mmc$), 6.98, 11.94, 6.18, Z = 4 [not 2], R = 0.079. Cu in 4(a): 0,0,0; Rb in 4(c): 0,0.3347,1/4; Cl(1) in 4(c): 0,0.1536,3/4; Cl(2) in 8(d): 0.2403,0.0816,0.1949. Anionic chains of face-sharing distorted octahedra (Cu...Cu = 3.09 A), linked by Rb ions.

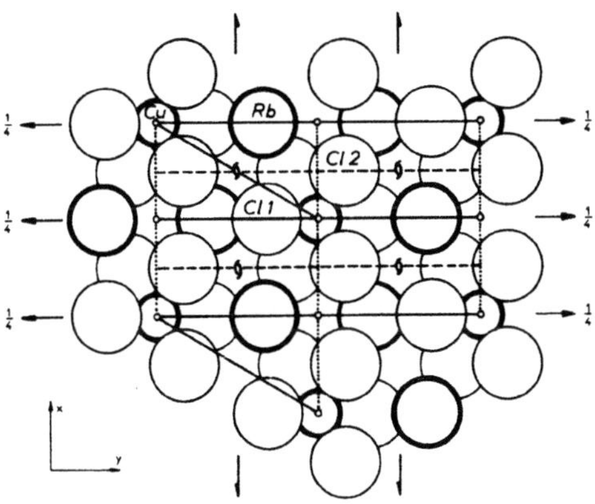

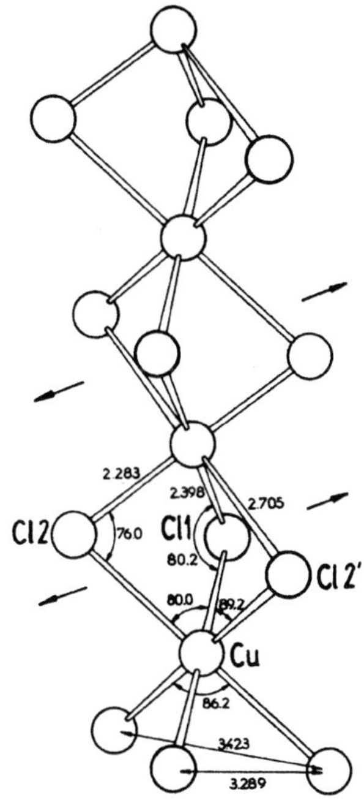

RUBIDIUM COPPER(I) CHLORIDES
$Rb_{18}Cu_{31}Cl_{49}$, $Rb_4Cu_5Cl_9$

J. Solid State Chem., **63**, 316-325 and 326-335.

$Rb_{18}Cu_{31}Cl_{49}$, $I\bar{4}3d$, 21.220, Z = 4, R = 0.035. $CuCl_4$ tetrahedra sharing corners and edges, with 9- and (7+2)-coordinate Rb.

$Rb_4Cu_5Cl_9$, Pa, 15.694, 8.544, 14.660, 106.59, Z = 4, R = 0.034. Double chains along **b** of corner-sharing $CuCl_4$ tetrahedra; most tetrahedra have at least one long Cu-Cl distance, and one Cu is 3-coordinate. Rb ions have 7-, (7+1)-, and 8-coordinations.

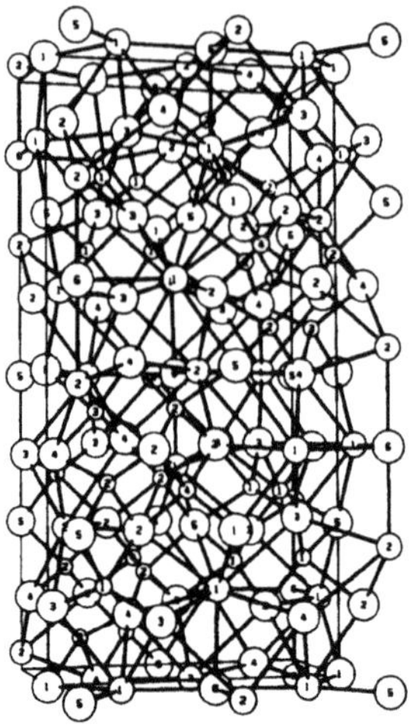

$Rb_{18}Cu_{31}Cl_{49}$

		x	y	z
Rb1	48e	0.8781	0.0031	0.0944
Rb2	24d	0.0281	0	$\frac{1}{4}$
Cu1	48e	0.0530	0.1895	0.3319
Cu2	48e	0.1764	0.0617	0.1725
Cu3	16c	0.0677	=x	=x
Cu4[a]	16c	0.1891	=x	=x
Cl1	16c	−0.0010	=x	=x
Cl2	48e	0.2516	0.1260	0.1195
Cl3	48e	0.3645	−0.0167	0.1175
Cl4	48e	−0.0096	0.1081	0.1343
Cl5	24d	0.2516	0	$\frac{1}{4}$
Cl6	12b	$\frac{7}{8}$	0	$\frac{1}{4}$

[a] There are only 12 Cu^+ ions distributed over the 16 sites.

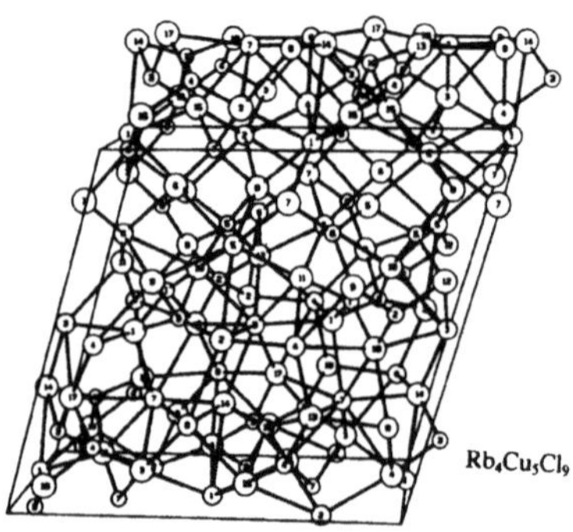

$Rb_4Cu_5Cl_9$

	x	y	z		x	y	z
Rb1	0.000	0.3344	0.000	Cl1	−0.3227	0.3448	−0.4[illegible]
Rb2	0.2967	0.1561	0.0135	Cl2	−0.0816	0.1300	−0.4[illegible]
Rb3	0.4792	0.4860	−0.4699	Cl3	−0.1953	0.1745	−0.0[illegible]
Rb4	0.1103	0.0185	−0.4783	Cl4	0.4498	0.2974	−0.0[illegible]
Rb5	−0.1194	0.1648	0.2655	Cl5	−0.3135	−0.2904	0.1[illegible]
Rb6	−0.4180	0.3410	0.2540	Cl6	−0.3085	0.2076	0.1[illegible]
Rb7	−0.2334	0.4825	−0.2544	Cl7	−0.5062	0.0015	0.1[illegible]
Rb8	−0.6021	0.0168	−0.2620	Cl8	−0.1139	0.4979	0.1[illegible]
Cu1	−0.2852	−0.3982	0.4450	Cl9	−0.2906	0.0342	0.3[illegible]
Cu2	−0.1661	−0.1119	0.4429	Cl10	−0.2083	0.4470	0.3[illegible]
Cu3	−0.4557	0.3782	−0.1696	Cl11	−0.4303	−0.3182	0.3[illegible]
Cu4	−0.3310	0.0990	−0.1729	Cl12	−0.0703	−0.1943	0.3[illegible]
Cu5	−0.1730	−0.3737	0.2500	Cl13	−0.3024	−0.1587	−0.2[illegible]
Cu6	0.1201	0.1054	0.2248	Cl14	0.4608	−0.3732	−0.2[illegible]
Cu7	−0.2449	0.3642	0.0227	Cl15	−0.3317	−0.4709	−0.0[illegible]
Cu8	−0.4438	0.1512	0.0094	Cl16	−0.4165	−0.0521	−0.0[illegible]
Cu9	−0.2127	0.2447	0.4803	Cl17	−0.4327	0.2372	−0.2[illegible]
Cu10	−0.4095	−0.2570	−0.2153	Cl18	−0.1901	−0.2650	−0.4[illegible]

TRICHLOROTELLURIUM TETRACHLOROAURATE(III)
$TeCl_3^+AuCl_4^-$

Z. Naturforsch., **41**B, 1483-1484.

$P\bar{1}$, 7.564, 7.720, 8.964, 78.26, 88.84, 89.35, Z = 2, R = 0.041. Trigonal pyramidal $TeCl_3^+$ and square planar $AuCl_4^-$ ions, with two longer Te...Cl interactions giving $(TeCl_3.AuCl_4)_2$ dimers and square pyramidal coordination at Te.

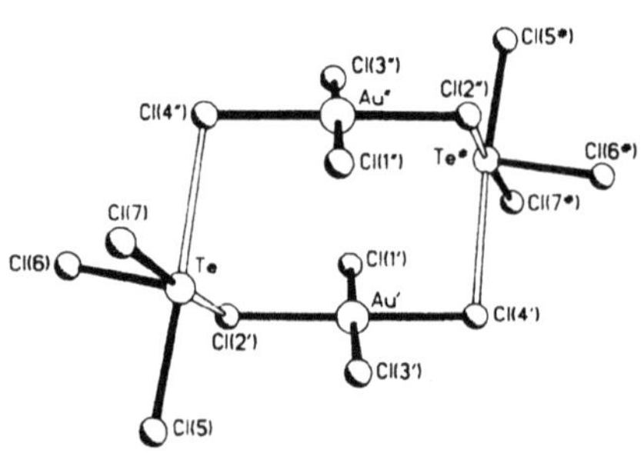

	x	y	z
	($\times 10^4$)		
Au	750	7870	4388.3
Cl(1)	3682	8377	4646
Cl(2)	464	6618	6911
Cl(3)	−2177	7477	4097
Cl(4)	1054	9141	1870
Te	3209	2326	2274
Cl(5)	4246	5061	2423
Cl(6)	2010	3319	− 76
Cl(7)	5784	1372	1287

Bond lengths (Å) and angles (deg.).

Au−Cl(1)	2.281(4)	Au−Cl(2)	2.278(4)
Au−Cl(3)	2.266(4)	Au−Cl(4)	2.279(4)
Te···Cl(2′)	3.005(5)	Te···Cl(4″)	3.052(5)
Cl(5)−Te	2.291(6)	Cl(6)−Te	2.292(4)
Cl(7)−Te	2.298(5)		
Cl(1)−Au−Cl(2)	91.4(2)	Cl(1)−Au−Cl(3)	177.8(3)
Cl(2)−Au−Cl(3)	89.8(2)	Cl(1)−Au−Cl(4)	88.2(2)
Cl(2)−Au−Cl(4)	179.5(2)	Cl(3)−Au−Cl(4)	90.6(2)
Au−Cl(2)···Te′	103.5(2)	Au−Cl(4)···Te″	97.2(2)
Cl(5)−Te−Cl(6)	93.4(3)	Cl(5)−Te−Cl(7)	95.6(3)
Cl(6)−Te−Cl(7)	93.6(2)	Cl(5)−Te···Cl(2′)	90.7(2)
Cl(6)−Te···Cl(2′)	77.9(2)	Cl(7)−Te···Cl(2′)	169.7(2)
Cl(5)−Te···Cl(4″)	167.0(2)	Cl(6)−Te···Cl(4″)	78.1(2)
Cl(7)−Te···Cl(4″)	94.7(2)	Cl(2′)···Te···Cl(4″)	78.1(2)

Symmetry operators: (′) −*x*, 1−*y*, 1−*z*; (″) *x*, −1+*y*, *z*.

TERNARY CADMIUM HALIDES
$(NH_4)_4CdCl_6$, K_4CdCl_6, Rb_4CdCl_6, In_4CdI_6, Tl_4CdI_6

Z. anorg. Chem., **539**, 7-17.

NH_4, K, Rb chlorides, R$\bar{3}$c, a = 12.335, 12.105, 12.439, c = 15.553, 14.909, 15.570 A, Z = 6, R = 0.037, 0.021, 0.056. NH_4, K, Rb(1) in 6(a): 0,0,1/4; NH_4, K, Rb(2) in 18(e): x,0,1/4, x = 0.3725, 0.3757, 0.3728; Cd in 6(b): 0,0,0; Cl in 36(f): -0.02846,0.15922,0.09824; -0.03154,0.15956,0.10229; -0.0278,0.1583,0.0983. K_4CdCl_6-type structure (**15**, 153; **18**, 414), with $CdCl_6$ octahedra, linked by K ions.

In, Tl iodides, P4/mnc, a = 9.060, 9.222, c = 9.754, 9.603 A, Z = 2, R = 0.045, 0.065. In, Tl in 8(g): x,1/2+x,1/4, x = 0.3583, 0.3534; Cd in 2(a): 0,0,0; I(1) in 4(e): 0,0,z, z = 0.7105, 0.7114; I(2) in 8(h): x,y,0, x = 0.3095, 0.3154, y = 0.1499, 0.1410. Tl_4HgBr_6-type structure (**39**A, 185), with CdI_6 octahedra, linked by 8-coordinate In or Tl ions.

LITHIUM THULIUM CHLORIDE
$LiTm_2Cl_5$

J. Less-Common Metals, **116**, 187-197.

C2/c, 15.288, 6.506, 7.150, 95.50, Z = 4, R = 0.059. Isostructural with the Dy compound (**50**A).

		x	*y*	*z*
Li	4c	0.25	0.25	0
Tm	8f	0.39644	0.27253	0.43239
Cl1	8f	0.2196	0.0695	0.3148
Cl2	8f	0.4086	0.1226	0.0733
Cl3	4e	0	0.0682	0.25

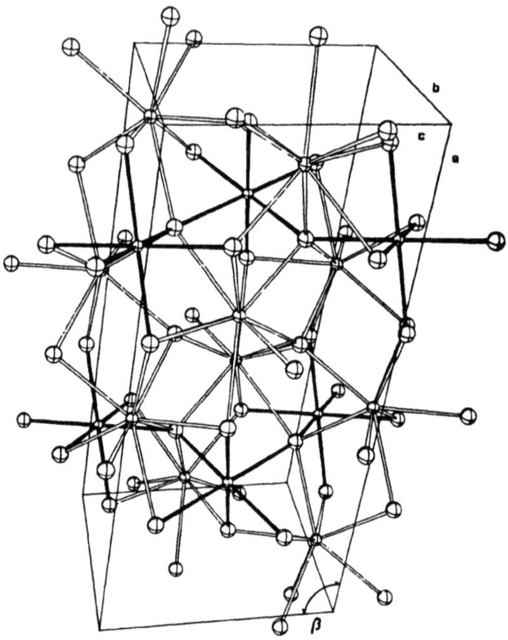

CAESIUM LITHIUM LUTETIUM HEXACHLORIDE
CAESIUM POTASSIUM SCANDIUM HEXACHLORIDE
$Cs_2LiLuCl_6$, Cs_2KScCl_6

Z. anorg. Chem., **535**, 208-212.

Fm3m, 10.409, 10.873, Z = 4, R = 0.055, 0.078. Cs in 8(c); Li or K in 4(b); Lu or Sc in 4(a); Cl in 24(e): x = 0.2483, 0.2263. Elpasolite structures (**2**, 498; **38**A, 213).

BARIUM BROMIDE CHLORIDE
$BaBr_{1.333}Cl_{0.667}$

Acta Cryst., C**42**, 915-917.

Pnma, 8.2031, 4.8606, 9.6359, Z = 4, neutron powder data. Atoms in 4(c): x,1/4,z, x = 0.2603, 0.0293, 0.1446, z = 0.1203, 0.8292, 0.4329, for Ba, Br, Br/Cl. $PbCl_2$-type structure, as for related materials (**2**, 16; **28**, 77; **50**A).

DIGALLIUM TRIBROMIDE
Ga_2Br_3

J. Solid State Chem., **61**, 171-180.

Pnma, 12.537, 15.474, 12.534, Z = 16, R = 0.062. $Ga(III)_2Br_6^{2-}$ anions, linked by Ga^+ ions with cubic and tricapped trigonal prismatic coordinations (some Ga^+ vacancies). Ga-Ga = 2.43, Ga-Br = 2.41, Ga(I)-Br = 2.94-3.92 A.

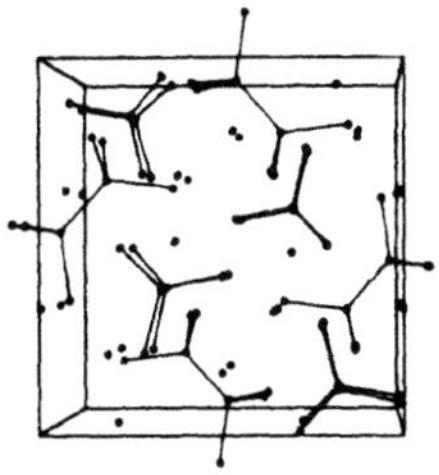

		x	y	z
Ga(1)	4c	0.1579	¼	0.3154
Ga(2)	4c	0.0266	¼	0.4577
Ga(3)	8d	0.3197	0.3288	0.6130
Ga(4)	4c	0.6897	¼	0.5158
Ga(5)[b]	8d	−0.0005	0.4440	0.6630
Ga(6)[b]	8d	0.3764	0.4375	0.3106
Br(1)	4c	0.3486	¼	0.3385
Br(11)	8d	0.6331	0.3727	0.3001
Br(2)	4c	0.0522	¼	0.6467
Br(22)	8d	0.9068	0.3707	0.4404
Br(31)	8d	0.2151	0.0778	0.5034
Br(32)	8d	0.4956	0.1146	0.5804
Br(33)	8d	0.2763	0.1155	0.7878

[b] The occupancy factors are Ga(5): 0.74(2) and Ga(6): 0.74(2).

LITHIUM GALLIUM BROMIDES
$LiGaBr_4$, $LiGaBr_3$

Z. Naturforsch., **41**B, 1391-1398.

$LiGaBr_4$, $P2_1/a$, 13.696, 6.854, 7.416, 93.14, Z = 4, R = 0.064. Isostructural with $LiAlCl_4$ (**43**A, 135), with $GaBr_4^-$ tetrahedra linked by octahedrally coordinated Li^+ ions. Ga-Br = 2.301-2.346, Li-Br = 2.605-2.909 A.

$LiGaBr_3$, $P2_1/m$, 8.007, 10.489, 6.535, 93.24, Z = 4, R = 0.046. The compound can be formulated as $Li_2[Ga_2Br_6]$, with $Br_3Ga\text{-}GaBr_3^{2-}$ ions linked by octahedrally coordinated Li^+ ions. Ga-Ga = 2.404, Ga-Br = 2.382-2.401, Li-Br = 2.675-2.892 A.

$LiGaBr_4$

	x	y	z
Ga	0,8998	0,3223	0,7044
Br1	0,0513	0,1766	0,6905
Br2	0,9306	0,6402	0,8084
Br3	0,8142	0,1775	0,9351
Br4	0,8073	0,3102	0,4361
Li	0,3731	−0,0118	0,1569

$LiGaBr_3$

	x	y	z
Ga1	0,2464	1/4	0,8639
Ga2	0,9636	1/4	0,9693
Br11	0,4565	1/4	0,1369
Br1	0,3201	0,0715	0,6621
Br22	0,9300	1/4	0,3293
Br2	0,8117	0,0702	0,8274
Li1	0	0	1/2
Li2	1/2	0	0

INDIUM TIN BROMIDE
$InSn_2Br_5$

INDIUM TIN IODIDE
$InSn_2I_5$

Z. anorg. Chem., **536**, 45-52.

I4/mcm, a = 8.285, 8.810, c = 14.438, 15.240 A, Z = 4, R = 0.054, 0.063. In in 4(a): 0,0,1/4; Sn in 8(h): x,1/2+x,0, x = 0.1735, 0.1705; Br or I(1) in 4(c): 0,0,0; Br or I(2) in 16(ℓ): x,1/2+x,z, x = 0.1600,0.1580, z = 0.3618, 0.3592. Anti-Tl_5Se_2I type (**51**A, 173).

POTASSIUM TRIBROMOPLUMBATE(II) MONOHYDRATE
$KPbBr_3.H_2O$

J. Less-Common Metals, **125**, 215-221.

$Cmc2_1$, 4.414, 19.728, 8.904, Z = 4, R = 0.061. $PbBr_6$ distorted octahedra and KO_2Br_4 trigonal prisms with three capping Br. Pb-Br = 2.77-3.45, K-O = 2.79, K-Br = 3.38-3.70(1) A.

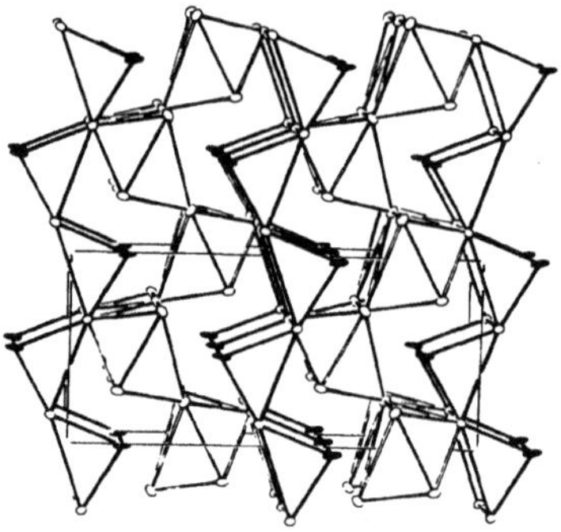

	x	y	z
Pb	0	0,1662	0
Br1	0	0,6139	0,1970
Br2	0	0,0410	0,8602
Br3	0	0,7146	0,7830
K	0	0,5524	0,5736
O	0	0,8639	0,023

CAESIUM TRIBROMOHEXACHLORODICHROMATE(III)
$Cs_3Cr_2Br_3Cl_6$

Acta Cryst., C**42**, 769-771.

$P6_3/mmc$, 7.336, 18.181, Z = 2, R = 0.050. Isostructural with the enneachloro compound (**21**, 277), with Br preferentially in terminal halogen positions.

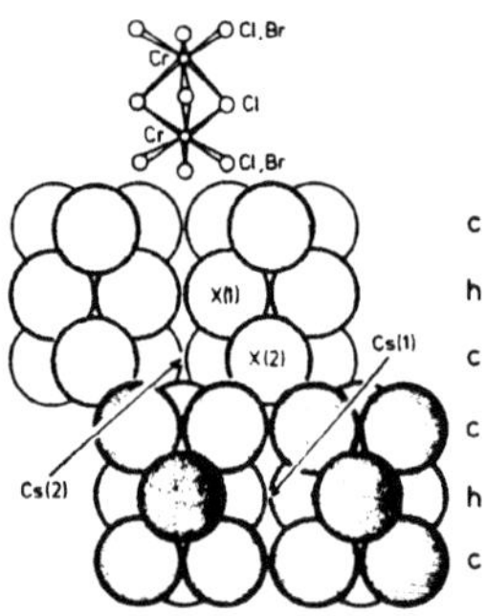

	x	y	z
Cs(1)	0	0	1/4
Cs(2)	1/3	2/3	0·0770
Cr	1/3	2/3	0·8373
X(1)*	0·5164 (4)	0·0328	1/4
X(2)*	0·8241 (2)	0·6482	0·0914

* X(1) = 0·94 (2) Cl + 0·06 Br, X(2) = 0·56 (1) Cl + 0·44 Br.

POTASSIUM TETRABROMOAURATE(III) DIHYDRATE
$KAuBr_4.2H_2O$

Acta Cryst., C**42**, 1091-1092.

$P2_1/n$, 9.579, 12.006, 8.480, 94.31, Z = 4, R = 0.041. Structure differs from that described previously (**4**, 50, 180) and from that of the chloride (**46**A, 186). Square-planar $AuBr_4^{2-}$ ions.

	x	y	z
Au	0·5026	0·5067	0·2519
Br(1)	0·6395	0·3371	0·2460
Br(2)	0·3679	0·6778	0·2539
Br(3)	0·3130	0·4166	0·0922
Br(4)	0·6901	0·5954	0·4134
K	0·5181	0·9723	0·2510
O(1)	0·0613	0·6353	0·4770
O(2)	0·5512	0·8640	0·5341

Interatomic distances (Å) *and angles* (°)

Au–Br(1)	2·425 (2)	K–O(2)	2·73 (2)
Au–Br(2)	2·426 (2)	K–O(2)	2·79 (2)
Au–Br(3)	2·436 (2)	K–Br(1)	3·657 (8)
Au–Br(4)	2·422 (2)	K–Br(2)	3·819 (8)
K–O(1)	2·72 (2)	K–Br(3)	2·593 (8)
K–O(1)	2·82 (2)	K–Br(4)	3·539 (8)
Br(1)–Au–Br(2)	179·0 (1)	Br(2)–Au–Br(3)	90·3 (1)
Br(1)–Au–Br(3)	89·8 (1)	Br(2)–Au–Br(4)	89·8 (1)
Br(1)–Au–Br(4)	90·1 (1)	Br(3)–Au–Br(4)	179·3 (1)
Au–Au	4·216 (2); 4·270 (2)	K–K	4·288 (2); 4·325 (3)

HEXAAQUOMANGANESE(II) TETRABROMOZINCATE
$[Mn(H_2O)_6][ZnBr_4]$

Acta Cryst., C**42**, 771-772.

Pbam, 12.444, 12.952, 8.002, Z = 4, R = 0.052. $Mn(H_2O)_6{}^{2+}$ octahedra and $ZnBr_4{}^{2-}$ tetrahedra, linked by hydrogen bonds.

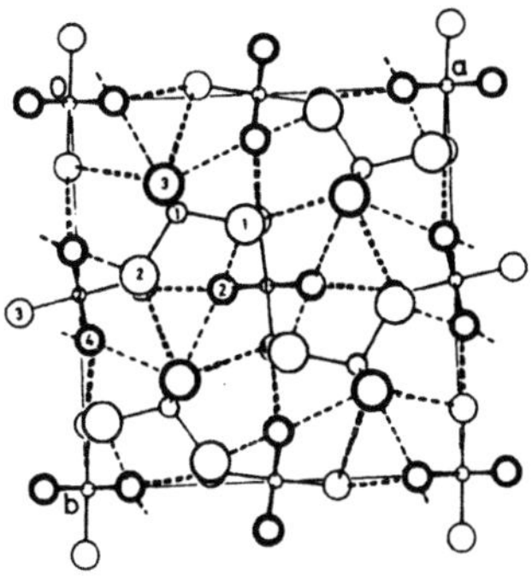

Atomic positions **(×10⁴)**

	x	*y*	*z*
Mn(1)	5000	5000	5000
Mn(2)	0	5000	5000
Zn(1)	2641	3029	0
Br(1)	4557	3421	0
Br(2)	1550	4567	0
Br(3)	2348	2206	2637
O(1)	5019	6672	5000
O(2)	3770	4968	3080
O(3)	−1728	5427	5000
O(4)	212	6206	3118

***Interatomic distances* (Å) *and angles* (°)**

Mn(1)–O(1)	2·166 (13) (×2)	O(1)–Mn(1)–O(2)	91·5 (4)
Mn(1)–O(2)	2·169 (8) (×4)	O(2)–Mn(1)–O(2)	90·2 (3)
Mn(2)–O(3)	2·220 (13) (×2)	O(3)–Mn(2)–O(4)	93·5 (3)
Mn(2)–O(4)	2·186 (9) (×4)	O(4)–Mn(2)–O(4)	92·9 (3)
Zn–Br(1)	2·437 (3)	Br(1)–Zn–Br(2)	112·3 (1)
Zn–Br(2)	2·411 (3)	Br(1)–Zn–Br(3)	104·0 (1)
Zn–Br(3)	2·392 (1) (×2)	Br(2)–Zn–Br(3)	106·4 (1)
		Br(3)–Zn–Br(3)	123·8 (1)

Possible H-bonded distances

O(2)–H···Br(3)	3·24 (1)	O(1)–H···Br(3)	3·57 (1) (×2)
O(2)–H···Br(1)	3·32 (1)	O(1)–H···O(4)	3·14 (1) (×2)
O(4)–H···Br(3)	3·32 (1)	O(3)–H···Br(3)	3·68 (1) (×2)
O(4)–H···Br(2)	3·47 (1)	O(3)–H···O(2)	3·01 (1) (×2)
O(4)–H···Br(2)	3·67 (1)		

CALCIUM HEXABROMODICADMATE HEPTAHYDRATE
$[Ca(H_2O)_7][Cd_2Br_6]$

Acta Cryst., C**42**, 773-774.

$P2_1/c$, 8.079, 18.677, 12.176, 96.60, Z = 4, R = 0.065. Infinite chains of edge- and corner-sharing $CdBr_6$ octahedra and $CdBr_4$ tetrahedra, hydrogen bonded to $Ca(H_2O)_7$ polyhedra.

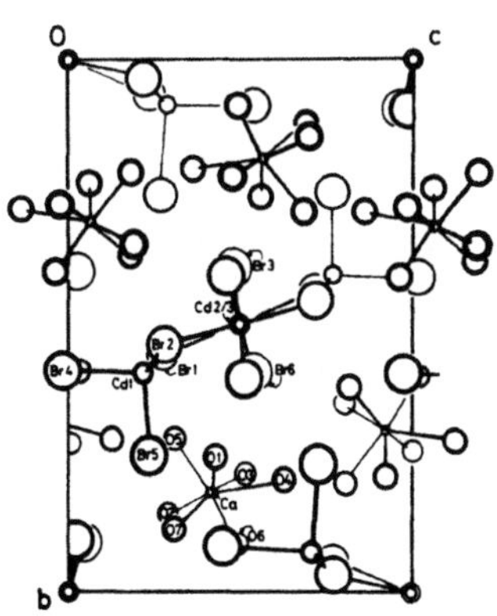

Atomic positions (×10⁴)

	x	*y*	*z*
Ca	2083	8081	4434
Cd(1)	6910	5937	2106
Cd(2)	5000	5000	5000
Cd(3)	0	5000	5000
Br(1)	−225	5339	2742
Br(2)	4369	5327	2771
Br(3)	2562	5967	5436
Br(4)	6841	5866	−23
Br(5)	7082	7311	2445
Br(6)	2429	4002	4718
O(1)	4665	7413	4454
O(2)	31	8555	3051
O(3)	−422	7755	5203
O(4)	3007	7825	6320
O(5)	1241	7061	3204
O(6)	1761	9203	5275
O(7)	3621	8696	3182

Interatomic distances (Å) *and angles* (°)

Environment of Ca

Ca–O(6)	2·36 (2)	Ca–O(3)	2·40 (2)
Ca–O(7)	2·37 (2)	Ca–O(1)	2·43 (1)
Ca–O(4)	2·38 (1)	Ca–O(5)	2·47 (1)
Ca–O(2)	2·39 (1)		

Tetrahedral environment of Cd(1)

Cd(1)–Br(2)	2·559 (2)	Cd(1)–Br(5)	2·600 (2)
Cd(1)–Br(4)	2·590 (3)	Cd(1)–Br(1)	2·606 (2)
Br(4)–Cd(1)–Br(2)	111·6 (1)	Br(2)–Cd(1)–Br(5)	114·8 (1)
Br(4)–Cd(1)–Br(5)	101·8 (1)	Br(2)–Cd(1)–Br(1)	115·6 (1)
Br(4)–Cd(1)–Br(1)	101·0 (1)	Br(5)–Cd(1)–Br(1)	110·3 (1)

Octahedral environment of Cd(2) (site symmetry C_i)

Cd(2)–Br(3)	2·768 (2)	Br(3)–Cd(2)–Br(2)	90·6 (1)
Cd(2)–Br(2)	2·771 (2)	Br(3)–Cd(2)–Br(6)	94·4 (1)
Cd(2)–Br(6)	2·783 (2)	Br(2)–Cd(2)–Br(6)	91·4 (1)

Octahedral environment of Cd(3) (site symmetry C_i)

Cd(3)–Br(3)	2·752 (2)	Br(3)–Cd(3)–Br(6)	93·6 (1)
Cd(3)–Br(6)	2·756 (2)	Br(3)–Cd(3)–Br(1)	90·2 (1)
Cd(3)–Br(1)	2·807 (2)	Br(6)–Cd(3)–Br(1)	90·1 (1)

Possible H bonds

X	O···*X*	Ca–O···*X*	*X*	O···*X*	Ca–O···*X*
O(1)···Br(5)	3·31 (1)	127 (1)	O(5)···Br(5)	3·41 (1)	105 (1)
Br(3)	3·47 (1)	87 (1)	Br(1)	3·45 (1)	152 (1)
Br(6)	3·60 (1)	156 (1)	Br(3)	3·47 (1)	87 (1)
Br(4)	3·68 (1)	87 (1)	O(6)···Br(4)	3·33 (2)	137 (1)
O(2)···Br(2)	3·35 (1)	108 (1)	Br(2)	3·60 (2)	120 (1)
Br(1)	3·48 (1)	120 (1)	Br(1)	3·67 (2)	130 (1)
O(3)···Br(4)	3·39 (2)	110 (1)	O(7)···Br(4)	3·30 (1)	97 (1)
Br(5)	3·58 (2)	152 (1)	Br(3)	3·41 (1)	129 (1)
Br(6)	3·67 (2)	129 (1)	Br(2)	3·70 (1)	152 (1)
O(4)···O(5)	2·85 (2)	129 (1)			
Br(5)	3·43 (1)	126 (1)			
Br(3)	3·64 (1)	84 (1)			

RUBIDIUM TRIIODIDE
RbI_3

THALLIUM(I) TRIIODIDE
TlI_3

Acta Cryst., C**42**, 1675-1678.

Pnma, a = 10.908, 10.599, b = 6.655, 6.419, c = 9.711, 9.436 A, Z = 4, R = 0.027, 0.042. Isostructural with the Cs salt (**19**, 321; **38**A, 219), with asymmetric, slightly bent triiodide anions.

	x	*y*	*z*
RbI₃			
I(1)	0,15807	¼	0,35086
I(2)	0,37722	¼	0,54612
I(3)	0,57527	¼	0,73483
Rb	0,33091	¼	0,03098
TlI₃			
I(1)	0,1510	¼	0,3431
I(2)	0,3787	¼	0,5429
I(3)	0,5826	¼	0,7358
Tl	0,3350	¼	0,0295

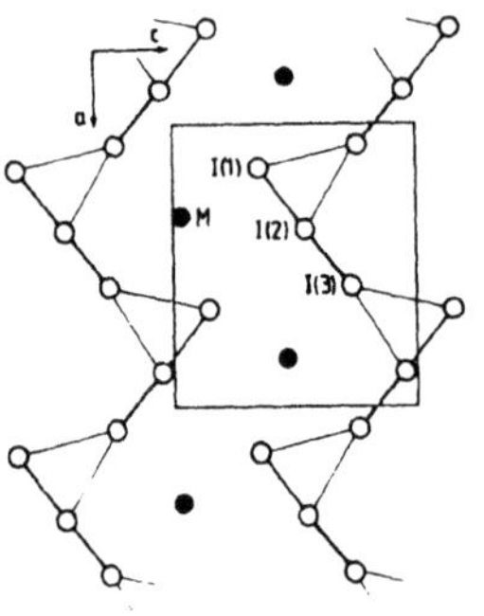

	TlI₃	**RbI₃**	**CsI₃**	**NH₄I₃**
T (K)	300	300	113	293
r(M^+)	1,40	1,48	1,69	1,48
I(1)–I(2)	3,063 (3)	3,051 (1)	3,038 (2)	3,114 (2)
I(2)–I(3)	2,826 (3)	2,833 (1)	2,842 (2)	2,797 (2)
I(1)–I(2)–I(3)	177,89 (9)	178,11 (3)	178,00 (4)	178,55 (5)
I(1)···I(3)	4,039 (3)	4,124 (1)	4,160 (2)	4,096 (2)
I(2)···I(3)	3,770 (3)	3,921 (1)	3,972 (2)	3,881 (2)
I(1)···I(3)···I(2)	46,02 (5)	44,49 (1)	43,80 (1)	45,84 (2)
I(2)–I(3)···I(2)	173,53 (8)	172,55 (2)	171,34 (4)	172,40 (4)
M···I(1)	3,544 (3)	3,634 (1)	3,709 (2)	3,642 (10)
I(1)	3,559 (3)	3,749 (1)	3,830 (2)	3,688 (11)
I(1) (2*)	3,663 (2)	3,761 (1)	3,805 (2)	3,731 (5)
I(3)	3,664 (2)	3,800 (1)	3,901 (2)	3,751 (10)
I(3)	3,817 (3)	3,921 (1)	3,989 (2)	3,838 (10)
I(2) (2*)	3,930 (2)	4,031 (1)	4,069 (2)	4,070 (6)
I(3) (2*)	3,996 (2)	4,158 (1)	4,220 (2)	4,129 (6)

STRONTIUM IODIDE DIHYDRATE
$SrI_2.2H_2O$

Acta Cryst., C**42**, 651-653.

Pcmn, 15.5805, 4.3403, 9.8814, Z = 4, R = 0.024. Columns along **b** of face-sharing $SrI_5(H_2O)_4$ tricapped trigonal prisms; Sr-I = 3.372-3.842(1), Sr-O = 2.682-2.731(2), O-H...I = 3.48-3.80 A.

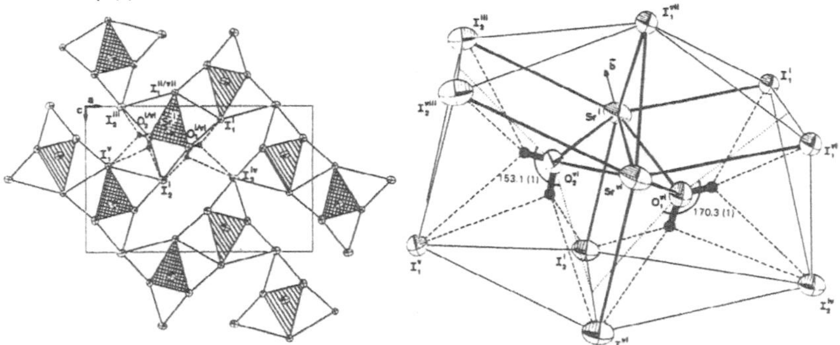

	x	y	z
Sr	0·38154	0·75	0·16797
I(1)	0·59933	0·75	0·09152
I(2)	0·34260	0·75	0·50591
O(1)	0·4599	0·25	0·2813
O(2)	0·2842	0·25	0·2114

ZIRCONIUM(IV) IODIDE
γ-ZrI_4

Kristallografija, **31**, 446-449 [Soviet Physics - Crystallography, **31**, 261-263].

$I\bar{1}$, 8.348, 8.343, 23.28, 90.59, 90.50, 90.04, Z = 8, R = 0.041. Distorted cubic-close-packed I, with Zr in 1/4 of the octahedral holes, giving zigzag chains of edge-sharing ZrI_6 octahedra. Zr-I = 2.69-3.03 A. Monoclinic form is described in **45**A, 182.

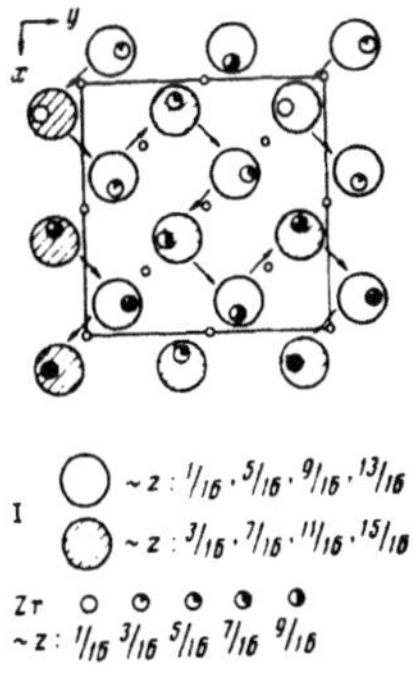

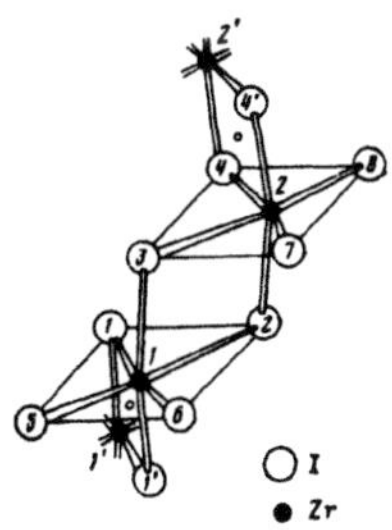

	x	y	z
Zr1	1243	8438	621
Zr2	4060	1260	1878
I1	8802	1125	599
I2	3564	1224	652
I3	1283	8934	1848
I4	1374	3697	1902
I5	8807	6323	686
I6	3706	6369	574
I7	6130	8788	1934
I8	6177	3686	1818

HAFNIUM(IV) IODIDE
HfI_4

Ž. Neorg. Khim., **31**, 1876-1879 [Russ. J. Inorg. Chem., **31**, 1080-1081].

$I\bar{1}$, 8.337, 8.337, 23.157, 90.58, 90.58, 90.09, Z = 8, R = 0.048. Isostructural with one of the forms of ZrI_4 (preceding report); compare the cubic and monoclinic descriptions (**24**, 298; **46**A, 198). Zigzag chains of edge-sharing HfI_6 octahedra; Hf-I = 2.67-2.68 (terminal), 2.85-2.86 and 3.00 (bridging) A.

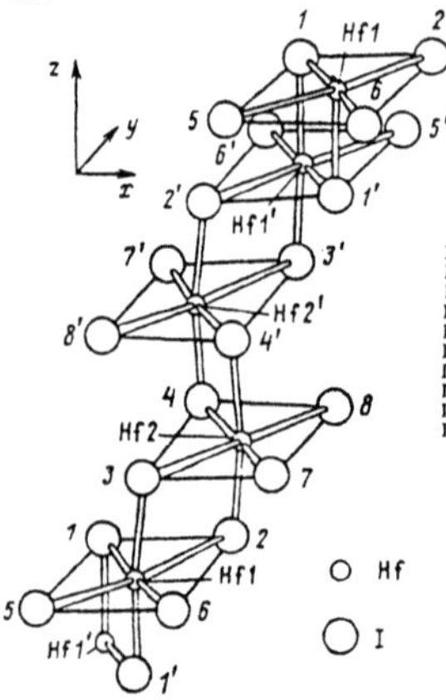

	x	y	z
Hf1	1235	8454	622
Hf2	4044	1257	1879
I1	8819	1119	597
I2	3560	1209	656
I3	1289	8944	1842
I4	1380	3688	1901
I5	8819	6343	691
I6	3686	6393	568
I7	6112	8820	1936
I8	6157	3678	1817

PLATINUM IODIDES
α- and β-PtI_2, Pt_3I_8

Z. anorg. Chem., **539**, 141-153.

α-PtI_2, cubic, a = 11.09 A. No structure determination.

β-PtI_2, $P2_1/c$, 6.588, 8.715, 6.889, 102.76, Z = 4. Two PtI_4 square-planes share an edge to form Pt_2I_6 groups, which share corners to give puckered layers. Pt-I = 2.58-2.61(1) A.

Pt_3I_8, $P4_12_12$, 11.664, 10.682, Z = 4. Three-dimensional framework of edge-sharing Pt(IV)I_6 octahedra and Pt(II)I_4 square-planes. Pt(IV)-I = 2.65-2.75, Pt(II)-I = 2.57-2.69(1) A.

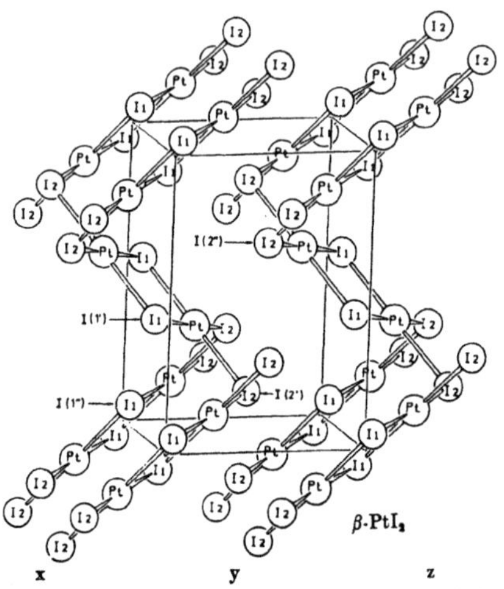

β-PtI_2

	x	y	z
Pt	0,2368	0,1275	−0,0038
I(1)	−0,0151	0,0783	0,2283
I(2)	0,4596	0,3273	0,2328

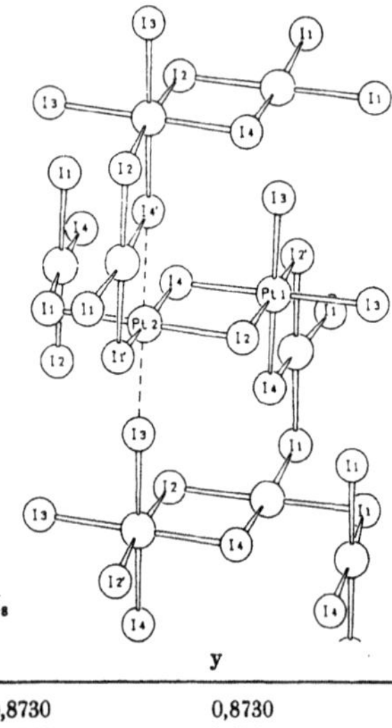

Pt_3I_8

	x	y	z
Pt(1)	0,8730	0,8730	0,0
Pt(2)	0,3746	0,8731	−0,0049
I(1)	0,3806	0,1050	−0,0157
I(2)	0,3752	0,6486	−0,0035
I(3)	0,1022	0,8749	0,0168
I(4)	0,6373	0,8717	0,0025

SAMARIUM DIIODIDE (HIGH-TEMPERATURE)
SmI_2

New Front. Rare Earth Sci. Appl., **1**, 388-341.

Pbca, 15.13, 8.15, 7.84, Z = 8, powder data. Layer structure with 7-coordinate Sm.

CAESIUM TETRAIODOTHALLATE(III)
$CsTlI_4$

Z. Naturforsch., **41**B, 269-272.

$P2_1/c$, 7.3110, 18.7042, 8.4770, 105.14, Z = 4, R = 0.049. Tetrahedral TlI_4^- anions linked by 11-coordinate Cs^+ cations. Tl-I = 2.736-2.770(3) A, I-Tl-I = 105-118°, Cs-I = 3.97-4.46 A.

CAESIUM ENNEAIODODICHROMATE(III)
$Cs_3Cr_2I_9$

J. Solid State Chem., **64**, 90-101.

Form I, $P6_3/mmc$, 8.044, 20.15, Z = 2, neutron powder data. Cs(1) in 2(b): 0,0,1/4; Cs(2) in 4(f): 1/3,2/3,0.0704; Cr in 4(f): 1/3,2/3,0.8292; I(1) in 6(h): -0.4978,2x,1/4; I(2) in 12(k): -0.1824,2x,0.0896. Isostructural with the chloride and bromide (**21**, 277; **37**A, 206), for which parameters are also given.

Form II, $R\bar{3}c$, 14.07, 20.82 A, Z = 6, neutron powder data. Cs in 18(e): 0.6732,0,1/4; Cr in 12(c): 0,0,0.8287; I(1) in 18(e): 0.1621,0,1/4; I(2) in 36(f): 0.4968,0.3057,0.4263. Dimeric anions as in form I. Both forms present in the same sample.

AMMONIUM SILVER IODIDE MONOHYDRATE
$(NH_4)_2AgI_3.H_2O$

Aust. J. Chem., **39**, 171-175.

Pbnm, 21.20, 10.84, 4.629, Z = 4, R = 0.032. Isostructural with K_2AgI_3 (**15**, 169; **41**A, 184), which might therefore also be a monohydrate. Polymeric chains along c of corner-sharing AgI_4 tetrahedra; Ag-I = 2.827-2.865(3) A, I-Ag-I = 105.9-111.9, Ag-I-Ag = 108.1^o. A refinement of the structure of $(NMe_4)Ag_2I_3$ (**28**, 392) is also described.

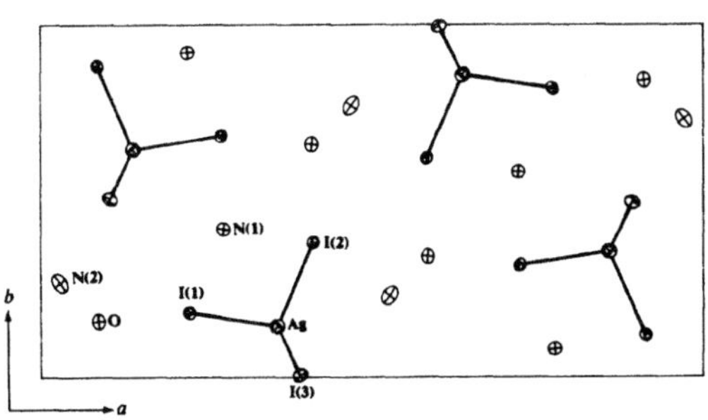

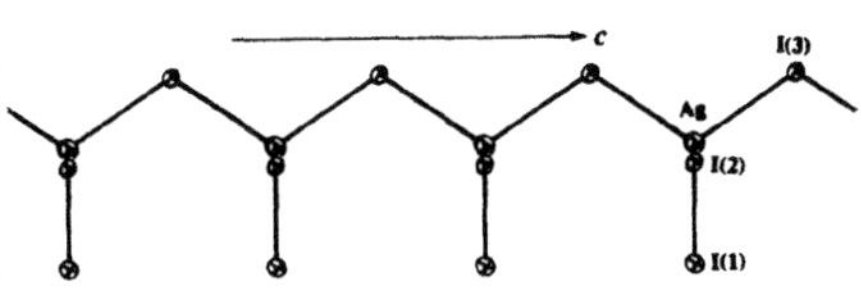

	x	y	z
Ag	0·35951	0·14433	1/4
I(1)	0·22586	0·18318	1/4
I(2)	0·41441	0·38198	1/4
I(3)	0·39460	0·00547	3/4
N(1)	0·2771	0·4198	−1/4
N(2)	0·0289	0·2687	1/4
O	0·0878	0·1590	−1/4

LEAD YTTRIUM OXYFLUORIDE
$Pb_8Y_6F_{32}O$

J. Solid State Chem., **64**, 148-161.

$R\bar{3}$, 10.817, 19.942, Z = 3, R = 0.059. Fluorite-related structure, with Y_6F_{36} clusters of YF_8 square antiprisms, and Pb coordination polyhedra derived from distorted PbF_8 cubes; Pb(2) forms Pb_6F_{32} blocks which enclose a cube of F ions with the oxide ion at its centre.

		x	y	z
Pb(1)	6(c)	0	0	0.26121
Pb(2)	18(f)	0.22784	0.19146	0.08341
Y	18(f)	0.08649	0.47275	0.08209
F(1)	18(f)	0.0357	0.2525	0.0414
F(2)	18(f)	0.4309	0.1386	0.1161
F(3)	18(f)	0.4751	0.3940	0.0365
F(4)	18(f)	0.2959	0.5008	0.0509
F(5)	18(f)	0.1255	0.3621	0.1679
F(6)	6(c)	0	0	0.1358
O	3(a)	0	0	0

POTASSIUM PENTAFLUOROPEROXOTITANATE
$K_3Ti(O_2)F_5$

Z. anorg. Chem., **537**, 175-188.

Fm3m, 8.836, Z = 4, R = 0.020. Disordered cryolite structure.

			x	y	z
Ti	4a	1	0	0	0
K1	24e	1/6	0,5298	0	0
K2	32f	1/4	0,2660	0,2660	0,2660
F1	96j	1/6	0,2127	0,034	0
F2	96k	1/24	0,210	0,029	0,029
O1	96j	1/12	0,193	0,116	0

AMMONIUM TETRAFLUORODIOXOVANADATE(V)
$(NH_4)_3VO_2F_4$

J. Solid State Chem., **65**, 260-264.

Immm (or I222), 9.126, 18.818, 6.264, Z = 6, R = 0.07. Two independent octahedral anions, one with cis-oxo ligands, the other with disordered O/F sites. Cryolite type above 418K.

		S.O.F.	x	y	z
V(1)	2a	1.00	0.0	0.0	0.0
V(2)	4g	1.00	0.0	0.3146	0.0
O(1)	8l	1.00	0.0	0.2569	0.2090
F(1)	8n	1.00	0.2027	0.3254	0.0
F(2)	8l	1.00	0.0	0.3965	0.2090
O/F(3)	4g	0.50	0.0	0.0992	0.0
O/F(4)	16o	0.125	0.0347	0.0949	0.1005
O/F(5)	4e	0.50	0.2090	0.0	0.0
O/F(6)	8m	0.25	0.1454	0.0	0.2063
O/F(7)	4i	0.50	0.0	0.0	0.2948
O/F(8)	16o	0.125	0.1425	0.0362	0.1868
N(1)	4h	1.00	0.0	0.1357	0.5
N(2)	8n	1.00	0.2710	0.1702	0.0
N(3)	16o	0.25	0.2009	0.5	0.0607
N(4)	4e	0.50	0.4287	0.0	0.0

SODIUM PENTAFLUOROXOVANADATE(V)
Na_2VF_5O

POTASSIUM catena-mu-FLUORO-DIFLUORODIOXOVANADATE(V)
$K_2VF_3O_2$

Acta Chem. Scand., A**40**, 325-330.

Na, Fm3m, 8.454, Z = 4, R = 0.052. 4V in 24(e): 0.0315,0,0; 4Na(1) in 4(b): 1/2,0,0; 4Na(2) in 8(c): 1/4,1/4,1/4; 24F,O in 24(e): 0.2216,0,0. Disordered $V(F,O)_6$ distorted octahedra, linked by Na ions.

K, Pnma, 7.383, 5.646, 11.426, Z = 4, R = 0.023. Infinite chains of cis-F-bridged VF_4O_2 octahedra, linked by K ions, as previously described (**37**A, 185).

AMMONIUM PENTAFLUOROPEROXOZIRCONATE(IV)
$(NH_4)_3Zr(O_2)F_5$

Z. anorg. Chem., **535**, 135-142.

Fm3m, 9.4401, Z = 4, R = 0.014. 4 Zr in 4(a): 4 N(1) in 4(b) (or 24(e): x = 0.494); 8 N(2) in 8(c); 16 F(1) in 96(j): 0.2117,0.0451,0; 4 F(2) in 96(k): 0.206,0.040,0.040; 8 O in 96(j): 0.1841,0.1177,0. Elpasolite structure, as for the Ti compound (**8**, 163; **43**A, 347; **44**A, 167), with a new disorder model.

CHROMIUM OXIDE TRIFLUORIDE
$CrOF_3$

Inorg. Chem., **25**, 426-427.

$P4_1$, 5.174, 8.921, Z = 4, R = 0.053. Three-dimensional array of corner-sharing $CrOF_5$ octahedra.

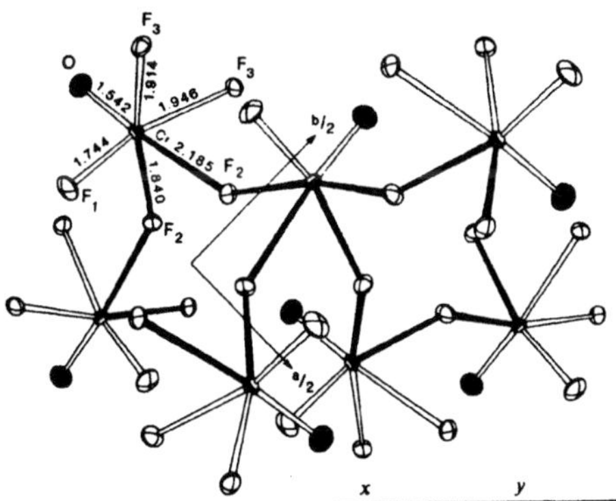

Interatomic Distances (Å) and Angles (deg) for Chromium(V) Oxide Trifluoride

Cr–O	1.542 (5)	Cr–F(3)[a]	1.914 (3)
Cr–F(1)	1.744 (4)	Cr–F(3)	1.946 (3)
Cr–F(2)	1.840 (3)	Cr–F(2)[b]	2.185 (3)
O–Cr–F(1)	101.7 (2)	F(2)–Cr–F(3)[a]	165.4 (2)
O–Cr–F(2)	98.3 (2)	F(2)–Cr–F(3)	88.8 (2)
O–Cr–F(3)[a]	96.1 (2)	F(2)–Cr–F(2)[b]	83.4 (2)
O–Cr–F(3)	96.3 (2)	F(3)–Cr–F(3)[a]	87.1 (2)
O–Cr–F(2)[b]	174.8 (2)	F(3)–Cr–F(2)[a,b]	82.1 (1)
F(1)–Cr–F(2)	91.0 (2)	F(3)–Cr–F(2)[b]	78.9 (1)
F(1)–Cr–F(3)[a]	88.5 (2)	Cr–F(2)–Cr[c]	147.3 (2)
F(1)–Cr–F(3)	161.8 (2)	Cr–F(3)–Cr[d]	144.3 (2)
F(1)–Cr–F(2)[b]	83.1 (2)		

[a] y, −1 − x, 1/4 + z. [b] y, −x, 1/4 + z. [c] −y, x, −1/4 + z. [d] −1 − y, x, −1/4 + z.

	x	y	z
Cr	−0.1096	−0.4171	0.00000
F(1)	0.1563	−0.4272	0.1201
F(2)	0.0617	−0.1960	−0.1274
F(3)	−0.4333	−0.3136	−0.0934
O	−0.0745	−0.6686	−0.0905

SODIUM TRIOXOFLUOROMOLYBDATE(VI)
$NaMoO_3F$

Rev. Chim. Minér., **23**, 27–34; J. Solid State Chem., **65**, 27–39.

$P2_1/m$, 7.138, 3.715, 6.510, 110.08, Z = 2, R = 0.054. Chains of edge-sharing $Mo(O,F)_6$ octahedra, linked by octahedrally-coordinated Na ions. Nmr and Raman data, and energy calculations suggest ordering of O and F.

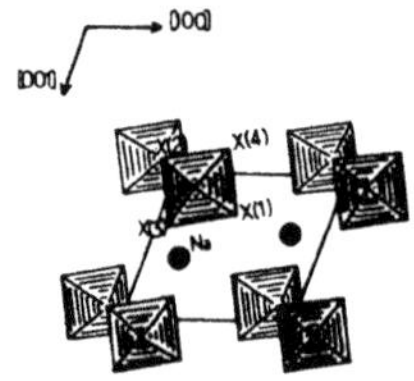

COPPER(II) DIFLUOROTRIOXOTUNGSTATE(VI)
$CuWO_3F_2$

Z. anorg. Chem., **539**, 165–174.

$P2_1/m$, 5.223, 9.599, 3.670, 106.26, Z = 2, R = 0.019. Chains along **c** of WO_4F_2 octahedra sharing opposite O corners (O/F are probably ordered, with X(1) = F); chains are linked by chains of edge-sharing CuO_2F_4 distorted octahedra. W–F = 2.09, W–O = 1.75–1.90, Cu–F = 1.96, 2.53, Cu–O = 1.90 A.

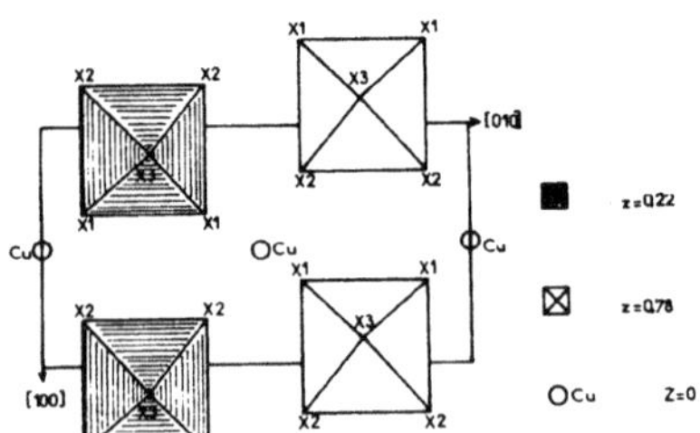

		X	Y	Z
2(e)	W	0,0369	0,25	0,2216
2(b)	Cu	0,5	0,0	0,0
4(f)	X(1)	0,3562	0,3902	0,3485
4(f)	X(2)	−0,1856	0,1100	0,1282
2(e)	X(3)	0,1198	0,25	−0,2498

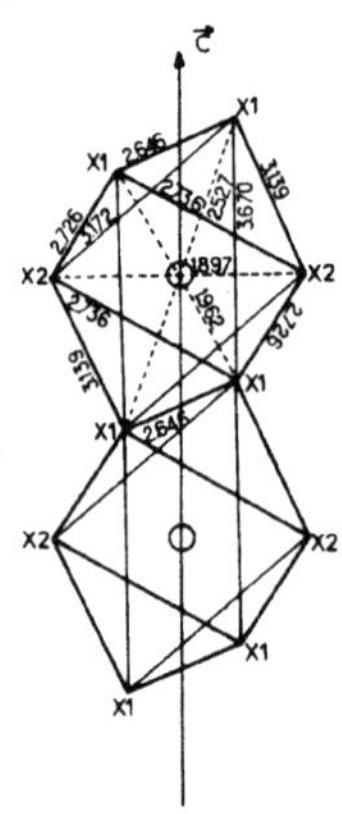

POTASSIUM PENTAFLUORODIOXOURANATE(VI)
$K_3UO_2F_5$

Radiokhimija, **28**, 682–685.

$I4_1/a$, 9.159, 18.170, Z = 8, R = 0.043. Structure as previously described (**18**, 405), with pentagonal bipyramidal $UO_2F_5^{3-}$ ions linked by K^+ ions.

THALLIUM URANIUM OXYFLUORIDE
TlU_3OF_{11}

Rev. Chim. Minér., **23**, 35-47.

Cm, 14.051, 8.106, 8.389, 90.00, R = 0.060. (001) double layers of edge-and corner-sharing UOF_8 distorted tricapped trigonal prisms, with Tl^+ ions in channels; ordered O/F distribution.

CUMENGEITE
$Pb_{21}Cu_{20}Cl_{42}(OH)_{40}$

Miner. Mag., **50**, 157-162.

I4/mmm, 15.065, 24.436, Z = 2, R = 0.071. Five Pb sites (one half-occupied) with various types of coordination, and two Cu sites with octahedral and square-pyramidal coordinations. The coordination polyhedra link to form columns along **c**.

	x	y	z
Pb(1)	0	0	0
Pb(2)	0	0	0.3199
Pb(3)	0.27634	0	0.23459
Pb(4)	0.19780	0.19780	0.36453
Pb(5)†	1/2	0	0.4104
Cu(1)	0.2599	0.0962	0.0910
Cu(2)	0.2610	0.2610	0
Cl(1)	0.1318	0.1318	0
Cl(2)	0.3815	0.1269	0.1786
Cl(3)	0	0	0.1187
Cl(4)	0.1696	0	0.3765
Cl(5)	0.1274	0.1274	0.2461
Cl(6)	0.3701	0	1/2
O(1)	0.3007	0.1762	0.4412
O(2)	0.1850	0.1850	0.1275
O(3)	0.2120	0	0.1357
O(4)	0.3272	0	0.0571

† position half-occupied

BISMUTH CHLORONIOBATE and TUNGSTATE
Bi_4NbO_8Cl (I), $PbBi_3WO_8Cl$ (II)

J. Solid State Chem., **62**, 92-104.

I. $P2_1cn$, 5.493, 5.496, 28.75, Z = 4, R = 0.095. Bi_2O_2, NbO_4, and Cl layers along **c**.

II. P4, 3.894, 14.34, Z = 1, R = 0.053. M_2O_2, WO_4, and Cl layers (M = 3/4Bi + 1/4Pb).

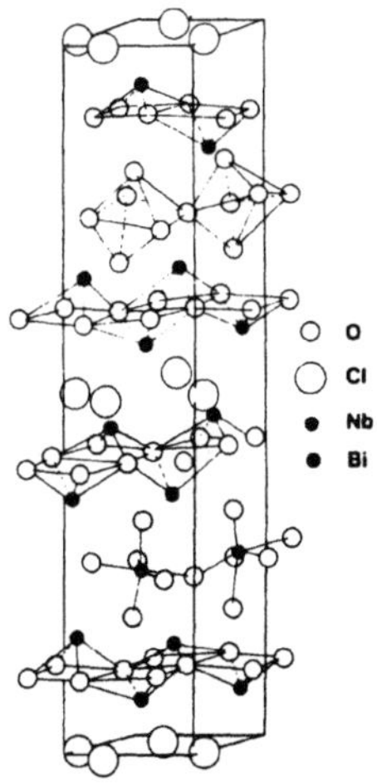

The structure of Bi_4NbO_8Cl, Nb omitted in upper half of cell.

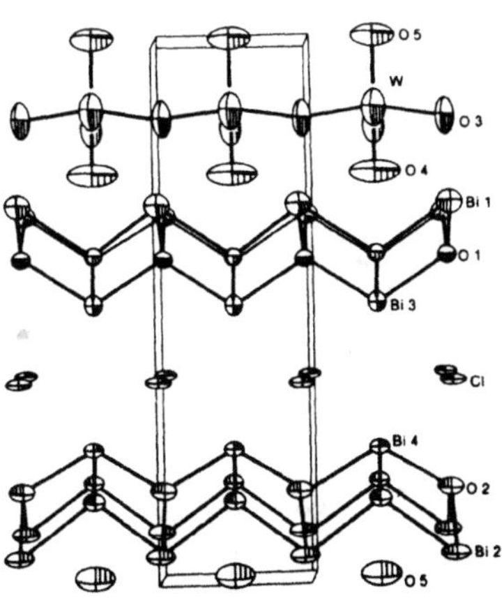

The structure of $PbBi_3WO_8Cl$; cell outlined.

IRON(III) OXIDE CHLORIDE
FeOCl

J. Amer. Chem. Soc., **108**, 7946-7951.

Pmnm, 3.7730, 7.9096, 3.3010, Z = 2, neutron powder data. Fe in 2(b): 1/4,0.11555,3/4; O in 2(a): 1/4,-0.04795,1/4; Cl in 2(a): 1/4,0.32961,1/4. Structure as previously described (**3**, 67, 376; **35**A, 181).

PALLADIUM(II) OXIDE CHLORIDE
Pd_2OCl_2

Z. Naturforsch., **41**B, 1363-1366.

$I4_1/amd$, 6.313, 9.872, Z = 4, R = 0.028. Pd in 8(c): 0,0,0; O in 4(a): 0,3/4,1/8; Cl in 8(e): 0,1/4,0.1722. Bands of edge-sharing PdO_2Cl_2 square planes, connected by sharing O atoms. Pd-Cl = 2.320, Pd-O = 2.003 Å.

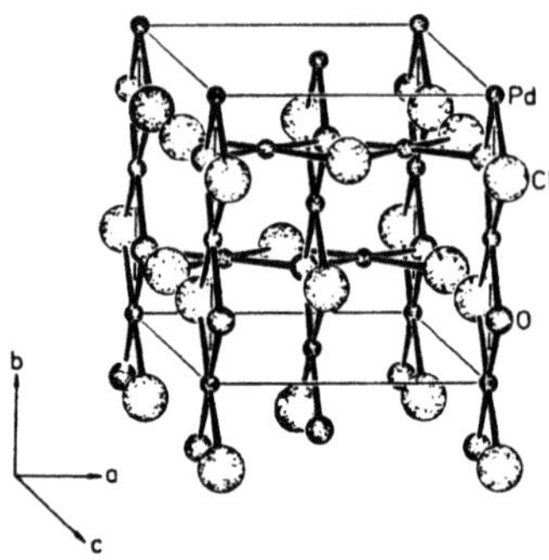

ATACAMITE
$Cu_2Cl(OH)_3$

Acta Cryst., C**42**, 1277-1280.

Pnma, 6.030, 6.865, 9.120, Z = 4, R = 0.029. Structure as previously described (**12**, 220), with edge-sharing $Cu(OH)_4Cl_2$ and $Cu(OH)_5Cl$ distorted octahedra. The relationship to the spinel structure is discussed.

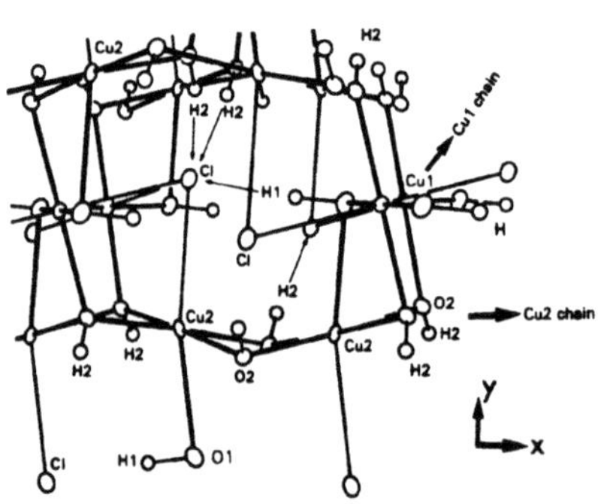

(× 10^4)

	x	y	z
Cu(1)	0	0	0
Cu(2)	1906	¼	2553
Cl	3518	¼	556
O(1)	1498	¼	−18
O(2)	4406	651	2879
H(1)	3049	¼	−148
H(2)	4331	−334	2279

Selected bond distances (Å) *and angles* (°) *for atacamite*

O(1)–Cu(1)	1·940 (2)	O(2)–Cu(2)	1·993 (2)
O(1^{i})–Cu(1)	1·940 (2)	O(2^{vi})–Cu(2)	1·993 (2)
O(2^{ii})–Cu(1)	2·017 (2)	O(2^{vii})–Cu(2)	2·010 (2)
O(2^{iii})–Cu(1)	2·017 (2)	O(2^{viii})–Cu(2)	2·010 (2)
Cl^{iv}–Cu(1)	2·776 (1)	O(1)–Cu(2)	2·358 (4)
Cl^{v}–Cu(1)	2·776 (1)	Cl^{ix}–Cu(2)	2·750 (1)
O(1)–Cu(1)–O(1^{i})	180	O(2)–Cu(2)–O(2^{vi})	79·1 (1)
O(1)–Cu(1)–O(2^{ii})	96·0 (1)	O(2)–Cu(2)–O(2^{vii})	177·3 (1)
O(1)–Cu(1)–O(2^{iii})	84·0 (1)	O(2)–Cu(2)–O(2^{viii})	101·3 (1)
O(1)–Cu(1)–Cl^{iv}	101·1 (1)	O(2)–Cu(2)–O(1)	103·1 (1)
O(1)–Cu(1)–Cl^{v}	78·9 (1)	O(2)–Cu(2)–Cl^{ix}	85·5 (1)
O(1^{i})–Cu(1)–O(2^{ii})	84·0 (1)	O(2^{vi})–Cu(2)–O(2^{vii})	101·2 (1)
O(1^{i})–Cu(1)–O(2^{iii})	96·0 (1)	O(2^{vi})–Cu(2)–O(2^{viii})	177·3 (1)
O(1^{i})–Cu(1)–Cl^{iv}	78·9 (1)	O(2^{vi})–Cu(2)–O(1)	103·1 (1)
O(1^{i})–Cu(1)–Cl^{v}	101·1 (1)	O(2^{vi})–Cu(2)–Cl^{ix}	85·5 (1)
O(2^{ii})–Cu(1)–O(2^{iii})	180	O(2^{vii})–Cu(2)–O(2^{viii})	78·3 (1)
O(2^{ii})–Cu(1)–Cl^{iv}	84·4 (1)	O(2^{vii})–Cu(2)–O(1)	74·1 (1)
O(2^{ii})–Cu(1)–Cl^{v}	95·6 (1)	O(2^{vii})–Cu(2)–Cl^{ix}	97·2 (1)
O(2^{iii})–Cu(1)–Cl^{iv}	95·6 (1)	O(2^{viii})–Cu(2)–O(1)	74·1 (1)
O(2^{iii})–Cu(1)–Cl^{v}	84·4 (1)	O(2^{viii})–Cu(2)–Cl^{ix}	97·2 (1)
Cl^{iv}–Cu(1)–Cl^{v}	180	O(1)–Cu(2)–Cl^{ix}	168·7 (1)
Hydrogen bonds			
H(1)···O(1)	0·95 (8)	H(2)···O(2)	0·87 (5)
H(1)···Cl^{x}	2·10 (5)	H(2)···Cl^{xi}	2·22 (5)
O(1)···Cl^{x}	3·044 (2)	O(2)···Cl^{xi}	3·075 (3)
O(2)–H(2)···Cl^{x}	167 (4)	O(2)–H(2)···Cl^{x}	168 (4)

Symmetry operators: (i) *px, py, pz*; (ii) $\frac{1}{2}-x$, *py*, $z-\frac{1}{2}$; (iii) $x-\frac{1}{2}$, *y*, $\frac{1}{2}-z$; (iv) *x*, $y-1$, *z*; (v) $-x$, $1-y$, *pz*; (vi) *x*, $\frac{1}{2}-y$, *z*; (ix) $\frac{1}{2}-x$, $y-\frac{1}{2}$, $z+\frac{1}{2}$; (x) $1-x$, $1-y$, $-z$; (xi) *x*, $y-1$, *z*.

NEODYMIUM and GADOLINIUM OXYCHLORIDE
NdOCl, GdOCl

Z. anorg. Chem., **533**, 181-185.

P4/nmm, a = 4.0249, 3.9495, c = 6.7837, 6.6708 A, Z = 2, R = 0.064, 0.053. Nd or Gd in 2(c): 0,1/2,z, z = 0.1749, 0.1720; O in 2(a): 0,0,0; Cl in 2(c): z = 0.6292, 0.6271. PbFCl-type structures (**2**, 45; **3**, 64), as for LaOCl (**11**, 312; **50**A).

CAESIUM NEPTUNYL(V,VI) CHLORIDE
$Cs_7(NpO_2)_3Cl_{12}$

J. Chem. Soc., Dalton, 1403-1404 (1986).

$I\bar{4}3d$, 15.084, Z = 4, R = 0.035. Np in 12(a): 1/8,1/2,1/4; Cs(1) in 16(c): x,x,x, x = 0.1020; Cs(2) in 12(b): 7/8,1/2,3/4; O in 24(d): 0.2346,1/2,1/4; Cl in 48(e): 0.1304,0.5598,0.0824. Trans-octahedral $NpO_2Cl_4{}^{n-}$ anions (disorder of two Np(VI) and one Np(V)), linked by 9- and 8-coordinate Cs^+ ions. Np-O = 1.790(15), Np-Cl = 2.685(5), Cs(1)-O = 3.17, Cs(1)-Cl = 3.73, 3.74, Cs(2)-Cl = 3.54 A.

CAESIUM AQUODICHLORODIOXONEPTUNATE(V)
$Cs[NpO_2Cl_2(H_2O)]$

Radiokhimija, **28**, 695-701.

$P2_1/n$, 11.71, 6.99, 8.76, $\alpha = 93.1^o$, Z = 4, R = 0.13. Infinite chain anions, with Np having distorted pentagonal bipyramidal coordination, linked by Cs^+ cations.

BISMUTH OXIDE BROMIDE
BiOBr

Acta Cryst., C**42**, 1098-1099.

P4/nmm, 3.9233, 8.105, Z = 2, R = 0.048. Bi in 2(c): 1/4,1/4,0.1540; O in 2(a): 1/4,3/4,0, Br in 2(c): 1/4,1/4,0.6568 [origin at 2/m]. BiOCl (PbFCl) structure type (**3**, 64, 370). Bi - 4 O = 2.325, Bi - 4 Br = 3.170 A.

LANTHANUM OXYBROMIDE
LaOBr

SAMARIUM OXYBROMIDE
SmOBr

Mater. Res. Bull., **21**, 1291-1294.

P4/nmm, a = 4.159, 3.945, c = 7.392, 7.904, Z = 2, R = 0.056, 0.085. Ln in 2(c):1/2,0,z, z = 0.1622, 0.1464; O in 2(a): 0,0,0; Br in 2(c): z = 0.6349, 0.6550. PbFCl-type structures, as previously described (**30**A, 289).

SCANDIUM CHLORIDE HYDRIDE
ScClH

Inorg. Chem., **25**, 4811-4818.

$R\bar{3}m$, a = 3.4785, c = 26.531 A, Z = 6, R = 0.090. Sc, Cl in 6(c): 0,0,z, z = 0.1192, 0.3912; H assumed to be in centre of Sc tetrahedra. ZrCl-type structure (**42**A, 186).

ZIRCONIUM BROMIDE HYDRIDE
Zr_2Br_2D

Inorg. Chem., **25**, 4709-4714.

C2/m, 19.437, 3.5253, 5.9036, 100.98, at 14K, Z = 4, neutron powder data. Atoms in 4(i): x,0,z, x = 0.1815, 0.3154, 0.0851, 0.9168, 0.7088, z = 0.954, 0.547, 0.221, 0.283, 0.621 for Zr(1), Zr(2), Br(1), Br(2), D, respectively. Br-Zr-D-D-Zr-Br layers, with D in zigzag chains of distorted Zr tetrahedra.

SCANDIUM CHLORIDE BORIDES and NITRIDES
$Sc_7Cl_{12}B$, $Sc_7Cl_{12}N$, Sc_4Cl_6B, Sc_4Cl_6N

J. Solid State Chem., **64**, 331-346.

$Sc_7Cl_{12}B$, $Sc_7Cl_{12}N$, $R\bar{3}$, a = 13.014, 12.990, c = 8.899, 8.835, Z = 3, R = 0.032, 0.042. Structures similar to that of Sc_7Cl_{12} (**49**A), with B or N in the centres of Sc_6Cl_{12} clusters which are linked by isolated Sc(I) ions.

Sc_4Cl_6B, Sc_4Cl_6N (2 samples), Pbam, a = 11.741, 11.634, 11.625, b = 12.187, 12.144, 12.118, c = 3.599, 3.550, 3.545 A, Z = 2, R = 0.035, 0.084, 0.077. Infinite chains of $Sc_6Cl_{12}X$ clusters (X = B, N) sharing trans Sc-Sc edges, similar to the chains in $NaMo_4O_6$ (**48**A, 226).

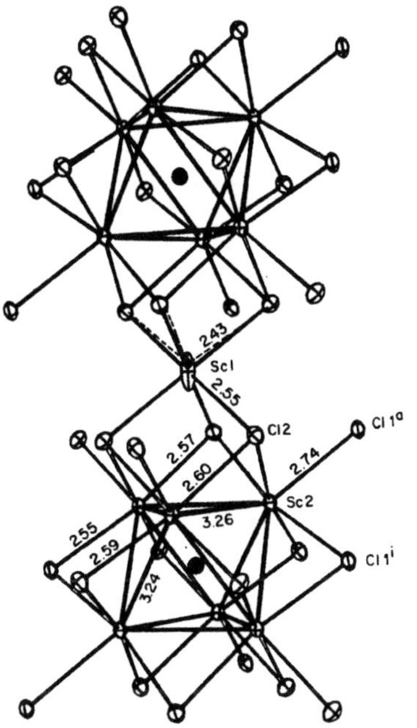

Two $Sc_6Cl_{12}N$ clusters, the bridging Cl^a atoms from other clusters and the intervening Sc(1) in $Sc_7Cl_{12}N$. (The threefold axis along *c* is vertical.)

	x	*y*	*z*
$Sc_7Cl_{12}B$			
Sc1	0.0	0.0	0.5
Sc2	0.1630	0.0439	0.1502
Cl1	0.3106	0.2295	0.0058
Cl2	0.1292	0.1791	0.3354
B	0.0	0.0	0.0
$Sc_7Cl_{12}N$			
Sc1	0.0	0.0	0.5
Sc2	0.1616	0.0438	0.1492
Cl1	0.3104	0.2292	0.0051
Cl2	0.1300	0.1803	0.3351
N	0.0	0.0	0.0

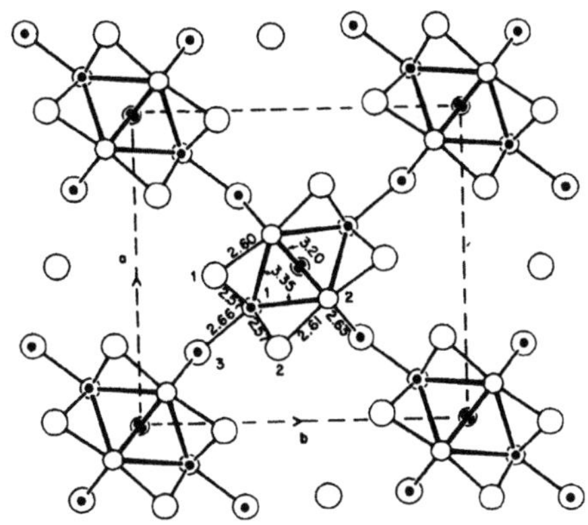

Projection of Sc_4Cl_6B along $[00\bar{1}]$

	x	*y*	*z*
Sc_4Cl_6B			
Sc1	0.3745	0.3521	0.5
Sc2	0.3948	0.5830	0.0
Cl1	0.4794	0.2407	0.0
Cl2	0.2400	0.4297	0.0
Cl3	0.2326	0.1814	0.5
B	0.5	0.5	0.5
$Sc_4Cl_6N(1)$			
Sc1	0.3770	0.3522	0.5
Sc2	0.3950	0.5799	0.0
Cl1	0.4844	0.2409	0.0
Cl2	0.2401	0.4270	0.0
Cl3	0.2370	0.1781	0.5
N	0.5	0.5	0.5
$Sc_4Cl_6N(2)$			
Sc1	0.3767	0.3537	0.5
Sc2	0.3970	0.5803	0.0
Cl1	0.4828	0.2416	0.0
Cl2	0.2404	0.4294	0.0
Cl3	0.2340	0.1790	0.5
N	0.5	0.5	0.5

SCANDIUM CARBIDE IODIDE and BROMIDE
$Sc_7I_{12}C$, $Sc_7Br_{12}C$

Inorg. Chem., 25, 3434-3438.

R3, a = 14.717, 13.628, c = 9.847, 9.203 A, Z =3, R = 0.045, 0.034. $Sc_6X_{12}C$ clusters, linked by an additional Sc.

	x	*y*	*z*
I(1)	0.0773	0.3065	0.0008
I(2)	0.1759	0.051	0.66735
I(3)	0.3063	0.2269	0.0034
I(4)	0.1339	0.1888	0.3317
Sc(1)	0	0	0.4752
Sc(2)	0.1045	0.1462	0.8708
Sc(3)	0.1419	0.0406	0.1374
C	0	0	−0.004

Br(1)	0.0772	0.3088	0
Br(2)	0.1764	0.0495	0.6668
Br(3)	0.3100	0.2276	0.0062
Br(4)	0.1336	0.1882	0.3344
Sc(1)	0	0	0.477
Sc(2)	0.1124	0.1566	0.865
Sc(3)	0.1518	0.0425	0.149
C	0	0	0.003

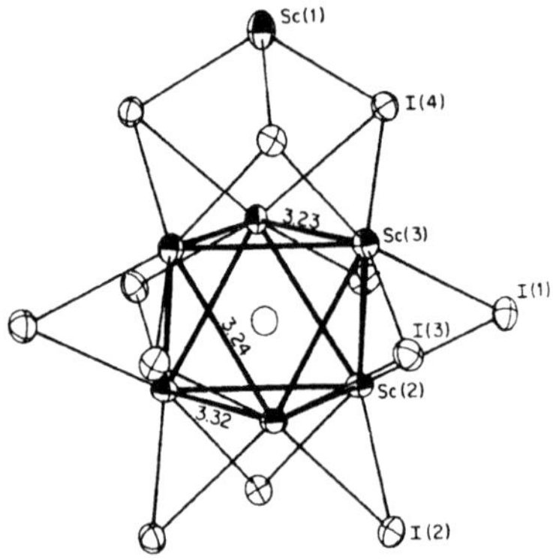

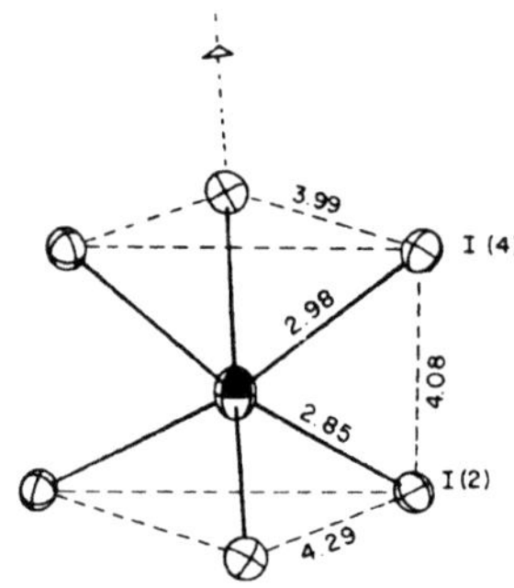

SCANDIUM CARBIDE CHLORIDE
SCANDIUM NITRIDE CHLORIDE
Sc_2Cl_2C, Sc_2Cl_2N

Inorg. Chem., **25**, 283-287.

$P\bar{3}m1$, a = 3.3997, 3.3495, c = 8.858, 8.808 A, Z = 1, R = 0.032, 0.050. Sc in 2(d): 1/3,2/3,z, z = 0.3629, 0.3656; Cl in 2(d): z = 0.8249, 0.8242; C or N in 1(b): 0,0,1/2. Isostructural with Ta_2S_2C (**38**A, 66), with slabs of Cl-Sc-Sc-Cl hcp layers, as in ZrCl and ZrBr (**42**A, 186; **43**A, 139), and with C or N in octahedral interstices. A 3R-form (**38**A, 66) of Sc_2Cl_2C is also found.

GADOLINIUM NITRIDE CHLORIDE
Gd_3Cl_6N

J. Less-Common Metals, **116**, 279-292.

$P\bar{1}$, 7.107, 8.156, 9.707, 75.37, 109.10, 114.64, Z = 2, R = 0.03. Gd_6N_2 units of two edge-sharing Gd_4N tetrahedra; Gd ions have tricapped trigonal prismatic coordinations.

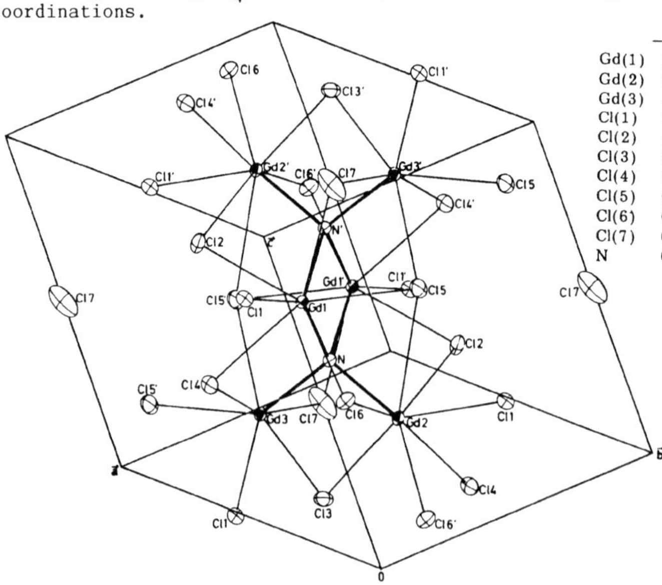

	x	y	z
Gd(1)	0.25905	0.41670	0.37381
Gd(2)	0.72050	0.57234	0.22018
Gd(3)	0.58850	0.91553	0.31231
Cl(1)	0.8246	0.2821	0.2025
Cl(2)	0.0	0.5	0.5
Cl(3)	0.7109	0.8981	0.0853
Cl(4)	0.1638	0.7397	0.2084
Cl(5)	0.4687	0.1772	0.3951
Cl(6)	0.3013	0.4166	0.0898
Cl(7)	0.0	0.0	0.5
N	0.5921	0.6330	0.3763

AMMONIUM NIOBIUM NITRIDE BROMIDE
$(NH_4)_3Nb_2NBr_{10}$

AMMONIUM TANTALUM NITRIDE IODIDE
$(NH_4)_3Ta_2NI_{10}$

Z. Naturforsch., **41**B, 423-428.

Nb compound, P4/mnc, 7.278, 19.328, Z = 2, R = 0.031. NH_4(1) in 2(b): 0,0,1/2; NH_4(2) in 4(d): 0,1/2,1/4; Nb in 4(e): 0,0,0.0955; Br(1) in 16(i): 0.1869,0.2919,0.1047; Br(2) in 4(e): 0,0,0.2387; N in 2(a): 0,0,0.

Ta compound, I4/mmm, 7.625, 19.902, Z = 2, R = 0.19. NH_4(1) in 2(b): 0,0,1/2; NH_4(2) in 4(d): 0,1/2,1/4; Ta in 4(e): 0,0,0.0928; 0.5 I(1) in 32(o): 0.190,0.304,0.0975; I(2) in 4(e): 0,0,0.2382; N in 2(a): 0,0,0.

D_{4h}-symmetry anions of two MX_5N octahedra sharing the N atom. M-N = 1.85, M-X = 2.53, 2.77; 2.74, 2.89 A.

CALCIUM PHOSPHIDE CHLORIDES
$Ca_{1.85}P_{0.85}Cl_{1.15}$ (I), Ca_3PCl_3 (II)

J. Less-Common Metals, **124**, 93-103.

I, Fm3m, 5.82, Z = 2, powder data. 3.7 Ca in 4(b): 1/2,1/2,1/2; 1.7 P + 2.3 Cl in 4(a): 0,0,0. NaCl-type structure.

II, Pm3m, 5.683, Z = 1, powder data. Ca in 3(d): 1/2,0,0; P in 1(a): 0,0,0; Cl in 3(c): 0,1/2,1/2. Mg_3NF_3-type structure (**34**A, 189; **35**A, 153).

COPPER PHOSPHIDE IODIDE
$Cu_2P_3I_2$

J. Solid State Chem., **65**, 178-189.

$P2_1/c$, 15.343, 12.925, 15.260, 116.38, Z = 16, R = 0.068. Tubes of condensed four- and five-membered rings of P atoms, surrounded by partially-occupied Cu sites, which are in turn surrounded by iodine atoms which form channels.

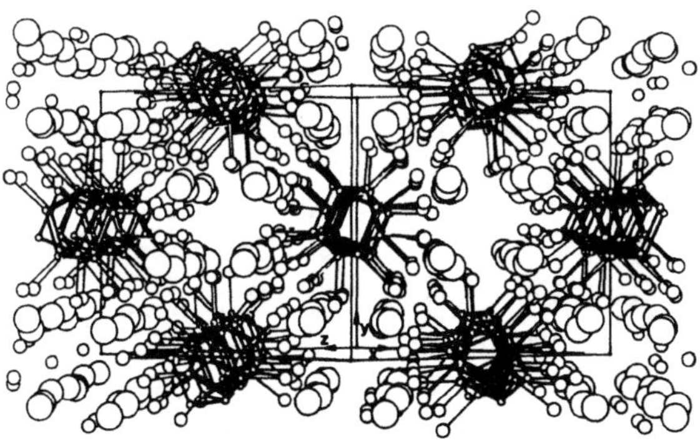

EUROPIUM PHOSPHIDE HALIDES
Eu_2PX (X = Cl, Br, I)

J. Less-Common Metals, **123**, 25-35.

Cl, Fm3m, 6.106, Z = 2, R = 0.070. Eu in 4(a): 0,0,0; P+Cl in 4(b): 1/2,1/2,1/2. NaCl-type structure.

Br, I, $R\bar{3}m$, a = 4.385, 4.445, c = 21.90, 23.126 A, Z = 3, R = 0.045, 0.058. Eu in 6(c): 0,0,z, z = 0.23728, 0.23266; P in 3(b): 0,0,1/2; Br, I in 3(a): 0,0,0. Anti-$NaFeO_2$-type structures (**3**, 75, 392).

ANTIMONY THIOBROMIDE
SbSBr

Z. Kristallogr., **175**, 305-315.

Pnam, 8.2424, 9.7556, 3.9680, Z = 4, R = 0.089. Sb, S, Br in 4(c): x,y,1/4, x = 0.1210, 0.8381, 0.5135, y = 0.1323, 0.0477, 0.8230. Sbis coordinated to 4 Br and 3 S, with the polyhedra linked into chains of pairs along **c**; each chain is connected to four neighbouring chains.

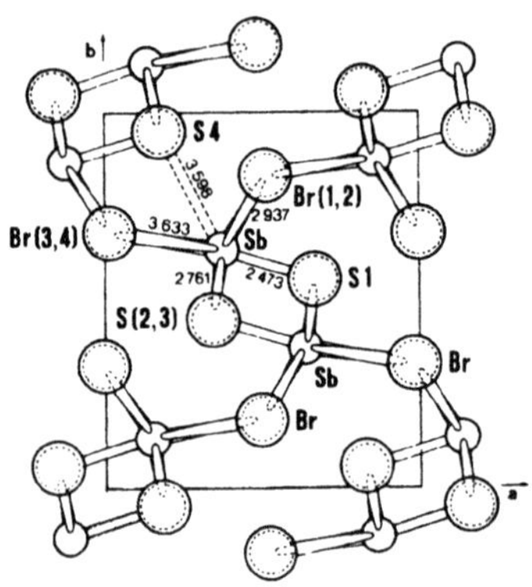

Bond distances (Å) and selected bond angles (°)

Sb – S(1)	2.473(8)	S(1) – Sb – S(2)	84.24(9)
– S(2)	2.671(5)	– Br(1)	81.67(9)
– S(3)	2.671(5)	– Br(3)	145.25(7)
Sb – Br(1)	2.937(3)	S(2) – Br – S(3)	95.95(9)
– Br(2)	2.937(3)	– Br(3)	112.34(9)
– Br(3)	3.633(4)	– Br(4)	64.46(8)
– Br(4)	3.633(4)	– Br(2)	87.84(8)
		– Br(1)	164.97(9)
		Br(3) – Sb – Br(4)	66.19(6)
		– Br(2)	127.28(8)
		– Br(1)	82.42(8)
		Br(2) – Sb – Br(1)	84.97(9)

NIOBIUM SELENOBROMIDE
$Nb_6Se_{20}Br_6$

J. Solid State Chem., **61**, 90-97.

$P2_1/c$, 19.370, 13.026, 13.598, 90.062, Z = 4, R = 0.039. Chains along **a**.

	x	y	z
Br1	0.0434	0.1577	0.1085
Br2	0.2467	0.1608	0.1018
Br3	0.5431	0.1752	0.6356
Br4	0.7441	0.1748	0.6448
Br5	0.1475	0.0901	0.3070
Br6	0.6436	0.0979	0.4419
Nb1	0.0447	0.2298	0.2940
Nb2	0.2468	0.2363	0.2852
Nb3	0.3956	0.2484	0.3656
Nb4	0.5436	0.2444	0.4496
Nb5	0.7471	0.2385	0.4579
Nb6	0.8961	0.2358	0.3802
Se1	0.1421	0.3533	0.2171
Se2	0.1467	0.3335	0.3872
Se3	0.3066	0.4046	0.3468
Se4	0.3500	0.3368	0.2026
Se5	0.2897	0.1614	0.4539
Se6	0.3251	0.0805	0.3087
Se7	0.4391	0.3475	0.5245
Se8	0.4827	0.4066	0.3763
Se9	0.4667	0.0874	0.4310
Se10	0.5033	0.1603	0.2824
Se11	0.6479	0.3369	0.3519
Se12	0.6456	0.3661	0.5198
Se13	0.8223	0.0765	0.4470
Se14	0.7891	0.1484	0.2957
Se15	0.8108	0.3970	0.3873
Se16	0.8519	0.3384	0.5366
Se17	0.9385	0.3264	0.2165
Se18	0.9836	0.3927	0.3607
Se19	0.9688	0.0709	0.3236
Se20	0.0059	0.1554	0.4667

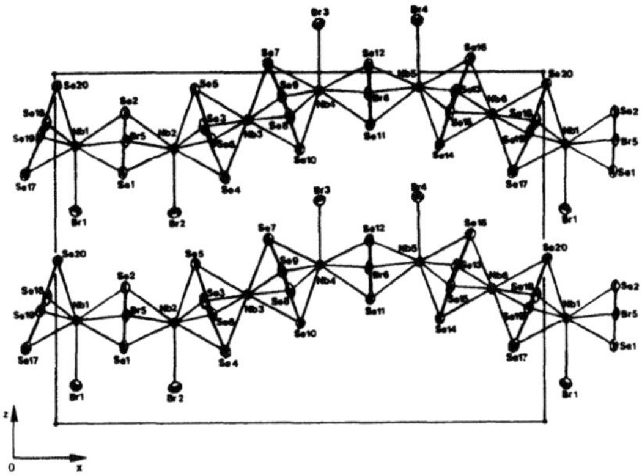

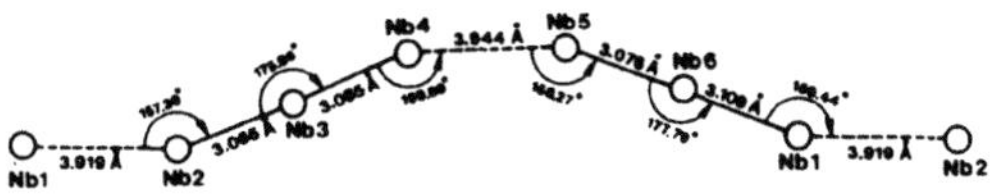

SILVER GERMANIUM SULPHIDE IODIDE
Ag_7GeS_5I

Rev. Chim. Minér., **23**, 796-809 (1986, publ. 1987).

$F\bar{4}3m$, 10.7116 A at 173K, Z = 4, R = 0.075. Partial ordering of Ag relative to room-temperature structure.

LEAD(II) MERCURY(II) THIOIODIDE
$Hg_2PbI_2S_2$

Acta Cryst., C**42**, 515-517.

P4/mbm, 13.501, 4.593, Z = 4, R = 0.039.

The Hg and S atoms form planar Hg_4S_4 rings, packed into columns running along the [001] direction. The columns are connected by Pb atoms. The I atoms occupy three different types of holes. $Hg_2SnI_2S_2$ is related to $Hg_3I_2S_2$ and has space group *Imma*, $a = 9{\cdot}282\ (3)$, $b = 9{\cdot}565\ (2)$ and $c = 13{\cdot}460\ (5)$ Å.

	Wyckoff position	*x*	*y*	*z*
Hg	8(*i*)	0·16896	0·08649	0
Pb	4(*h*)	0·6332	0·1332	0·5
I(1)	2(*b*)	0	0	0·5
I(2)	2(*d*)	0	0·5	0
I(3)	4(*h*)	0·8130	0·3139	0·5
S	8(*i*)	0·0763	0·2370	0

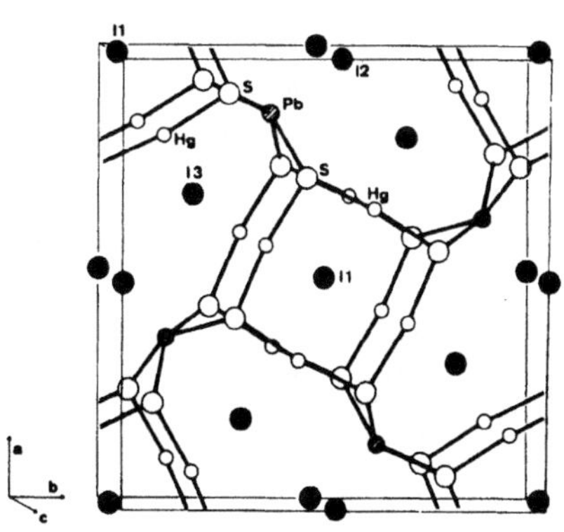

Relevant bond distances (Å) *and angles* (°)

Hg–S	2·382 (2)	Pb–S (4×)	2·989 (1)
	2·386 (2)	Pb–I(2) (2×)	3·427 (1)
Hg–I(1) (2×)	3·441	Pb–I(3) (2×)	3·431 (1)
Hg–I(3) (2×)	3·301 (1)	Pb–I(3) (2×)	4·385 (2)
Hg–Hg	3·624 (1)		
S–Hg–S	171·06 (8)	S_s–Pb–Sb_s*	73·71 (7)
Hg–S–Hg	98·94 (8)	S_s–Pb–Sb_a	100·40 (7)

* Subscript *s*: atom in the same column; subscript *a*: atom in an adjacent column.

LANTHANON THIOIODIDES
α-CeSI, SmSI, GdSI

Z. anorg. Chem., **535**, 229-239.

α-CeSI, Pbca, 7.330, 7.050, 14.350, Z = 8. Structure in **34**A, 214.

SmSI, R$\bar{3}$m, 4.547, 32.630 A, Z = 6. Structure in **39**A, 201.

GdSI, P$\bar{6}$, 10.737, 4.179 A, structure proposed. GdSI and SmSI types transform to α-CeSI type at high pressure.

HEXAAMMINECHROMIUM(III) HEXACYANOCHROMATE(III)
$[Cr(NH_3)_6][Cr(CN)_6]$

Acta Cryst., B**42**, 359-364.

R$\bar{3}$, a = 7.4635 A, α = 97.830^o, Z = 1, synchrotron radiation, R = 0.029. Isostructural with related compounds (**39**A, 205; **43**A, 161). Electron-density maps show bond and lone-pair electrons.

POTASSIUM NITROPRUSSIDE HYDRATE
$K_2[Fe(CN)_5NO].1.25H_2O$

J. Cryst. Spect. Res., **16**, 537-555.

$P2_12_12_1$, 9.706, 18.826, 25.258, Z = 16, R = 0.058. Octahedral anions, linked by K ions and water molecules, one of which is disordered. Fe-C = 1.88-1.96, Fe-N = 1.59-1.63(2) A.

cis-DIAMMINEDICYANOPALLADIUM(II)
$Pd(CN)_2(NH_3)_2$

Acta Cryst., C**42**, 1283-1284.

$P2_1/n$, 6.825, 12.733, 6.779, 111.82, Z = 4, R = 0.030. Square-planar molecules, linked by N-H...N hydrogen bonds.

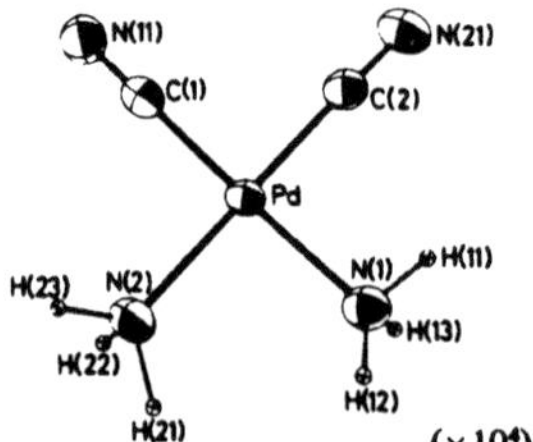

($\times 10^4$)

	x	y	z
Pd	2480	−138·0	136
C(1)	3455	1211	1580
C(2)	2120	570	−2526
N(1)	1491	−1573	−1425
N(2)	2900	−837	3029
N(11)	4034	1973	2466
N(21)	1899	983	−4093

Interatomic distances (Å) *and angles* (°)

Pd–C(1)	1·966 (5)	N(1)–H(11)	0·93 (7)
Pd–C(2)	1·949 (6)	N(1)–H(12)	1·10 (10)
Pd–N(1)	2·093 (5)	N(1)–H(13)	0·81 (11)
Pd–N(2)	2·075 (5)	N(2)–H(21)	0·83 (8)
C(1)–N(11)	1·133 (7)	N(2)–H(22)	0·91 (8)
C(2)–N(21)	1·144 (7)	N(2)–H(23)	0·89 (8)
C(1)–Pd–C(2)	87·7 (2)	H(11)–N(1)–H(12)	94 (6)
N(1)–Pd–N(2)	90·5 (2)	H(11)–N(1)–H(13)	81 (8)
N(1)–Pd–C(2)	91·7 (2)	H(12)–N(1)–H(13)	119 (9)
N(2)–Pd–C(1)	90·1 (2)	H(21)–N(2)–H(22)	106 (7)
N(1)–Pd–C(1)	178·9 (2)	H(21)–N(2)–H(23)	98 (6)
N(2)–Pd–C(2)	177·8 (2)	H(22)–N(2)–H(23)	116 (7)
Pd–C(1)–N(11)	178·0 (5)		
Pd–C(2)–N(21)	179·7 (3)		

POTASSIUM TETRACYANOMERCURATE(II)
$K_2Hg(CN)_4$

J. Chem. Phys., **85**, 6004-6009.

298K, Fd3m, 12.7835, Z = 8, neutron powder data. K in 16(d): 5/8,5/8,5/8; Hg in 8(a): 0,0,0; C in 32(e): x,x,x, x = 0.3472; N in 32(e): x = 0.3991. Spinel-type structure, with ordered CN groups. Hg has tetrahedral coordination to four C atoms.

Below 111K, $R\bar{3}c$, a = 15.7230 A, α = 32.803° (hexagonal cell, a = 8.879, c = 44.591 A), at 5.9K, Z = 4, neutron powder data at 113, 107, 80, 5.9K. $Hg(CN)_4$ tetrahedra are rotated by 7°, and K shifted by 0.1 A, from the positions in the high-temperature form.

BARIUM HEXA(THIOCYANATO-S)CADMATE HEPTAHYDRATE
$Ba_2Cd(SCN)_6.7H_2O$

Z. Naturforsch., **41**B, 439-443.

$P2_1/c$, 9.436, 14.697, 18.271, 109.02, Z = 4, R = 0.035. $Cd(SCN)_6$ octahedra, linked by two Ba ions which are coordinated to 5 N and 4 H_2O. Cd-S = 2.68-2.75(1) A, Cd-S-C = 101-108°, Ba-N = 2.86-3.13, Ba-O = 2.66-2.93 A.

RUBIDIUM OZONIDE
RbO_3

Z. anorg. Chem., **532**, 37-46.

$P2_1/c$, 6.452, 6.022, 8.763, 122.34, at 253K, Z = 4, R = 0.038. Bent ozonide anions (close to C_{2v} symmetry) and Rb^+ cations, in a CsCl-like arrangement. O-O = 1.34(1) A, O-O-O = 114°.

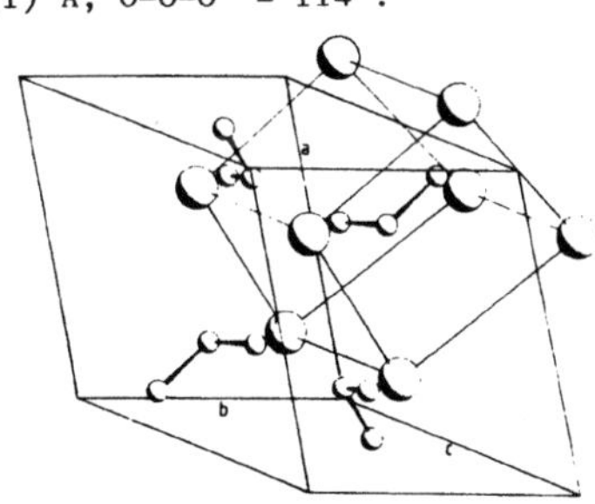

	x	y (×10⁴)	z
Rb	2286	2437	850
O1	2838	7858	3907
O2	3238	7433	5533
O3	1033	9329	2951

BROMELLITE
BeO

J. Appl. Phys., **59**, 3728-3733.

$P6_3mc$, 2.675-2.718, 4.341-4.408, at 298-1183K and/or 0.0001-5.7 GPa, Z = 2, R = 0.015-0.026. Be and O in 2(b): 1/3,2/3,z, z(Be) = 0, z(O) = 0.377-0.379 (no change in **z** with pressure, but slight increase with increasing temperature). Structure as previously described (**1**, 79, 115; **20**, 261; **27**, 473; **29**, 296; **34**A, 238).

CARBON MONOXIDE
ε-CO

J. Chem. Phys., **84**, 2837-2845.

$R\bar{3}c$, a = 6.059 A, α = 85.73° (hexagonal cell, a = 8.244, c = 11.250 A), at 100K and 5.5 GPa, Z = 8 (24 in hexagonal cell), powder data. C/O(1) in 12(c): 0,0,0.0501; C/O(2) in 36(f): parameters not given [compare N_2, this volume, p. 69].

ANTIMONY(III,V) OXIDE (V and Mo DOPED)
β-Sb_2O_4

VANADIUM ANTIMONY OXIDE
$VSbO_4$

VANADIUM PENTOXIDE
V_2O_5

J. Solid State Chem., **64**, 249-260.

Sb_2O_4, C2/c, a = 12.058, 12.057, b = 4.831, 4.834, c = 5.387, 5.384 A, β = 104.54, 104.58°, for V, Mo doped samples, Z = 4, neutron powder data. Sb(1) in 4(c): 3/4,1/4,1/2; Sb/V(2) in 4(e): 1/2,y,3/4, y = 0.2852 for Sb, 0.3170 for V; O(1) in 8(f): 0.6916,0.0518,0.1759; O(2) in 8(f): 0.4064,0.5878,0.5349 (for V-doped sample; similar values for the Mo-doped specimen). Structure as previously described (**29**, 441), with corrugated layers of $Sb(V)O_6$ octahedra, separated by sheets of $Sb(III)O_4E$ trigonal bipyramids (E = lone-pair). V substitutes for Sb(III) ions, but Mo is found in channels with two nearest Sb(III) ions sites vacant.

$VSbO_4$, $P4_2/mnm$, a = 4.632, 4.622, c = 3.034, 3.048 A, for two samples, Z = 1, neutron powder data. V/Sb in 2(a): 0,0,0; O in 4(f): x,x,0, x = 0.2958, 0.3023. Rutile structure (**15**, 290).

V_2O_5, $Pmn2_1$, 11.555, 4.371, 3.557, Z = 2, neutron powder data. V in 4(b): 0.1487,0.1086,0; O(1) in 4(b): 0.1468,0.4666,0.0320; O(2) in 4(b): 0.1813,-0.0056,0.5069; O(3) in 2(a): 0,-0.0030,0.0020. Structure essentially as previously described (**26**, 353). [See also this volume, p. 125.]

BISMUTH(III) OXIDE
δ-Bi_2O_3

J. Phys. C, **16**, L561-L566 (1983).

Fm3m, 5.648, Z = 2, neutron powder data. 4 Bi in 4(a): 0,0,0; 3.4 O(1) in 8(c): 1/4,1/4,1/4; 2.6 O(2) in 32(f): x,x,x, x = 0.354. Defect fluorite structure, as previously described (**27**, 476; **44**A, 342).

DICHLORINE HEXAOXIDE (CHLORYL PERCHLORATE)
Cl_2O_6 $ClO_2^+ClO_4^-$

Angew Chem., **98**, 994-995.

Cc, 5.563, 8.613, 9.785, 100.56, at -70°C, Z = 6, R = 0.028. Bent ClO_2^+ and tetrahedral ClO_4^- ions in a CsCl-type packing. Cl-O = 1.408 (cation), 1.417-1.467(2) A (anion), O-Cl-O = 118.9 (cation), 108.0-110.8° (anion).

BADDELEYITE
ZrO_2

Phys. Chem. Miner., **13**, 233-237.

Below 35kbar, $P2_1/c$, a = 5.168, 5.120, b = 5.232, 5.216, c = 5.341, 5.281 A, β=99.25, 99.01^o, at 0.001, 15kbar, Z = 4, R = 0.073 at 15kbar. Zr: 0.2754,0.0395,0.2086; O(1): 0.0710,0.3396,0.3380; O(2): 0.4551,0.7607,0.4855. Structure as previously described (**4**, 9; **23**, 341; **30**A, 307).

Above 35kbar, Pbcm, a = 5.005, 4.992, b = 5.235, 5.229, c = 5.051, 5.046 A, at 39, 51 kbar, Z = 4, R = 0.034, 0.045. Zr: 0.2701,0.0331,1/4; 0.5 O(1): 0.0716,0.3734,0.1337; O(2): 0.4570,3/4,1/2, at 51 kbar (parameters at 39 kbar also given). Disorder of the O(1) atoms results in the higher symmetry relative to the normal pressure phase.

VANADIUM PENTOXIDE
V_2O_5

Acta Cryst., C**42**, 1467-1469.

Pmmn, 11.512, 3.564, 4.368, Z = 2, R = 0.020. Structure as previously described (**4**, 22, 134; **13**, 232; **26**, 353). Layers of VO_5 square pyramids sharing edges and corners, with layers connected by a weak V-O interaction; V-O = 1.58, 1.78-2.02, 2.79 A. [See also, this volume, p. 124.]

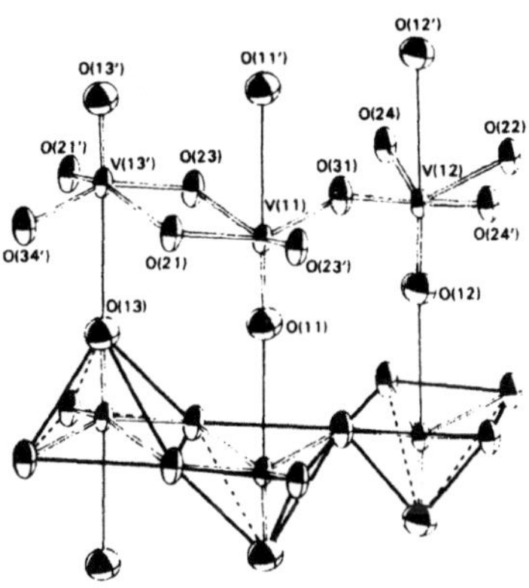

	x	y	z
V	0·10118	$\frac{1}{4}$	0·8917
O(1)	0·1043	$\frac{1}{4}$	0·531
O(2)	−0·0689	$\frac{1}{4}$	0·003
O(3)	$\frac{1}{4}$	$\frac{1}{4}$	0·001

MAGNETITE
Fe_3O_4

J. Solid State Chem., **62**, 75-82.

Fd3m is confirmed (**48**A, 195; **49**A; **51**A, 187).

SILVER(I,III) OXIDE (TETRAGONAL)
AgO

J. Solid State Chem., **65**, 225-230.

$I4_1/a$, 6.833, 9.122, Z = 16, neutron powder data. Ag(I) in 8(c): 0,0,0; Ag(III) in 8(d): 0,0,1/2; O in 16(f): 0.0798,0.0130,0.2128. Ag(I) with linear coordination, Ag-O = 2.18 A, and Ag(III) with square-planar coordination, Ag-O = 2.03 A, as in the monoclinic form (**22**, 302; **23**, 328; **24**, 327).

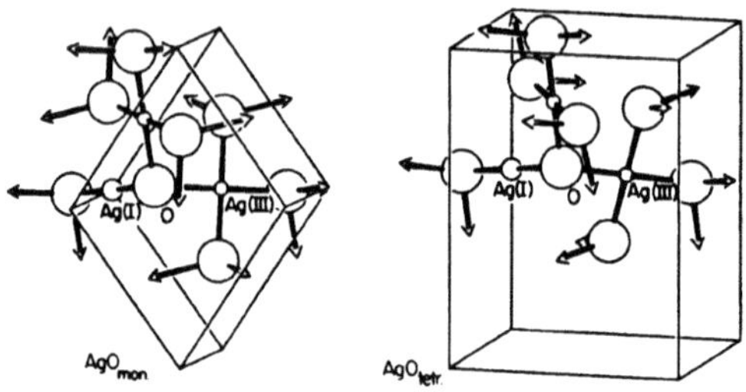

SILVER (III) OXIDE
Ag_2O_3

Z. anorg. Chem., **535**, 39-46.

Fdd2, 12.869, 10.490, 3.6638, Z = 8, R = 0.012. Ag in 16(b): -0.0467,0.1325,0.5; O(1) in 16(b): 0.0830,0.2333,0.6153; O(2) in 8(a): 0,0,0.8402. Isostructural with Au_2O_3 (**45**A, 220), with square-planar AgO_4 groups connected to give a three-dimensional framework; AgO = 1.96-2.05(1) A. [Compare **23**, 335.]

URANIUM OXIDE
β-U_4O_{9-x} (x = 0.062)

J. Solid State Chem., **61**, 1-7.

$I\bar{4}3d$, 21.9, Z = 64, neutron radiation at 230 and 500°C, R ∿ 0.1. Anion-excess fluorite-related structure proposed.

POTASSIUM β-ALUMINA
$K_{1.50}Al_{11.0}O_{17.25}$

J. Solid State Chem., **61**, 81-89.

$P6_3/mmc$, 5.598, 22.732, Z = 2, R = 0.029. β-alumina structure (**5**, 72; **33**A, 275; **37**A, 226; **42**A, 239; **43**A, 179).

		Number per unit cell	*x*	*z*
K(1)	6*h*	1.23	0.6786	¼
K(2)	6*h*	1.45	0.8830	¼
K(3)	6*h*	0.26	0.9672	¼
Al(1)	12*k*	10.70	0.8318	0.10426
Al(2)	4*f*	4	⅓	0.02430
Al(3)	4*f*	4	⅓	0.17454
Al(4)	2*a*	2	0	0
Al(5)	12*k*	0.89	0.8377	0.1748
O(1)	12*k*	12	0.1575	0.04953
O(2)	12*k*	12	0.5033	0.14626
O(3)	4*f*	4	⅔	0.05503
O(4)	4*e*	4	0	0.14138
O(5)	6*h*	2	0.318	¼
O(6)	6*h*	0.5	0.799	¼

POTASSIUM MAGNESIUM β-ALUMINA
BARIUM MAGNESIUM β-ALUMINA
$(K,Ba,Mg)_3Al_{10}O_{17}$, $(Ba,Mg)_2Al_{10}O_{17}$ (idealized)

J. Solid State Chem., **61**, 236-244.

P6$_3$/mmc, a = 5.6408, 5.6275, c = 22.645, 22.658 A, Z = 2, R = 0.034, 0.031. β-Alumina structures (**5**, 72; **42**A, 239; **43**A, 179; **51**A, 191).

		Number per unit cell	*x*	*z*
K(1)	2*d*	1.34	$\frac{2}{3}$	$\frac{1}{4}$
K(2)	6*h*	1.32	0.9112	$\frac{1}{4}$
K(3)	2*b*	0.39	0	$\frac{1}{4}$
K(4)	6*h*	0.70	0.821	$\frac{1}{4}$
Ba	6*h*	0.044	0.617	$\frac{1}{4}$
Al(1)	i2*k*	12	0.8330	0.10521
Al(2)	4*f*	4	$\frac{1}{3}$	0.02481
Al(3)	4*f*	4	$\frac{1}{3}$	0.17345
Al(4)	2*a*	2	0	0
O(1)	12*k*	12	0.1533	0.05150
O(2)	12*k*	12	0.5035	0.14714
O(3)	4*f*	4	$\frac{2}{3}$	0.05815
O(4)	4*e*	4	0	0.14389
O(5)	2*c*	2	$\frac{1}{3}$	$\frac{1}{4}$

		Number per unit cell	*x*	*z*
Ba	2*d*	1.912	0.6678	$\frac{1}{4}$
Al(1)	12*k*	12	0.8343	0.10544
Al(2)	4*f*	4	$\frac{1}{3}$	0.02400
Al(3)	4*f*	4	$\frac{1}{3}$	0.17416
Al(4)	2*a*	2	0	0
O(1)	12*k*	12	0.1534	0.05152
O(2)	12*k*	12	0.5042	0.14799
O(3)	4*f*	4	$\frac{2}{3}$	0.05901
O(4)	4*e*	4	0	0.14437
O(5)	2*c*	2	$\frac{1}{3}$	$\frac{1}{4}$

β- and β"-ALUMINA
$M_xAl_{11}O_{17}$ (M = Sr, Cd, Pb, Na)

J. Phys. Chem. Solids, **47**, 693-706; Phys. Rev., B, **34**, 5838-5849.

β, P6$_3$/mmc, a = 5.620, 5.607, c = 22.19, 22.52 A, for M = Cd, Pb, Z = 2, R = 0.032, 0.040. β", R3m, a = 5.622, 5.616, 5.612, c = 33.50, 33.21, 33.78 A, for M = Sr, Cd, Pb, Z = 3, R = 0.029, 0.038, 0.041. Structures as previously described (**5**, 72; **33**A, 275, 277), with divalent Pb in 2(d) in β-alumina, and other divalent cations [and Na] distributed over 2(d) and 6(h) sites in β-alumina, and 6(c) and 9(d) [18(h)] sites in β"-alumina.

β"-ALUMINA (COBALT STABILIZED)
$Na_2CoAl_{10}O_{17}$ (idealized)

J. Solid State Chem., **62**, 26-34.

R$\bar{3}$m, 5.6153, 33.712, Z = 3, R = 0.064. β"-Alumina structure (**33**A, 277), with Co(II) in the Al(2) site.

Atom	Al(1)	Al(2)	Co	Al(3)	Al(4)	O(1)	O(2)	O(3)	O(4)	O(5)	Na(1)	Na(2)
Wyckoff position	18*h*	6*c*		6*c*	3*a*	18*h*	18*h*	6*c*	6*c*	3*b*	6*c*	18*h*
Theor. occup.	0.5	0.16667	—	0.16667	0.08333	0.5	0.5	0.16667	0.16667	0.08333	0.16667	0.5
Refined occup.	—	0.093(1)	0.0584(6)	—	—	—	—	—	—	0.078(1)	0.051(2)	0.100(2)
x	0.3347(3)	0		0	0	0.1535(4)	0.1631(3)	0	0	0	0	−0.063(2)
z	0.0711(1)	0.3500(1)		0.4502(1)	0	0.0343(1)	0.2353(1)	0.2958(2)	0.966(2)	0.5	0.176(1)	0.1713(3)

LITHIUM NICKEL ALUMINATE
$LaNiAl_{11}O_{19}$

J. Solid State Chem., **65**, 301-308.

$P6_3/mmc$, 5.57, 22.01, Z = 2, R = 0.053. Magnetoplumbite-type structure (**6**, 74), with split Al(5) site and some La in a 6(h) site.

		Occupancy factor	x (× 10⁴)	y (× 10⁴)	z (× 10⁴)
La_1	2d	0.76	6667	3333	2500
La_2	6h	0.08	7199	4398	2500
Al_1	2a	1	0	0	0
Al_2	4f	1	3333	6667	274
Al_3	4f	1	3333	6667	1897
Al_4	12k	1	8318	6637	1078
Al_5	4e	0.5	0	0	2391
O_1	6h	1	1815	3631	2500
O_2	12k	1	1540	3081	523
O_3	12k	1	5043	86	1502
O_4	4e	1	0	0	1488
O_5	4f	1	6667	3333	562

La(1)	O(1) × 6	2,6940 Å
	O(3) × 6	2,7867
La(2)	O(1) × 2	2,3725
	O(1) × 2	2,8075
	O(3) × 2	2,5918
Al(1)	O(2) × 6	1,8792
Al(2)	O(2) × 3	1,8128
	O(5)	1,8402
Al(3)	O(3) × 3	1,8631
	O(1) × 3	1,9739
Al(4)	O(3) × 2	1,8339
	O(4)	1,8556
	O(5) × 3	1,9545
Al(5)	O(1) × 3	1,7666
	O(4)	1,9837
	O(4)	2,4617

DELAFOSSITE-TYPE OXIDES
$CuMO_2$ (M = Al, Ga, Sc, Y)

Z. anorg. Chem., **543**, 73-80.

R-type, Al, Ga, Y, $R\bar{3}m$, a = 2.858, 2.977, 3.353, c = 16.958, 17.171, 17.136 A, Z = 3, R = 0.032, 0.045, 0.048. Cu in 3(a): 0,0,0; M in 3(b): 0,0,1/2; O in 6(c): 0,0,z, z = 0.1099, 0.1076, 0.1066. Structures as previously described (**3**, 392; **10**, 110; **19**, 372; **20**, 270; **31**A, 137; **37**A, 262).

2H-type, Al, Sc, $P6_3/mmc$, a = 2.863, 3.223, c = 11.314, 11.413 A, Z = 2, R = 0.019, 0.019, Cu in 2(c): 1/3,2/3,1/4; M in 2(a): 0,0,0; O in 4(f): 1/3,2/3,z, z = 0.0851, 0.0893. Structures as previously described (**50**A).

SODIUM NEODYMIUM ALUMINATE
$Nd_{0.9}Na_{1.3}Al_{23}O_{36.5}$

J. Solid State Chem., **64**, 102-107.

$P\bar{6}m2$, 5.570, 22.257, Z = 1, R = 0.049. Alternate stacking of β-alumina and magnetoplumbite structures, with ordering of Na and Nd between the layers.

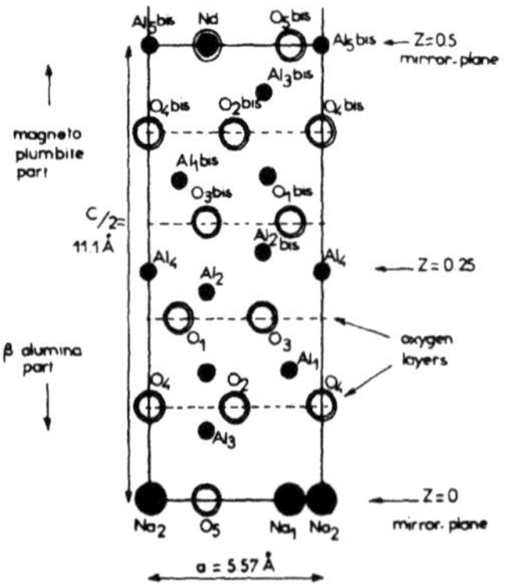

	$P\bar{6}m2$	$P6_3/mmc$ corresp.	Occup.	x = −y (10⁴)	z
Al_1	6n	12k	1	−1686	1470
Al_{1bis}	6n		1	1685	−6390
Al_2	2h	4f	1	3334	2284
Al_{2bis}	2h		1	6667	7203
Al_3	2h	4f	1	3334	755
Al_{3bis}	2h		1	6667	5595
Al_4	2g	2a	1	0	2533
Al_{5bis}	(1b)/2g	(2b)/4e	1	0	5086
O_1	6n	12k	1	1564	2043
O_{1bis}	6n		1	−1557	6936
O_2	6n	12k	1	−4986	1053
O_{2bis}	6n		1	−5050	5971
O_3	2i	4f	1	6667	2001
O_{3bis}	2i		1	3334	6907
O_4	2g	4e	1	0	1104
O_{4bis}	2g		1	0	5986
O_5	(1c)/3j	(2c)/6h	1.15	2880	0
O_{5bis}	3k	6h	1	−1803	5000
Nd	(1d)/3k	(2d)/6h	0.9	3197	5000
Na_1	3j	6h	0.77	−1931	0
Na_2	(1e)/3j	(2b)/6h	0.53	9494	0

CALCIUM GALLATE
$Ca_5Ga_6O_{14}$

Kristallografija, **31**, 1217-1219 [Soviet Physics - Crystallography, **31**, 719-720].

$Cmc2_1$, 11.511, 11.282, 10.489, Z = 4, R = 0.055. Isostructural with $Ca_5Al_6O_{14}$ (**44**A, 189). Ga-4 O = 1.78-1.90, Ca-6 O = 2.26-2.63 A.

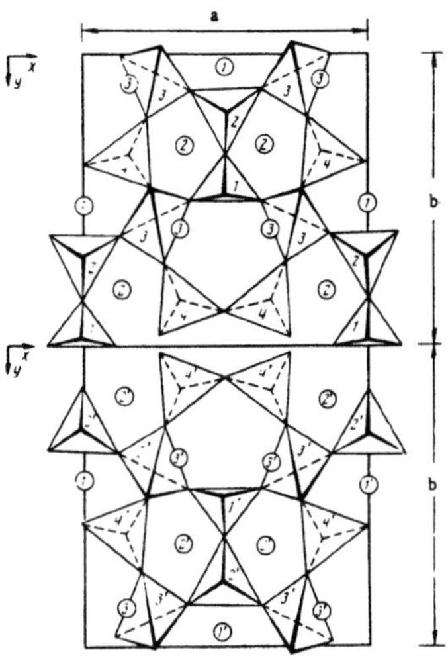

	x	y	z
Ca_1	0	0.4737	0.4179
Ca_2	0.1431	0.1837	0.3748
Ca_3	0.1582	−0,1012	0.4234
Ga_1	0	−0.0008	0 1370
Ga_2	0	0.2999	0.1120
Ga_3	−0.2438	0.5974	0.6483
Ga_4	0.1435	0.3540	0.6805
O_1	0	0.0114	0.3140
O_2	0	0.1478	0.0552
O_3	0	0.2962	0.2903
O_4	0	0.6158	0.2483
O_5	−0.1254	0.6255	0.5417
O_6	0.1255	−0.0645	0.0558
O_7	0.1600	0.3494	0.5121
O_8	0.2677	0.0286	0.2667
O_9	0.2856	0.2772	0.2588

BARIUM HEXAGALLATE
β-$Ba_{0.65}Ga_{10.80}O_{16.84}$

Solid State Ionics, **21**, 143-149.

Hexagonal, $P6_3/mmc$, 5.859, 23.192 A, Z = 2, R = 0.018. β-Alumina type structure (**5**, 72; **43**A, 179; **52**A, 177) with Ba in a 6(h) site.

SODIUM PENTAINDATE(I)
$Na_{24}In_5O_{15}$

J. Less-Common Metals, **116**, 129-136.

$I\bar{4}3m$, 11.0759, Z = 2, R = 0.060. In(1) in 8(c): x,x,x, x = 0.1431; In(2) in 2(a): 0,0,0; Na(1) in 24(g): 0.4165,0.1330,0.1330; Na(2) in 24(g): 0.3912,0.3912,0.1749; O(1) in 6(b): 1/2,1/2,0; O(2) in 24(g): 0.2566,0.0506,0.2566. New structure type derived from that of Na_2O by doubling **a** and reordering cations and anions. In_5 clusters of indium-centred In_4 tetrahedra are present. In-In = 2.74, In-O = 2.05 A.

BARIUM NICKEL INDATE
$BaNi_2In_8O_{15}$

Z. anorg. Chem., **542**, 25-30.

Cmca, 12.2811, 11.0214, 9.9103, Z = 4, R = 0.088. Framework of InO_6 octahedra, with Ni statistically occupying one In site. In/Ni-6 O = 1.91-2.16, In-6 O = 2.11-2.49, Ba-12 O = 2.72-3.58(6) A.

		x	y	z
In_I	(8f)	0,0	0,3119	0,1385
In_{II}	(16g)	0,3692	0,0830	0,1482
Ni_I/In_{III}	(16g)	0,2620	0,1624	0,6128
Ba	(4a)	0,0	0,0	0,0
O_I	(8d)	0,760	0,5	0,5
O_{II}	(8e)	0,25	0,186	0,25
O_{III}	(16g)	0,858	0,911	0,256
O_{IV}	(8f)	0,0	0,637	0,272
O_V	(16g)	0,652	0,277	0,505
O_{VI}	(4b)	0,5	0,5	0,5

SODIUM THALLATE(I)
Na_3TlO_2

J. Less-Common Metals, **120**, 225-237.

Pnma, 11.4561, 6.7067, 5.3676, Z = 4, R = 0.073. Tl and Na ions have 3-coordinations; Tl-O = 2.31, 2.34, Na-O = 2.30-2.36 A.

	$x \times 10^4$	$y \times 10^4$	$z \times 10^4$
Tl(1) (4c)	0905	2500	2349
Na(1) (8d)	8337	4995	2495
Na(2) (4c)	0782	7500	2567
O(1) (4a)	0000	5000	0000
O(2) (4c)	2558	2500	9918

POTASSIUM DIGERMANATE
$K_6Ge_2O_7$

Mh. Chem., **117**, 969-976.

Pc, 6.549, 9.094, 11.426, β = 126.78, Z = 2, R = 0.131. The structure is very similar to that of $K_6Co_2O_7$, which is reported as $P2_1/c$, with linear bridging oxygens (**40**A, 212). In the Pc description of the present compound, Ge-O-Ge = 157°.

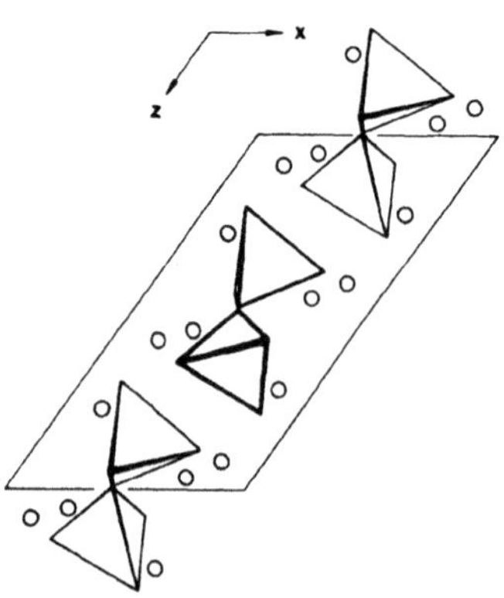

	x	y	z
K(1)	0.849	0.0383	0.7219
K(2)	0.817	0.3988	0.4196
K(3)	0.715	0.2224	0.9646
K(4)	0.309	0.2717	0.0527
K(5)	0.198	0.0969	0.5844
K(6)	0.166	0.4742	0.2752
Ge(1)	0.4212	0.1134	0.3710
Ge(2)	0.5788	0.3901	0.6290
O(1)	0.164	0.163	0.198
O(2)	0.701	0.103	0.389
O(3)	0.379	−0.033	0.448
O(4)	0.441	0.250	0.495
O(5)	0.654	0.548	0.575
O(6)	0.335	0.421	0.644
O(7)	0.854	0.324	0.789

TRICALCIUM GERMANATE
Ca_3GeO_5

Z. Kristallogr., **176**, 303-317.

3M form, $P2_1/m$, 12.546, 7.196, 9.515, 116.43, Z = 6, R = 0.039. 9R form, R3m, a = 7.190, c = 25.243 A, Z = 9, R = 0.059. The structures are related to those of tricalcium silicate monoclinic (**52**A, 322) and rhombohedral (**51**A, 327) forms; 2-layer (**51**A, 198) and 24-layer (**52**A, 184) forms have also been described. The structures exhibit orientational disorder of GeO_4 tetrahedra.

3M

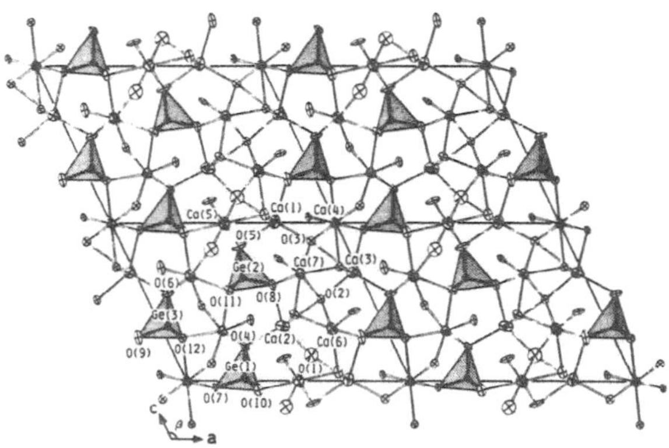

9R

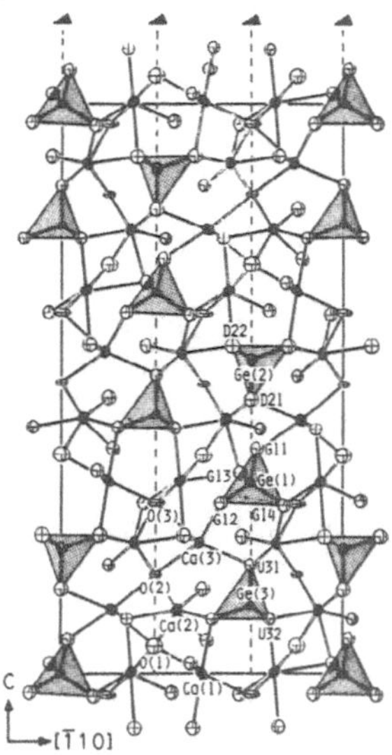

BARIUM HYDROXYGERMANATE
$Ba_9Ge_{23}O_{53}(OH)_4$

Z. Kristallogr., **175**, 125-130.

$P6_3cm$, 11.6784, 19.2545, Z = 2, R = 0.034. Ge atoms have tetrahedral, square-pyramidal, and octahedral coordination, and form Ge_3O_{10} and $Ge_3(O,OH)_{11}$ rings; Ba ions have 9-, 10-, and 11-coordinations; OH is in the O(7) site and partially in the O(4) site.

	x	y	z
Ba(1)	0	32357	73732
Ba(2)	0	31161	94781
Ba(3)	0	30233	53016
Ge(1)	0	0	0
Ge(2)	33333	66666	72607
Ge(3)	33333	66666	97635
Ge(4)	0	16165	14466
Ge(5)	0	16615	35601
Ge(6)	15277	49014	34992
Ge(7)	15488	48501	10705
O(1)	0	13261	44998
O(2)	0	13133	67024
O(3)	0	14217	5265
O(4)	0	14188	83660
O(5)	13759	34010	14328
O(6)	25358	51569	78506
O(7)	0	21970	26259
O(8)	0	45024	7014
O(9)	0	46266	31618
O(10)	25243	51114	3396
O(11)	12179	34416	38585
O(12)	18431	57683	66816
O(13)	18179	58670	92248

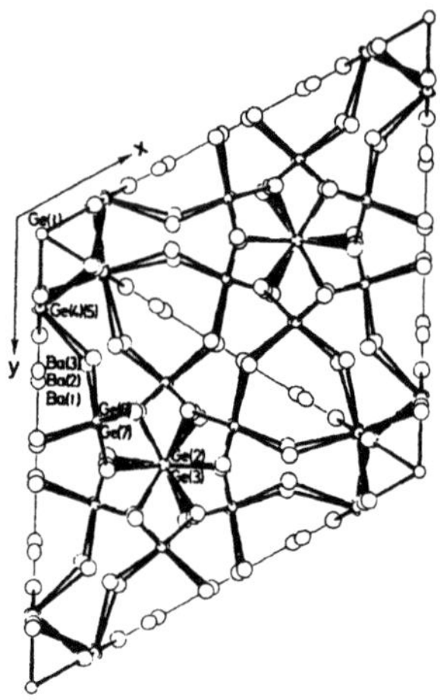

ALUMINUM DIGERMANATE
$Al_2Ge_2O_7$

J. Solid State Chem., **62**, 402-404.

C2/c, 7.132, 7.741, 9.702, 110.62, Z = 4, R = 0.051. Digermanate anions, linked by AlO_3 chains of edge-sharing AlO_5 trigonal bipyramids.

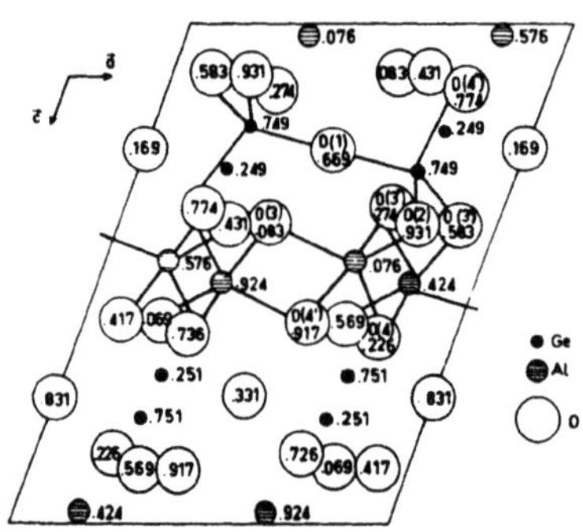

	X	*Y*	*Z*
Ge	0.2414	0.2488	0.2945
Al	0.3290	0.0764	0.0211
O(1)	0	0.1694	0.25
O(2)	0.2891	0.4310	0.3982
O(3)	0.3972	0.0834	0.3973
O(4)	0.2887	0.2739	0.1292

MAIN INTERATOMIC DISTANCES (Å) AND ANGLES (°) IN THE Ge_2O_7 AND AlO_5 POLYHEDRA

Ge–O(2) = 1.696(8)	
Ge–O(1) = 1.734(5)	Ge–O(1)–Ge = 138.5(4)
Ge–O(3") = 1.757(7)	Ge–O(2)–Al = 140.7(5)
Ge–O(4") = 1.762(8)	
Average d_{Ge-O} = 1.74	
	O(2)–Al–O(4) = 127.6(6) }
	O(2)–Al–O(3) = 115.7(5) } Σ = 359.9
Al–O(2) = 1.748(9)	O(3)–Al–O(4) = 116.6(5) }
Al–O(4) = 1.820(7)	O(3')–Al–O(4') = 167.8(6)
Al–O(3) = 1.831(7)	O(4)–Al–O(4') = 79.2(6)
Al–O(3') = 1.903(9)	O(3)–Al–O(4') = 94.7(5)
Al–O(4') = 1.932(9)	O(2)–Al–O(4') = 94.8(6)
Average d_{Al-O} = 1.85	O(4)–Al–O(3') = 93.4(6)
	O(3)–Al–O(3') = 79.7(5)
	O(2)–Al–O(3') = 97.5(6)

SODIUM ALUMINOGERMANATE (C-PHASE)
$NaAlGeO_4$

POTASSIUM ALUMINOGERMANATE (D-PHASE)
$KAlGeO_4$

Kristallografija, **31**, 883-891 [Soviet Physics - Crystallography, **31**, 522-527].

C-Phase, $P2_1/n$, 8.783, 15.432, 8.252, γ = 90.09°, Z = 12, R = 0.033. Beryllonite-type structure (**26**, 456; **27**, 580; **39A**, 283). D-Phase, $P6_3$ (or perhaps P3), 18.429, 8.599, Z = 24, R = 0.057. Aluminogermanate tetrahedral framework, with 6- and 9-coordinate K ions.

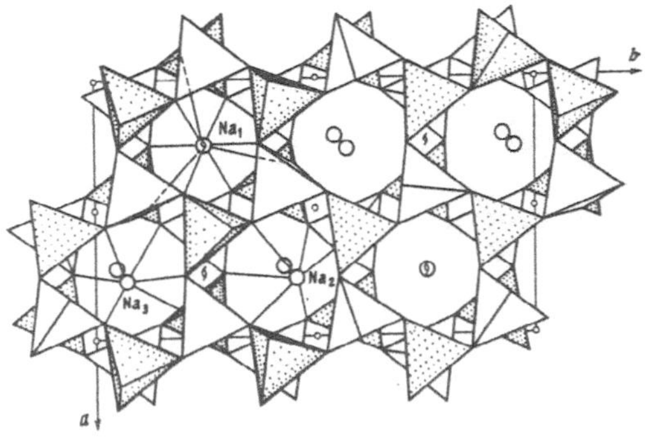

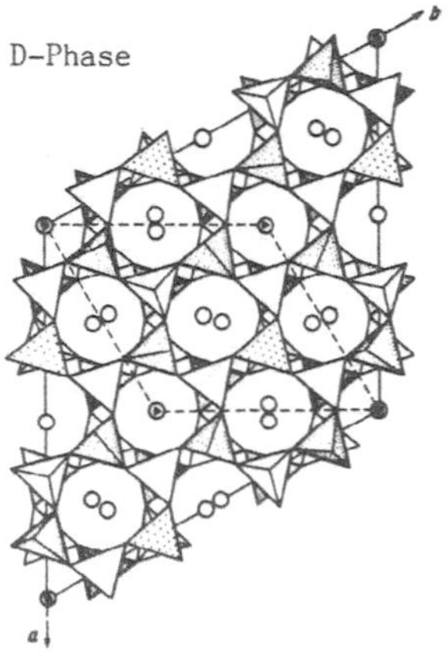

POTASSIUM ALUMINOGERMANATE
$KAlGeO_4$

Z. Kristallogr., **176**, 29-33.

$P6_3$, 18.413, 8.590, Z = 24, R = 0.035 (twinned crystal). Ordered Al/Ge tetrahedral framework, with channels along **c** which contain the K ions. [See also preceding report.]

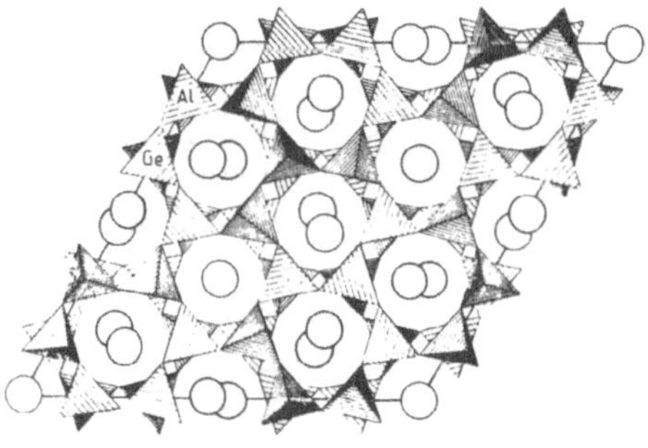

CANCRINITE (GERMANIUM)
$Na_8[Al_6Ge_6O_{24}]Ge(OH)_6.2H_2O$

Kristallografija, **31**, 874-878 [Soviet Physics - Crystallography, **31**, 516-519].

$P6_3$, 13.023, 5.024, Z = 1, R = 0.055. Cancrinite structure (**3**, 150, 524; **19**, 480; **30A**, 428; **35A**, 458), with the large channels along **c** filled by $Ge(OH)_6$ octahedra.

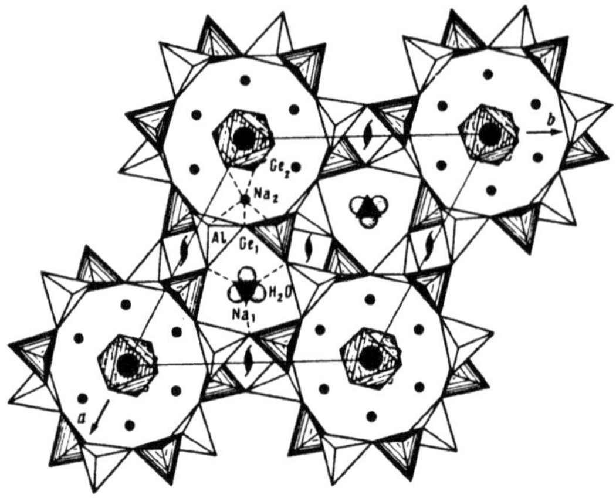

GALLIUM GERMANATE
α-Ga_4GeO_8

J. Solid State Chem., **65**, 377-382.

C2/m, 12.169, 3.005, 9.414, 125.06, Z = 2, R = 0.029. Intergrowth of GeO_2 rutile elements in a β-gallia network, with large hexagonal tunnels. Ge in octahedral sites, and Ga in both octahedral and tetrahedral sites.

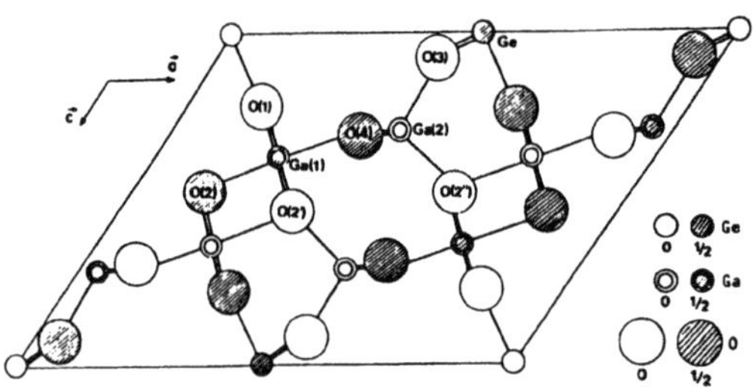

		x	y	z
M(1) = Ge	2a	0	0	0
M(2) = Ga(1)	4i	0.7576	0	0.3596
M(3) = Ga(2)	4i	0.4621	0	0.2809
O(1)	4i	0.1627	0	0.2183
O(2)	4i	0.6446	0	0.4618
O(3)	4i	0.4403	0	0.0738
O(4)	4i	0.8912	0	0.3064

Tetrahedral site		
M(3)–O(3)	1.810(8)	
M(3)–O(4)	2 × 1.816(5)	
M(3)–O(2)	1.867(4)	
Mean value = 1.83	in tetrahedra	Ge – O = 1.74; Ga – O = 1.82
Octahedral sites		
M(1)–O(1)	2 × 1.862(4)	
M(1)–O(3)	4 × 1.960(6)	
Mean value = 1.93	in octahedra	Ge – O = 1.90; Ga – O = 2.00
M(2)–O(1)	2 × 1.896(3)	
M(2)–O(4)	1.956(9)	
M(2)–O(2')	2 × 2.050(3)	
M(2)–O(2)	2.083(8)	
Mean value = 1.99		
Vacant site in tunnels		
□-O(1)	2.32	
□-O(3)	3 × 2.32	
□-O(4)	2 × 2.59	
□-O(4)	2.68	

BISMUTH(III) GERMANATE
Bi_2GeO_5

Acta Chem. Scand., **18**, 1555-1557 (1964) [not previously reported, since a fuller account was promised (**29**, 442)].

$Cmc2_1$, 15.697, 5.4928, 5.3836, Z = 4, R = 0.14. Bi, 0.1676,0.2174,0.2500; Ge, 0,0.6857,0.2000; O(1), 0,0.5803,0.4908; O(2), 0.0949,0.1368,0.6482; O(3), 0.2541,0.4777,0.4621. GeO_3^{2-} chains of corner-sharing tetrahedra and $Bi_2O_2^{2+}$ layers. Ge-O = 1.7-1.8, Bi-O = 2.2-2.7 A.

Ž. Neorg. Khim., **31**, 1391-1396 [Russ. J. Inorg. Chem., **31**, 795-797].

Cc, 15.686, 5.493, 5.376, 90.01, Z = 4, R = 0.055. Structure approximately as previously described in $Cmc2_1$ (above). $Bi_2O_2^{2+}$ (100) layers alternating with layers of GeO_3^{2-} chains. Ge-4 O = 1.68-1.76, Bi-(6+5) O = 2.05-2.61, 3.06-3.83 A.

LITHIUM SCANDIUM GERMANATE
$LiScGeO_4$

Ž. Strukt. Khim., **27**, No. 3, 167-168 [J. Struct. Chem., **27**, 488-489].

Pnma, 10.673, 5.9926, 4.9667, Z = 4, R = 0.054. Li in 4(a): 0,0,0; Sc, Ge, O(1), O(2) in 4(c): x,1/4,z, x = 0.2715, 0.0901, 0.0973, 0.4426, z = 0.0029, 0.4474, 0.7958, 0.1988; O(3) in 8(d): 0.1730,0.0300,0.2860. Olivine-type structure (**1**, 352). Li-6 O = 2.09-2.34, Sc-6 O = 2.07-2.20, Ge-4 O = 1.73-1.78 A.

HAFNIUM GERMANATE
$HfGeO_4$

THORIUM GERMANATE
α- and β-$ThGeO_4$

J. Less-Common Metals, **124**, 105-109.

Hf, α-Th, $I4_1/a$, a = 4.862, 5.145, c = 10.497, 10.531, Z = 4, R = 0.036, 0.036. Ge in 4(a): 0,0,0; Hf or Th in 4(b): 0,0,1/2; O in 16(f): 0.2678,0.1739,0.0831 and 0.2548,0.1493,0.0787 (origin at $\bar{4}$). Scheelite-type structure (**1**, 347; **29**, 331).

β-Th, $I4_1/amd$, 7.230, 6.539, Z = 4, R = 0.038. Ge in 4(b): 0,0,1/2; Th in 4(a): 0,0,0; O in 16(h): 0,0.1803,0.3214 (origin at $\bar{4}m2$). Zircon-type structure (**1**, 345).

IRON(III) TRIGERMANATE
$Fe_8Ge_3O_{18}$

J. Solid State Chem., **62**, 397-401.

$P2_1/c$, 8.754, 5.110, 14.280, 101.80°, Z = 2, R = 0.039. Close-packed oxygen layers, ABACBC, with Fe in octahedral sites and Ge distributed in octahedral and tetrahedral sites.

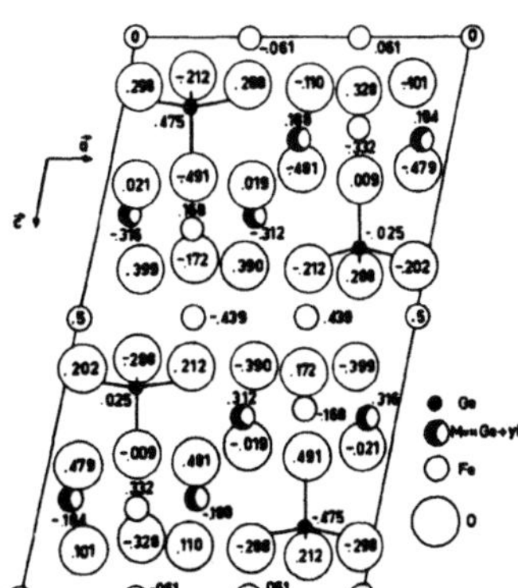

	x	y	z
Oct(1)—Fe (1)	0	0	0
Oct(2)—*M* (2) (2.7 Fe + 1.3 Ge)	0.5388	0.1878	0.1801
Oct(3)—*M* (3) (3.3 Fe + 0.7 Ge)	0.9134	0.1841	0.1824
Oct(4)—Fe (4)	0.3370	−0.0614	0.0024
Oct(5)—Fe (5)	0.2793	0.1678	0.3393
Tetr. —Ge	0.7900	−0.0252	0.3791
O (1)	0.1912	−0.2119	0.0728
O (2)	0.3558	0.2882	0.0846
O (3)	0.0378	0.2982	0.0885
O (4)	0.5401	−0.1101	0.0880
O (5)	0.6931	0.3280	0.1007
O (6)	0.8484	−0.1006	0.0842
O (7)	0.0843	0.0214	0.2637
O (8)	0.2523	0.5086	0.2489
O (9)	0.4267	0.0191	0.2618

CALCIUM COPPER(II) DIGERMANATE
$CaCuGe_2O_6$

Z. Kristallogr., **176**, 205-217.

$P2_1/c$, 10.198, 9.209, 5.213, 105.73, Z = 4, R = 0.055. Distorted pyroxene structure, with two types of Ge/O layers. Cu has distorted octahedral and Ca irregular (7+1) coordination. Ge-4 O = 1.71-1.79, Ge-5 O = 1.77-1.89, 2.09, Cu-O = 1.95-2.10, 2.28, 2.45, Ca-O = 2.33-2.63, 3.03 A.

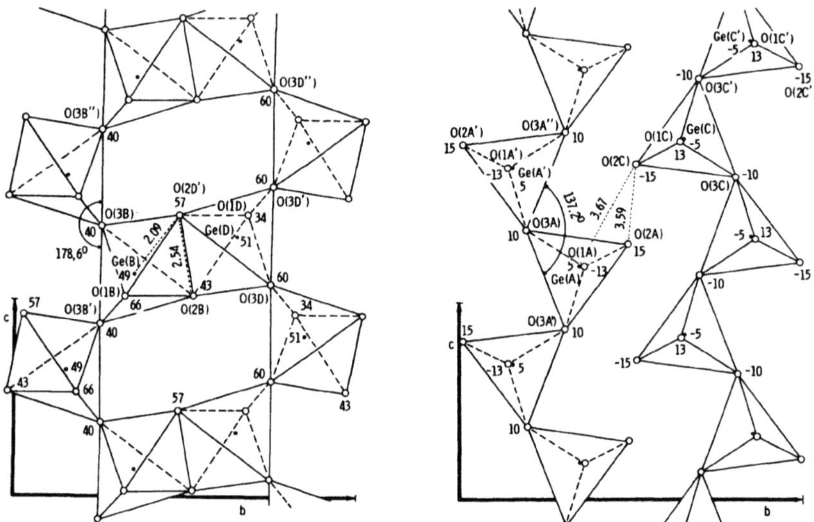

	x	y	z
Ca	0.7170	0.5475	0.1916
Cu	0.7727	0.1617	0.2436
Ge(A1)	0.0490	0.3492	0.2099
Ge(B1)	0.4910	0.3510	0.4008
O(1A1)	0.8728	0.3608	0.1158
O(2A1)	0.1459	0.4893	0.3762
O(3A1)	0.1016	0.1944	0.4194
O(1B1)	0.6581	0.3206	0.3764
O(2B1)	0.4321	0.5223	0.2603
O(3B1)	0.4006	0.2517	0.5903

SODIUM GADOLINIUM GERMANATE
SODIUM YTTRIUM GERMANATE
SODIUM LUTETIUM GERMANATE
$NaLnGeO_4$ (Ln = Gd, Y, Lu)

Phys. Status Solidi, A, **93**, 495-502.

Pnma, a = 11.539, 11.423, 11.277, b = 6.535, 6.45, 6.354, c = 5.296, 5.267, 5.241 A, Z = 4, R = 0.016, 0.024, 0.019. Olivine-like structures (**35A**, 315).

CALCIUM STANNATE(IV)
$CaSnO_3$

STRONTIUM STANNATE(IV)
$SrSnO_3$

Acta Cryst., B**42**, 167-172.

Pbnm, a = 5.532, 5.707, b = 5.681, 5.707, c = 7.906, 8.064 A (twinning in the Sr compound results in an apparently cubic cell, a = 8.068 A (1, 303, 333)), R = 0.031 for Ca compound, interpretation of twinned crystal data for the Sr compound. Ca or Sr in 4(c): x,y,1/4, x = -0.0131, -0.0009, y = 0.0509, 0.0124; Sn in 4(b): 0,1/2,0; O(1) in 4(c): x = 0.0994, 0.0736, y = 0.4615, 0.4896; O(2) in 8(d): x,y,z, x = 0.6992, 0.7132, y = 0.2981, 0.2853, z = 0.0514, 0.0368. Orthorhombic perovskites.

BISMUTH LEAD OXIDE
$Bi_{12}PbO_{19}$

J. Solid State Chem., **62**, 290-296.

5 and 295K, I23, 10.2401, 10.2707, Z = 2. Neutron powder data. $Bi_{12}GeO_{20}$-type structure (**32**A, 292; **45**A, 233).

983K, Fm3m, 5.64, neutron powder data. Fluorite structure related to that of δ-Bi_2O_3 (this volume, p. 124 and following report).

BISMUTH YTTRIUM OXIDE
$(Bi,Y)_2O_3$

J. Phys. C. **16**, L561-L566 (1983); J. Solid State Chem., **63**, 8-15 (1986).

Fm3m, 5.6485-5.4469 for 0-40% Y_2O_3, Z = 2, neutron powder data. 4 Bi,Y in 24(e): x,0,0, x = 0-0.046; 3.4-4.6 O(1) in 8(c): 1/4,1/4,1/4; 2.6-1.0 O(2) in 32(f): x,x,x, x = 0.354-0.320; 0-0.2 O(3) in 48(i): 1/2,x,x, x = 0.390-0.288. δ-Bi_2O_3-type structure (this volume, p. 124), with additional displacement of cations to 24(e) sites, and some interstitial oxygen in 48(i) sites.

BISMUTH CADMIUM OXIDE
$Bi_{0.79}Cd_{0.21}O_{1.395}$

J. Solid State Chem., **63**, 160-165.

Im3m, 4.281, Z = 2, R = 0.037. 2 Bi/Cd in 2(a): 0,0,0; 2.79 O in 12(d): 1/2,1/4,0. Anti-α-AgI type structure (**3**, 8; **45**, 380), with oxygen ions in tetrahedral sites in the b.c.c. cation lattice.

BARIUM NICKEL SCANDIUM OXIDE
$BaNi_2Sc_8O_{15}$

J. Less-Common Metals, **124**, 229-234.

Cmca, 12.008, 10.740, 9.750 A, Z = 4, R = 0.068. The material is metastable, and the structure contains a framework of ScO_6 and NiO_6 octahedra (one site statistically occupied), with 12-coordinate Ba.

		x	y	z
Sc1	(8f)	0,0	0,3126	0,1370
Sc2	(16g)	0,3680	0,0840	0,1468
Ni + Sc3	(16g)	0,2599	0,1623	0,6135
Ba	(4a)	0,0	0,0	0,0
01	(8d)	0,759	0,5	0,5
02	(8e)	0,25	0,186	0,25
03	(16g)	0,869	0,912	0,258
04	(8f)	0,0	0,642	0,262
05	(16g)	0,632	0,268	0,501
06	(4b)	0,5	0,5	0,5

BARIUM CALCIUM YTTRIUM OXIDE
$BaCa_2Y_6O_{12}$

CALCIUM MAGNESIUM YTTRIUM OXIDE
$CaMg_3Y_8O_{16}$ (3 x $Ca_{0.33}MgY_{2.67}O_{5.33}$)

J. Less-Common Metals, **116**, 211-218.

$BaCa_2Y_6O_{12}$, $P6_3/m$, 10.221, 3.361, Z = 1, R = 0.092. $SrCa_2Sc_6O_{12}$-type structure (**41**A, 230).

$CaMg_3Y_8O_{16}$, Pm, 6.516, 10.620, 5.866, 92.4, Z = 1, R = 0.103.

$BaCa_2Y_6O_{12}$

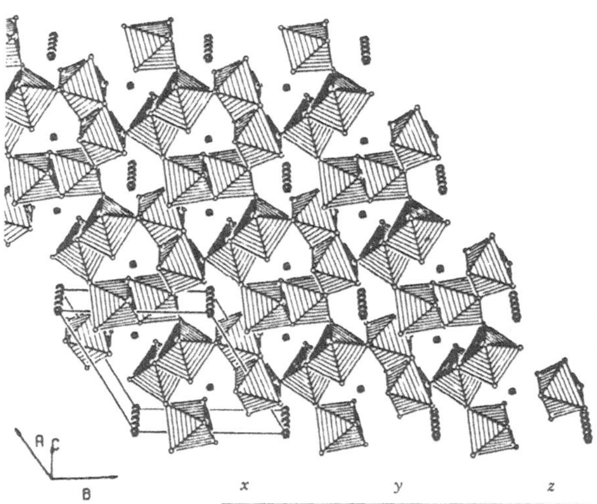

		x	y	z
$\frac{1}{4}Ba_I$	2b	0.0	0.0	0.0
$\frac{1}{4}Ba_{II}$	2a	0.0	0.0	0.25
Ca	2d	0.3333	0.3333	0.25
Y	6h	0.3470	0.9993	0.25
O_I	6h	0.187	0.297	0.25
O_{II}	6h	0.528	0.395	0.25

$CaMg_3Y_8O_{16}$

		x	y	z
M_I	2c	0.1380	0.1639	0.2312
M_{II}	2c	0.8823	0.1654	0.8160
M_{III}	2c	0.3144	0.3320	0.7813
M_{IV}	2c	0.6249	0.3331	0.2661
M_V	1b	0.1885	0.5	0.2963
M_{VI}	1b	0.8422	0.5	0.7414
M_{VII}	1a	0.2766	0.0	0.7142
M_{VIII}	1a	0.6800	0.0	0.2540
O_I	2c	0.502	0.187	0.096
O_{II}	2c	0.831	0.156	0.224
O_{III}	2c	0.179	0.174	0.620
O_{IV}	2c	0.081	0.347	0.013
O_V	2c	0.654	0.343	0.605
O_{VI}	2c	0.326	0.343	0.388
O_{VII}	1b	0.434	0.5	0.042
O_{VIII}	1b	0.864	0.5	0.393
O_{IX}	1b	0.138	0.5	0.716
O_X	1a	0.012	0.0	0.999
O_{XI}	1a	0.654	0.0	0.853
O_{XII}	1a	0.407	0.0	0.457

HOLLANDITES
$K_{1.54}Mg_{0.77}Ti_{7.23}O_{16}$

J. Chem. Phys., **85**, 474-484.

I4/m, 10.138-10.197, 2.9685-2.992, at 133-919K, Z = 1, R = 0.010-0.025. Hollandite structure (**13**, 190; **45**A, 237), with four K sites in the conduction tunnel.

$Ba_x(M,Ti)_8O_{16}$ (M = Ga, Al, Al/Fe)

Acta Cryst., B**42**, 229-236.

I4/m, 10.0, 2.9, for 3 compounds, Z = 1; I2/m, 10.1, 2.9, 10.0, 90.0-91.0, for 4 compounds, Z = 1, neutron powder data. Hollandite structures (**13**, 190; **22**, 339; **42**A, 269; **44**A, 208; **46**A, 243; **49**A).

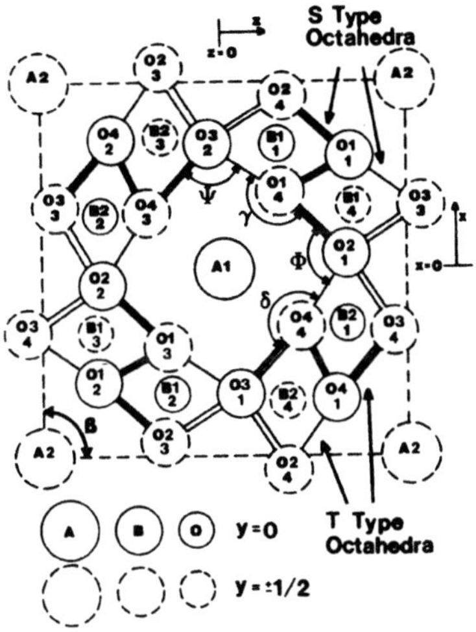

MANNARDITE
$Ba(Ti,V,Cr)_8O_{16}.H_2O$ (idealized)

Canad. Miner., **24**, 67-78.

$I4_1/a$, 14.357, 5.908, Z = 4, R = 0.027. Hollandite-type structure (**13**, 190), with √2a x 2c cell; Ba and H_2O are disordered in tunnels.

ZIRCONIUM TITANATES
$ZrTiO_4$ (I), $Zr_5Ti_7O_{24}$ (II)

IRON NIOBATE
$FeNb_2O_6$ (III)

J. Solid State Chem., **64**, 30-46.

I. Pbcn, 4.8042, 5.4825, 5.0313, Z = 2, neutron powder data. Zr,Ti in 4(c): 0,0.265,1/4; O in 8(d): 0.2704,0.1004,0.0700. α-PbO_2 structure type (**13**, 186; **16**, 224; **17**, 388; **23**, 345), with random distribution of cations.

II. Pbcn, 14.3574, 5.3247, 5.0200, Z = 1, neutron powder data. Ordered α-PbO_2 superstructure of fersmite type (**24**, 367).

III. Pbcn, 14.2661, 5.7334, 5.0495, Z = 4, neutron powder data. Ordered α-PbO_2 superstructure of niobite type (**2**, 55, 337; **37**A, 243).

$Zr_5Ti_7O_{24}$

		x	y	z	n
Zr(1)	4c2	0	0.2629	¼	0.484
Ti(1)	4c2	0	0.2629	¼	0.016
Zr(2)	8d1	0.3324	0.3335	0.1798	0.111
Ti(2)	8d1	0.3324	0.3335	0.1798	0.889
O(1)	8d1	0.0873	0.0668	−0.0234	1
O(2)	8d1	0.4194	0.0824	0.0599	1
O(3)	8d1	0.7600	0.1226	0.1402	1

$FeNb_2O_6$

		x	y	z
Fe	4c2	0	0.3311	¼
Nb	8d1	0.3389	0.3191	0.2506
O(1)	8d1	0.0963	0.1041	0.0727
O(2)	8d1	0.4189	0.1163	0.0990
O(3)	8d1	0.7560	0.1236	0.0793

REDLEDGEITE
$Ba(Cr,Ti)_8O_{16}$

Miner. Mag., **50**, 709-715.

I2/m, 10.129, 2.95, 10.135, 90.05, Z = 1, R = 0.063. Monoclinic hollandite-type phase. Weak superlattice reflexions with a period of 2.24**b** suggest tunnel cation ordering.

STRONTIUM IRON TITANIUM OXIDE
$Sr_2FeTiO_{5.5}$

Mater. Res. Bull., **21**, 823-833.

Pm3m, 3.90, Z = 1/2, neutron powder data. Sr in 1(b); 1/2,1/2,1/2; 0.25 Fe(1) + 0.5 Ti in 1(a): 0,0,0; 0.25 Fe(2) in 12(i): 0.04,0.04,0; 1.75 O(1) in 3(d): 1/2,0,0; 1.0 O(2) in 12(j): 1/2,0.05,0.05. Perovskite structure, with oxygen vacancies which generate 4- and 5-coordinate Fe, and with Fe and O displacements from ideal positions.

LANTHANUM TITANATE
$LaTiO_3$

J. Less-Common Metals, **116**, 95-104.

Pbnm, a = 5.630-5.625, b = 5.584-5.607, c = 7.901-7.918 A, at 10-298K, Z = 4, neutron powder data at five temperatures. La in 4(c): 0.9927,0.0471,1/4; Ti in 4(b): 0,1/2,0; O(1) in 4(c): 0.0790,0.4934,1/4; O(2) in 8(d): 0.7094,0.2924,0.0420, at 10K. Orthorhombic perovskite, as previously described (**45**A, 235).

MAGNESIUM NEODYMIUM TITANATE
$MgNd_2TiO_6$

Acta Cryst., C**42**, 1472-1475.

$P2_1/n$, 5.4661, 5.5905, 7.7768, 90.010, Z = 2, neutron powder data. Monoclinic perovskite with MgO_6 and TiO_6 octahedra, and 8-coordinate Nd.

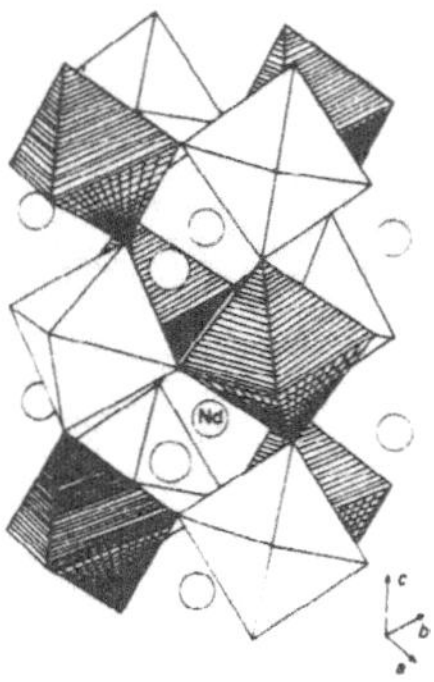

	x	y	z	n
Nd	0·5103	0·5472	0·2503	100
Mg(1)	0·5	0	0	96·0
Mg(2)*	0	0·5	0	4·0
Ti(1)	0	0·5	0	96·0
Ti(2)*	0·5	0	0	4·0
O(1)	0·196	0·210	−0·0494	100
O(2)	0·2852	0·698	−0·042	100
O(3)	0·4130	0·9746	0·2559	100

* 4% of the Mg and Ti ions are disordered, and are indicated as Mg(2) and Ti(2).

MgO_6 octahedra		
Mg–O(1)	2·071 (6)	2×
Mg–O(2)	2·083 (6)	2×
Mg–O(3)	2·051 (7)	2×
TiO_6 octahedra		
Ti–O(1)	1·981 (6)	2×
Ti–O(2)	1·939 (5)	2×
Ti–O(3)	1·962 (7)	2×
NdO_8 bicapped prism		
Nd–O(1)	2·619 (7)	1×
	2·351 (7)	1×
	2·742 (8)	1×
Nd–O(2)	2·718 (8)	1×
	2·627 (8)	1×
	2·397 (8)	1×
Nd–O(3)	2·448 (3)	1×
	2·350 (5)	1×

CALZIRTITE

$Ca_2Zr_5(Ti,Nb,Fe)_2O_{16}$

Amer. Min., **71**, 815-818.

$I4_1/acd$, 15.094, 10.043, Z = 8, R = 0.040. Anion-deficient fluorite superstructure, as previously described (**26**, 379; **49**A).

LITHIUM DIVANADATE

δ-LiV_2O_5

J. Solid State Chem., **65**, 63-71.

Amam, 11.242, 9.9054, 3.6018, Z = 4, neutron and X-ray powder data. Li in 4(c): 3/4,0.109,0; V in 8(f): 0.401,0.206,0; O(1) in 8(f): 0.1269,0.0464,0; O(2) in 8(f): 0.0729,0.2626,1/2; O(3) in 4(c): 1/4,0.2855,0. Insertion of Li into V_2O_5 results in separation, shifting, and puckering of layers to accommodate Li, with a long V-O bond being broken. V has square-pyramidal coordination, V-O = 1.62, 1.87-1.96(2) A, and Li has tetrahedral coordination, Li-O = 2.04, 2.10(3) A (two longer Li-O, 2.38 A).

VANADIUM HOLLANDITES

$K_{0.78}V_{10.44}O_{17.22}$, $K_{0.92}V_{10.16}O_{17.08}$, $Rb_{0.9}V_{10.2}O_{17.1}$

Z. Kristallogr., **176**, 113-124.

I2/m (**c** unique), a = 10.979, 10.979, 10.895, b = 9.383, 9.383, 9.385, c = 2.922, 2.902, 2.909 A, γ = 92.26, 92.26, 92.25°, Z = 1, R = 0.068, 0.074, 0.061. Hollandite-type structures (**13**, 190), but with some O in K sites creating additional oxygen octahedra (which are occupied by extra V).

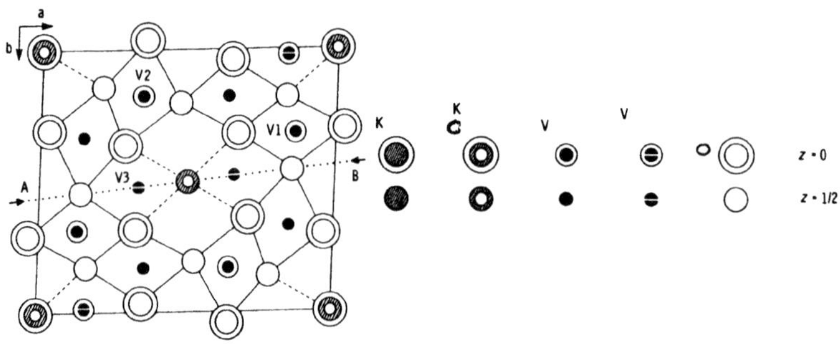

			$K_{0.78}V_{10.44}O_{17.22}$	$K_{0.92}V_{10.16}O_{17.08}$		$Rb_{0.9}V_{10.2}O_{17.1}$	
K, Rb/O in 2a (0,0,0)							
		S.O.F.	= 0.201 ≙ 80.4%	0.199	≙ 79.6%	0.171	≙ 68.4%
V(1) in 4i (x,y,0)	10^4	x	= 1348	1354		1351	
		y	= 6792	6789		6784	
V(2) in 4i		x	= 6535	6537		6538	
		y	= 8341	8346		8349	
						0.65	
V(3) in 4i		x	= 1641	1650		1644	
		y	= 9821	981C		9819	
		S.O.F.	= 0.305 ≙ 61.0%	0.270	≙ 54.0%	0.272	≙ 54.4%
O(1) in 4i		x	= 328	320		329	
		y	= 3040	3030		3057	
O(2) in 4i		x	= 7127	7140		7127	
		y	= 6356	6360		6374	
O(3) in 4i		x	= 3300	3280		3302	
		y	= 6755	6780		6773	
O(4) in 4i		x	= 3551	3580		3542	
		y	= 9595	9610		9594	

$K_{0.72}V_{10}O_{16}$

Mater. Res. Bull., **21**, 567-574.

C2/m, 14.67, 2.889, 10.93, 140.4, Z = 1, R = 0.051. Hollandite-type structure, as for $K_2Cr_8O_{16}$ (**42A**, 269), but with K deficiency and V excess.

			x	y	z
V(1)	4i	1	0.166	0	0.513
V(2)	4i	1	0.320	0	0.186
V(3)	4i	0.5	0.017	0	0.853
K	2a	0.36	0	0	0
O(1)	4i	1	0.325	0	0.997
O(2)	4i	1	0.041	0	0.686
O(3)	4i	1	0.305	0	0.337
O(4)	4i	1	0.363	0	0.649

ARSENIC VANADIUM OXIDE
$As_2V_4O_{13}$

Acta Cryst., C**42**, 1465-1467.

P1, 4.9441, 4.990, 8.8156, 90.12, 100.57, 80.72, Z = 1, R = 0.020. Cubic close-packed O, with four V^{4+} in octahedra sharing edges and corners, and two As^{5+} in tetrahedral sites sharing corners with VO_6 octahedra. V-O = 1.64, 1.92-2.27, As-O = 1.66-1.73 A.

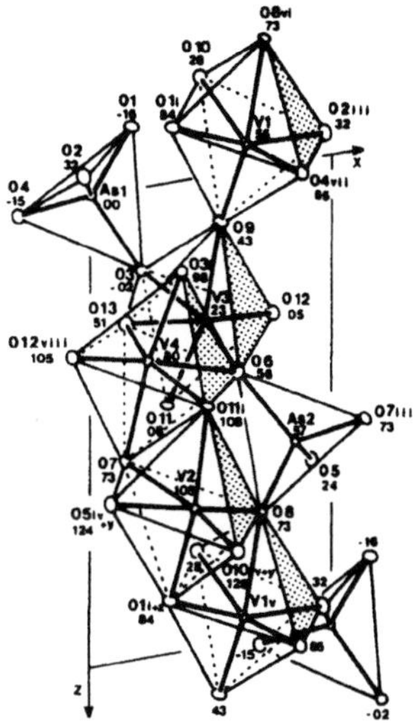

	x	*y*	*z*
As(1)	0·000	0·000	0·000
As(2)	0·75608	0·56610	0·58753
V(1)	0·5484	0·5478	−0·05821
V(2)	0·2606	0·0457	0·68320
V(3)	0·4367	0·2328	0·30307
V(4)	0·1057	0·7989	0·35768
O(1)	0·1916	−0·1581	−0·1275
O(2)	−0·0866	0·3286	−0·0479
O(3)	0·2143	−0·0208	0·1791
O(4)	−0·2715	−0·1466	0·0190
O(5)	0·8780	0·2423	0·6377
O(6)	0·5203	0·5638	0·4191
O(7)	0·0226	0·7274	0·5632
O(8)	0·5865	0·7303	0·7196
O(9)	0·4635	0·4311	0·0973
O(10)	0·3984	0·2819	−0·2141
O(11)	0·3076	0·0756	0·4701
O(12)	0·7489	0·0511	0·3158
O(13)	0·0588	0·5094	0·2743

BISMUTH VANADATE (HIGH-PRESSURE)
$BiVO_4$

Phase Transitions, **6**, 165-173.

$I4_1/a$, a = 5.105, 5.080, c = 11.577, 11.490 A, at 1.6, 4.3 GPa, Z = 4, R = 0.048, 0.051. Scheelite structure, as previously described for a high-temperature form, the normal pressure and temperature form being monoclinic (**45**A, 239). Increasing pressure results mainly in Bi-O polyhedral compression.

POTASSIUM TRIDECAVANADOMANGANATE(IV) and -NICKELATE OCTADECAHYDRATES
$K_7MV_{13}O_{38}.18H_2O$ (M = Mn, Ni)

Chem. Lett., 1267-1270 (1986); ibid., 1123-1124 (1975).

I4/mmm, a = 11.366, 11.37, c = 19.638, 19.56 A, Z = 2, R = 0.097, 0.089. Heteropolyanions of condensed octahedra, with disorder resulting in apparent higher symmetry (V(2) and O(2) occupancies of 3/4; partial occupancies for K, H_2O).

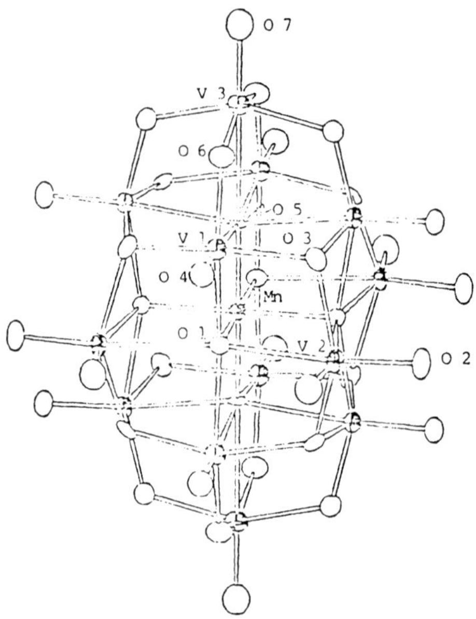

BARIUM COBALT(II) VANADATE(V)
BARIUM MAGNESIUM VANADATE(V)
$BaCo_2V_2O_8$, $BaMg_2V_2O_8$

Z. anorg. Chem., **534**, 153-158.

$I4_1/acd$, a = 12.4441, 12.4189, c = 8.4153, 8.4657 A, Z = 8, R = 0.047, 0.050. VO_4 tetrahedra, CoO_6 or MgO_6 octahedra, and 12-coordinate Ba. V-O = 1.69-1.74, Co-O = 2.04-2.17, Mg-O = 2.05-2.16, Ba-O = 2.76-3.11(1) A.

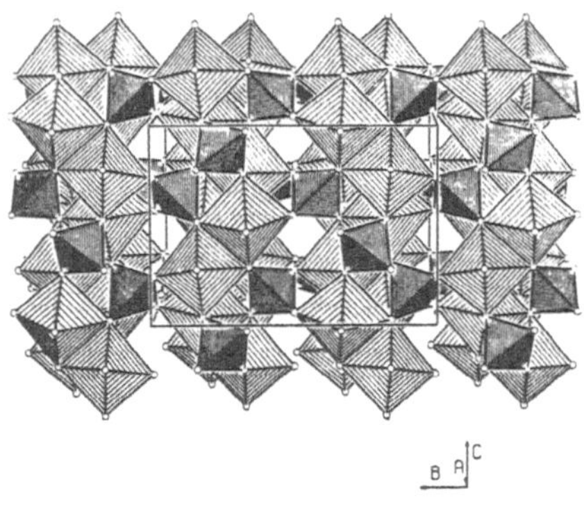

		x	y	z
Ba	(8a)	0,0	0,25	0,375
V	(16e)	0,0800	0,0	0,25
Co	(16f)	0,1689	0.4189	0,125
O_I	(32g)	0,157	0,074	0,383
O_{II}	(32g)	0,498	0,087	0,347
Ba	(8a)	0,0	0,25	0,375
V	(16e)	0,0810	0,0	0,25
Mg	(16f)	0,1682	0,4182	0,125
O_I	(32g)	0,158	0,074	0,384
O_{II}	(32g)	0,499	0,087	0,346

STRONTIUM NICKEL VANADATE
$SrNi_2V_2O_8$

Rev. Chim. Minér., **23**, 1-7.

$I4_1cd$, 12.1617, 8.3247, Z = 8, R = 0.076. Not isostructural with the Ba compound (**52**A, 202), but a new structure type, with NiO_6 octahedra, VO_4 tetrahedra, and 12-coordinate Ba. Ni-O = 1.98-2.25, V-O = 1.63-1.78, Ba-O = 2.65-3.20(2) A.

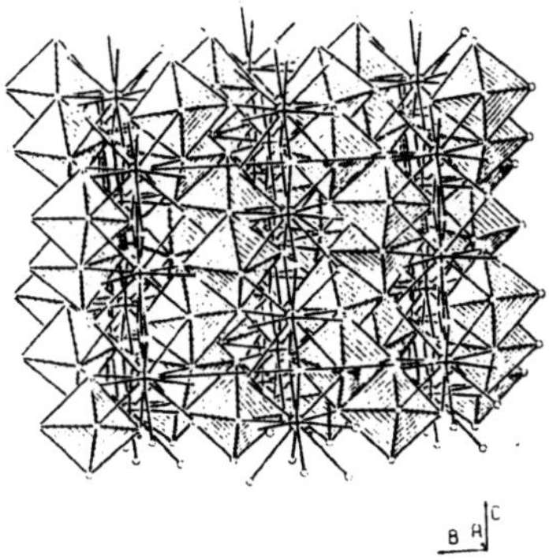

		x	*y*	*z*
Sr	(8 *a*)	0,0	0.0	0,0
V	(16 *b*)	0,2596	0.0802	0,1015
Ni	(16 *b*)	0,3283	0.3345	0,2294
O_I	(16 *b*)	0,153	0.499	0,013
O_{II}	(16 *b*)	0,350	0,672	0,488
O_{III}	(16 *b*)	0,164	0,678	0,725
O_{IV}	(16 *b*)	0.332	0.497	0,204

SILVER THALLIUM(I) METAVANADATE

$AgTl(VO_3)_2$

Mater. Res. Bull., **21**, 225-229.

C2/c, 10.736, 10.227, 5.8299, 102.32, Z = 4, R = 0.078. Diopside structure (**2**, 130), with VO_3 chains of corner sharing VO_4 tetrahedra, octahedrally coordinate Ag, and 8-coordinate Tl. V-O = 1.64-1.81, Ag-O = 2.44-2.47, Tl-O = 2.72-3.06 A.

ZINC DIVANADATE(V)

$Zn_4V_2O_9$

Mh. Chem., **117**, 131-138.

$P2_1$, 10.488, 8.198, 9.682, 118.66, Z = 4, R = 0.085. The structure contains a Zn/O framework with (4+1)-, and 5-, and (5+1)-coordinated Zn, incorporating tetrahedrally-coordinated V. V-O = 1.65-1.84, Zn-O = 1.93-2.70(2) A.

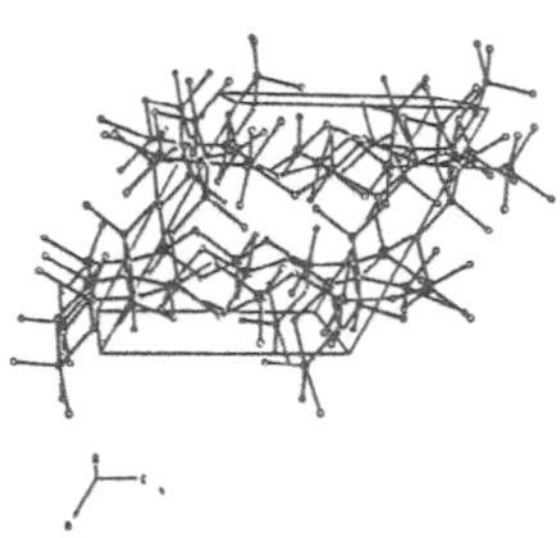

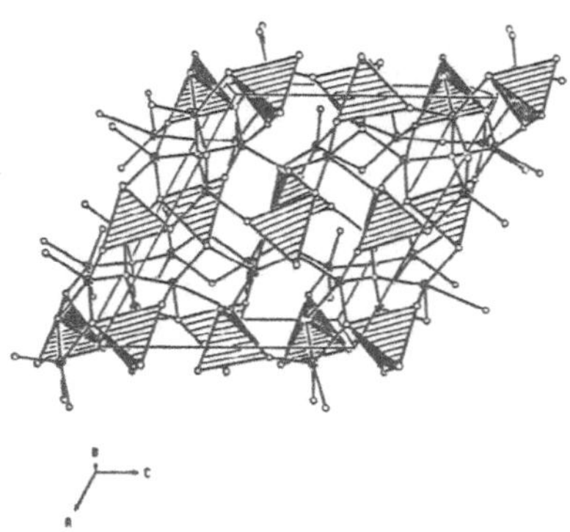

	x	y	z				
Zn_I	0,2245	0,4462	0,1637	O_{VI}	0,764	0,486	0,987
Zn_{II}	0,2581	0,9476	0,5082	O_{VII}	0,639	0,467	0,247
Zn_{III}	0,7871	0,3926	0,1880	O_{VIII}	0,396	0,143	0,527
Zn_{IV}	0,2596	0,3226	0,5154	O_{IX}	0,080	0,632	0,066
Zn_V	0,2301	0,8296	0,1611	O_X	0,747	0,832	0,004
Zn_{VI}	0,3901	0,1471	0,1025	O_{XI}	0,512	0,161	0,328
Zn_{VII}	0,2811	0,4682	0,8429	O_{XII}	0,152	0,670	0,829
Zn_{VIII}	0,9147	0,6825	0,1057	O_{XIII}	0,605	0,816	0,223
V_I	0,4800	0,1293	0,8394	O_{XIV}	0,539	0,119	0,038
V_{II}	0,9702	0,1186	0,5025	O_{XV}	0,952	0,571	0,301
V_{III}	0,5564	0,1463	0,5153	O_{XVI}	0,872	0,764	0,794
V_{IV}	0,9987	0,1215	0,1488	O_{XVII}	0,122	0,795	0,535
O_I	0,631	0,136	0,820	O_{XVIII}	0,937	0,428	0,801
O_{II}	0,156	0,469	0,604				
O_{III}	0,670	0,976	0,595				
O_{IV}	0,660	0,321	0,605				
O_V	0,890	0,635	0,545				

CADMIUM PYROVANADATE
$Cd_2V_2O_7$

Kristallografija, **31**, 1222-1223 [Soviet Physics - Crystallography, **31**, 722-723].

C2/m, 7.077, 8.996, 4.976, 103.20, Z = 2, R = 0.049. Structure as previously described (**32**A, 274), with linear O_3V-O-VO_3 groups, linked by CdO_6 octahedra. V-O = 1.757 (bridging), 1.712 (terminal), Cd-O = 2.278-2.398(4) A.

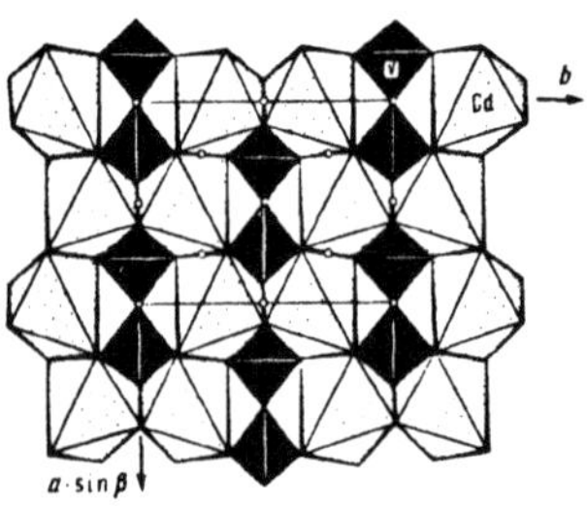

	x	y	z
Cd	0	0.30480	0.5
V	0.2259	0	0.9095
O_1	0	0	0
O_2	0.3966	0	0.213
O_3	0.2375	0.1559	0.7231

FRANCEVILLITE
$(Ba,Pb)[(UO_2)_2V_2O_8].5H_2O$

Neues Jb. Miner., Mh., 552-560.

Pcan, a = 10.419, 10.479, b = 8.510, 8.450, c = 16.763, 16.645 A, for two specimens with Ba:Pb = 96:4 and 69:31, respectively, Z = 4, R = 0.025, 0.027. Structure essentially as previously described for curienite and francevillite (**37**A, 239; **41**A, 410).

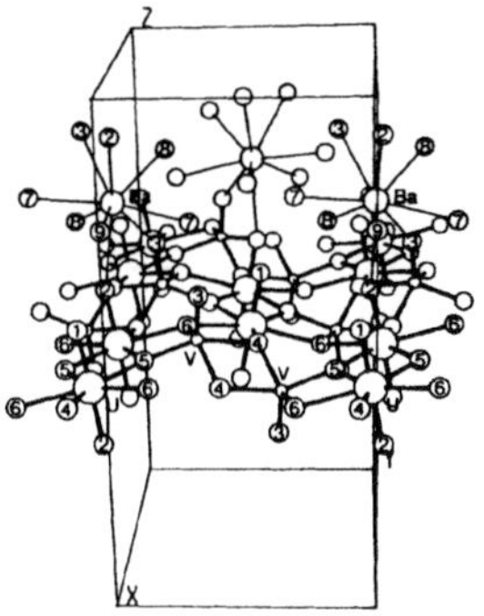

	x	y	z
Ba,Pb	0.33230	0	1/4
	0.33909	0	1/4
U	0.18504	0.02194	−0.00376
	0.18468	0.02088	−0.00357
V	0.03121	0.34636	0.04243
	0.03180	0.34726	0.04300
O(1)	0.1704	0.0604	−0.1084
	0.1695	0.0606	−0.1087
O(2)	0.2031	−0.0212	0.1008
	0.2023	−0.0235	0.1024
O(3)	−0.0056	0.3596	0.1355
	−0.0030	0.3591	0.1363
O(4)	−0.0976	0.4379	−0.0203
	−0.0980	0.4388	−0.0188
O(5)	−0.0146	0.1524	0.0127
	−0.0129	0.1524	0.0130
O(6)	0.1997	0.2902	0.0314
	0.1997	0.2904	0.0333
W(7)	0.2551	0.3033	0.2251
	0.2560	0.3004	0.2251
W(8)	0.5199	0.1868	0.3236
	0.5210	0.1900	0.3225
W(9)	0.0372	0	1/4
	0.0323	0	1/4

LITHIUM NIOBATE
$LiNbO_3$

Acta Cryst., B**42**, 61-68.

R3c, a = 5.14739, 5.15052, c = 13.85614, 13.86496 A, at 298K, for stoichiometric and congruent compositions, Z = 6, R = 0.012, 0.012. Structure as previously described (**31**A, 147). For the congruent composition, 6% Li is missing from the Li sites and replaced by Nb, with compensating vacancies at the Nb site.

Atomic coordinates ($\times 10^5$)
stoichiometric (*upper*) *and congruent* (*lower*)

	x	*y*	*z*
Li	0	0	27 872
	0	0	27 909
Nb	0	0	0
	0	0	0
O	4757	34 328	6336
	4790	34 299	6385

POTASSIUM HEPTANIOBATE
$K_3Nb_7O_{19}$

J. Solid State Chem., **61**, 181-187.

$P\bar{1}$, 14.143, 9.948, 6.463, 106.45, 95.82, 97.29, Z = 2, R = 0.049. Edge- and corner-sharing NbO_6 octahedra, with tunnels containing K ions.

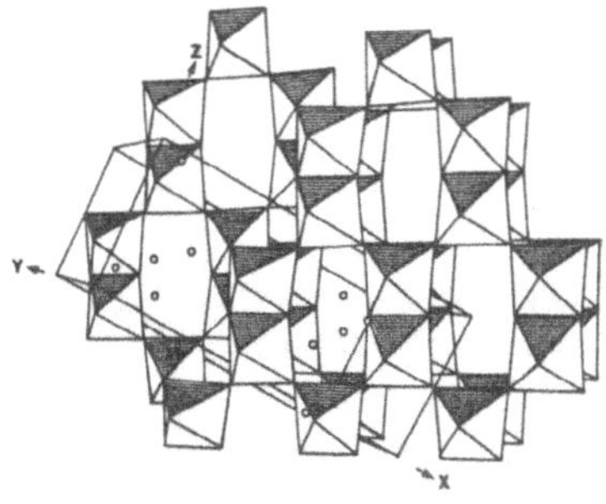

STRONTIUM DINIOBATE
$SrNb_2O_6$

STRONTIUM HEXANIOBATE
$SrNb_6O_{16}$

Acta Chem. Scand., A**40**, 467-475.

$SrNb_2O_6$, $P2_1/c$, 7.7223, 5.5944, 10.9862, 90.372, Z = 4, powder data. Almost isostructural with orthorhombic $CaTa_2O_6$ (**28**, 201).

$SrNb_6O_{16}$, Pmm2, 3.9563, 10.1889, 14.794, Z = 2, powder data for two-phase sample. Structure derived from comparison with $NaNb_6O_{15}F$ (**30**A, 335).

BARIUM TITANIUM NIOBIUM OXIDE
$BaTi_3Nb_4O_{17}$

Mater. Res. Bull., **21**, 1039-1044.

Cmcm, 6.639, 8.957, 21.145, Z = 4, powder data. Structure given derived from that of rutile.

BARIUM TITANIUM(IV) NIOBATE(V)
$Ba_6TiNb_4O_{18}$

Acta Cryst., C**42**, 266-268.

$R\bar{3}m$, 5.7852, 42.4886, Z = 3, neutron powder data. 18L structure with BaO_3 layers stacked $(hhcccc)_3$ to give a perovskite lattice with a shear plane every sixth layer; Ti and Nb are partially ordered in octahedral holes, with empty octahedra in the shear planes.

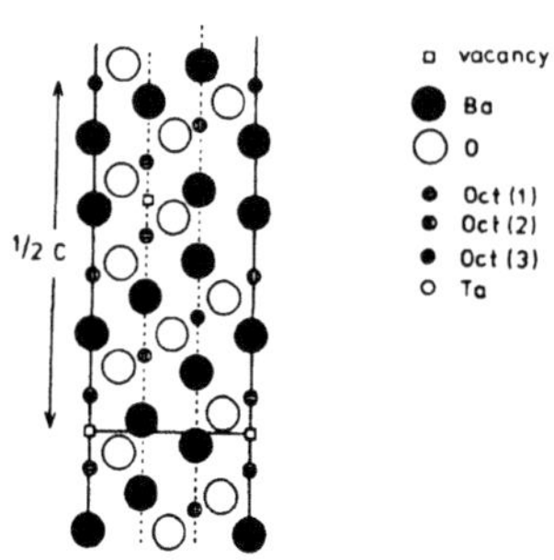

	$x\ (= -y)$	z	n (%)
Ba(1)	0	0·1391	100
Ba(2)	0	0·3134	100
Ba(3)	0	0·4144	100
Ti(1)	0	0·0512	6·3
Nb(1)	0	0·0512	93·7
Ti(2)	0	0·2235	28·2
Nb(2)	0	0·2235	71·8
Ti(3)	0	0·5	30·7
Nb(3)	0	0·5	69·3
O(1)	0·4994	0·13993	100
O(2)	0·5030	0·30201	100
O(3)	0·4968	0·41950	100

$Ba_6Ti\square Nb_4O_{18}$	
Ba(1)—O(1)*a*	2·893 (4)
—O(1)	2·851 (4)
—O(2)	2·821 (3)
Ba(2)—O(2)*a*	2·901 (5)
—O(1)	3·004 (3)
—O(3)	2·714 (3)
Ba(3)—O(3)*a*	2·933 (8)
—O(3)	2·762 (3)
—O(2)	3·256 (4)
Oct(1)—O(3)	1·846 (2)
—O(2)	2·261 (3)
Oct(2)—O(1)	2·104 (4)
—O(2)	1·920 (3)
Oct(3)—O(1)	2·017 (2)
O(1)—O(11)*a*	2·897 (4)
—O(1)*a*	2·888 (4)
—O(1)	2·818 (3)
—O(2)	2·820 (2)
O(2)—O(2)*a*	2·954 (3)
—O(2)*a*	2·832 (3)
—O(3)	2·869 (2)
O(3)—O(3)*a*	2·944 (2)
—O(3)*a*	2·841 (2)
—O(3)	3·127 (2)

BARIUM IRON NIOBATE
$Ba_6FeNb_9O_{30}$

Mh. Chem., **117**, 1239-1244.

P4bm, 12.597, 3.990, Z = 1, R = 0.053. Tetragonal tungsten bronze structure (**12**, 292; **39**A, 244), with Fe/Nb disorder in the octahedral framework.

		x	y	z
Ba_I	(2 a)	0.0	0.0	0.0
Ba_{II}	(4 c)	0.1721	0.6721	0.0060
Nb_I/Fe_I	(2 b)	0.5	0.0	0.5353
Nb_{II}/Fe_{II}	(8 d)	0.0754	0.2148	0.4919
O_I	(2 b)	0.5	0.0	0.031
O_{II}	(8 d)	0.079	0.208	0.021
O_{III}	(8 d)	0.343	0.008	0.494
O_{IV}	(4 c)	0.283	0.783	0.469
O_V	(8 d)	0.139	0.069	0.500

Ba_I—O_V	: 2.79 (3) (8 ×) Å
—O_{II}	: 2.80 (2) (4 ×)
Ba_{II}—O_{IV}	: 2.70 (4)
—O_{III}	: 2.85 (3) (2 ×)
—O_{III}	: 2.90 (3) (2 ×)
—O_{IV}	: 2.90 (4)
—O_I	: 3.07 (1)
—O_{II}	: 3.17 (2) (2 ×)
—O_V	: 3.35 (3) (2 ×)
—O_V	: 3.38 (3) (2 ×)
—O_{II}	: 3.50 (2) (2 ×)
Nb_I/Fe_I—O_I	: 1.98 (8)
—O_{III}	: 1.98 (2) (4 ×)
—O_I	: 2.01 (8)
Nb_{II}/Fe_{II}—O_{II}	: 1.88 (3)
—O_{III}	: 1.93 (2)
—O_{IV}	: 1.98 (2)
—O_V	: 2.00 (2)
—O_V	: 2.05 (2)
—O_{II}	: 2.11 (3)

NICKEL DINIOBATE (FORM II)

$Ni_4Nb_2O_9$

Z. anorg. Chem., **539**, 203-210.

Pcan, 5.0545, 8.7688, 14.3041, Z = 4, R = 0.054, 0.061, 0.077 for data from three individual parts of a triple crystal (which had previously suggested a cell with doubled **a** and **b** axes). A complex arrangement of NiO_6 and NbO_6 octahedra, which differs from that in the Fd2d form (**52A**, 206). Ni-O = 1.92-2.20, Nb-O = 1.79-2.29(4) A.

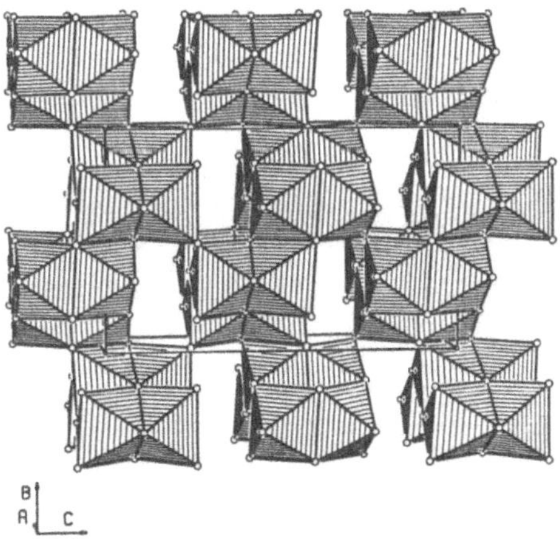

NiO_6

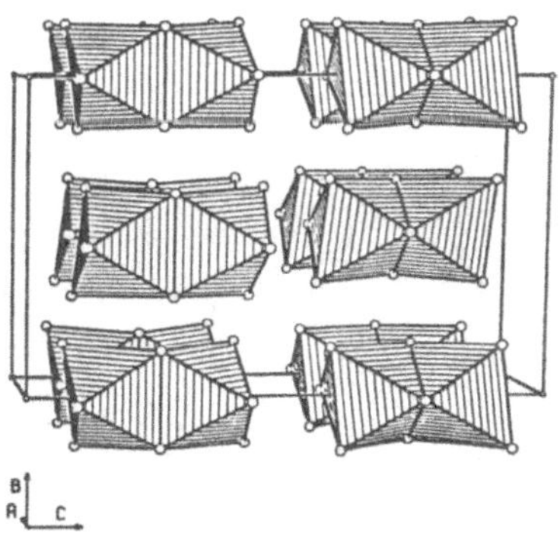

NbO_6

		x	y	z
Nb	(8d)	0,0018	0,0292	0,1447
Ni_I	(8d)	0,0013	0,3363	0,3123
Ni_{II}	(8d)	0,9797	0,3261	0,4987
O_I	(4c)	0,296	0,0	0,25
O_{II}	(8d)	0,169	0,167	0,074
O_{III}	(8d)	0,171	0,164	0,405
O_{IV}	(8d)	0,125	0,152	0,749
O_V	(8d)	0,169	0,507	0,083

ZINC NIOBATES and TANTALATES
$Zn(V,Nb)_2O_6$ (I), $Zn(V,Ta)_2O_6$ (II), $Zn(Nb,Ta)_2O_6$ (III)

Z. anorg. Chem., **542**, 31-36.

I. Pbcn, 14.113, 5.693, 5.025, Z = 4, R = 0.086. Niobite-type structure (**2**, 55, 337; **37**A, 243). V,Nb-6 O = 1.84-2.23, Zn-6 O = 2.03-2.17(1) A.

II. $P4_2/mnm$, 4.712, 9.191, Z = 2, R = 0.052. Trirutile-type structure. V,Ta-6 O = 1.98, 2.01, Zn-6 O = 2.05, 2.10(2) A.

III. Pbcn, 4.710, 17.103, 5.072, Z = 4, R = 0.107. Tri-α-PbO_2-type structure (**51**A, 226). Nb,Ta-6 O = 1.92-2.13, Zn-6 O = 2.00-2.30(3) A.

HYDROGEN TANTALATE HYDRATE
$(D_2O)_{0.38}D_2Ta_2O_6$

Solid State Comm., **59**, 569-573.

Fd3m, 10.5724, Z = 8, neutron powder data. Ta in 16(c): 0,0,0; O in 48(f): x,1/8,1/8, x = 0.3121; D in 48(f): x = 0.4105; 3.04 O(W) in 8(b): 3/8,3/8,3/8; 6.08 D(W) in 32(e): x,x,x, x = 0.322. Defect pyrochlore.

SODIUM TANTALATE
$Na_2Ta_4O_{11}$

Rev. Chim. Minér., **22**, 817-820 (1985).

$R\bar{3}c$, a = 6.198, c = 36.56 A, Z = 6, R = 0.048. TaO_7 pentagonal bipyramids connected by TaO_6 octahedra, as in $Na_2Nb_4O_{11}$ (**35**A, 257).

PARABARIOMICROLITE
$BaTa_4O_{10}(OH)_2.2H_2O$

Canad. Miner., **24**, 655-663.

$R\bar{3}m$, 7.4290, 18.505 A, Z = 3, powder data. 3Ba in 3(b): 0,0,1/2; 3 Ta(1) in 3(a): 0,0,0; 3Ta(2) in 9(d): 1/2,0,1/2; 18 O(1) in 18(h): x,-x,z, x = 0.1253, z = 0.1877; 12 O(2) + 6 OH in 18(h): x = 0.1253, z = 0.9377; 6 H_2O in 6(c): 0,0,3/8. Layer structure derived from the structure of pyrochlore. One layer has 6- and 3-membered rings of corner-sharing TaO_6 octahedra, with 8-coordinate Ba at the centre of the 6-membered rings; the second layer is sparsely occupied by TaO_6 octahedra.

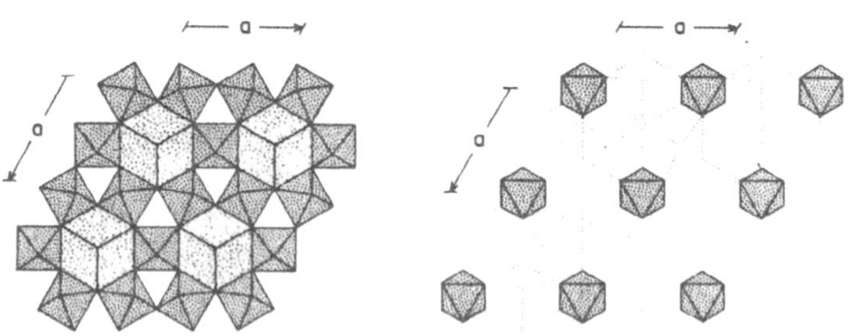

IRON(II) DITANTALATE(V)
$FeTa_2O_6$

J. Solid State Chem., **62**, 220-230.

$P4_2/mnm$, 4.749, 9.192, Z = 2, neutron powder data. Fe in 2(a): 0,0,0; Ta in 4(e): 0,0,0.333; O(1) in 4(f): 0.307,0.307,0; O(2) in 8(j): 0.297,0.297,0.322. Trirutile structure.

BARIUM IRON TANTALATE
$Ba_4FeTa_{10}O_{30}$

Z. anorg. Chem., **542**, 18-24.

Amm2, 3.906, 10.201, 14.990, Z = 1, R = 0.071. Framework of corner-sharing TaO_6 octahedra, with Ba and Fe(II) in tunnels. Ta-6 O = 1.78-2.08, Ba-13 or 15 O = 2.66-3.41, Fe-12 O = 2.35-3.36 A. Fe site is half occupied.

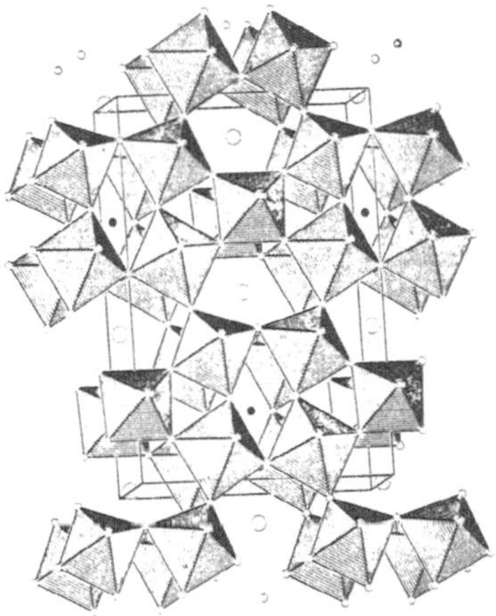

			x	y	z
	Ta_I	(2a)	0,0	0,0	0,2056
	Ta_{II}	(4d)	0,0	0,3167	0,0819
	Ta_{III}	(4d)	0,0	0,3154	0,3346
	Ba_I	(2b)	0,5	0,0	0,4064
	Ba_{II}	(2b)	0,5	0,0	0,0033
0,5	Fe	(2b)	0,5	0,0	0,7002
	O_I	(2b)	0,5	0,0	0,195
	O_{II}	(4e)	0,5	0,335	0,091
	O_{III}	(4e)	0,5	0,315	0,358
	O_{IV}	(4d)	0,0	0,125	0,113
	O_V	(4d)	0,0	0,151	0,292
	O_{VI}	(4d)	0,0	0,264	0,455
	O_{VII}	(4d)	0,0	0,367	0,209
	O_{VIII}	(2a)	0,0	0,0	0,882
	O_{IX}	(2a)	0,0	0,0	0,541

NICKEL DITANTALATE
$NiTa_2O_6$

Z. anorg. Chem., **536**, 15-23.

$P4_2/mnm$, 4.7219, 9.150, Z = 2, R = 0.074. Ni in 2(a): 0,0,0; Ta in 4(e): 0,0,0.3316; O(1) in 4(f): 0.307,0.307,0; O(2) in 8(j): 0.297,0.297,0.327. Trirutile-type. Ni-6 O = 2.05, 2.08, Ta-6 O = 1.98-2.01(3) A. Other AB_2O_6 compounds have partially-ordered tri-α-PbO_2 type structures.

LANTHANUM TANTALATE
$LaTa_7O_{19}$

Z. anorg. Chem., **543**, 117-128.

$P\bar{6}c2$, 6.236, 19.995, Z = 2, R = 0.068. Double (001) layers of TaO_7 pentagonal bipyramids, linked by La and Ta coordination polyhedra. La-O = 2.53, Ta-O = 1.90-2.46 A.

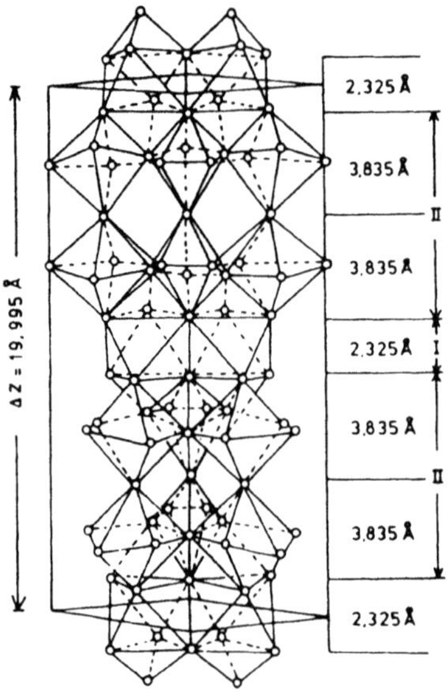

		x	y	z
La(1)	2c	1/3	2 3	0.0
Ta(1)	2e	2 3	1 3	0,0
Ta(2)	12l	0.3610	0.0020	0,1563
O(1)	12l	0,760	0,01	0,156
O(2)	4i	2 3	1 3	0,164
O(3)	4h	1 3	2 3	0.127
O(4)	6k	0.37	−0.041	1 4
O(5)	12l	0.423	0.056	0.061

NEODYMIUM TANTALATES
$NdTaO_4$, $NdTa_3O_9$

Kristallografija, **31**, 892-896 and 897-900 [Soviet Physics - Crystallography, **31**, 528-530 and 531-533].

$NdTaO_4$, Pa, 5.422, 5.596, 5.242, 96.76, Z = 2, R = 0.057. $NdTa_3O_9$, P4mm, 7.822, 3.883, Z = 4/3 (4 x $Nd_{0.33}TaO_3$), R = 0.067. The $NdTaO_4$ structure contains {110} sheets of edge-sharing NdO_8 polyhedra, linked by columns of edge-sharing TaO_6 octahedra. $NdTa_3O_9$ contains a framework of corner-sharing TaO_6 octahedra, with two partially-occupied Nd sites.

$NdTaO_4$

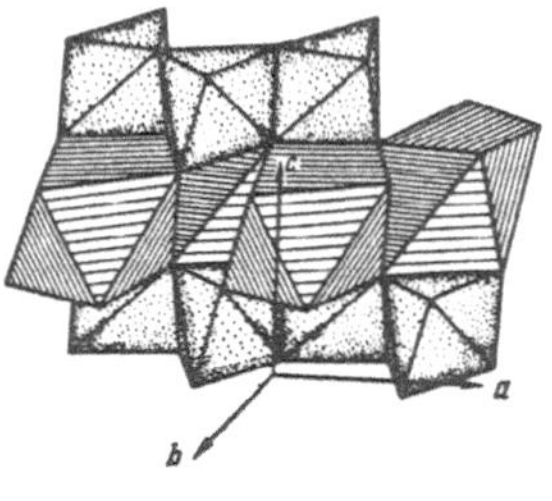

	x	y	z
Nd	0	0.2349	0
Ta	0.4946	0.3132	0.5059
O_1	0.142	0.904	0.748
O_2	0.252	0.563	0.283
O_3	0.363	0.086	0.262
O_4	0.262	0.439	0.728

$NdTa_3O_9$ ($Na_{0.33}TaO_3$)

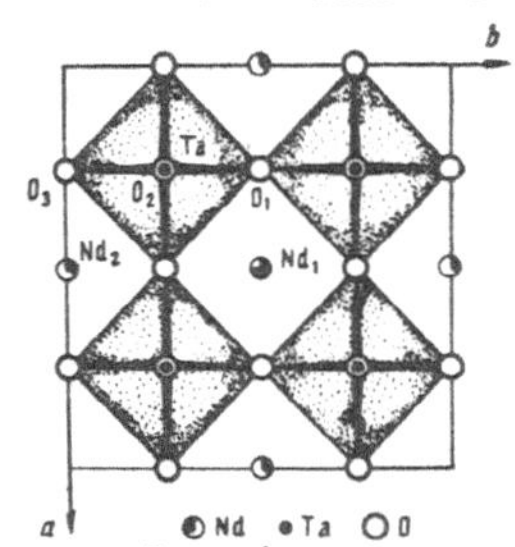

	Multiplicity of position	Degree of filling of position	x	y	z
Nd_1	1	0.736	0 5	0.5	0.5
Nd_2	2	0.296	0.5	0	0.545
Ta	4	1	0 2422	0.2422	0.021
O_1	4	1	0.266	0.5	0.025
O_2	4	1	0 256	0 256	0.504
O_3	4	1	0 272	0	−0.019

CALCIUM CHROMATE(V,VI)
$Ca_{10}(CrO_4)_6(CrO_4)$ (I)

CALCIUM CHROMATE(IV,VI)
$Ca_5(CrO_4)_2(CrO_4)$ (II)

Proc. 10th Conf. Coord. Chem., 131-137 (1985).

I, R3c, a = 14.148 A, α = 44.7°, Z = 2 (hexagonal cell, 10.776, 38.119, Z = 6), R = 0.031. II, Pnma, 6.742, 16.078, 10.345, Z = 4, R = 0.031. Differences in Cr-O distances indicate two oxidation states in each compound.

MONAZITE-TYPE STRUCTURES
$SrCrO_4$, $SrSeO_4$, $PbCrO_4$ (crocoite), $PbSeO_4$

Z.Kristallogr., **176**, 75-83.

$P2_1/n$, a = 7.065, 7.101, 7.127, 7.154, b = 7.375, 7.340, 7.438, 7.407, c = 6.741, 6.874, 6.799, 6.954 A, β = 103.08, 103.48, 102.43, 103.14°, Z = 4, R = 0.038, 0.033, 0.039, 0.039. Monazite-type structures (**9**, 236; **11**, 374; **13**, 316; **17**, 487; **23**, 455; **30**A, 320; **32**A, 358; **33**A, 402).

	x	*y*	*z*
Sr/Pb	0.22813	0.15869	0.39806
	0.22082	0.15773	0.39876
	0.22130	0.14545	0.39692
	0.21598	0.14932	0.40191
Cr/Se	0.19769	0.16487	0.88691
	0.19364	0.16510	0.88588
	0.20107	0.16364	0.88184
	0.19595	0.16362	0.88195
O(1)	0.2584	0.0055	0.0562
	0.2484	0.0052	0.0552
	0.2538	0.0042	0.0574
	0.2397	0.0057	0.0490
O(2)	0.1201	0.3373	0.0024
	0.1154	0.3375	−0.0004
	0.1245	0.3425	−0.0110
	0.1196	0.3396	−0.0142
O(3)	0.0256	0.1012	0.6981
	0.0192	0.1094	0.6935
	0.0295	0.0999	0.6858
	0.0252	0.1102	0.6851
O(4)	0.3776	0.2179	0.7881
	0.3752	0.2167	0.7872
	0.3859	0.2141	0.7819
	0.3806	0.2083	0.7881

HASHEMITE
$BaCrO_4$

Amer. Min., **71**, 1217-1220.

Pnma, 9.1028, 5.5276, 7.3314, Z = 4, R = 0.048. $BaSO_4$ structure type (**1**, 343; **43**A, 278; **44**A, 272). Cr-O = 1.63-1.65, Ba-O = 2.73-2.93 A.

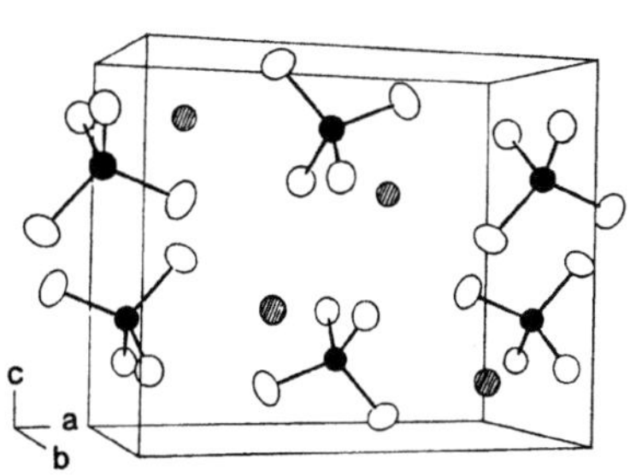

	x	*y*	*z*
Ba	0.81690	3/4	−0.15551
Cr	0.44225	3/4	0.19404
O(1)	0.6099	3/4	0.1146
O(2)	0.3185	3/4	0.0330
O(3)	0.4180	0.5078	0.3231

THALLIUM(III) CHROMATE
$Tl_2(CrO_4)_3$

Rev. Chim. Minér., **23**, 70-79.

Pbcn, 12.869, 8.716, 8.699, Z = 4, R = 0.049. CrO_4 tetrahedra linked by distorted TlO_6 octahedra.

AMMONIUM DODECAPEROXODECAMOLYBDATE HYDRATE
$(NH_4)_8[Mo_{10}O_{22}(O_2)_{12}].16H_2O$

Acta Chem. Scand., A**40**, 83-90.

$P2_1/c$, 12.726, 18.722, 12.530, 110.50, Z = 2, R = 0.035. The crystals are not fully stoichiometric. The complex anion contains edge- and corner-sharing MoO_7 pentagonal bipyramids.

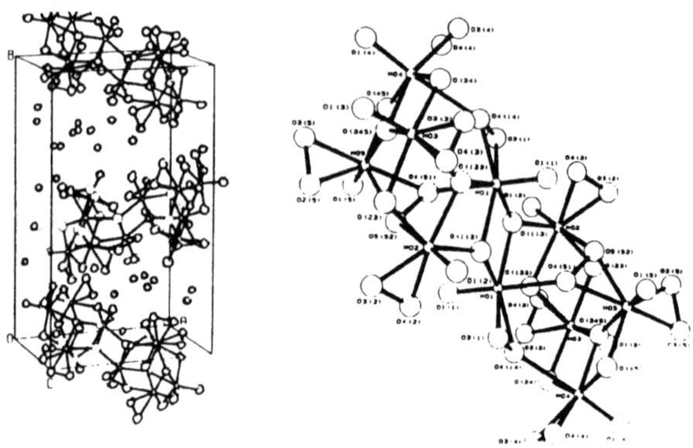

LITHIUM MOLYBDATE BRONZE
$Li_{0.33}MoO_3$

J. Solid State Chem., **64**, 47-56.

$P\bar{1}$, 13.079, 15.453, 7.476, 96.97, 106.56, 103.368, Z = 24, R = 0.032. Clusters of six edge- and corner-sharing MoO_6 octahedra, linked into a three-dimensional structure by further edge- and corner-sharing, and by LiO_6 octahedra.

LITHIUM OCTAMOLYBDATE
$LiMo_8O_{10}$

ZINC OCTAMOLYBDATE
$ZnMo_8O_{10}$

J. Solid State Chem., **64**, 347-358.

$I4_1md$, a = 5.8515, 5.8961, c = 24.783, 24.840 A, Z = 4, R = 0.042, 0.026. Infinite chains along **a** and **b** of edge-sharing Mo_6 octahedral clusters, cross-linked by Mo-O-Mo bonds. Mo-Mo = 2.59-3.31, Li-4 O = 1.70, 1.96, Zn-4 O = 1.88, 1.93 A.

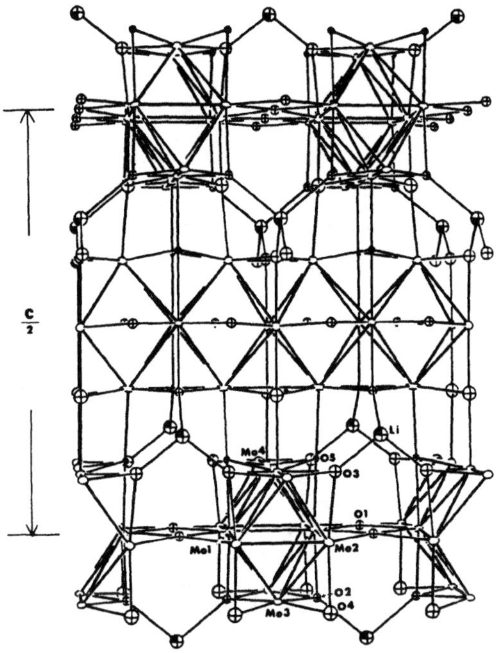

Positional Parameters for $LiMo_8O_{10}$

		Multiplier	x	y	z
Mo1	8b	0.5	0.5	0.2262	0.0
Mo2	8b	0.5	0.0	0.2353	0.00429
Mo3	8b	0.5	0.0	0.2480	0.32808
Mo4	8b	0.5	0.0	0.2760	0.17549
O1	8b	0.5	0.5	0.246	0.250
O2	8b	0.5	0.5	0.239	0.0811
O3	8b	0.5	0.5	0.230	0.4224
O4	8b	0.5	0.0	0.242	0.085
O5	8b	0.5	0.0	0.263	0.4144
Li	4a	0.25	0.5	0.5	0.131

Positional Parameters for $ZnMo_8O_{10}$

		Multiplier	x	y	z
Mo1	8b	0.5	0.5	0.2192	0.0
Mo2	8b	0.5	0.0	0.2389	0.00451
Mo3	8b	0.5	0.0	0.2715	0.32858
Mo4	8b	0.5	0.0	0.2752	0.17670
O1	8b	0.5	0.5	0.244	0.2538
O2	8b	0.5	0.5	0.226	0.0814
O3	8b	0.5	0.5	0.238	0.4236
O4	8b	0.5	0.0	0.266	-0.0876
O5	8b	0.5	0.0	0.276	0.4168
Zn	4a	0.25	0.5	0.5	0.12030

SODIUM TETRAMOLYBDATE
$NaMo_4O_6$

J. Amer. Chem. Soc., **101**, 3963-3964 (1979).

P4/mbm, 9.559, 2.860, Z = 2, R = 0.046. Molybdate chains with 8-coordinate Na ions in channels.

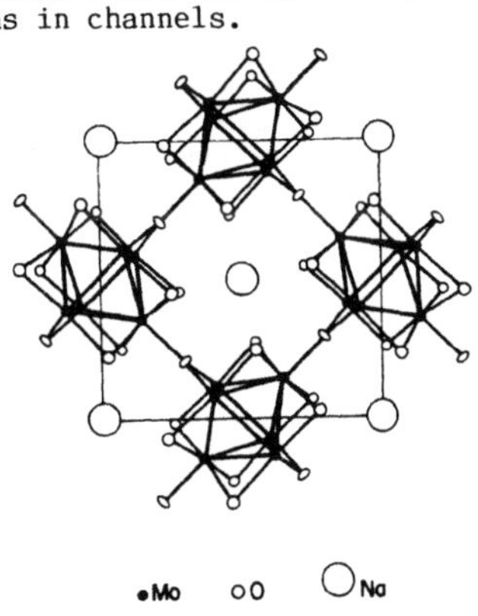

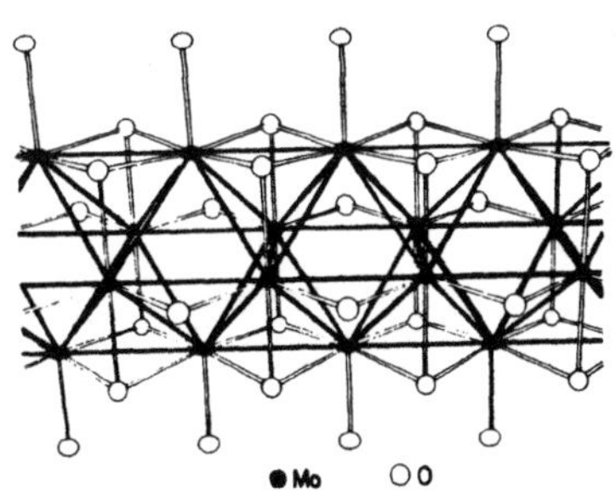

BARIUM TETRAMOLYBDATE
$Ba_5(Mo_4O_6)_8$

J. Less-Common Metals, **116**, 169-186.

Pbam, 9.517, 9.822, 22.813, Z = 16 (subcell with c' = c/8), R = 0.064. $NaMo_4O_6$ (preceding report) superstructure, with Ba ordering in five of eight sites in channels of infinite edge-shared octahedral molybdate chains.

POTASSIUM DECAOXOOCTAPEROXOPENTAMOLYBDATE(VI) PENTAHYDRATE
$K_6[Mo_5O_{10}(O_2)_8].5H_2O$

Acta Chem. Scand., A**40**, 1-7.

$P2_1/n$, a = 15.320, 15.364, b = 12.722, 12.784, c = 13.956, 13.97, β = 92.47, 92.44, at 173, 298K, Z = 4, R = 0.025 at 173K. The complex anion contains one Mo with tetrahedral coordination (Mo-O = 1.77-1.78 A) and four Mo with pentagonal bipyramidal coordinations (Mo-O = 1.69-2.43 A). Anions are linked by 6- to 10-coordinate K ions, and by hydrogen bonding via water molecules.

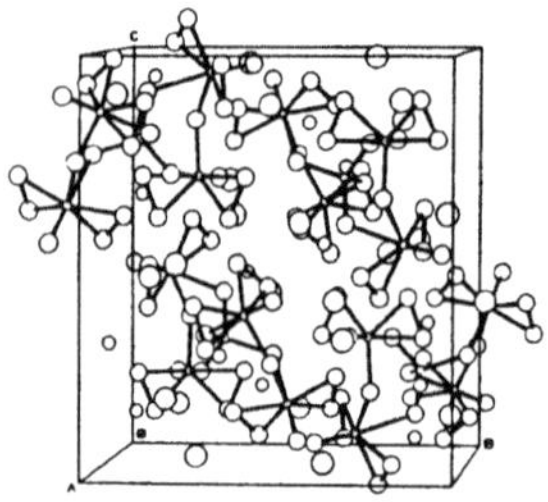

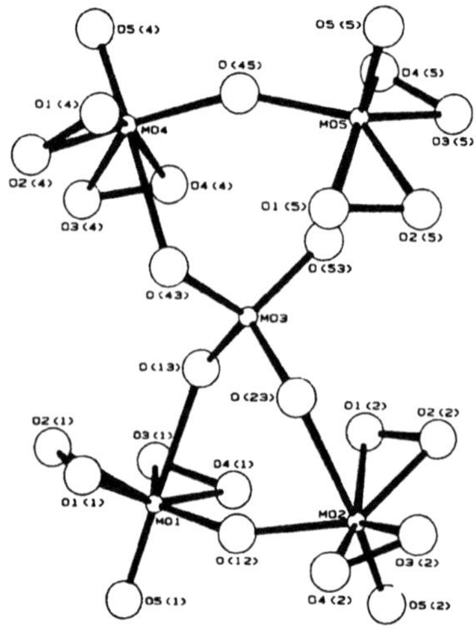

POTASSIUM DIPEROXOHEPTAMOLYBDATE(VI) HYDRATE
$K_6Mo_7O_{22}(O_2)_2.10H_2O$

Acta Chem. Scand., A**40**, 335-343.

$P2_1/c$, 16.629, 10.538, 19.704, 103.56, Z = 4, R = 0.043. The crystals are not fully stoichiometric. The complex anion contains 5 Mo with octahedral and 2 Mo with pentagonal bipyramidal coordinations.

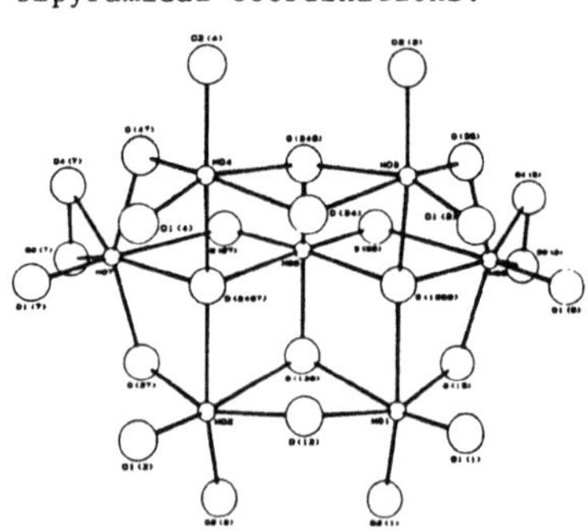

INDIUM MOLYBDATE
$In_{11}Mo_{40}O_{62}$

Inorg. Chem., **25**, 3428-3433.

$Pmc2_1$, 28.878, 9.513, 9.883, Z = 2, R = 0.038. Chains of four and five edge-sharing Mo_6 octahedra with oxygens above all free edges, with In_5^{7+} and In_6^{8+} chains between the Mo clusters. Mo-Mo = 2.64-3.12, In-In = 2.62-2.68 A.

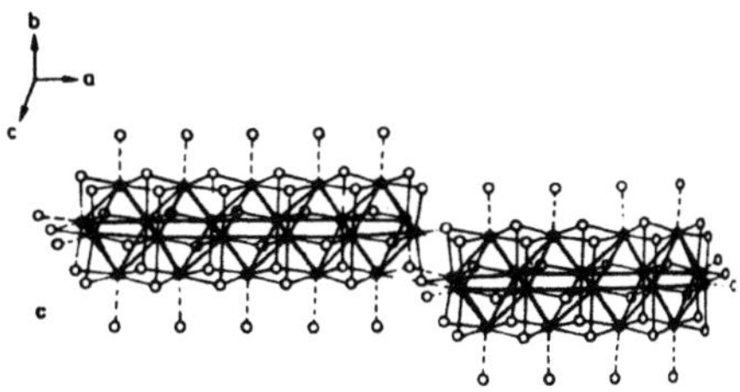

THALLIUM MOLYBDENUM BRONZE (RED)
$Tl_{0.33}MoO_3$

C.R. Acad. Sci, Paris, Ser. 2, **302**, 635-640.

C2/m, 14.541, 7.780, 6.414, 93.09, Z = 12, R = 0.021. Isostructural with the K compound (**30**A, 341; **43**A, 204). Mo-O = 1.69-2.38, Tl-O = 2.67-3.33 A.

THALLIUM HEXAMOLYBDATE
$TlMo_6O_{17}$

Solid State Comm., **59**, 137-142.

$P\bar{3}m1$, 5.557, 14.030, Z = 1, R = 0.020. Structure similar to that of the K compound (described in $P\bar{3}$, **50**A).

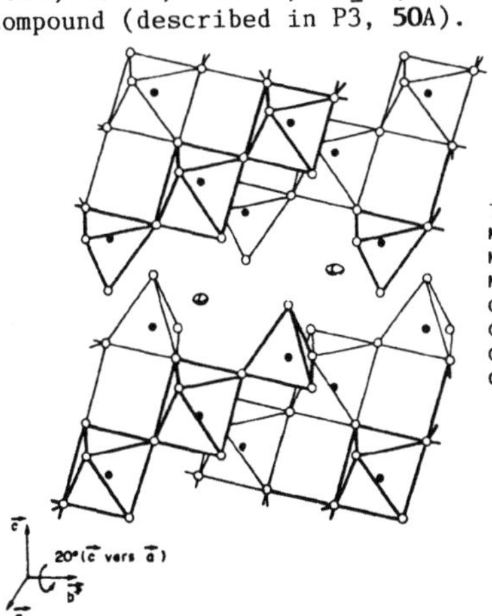

		x	y	z
Tl	1(b)	0	0	1/2
Mo1	2(d)	1/3	2/3	0.35887
Mo2	2(d)	1/3	2/3	0.93100
Mo3	2(c)	0	0	0.21426
O1	6(g)	0.1583	0.8417	0.3184
O2	6(g)	0.1637	0.8363	0.8453
O3	2(d)	1/3	2/3	0.4815
O4	3(e)	1/2	0	0

BISMUTH(III) MOLYBDATE
$Bi_2Mo_2O_9$

J. Solid State Chem., **63**, 70-75.

$P2_1/n$, 11.972, 10.813, 11.899, 90.13, Z = 8, R = 0.051. Previous study in **41**A, 404. The structure may be represented as $Bi(Bi_3O_2)(MoO_4)_4$, with Bi_3O_2 chains along **b**, MoO_4 tetrahedra, and additional Bi ions. Mo-4 O = 1.71-1.80, Bi-8 O = 2.13-2.92(2) A.

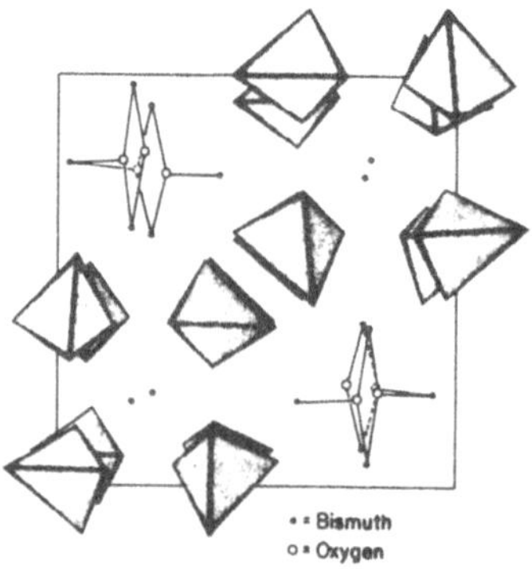

	x	*y*	*z*
Bi 1	−0.10020	0.1209	0.7564
Bi 2	0.60326	0.1229	0.7583
Bi 3	0.25585	0.1232	0.7585
Bi 4	0.25301	0.1228	0.4122
Mo 1	−0.0827	0.1230	0.8722
Mo 2	−0.0784	0.1250	0.4163
Mo 3	0.5893	0.1272	0.0814
Mo 4	0.5791	0.1321	0.4156
O1	0.252	0.261	0.285
O2	0.255	0.493	0.227
O3	0.409	0.257	0.689
O4	0.326	0.259	0.901
O5	0.105	0.437	0.447
O6	0.101	0.251	0.815
O7	0.405	0.444	0.048
O8	0.093	0.494	0.684
O9	0.194	0.260	0.596
O10	0.097	0.313	0.050
O11	0.403	0.308	0.454
O12	0.056	0.187	0.388
O13	0.044	0.046	0.112
O14	0.447	0.066	0.383
O15	0.399	0.495	0.820
O16	0.962	0.300	0.599
O17	0.177	0.489	0.904
O18	0.192	0.988	0.903

ZIRCONIUM MOLYBDATE (HIGH-TEMPERATURE)
$Zr(MoO_4)_2$

Acta Cryst., C**42**, 257-259.

$P\bar{3}1c$, 10.1391, 11.7084, Z = 6, R = 0.031. Distorted MoO_4 tetrahedra, linked by ZrO_6 octahedra. The Hf compound is isostructural, a previous description of that structure (**33**A, 338; **40**A, 201) being incorrect.

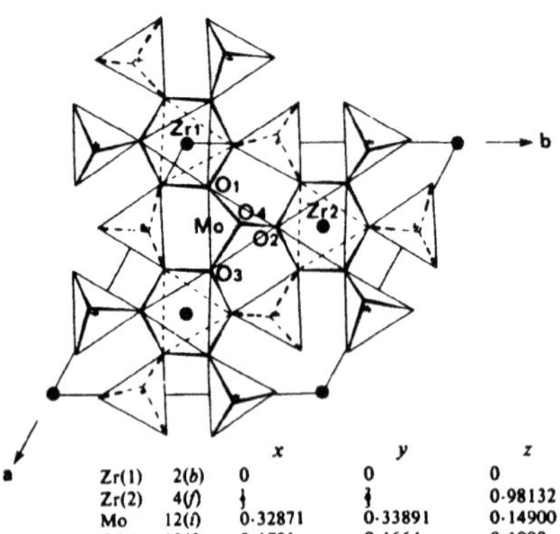

		x	*y*	*z*
Zr(1)	2(*b*)	0	0	0
Zr(2)	4(*f*)	1/3	2/3	0·98132
Mo	12(*i*)	0·32871	0·33891	0·14900
O(1)	12(*i*)	0·1731	0·1664	0·1008
O(2)	12(*i*)	0·3323	0·4959	0·0814
O(3)	12(*i*)	0·4979	0·3364	0·1178
O(4)	12(*i*)	0·3189	0·3522	0·2923

Bond lengths (Å), *bond angles* (°) *and bond strengths* (*s*)

MoO_4 tetrahedra

		s	Bond distribution		
Mo—O(1)	1·763 (7)	1·50	1σ + 0·5π	O(1)—O(2)	2·903 (17)
—O(2)	1·762 (5)	1·50	1σ + 0·5π	—O(3)	2·860 (16)
—O(3)	1·767 (5)	1·48	1σ + 0·5π	—O(4)	2·825 (12)
—O(4)	1·690 (3)	1·89	1σ + 1π	O(2)—O(3)	2·887 (12)
				—O(4)	2·836 (8)
				O(3)—O(4)	2·790 (9)

O(1)—Mo—O(2)	110·89 (4)	O(2)—Mo—O(3)	109·79 (3)
O(1)—Mo—O(3)	108·26 (3)	O(2)—Mo—O(4)	110·45 (3)
O(1)—Mo—O(4)	109·74 (4)	O(3)—Mo—O(4)	107·62 (2)

ZrO_6 polyhedra

			s
Zr(1)—O(1$^{i\ to\ vi}$)	(6×)	2·088 (8)	0·68
Zr(2)—O(2^{i,ix,x})	(3×)	2·087 (8)	0·68
O(3^{vi,vii,viii})	(3×)	2·081 (7)	0·69

O(1^{i})—Zr(1)—O(1^{ii})	(6×)	91·19 (3)	O(3^{vii})—Zr(2)—O(3^{vi})	(3×)	91·91 (3)
O(1^{i})—Zr(1)—O(1^{v})	(6×)	88·81 (3)	O(2^{i})—Zr(2)—O(3^{vii})	(3×)	87·92 (3)
O(2^{i})—Zr(2)—O(2^{ix})	(3×)	91·54 (3)	O(2^{i})—Zr(2)—O(3^{vi})	(3×)	88·62 (3)

O(4) coordination polyhedron

O(4)—O(1^{xi})	3·99 (1)	O(4)—O(1^{xii})	3·29 (2)*
—O(2^{xi})	3·88 (1)	—O(1^{xiii})	3·13 (2)*
—O(3^{xi})	4·06 (1)	—O(2^{xiii})	3·01 (1)*
—O(4^{xii})	4·03 (2)	—O(2^{xiv})	3·15 (2)*
—O(4^{xiii})	3·06 (1)	—O(3^{xii})	3·41 (1)*
—O(4^{xiv})	3·48 (2)	—O(3^{xiv})	3·21 (2)*

Symmetry code: (i) *x, y, z*; (ii) −*y*, *x*−*y*, *z*; (iii) *y*−*x*, −*x*, *z*; (iv) −*x*, −*y*, −*z*; (v) *y*, *y*−*x*, −*z*; (vi) *x*−*y*, *x*, −*z*; (vii) 1−*x*, 1−*y*, −*z*; (viii) *y*, 1+*y*−*x*, −*z*; (ix) 1−*y*, 1+*x*−*y*, *z*; (x) *y*−*x*, 1−*x*, *z*; (xi) *y*, *x*, ½+*z*; (xii) *x*, *x*−*y*, ½−*z*; (xiii) *y*−*x*, *y*, ½−*z*; (xiv) 1−*y*, 1−*x*, ½−*z*.

VANADOMOLYBDOSILICIC ACID HYDRATE
$H_5[SiMo_{11}VO_{40}].30\text{-}36H_2O$

Ž. Strukt. Khim., **27**, No. 2, 101-109 [J. Struct. Chem., **27**, 261-268].

P4/mnc, 12.792, 18.131, at 203K, Z = 2, R = 0.081. Heteropolyanion (**46**A, 266; 52A, 234) with a central SiO_4 tetrahedron, but disordered so that there are eight oxygen sites around Si, and with Mo/V disorder. The water molecules are involved in an extensive hydrogen bond system.

POTASSIUM MANGANESE(II) MOLYBDATE(VI)
$K_2Mn_2(MoO_4)_3$

Kristallografija, **31**, 440-445 [Soviet Physics - Crystallography, **31**, 258-261].

Pcca, 10.698, 12.292, 18.060, Z = 8, R = 0.068. Bands of edge-sharing MnO_6 octahedra with attached MoO_4 tetrahedra, linked by single MnO_6 octahedra, with channels containing K ions. Mo-O = 1.75-1.80, Mn-O = 2.11-2.29, K-O = 2.51-3.50 A.

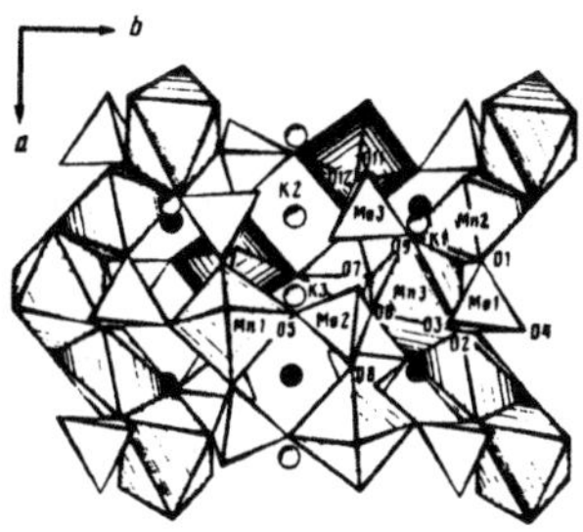

	x	y	z
Mo(1)	0.0475	0.0234	0.1161
Mo(2)	0.4241	0.3572	0.0871
Mo(3)	0.2195	0.3005	0.2952
Mn(1)	0.0845	0.3245	0.0974
Mn(2)	0.25	0	0.2856
Mn(3)	0.50	0.1673	0.25
K(1)	0.2747	0.1510	0.4620
K(2)	0.25	0.50	0.4439
K(3)	0.50	0.4964	0.25
O(1)	0.1168	0.0250	0.3740
O(2)	0.3917	0.0257	0.2026
O(3)	0.4116	0.0701	0.0466
O(4)	0.1065	0.1528	0.0946
O(5)	0.4411	0.4978	0.1015
O(6)	0.4387	0.2837	0.1718
O(7)	0.4636	0.3168	0.4790
O(8)	0.2723	0.3368	0.0492
O(9)	0.3021	0.1727	0.2976
O(10)	0.3253	0.4045	0.3190
O(11)	0.1071	0.3005	0.3671
O(12)	0.1501	0.3309	0.2096

RUBIDIUM MANGANESE MOLYBDATE
$Rb_4Mn(MoO_4)_3$

Izv. Akad Nauk SSSR, Ser. Fiz., **50**, 353-355.

$P\bar{6}2c$, 6.088, 23.676, Z = 2. High-temperature $K_4Zn(MnO_4)_3$ is isostructural, 6.16, 22.44 A, at 400°C. Powder data.

KAMIOKITE
$Fe_2Mo_3O_8$

Acta Cryst., C**42**, 9-11.

$P6_3mc$, 5.781, 10.060, Z = 2, R = 0.027. Layers of edge- and corner-sharing MoO_6 octahedra alternate along **c** with layers of FeO_6 octahedra and FeO_4 tetrahedra; there are triangular Mo clusters with Mo...Mo = 2.533(1) A.

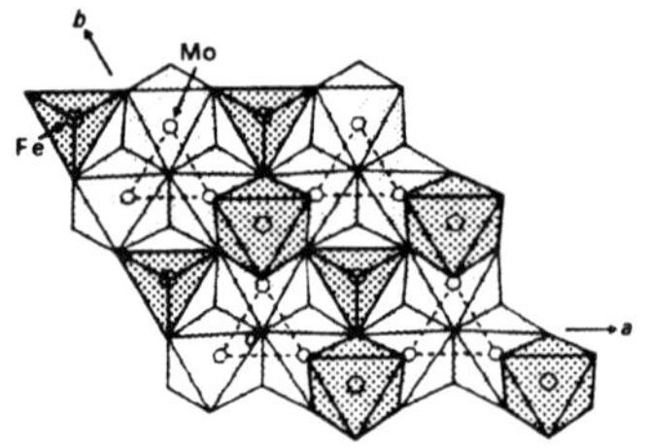

(a)

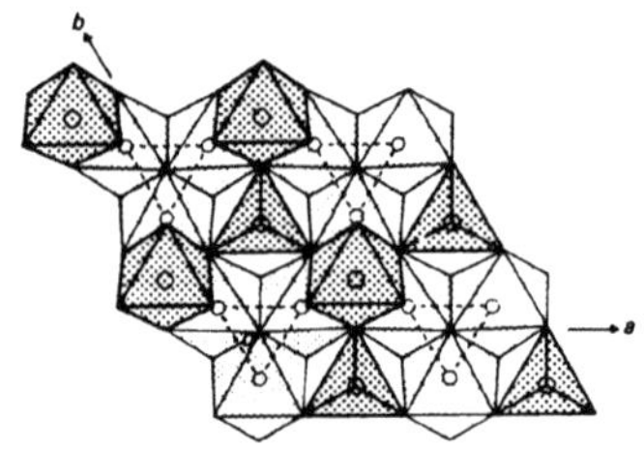

(b)

Projection of the crystal structure along **c** with the MoO_6, FeO_4 and FeO_6 polyhedra. The Mo–Mo bonding is shown by dotted lines. (*a*) Projection from $z = \sim 0{\cdot}15$ to $\sim 0{\cdot}65$. (*b*) Projection from $z = \sim 0{\cdot}65$ to $\sim 1{\cdot}15$.

		x	*y*	*z*
Fe(1)	2(*b*)	$\frac{1}{3}$	$\frac{2}{3}$	0·9504
Fe(2)	2(*b*)	$\frac{1}{3}$	$\frac{2}{3}$	0·5120
Mo	6(*c*)	0·14603	0·29206	$\frac{1}{4}$
O(1)	2(*a*)	0	0	0·3923
O(2)	2(*b*)	$\frac{1}{3}$	$\frac{2}{3}$	0·1486
O(3)	6(*c*)	0·4872	0·9744	0·3645
O(4)	6(*c*)	0·1665	0·3330	0·6352

Selected bond distances (Å) *and angles* (°)

Fe tetrahedron					
Fe(1)–O(2^{i})	1·994 (9)	O(2^{i})	–Fe(1)–	O(3^{v})	115·7 (1)
O(3^{v})	1·993 (3) (×3)	O(3^{v})		O(3^{vii})	102·6 (2)
Mean:	1·993				
Fe octahedron					
Fe(2)–O(3)	2·139 (4) (×3)	O(3)	–Fe(2)–	O(3^{ii})	77·2 (2)
O(4)	2·080 (4) (×3)	O(3)		O(4)	172·7 (2)
		O(3)		O(4^{ii})	97·1 (2)
Mean:	2·110	O(4)		O(4^{iii})	88·2 (2)
Mo octahedron					
Mo–O(1)	2·047 (6)	O(1)	–Mo–	O(2)	164·1 (3)
O(2)	2·135 (4)	O(1)		O(3^{iii})	89·5 (2)
O(3^{iii})	2·081 (4) (×2)	O(1)		O(4^{vi})	100·6 (2)
O(4^{vi})	1·953 (4) (×2)	O(2)		O(3^{iii})	78·3 (2)
		O(2)		O(4^{vi})	90·0 (2)
Mean:	2·042	O(3^{iii})		O(3^{iv})	79·8 (2)
		O(3^{iii})		O(4^{vi})	91·4 (1)
		O(3^{iii})		O(4^{viii})	166·6 (1)
		O(4^{vi})		O(4^{viii})	95·3 (2)
Mo–Mo^{ii}	2·5326 (5) (×2)				
Mo^{vi}	3·2486 (5) (×2)				

Symmetry code: none *x*, *y*, *z*; (i) *x*, *y*, *z* + 1; (ii) –*y*, –*x*, *z*; (iii) 1 – *y*, 1 – *x*, *z*; (iv) *x*, *x* – *y* + 1, *z*; (v) 1 – *x*, 2 – *y*, *z* + ½; (vi) *y*, *x*, *z* – ½; (vii) *y* – 1, *x*, *z* + ½; (viii) –*x*, *y* – *x*, *z* – ½.

LITHIUM IRON(II) MOLYBDATE
$Li_2Fe_2(MoO_4)_3$

J. Solid State Chem., **62**, 231-240; Mater. Res. Bull., **21**, 719-726.

Pbcn, 12.8947, 9.4957, 9.3477, Z = 4, neutron powder data. The structure is related to that of garnet and langbeinite, and positional parameters were derived from those of $Sc_2(WO_4)_3$ (**31**A, 158). It contains corner-sharing FeO_6 octahedra and MoO_4 tetrahedra, and tetrahedrally-coordinated Li.

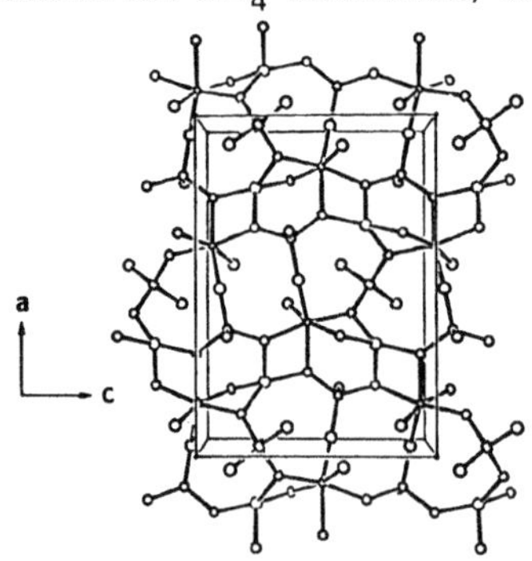

	x	y	z
Mo1	0.0	0.4605	0.25
Mo2	0.3564	0.4008	0.1063
Fe	0.3778	0.2457	0.4678
O1	0.1619	0.0875	0.1121
O2	0.0939	0.3528	0.1667
O3	0.2806	0.3300	0.9665
O4	0.4362	0.0671	0.3754
O5	0.4869	0.3604	0.0724
O6	0.3207	0.3266	0.2704
Li	0.1895	0.2208	0.2711

CAESIUM COBALT(II) MOLYBDATE
$Cs_2Co_2(MoO_4)_3$

Ž. Strukt. Khim., **27**, No. 6, 100-106 [J. Struct. Chem., **27**, 928-933].

$P2_12_12_1$, 10.832, 10.829, 10.830, Z = 4, R = 0.069. Langbeinite arrangement (**21**, 362; **45**A, 332) of corner-sharing MoO_4 tetrahedra and CoO_6 octahedra, with

Cs in large ellipsoidal interstices. Deviation from cubic symmetry results from small rotations of the tetrahedra.

SODIUM NICKEL MOLYBDATE HYDRATE
$NaNi_2(MoO_4)_2(OH).H_2O$

Acta Cryst., C**42**, 1667-1669.

C2/m, 9.3896, 6.2928, 7.5727, 116.080, at 15K, Z = 2, neutron powder data. Isostructural with the Zn and Mn salts (**42**A, 276; **52**A, 217). Mo-O = 1.75-1.83, Ni-O = 2.07-2.08(1) A.

COPPER MOLYBDATE
$Cu_4Mo_6O_{20}$

SODIUM COPPER HYDROXIDE MOLYBDATE
$NaCu(OH)(MoO_4)$

Inorg. Chem., **25**, 3782-3785.

$Cu_4Mo_6O_{20}$, P$\bar{1}$, 10.071, 9.796, 8.052, 104.98, 100.63, 82.91, Z = 2, R = 0.045. $NaCu(OH)(MoO_4)$, Pnma, 7.726, 5.968, 9.495, Z = 4, R = 0.034. $Cu_4Mo_6O_{20}$ contains chains of edge-sharing MoO_6 octahedra, with 4- and 5-coordinate Cu sites, some of which have partial occupancy. $NaCu(OH)MoO_4$ is isostructural with the Zn analogue (**43**A, 208; **46**A, 269).

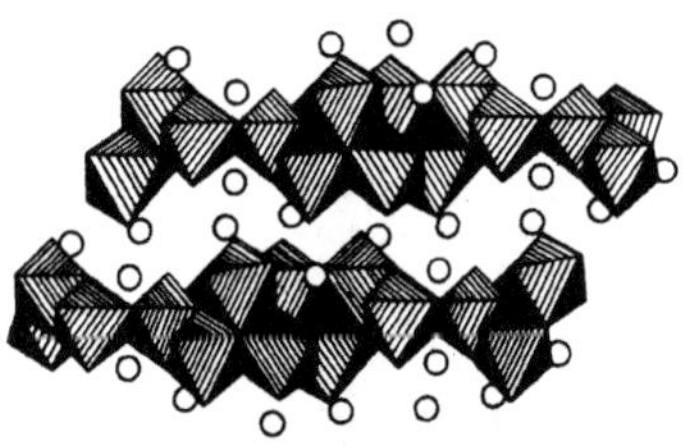

$Cu_4Mo_6O_{20}$

Parameters for $Cu_4Mo_6O_{20}$

		x	y	z
	Mo1	0.79930	0.79927	0.03380
	Mo2	1.08763	0.79079	−0.13225
	Mo3	0.51538	0.80411	−0.24815
	Mo4	0.63142	0.46538	−0.34954
	Mo5	0.23437	0.98518	0.24617
	Mo6	0.82174	0.67941	0.57696
	Cu1	1.0295	0.6232	0.2558
0.76 (1)	Cu2	0.5813	0.8505	0.3335
0.24 (1)	Cu3	0.8375	1.0973	0.4009
0.89 (1)	Cu3′	0.8897	0.0930	0.4127
0.11 (3)	Cu4	0.6369	0.5323	0.1391
	Cu4′	0.6054	0.4640	0.0877
	O1	1.1476	0.7557	−0.2959
	O2	0.3256	0.1653	0.4200
	O3	0.2739	0.4652	0.5433
	O4	0.3461	0.0824	0.0922
	O5	0.3836	0.8818	0.2863
	O6	0.4626	0.5572	0.1962
	O7	0.0374	0.4257	0.3069
	O8	0.6901	0.7368	0.1435
	O9	0.2263	0.6497	0.3250
	O10	0.8592	0.1279	0.9522
	O11	0.1377	0.9717	0.3994
	O12	0.8616	0.9438	0.1827
	O13	0.0456	0.3248	0.9684
	O14	0.5970	0.0708	0.3277
	O15	0.7939	0.4203	0.0742
	O16	0.4777	0.6326	0.5610
	O17	0.1148	0.1699	0.1895
	O18	0.3035	0.3526	0.2207
	O19	0.5618	0.2394	0.1012
	O20	0.9087	0.7209	0.4339

Parameters for $NaCu(OH)(MoO_4)$

	x	y	z
Mo	0.1314	−0.2500	0.16289
Cu	0.5000	0.0000	0.0000
Na	0.3697	0.2500	0.3056
O1	0.3259	−0.2500	0.0546
O2	0.1201	−0.4895	0.2686
O3	−0.0445	−0.2500	0.0469
O4	0.3681	0.2500	0.0612

COPPER(I) PENTAMOLYBDATE(VI)
$Cu_6Mo_5O_{18}$

J. Solid State Chem., **62**, 64-74.

I2/c, 14.702, 6.272, 15.264, 101.87, Z = 4, R = 0.027. Chains along **a** of edge-sharing MoO_6 octahedra, and (001) sheets of corner-sharing CuO_4 tetrahedra, with short Cu...Cu contacts (< 2.8 A).

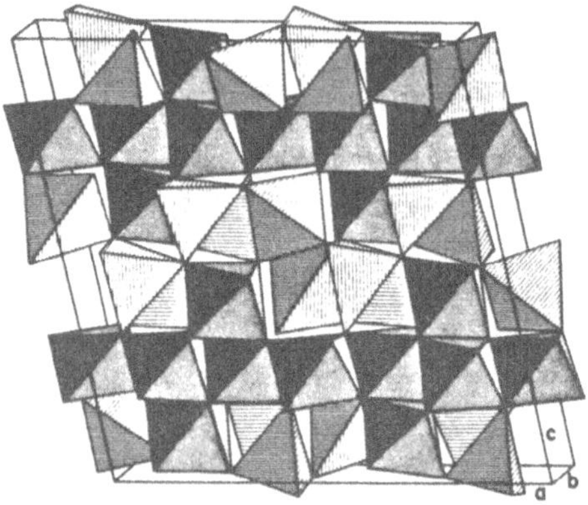

Fractional Coordinates (×10,000)			
	X	Y	Z
Mo(1)	5000	9320	2500
Mo(2)	3389.6	5820.5	2421.9
Mo(3)	1586.8	6356.9	711.9
Cu(1)	3885.8	6477.8	378.1
Cu(2)	2143.6	3799.2	3692.5
Cu(3)	5031.0	3043.1	919.8
O(1)	4018	3947	3107
O(2)	2598	6639	3295
O(3)	868	8497	187
O(4)	4201	6661	1724
O(5)	923	4076	348
O(6)	5947	8474	1871
O(7)	2452	6328	65
O(8)	4346	11043	1700
O(9)	2557	4243	1652

$Cu_4Mo_5O_{17}$

J. Solid State Chem., **65**, 215-224.

$P\bar{1}$, 9.573, 10.958, 6.782, 91.63, 111.01, 72.96, Z = 2, R = 0.026. Infinite stepped ribbons along **a** of edge-sharing MoO_6 octahedra, linked by three CuO_4 tetrahedra and one CuO_6 distorted octahedron.

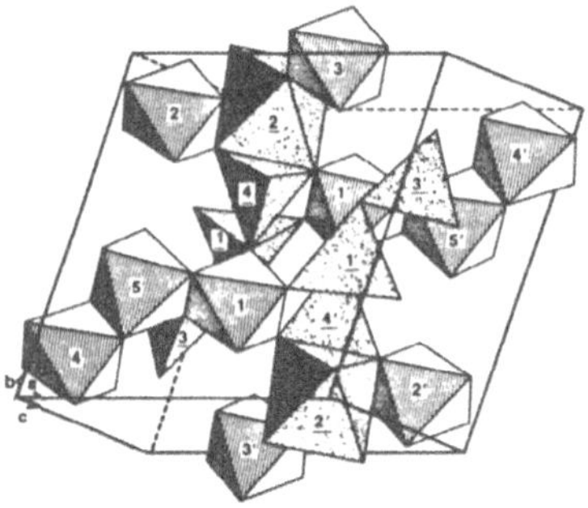

Atomic Coordinates (×10,000)			
	X	Y	Z
Mo(1)	5574.2	6,720.5	3499.7
Mo(2)	3768.8	10,272.0	1338.8
Mo(3)	2086.6	8,003.9	−170.3
Mo(4)	393.3	11,486.2	−2422.5
Mo(5)	8105.7	6,123.9	1276.2
Cu(1)	7777.5	4,441.3	6979.0
Cu(2)	7096.9	10,790.7	4552.0
Cu(3)	9006.2	7,053.8	6865.8
Cu(4)	5422.3	3,805.0	2262.0
O(1)	3288	6,545	1642
O(2)	4995	7,327	5576
O(3)	−44	13,260	−2732
O(4)	4176	8,449	1599
O(5)	8312	5,875	−1192
O(6)	2502	11,781	−515
O(7)	6149	5,030	4109
O(8)	2478	7,653	−2455
O(9)	3391	10,656	3591
O(10)	5691	10,363	1898
O(11)	7480	6,969	4075
O(12)	135	7,940	−811
O(13)	1645	9,889	−523
O(14)	9005	4,641	2662
O(15)	5997	6,319	552
O(16)	−1499	11,317	−2915
O(17)	814	11,002	−4605

LITHIUM LANTHANUM MOLYBDATE
$Li_{0.3125}La_{0.5625}MoO_4$

Solid State Ionics, **18 -19**, 922-928.

$I4_1/a$, 5.3350, 11.7584, Z = 4, neutron powder data. La in 4(b): 0,1/4,5/8; Mo in 4(a): 0,1/4,1/8; O in 16(f): 0.1413,0.0116,0.2074; Li position unknown, postulated to be at ∿0.05,0.1,0.65. Scheelite-type structure (**1**, 347; **29**, 331).

LANTHANUM MOLYBDENUM ALUMINUM OXIDE
$La_3Mo_{4.33}Al_{0.67}O_{14}$

J. Solid State Chem., **62**, 241-252.

Pnma, 17.750, 5.6600, 11.070, Z = 4, R = 0.034. Isostructural with $La_3Mo_4SiO_{14}$ (**51**A, 331), with Mo_3O_{13} clusters and chains of edge-sharing MoO_6 octahedra (Mo...Mo = 2.535 and 3.167 A, alternately), and $(Mo,Al)O_4$ tetrahedra.

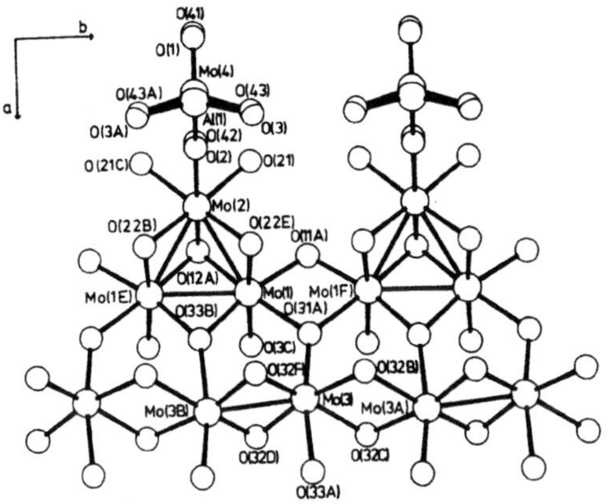

	x	y	z
La(1)	0.33565	0.25000	0.13199
La(2)	0.34657	0.25000	0.49908
La(3)	0.53414	0.25000	0.69746
Mo(1)	0.33906	0.02356	0.82069
Mo(2)	0.21638	0.25000	0.86281
Mo(3)	0.50834	0.2219	0.00788
Mo(4)	0.0548	0.25000	0.0968
Al(1)	0.0662	0.25000	0.1114
O(1)	−0.0246	0.25000	0.0549
O(2)	0.1318	0.25000	0.9949
O(3)	0.0850	−0.0149	0.1859
O(11)	0.2890	0.75000	0.7446
O(12)	−0.2206	0.25000	0.7918
O(21)	0.1526	−0.0001	0.8023
O(22)	−0.2709	0.5070	0.0339
O(31)	−0.0999	0.75000	0.5994
O(32)	0.0446	0.0010	0.6112
O(33)	0.0946	0.75000	0.4096
O(41)	−0.0304	0.25000	0.0120
O(42)	0.1241	0.25000	0.9792
O(43)	0.0809	−0.0084	0.1812

MOLYBDATES
$K_5Nd(MoO_4)_4$, $K_5Bi(MoO_4)_4$, $Rb_5Gd(MoO_4)_4$

Kristallografija, **31**, 237-243 [Soviet Physics - Crystallography, **31**, 138-142].

K/Nd, P2/m, 10.360, 17.943, 14.301, 103.98, Z = 6, R = 0.059. K/Bi, $P\bar{3}m1$, 6.019, 41.761, Z = 3, R = 0.048. Rb/Gd, P1, 10.604, 18.349, 14.725, 90.11, 104.41, 89.99, Z = 6, R = 0.067. Structures are proposed in a palmierite-like subcell with a = 6, b = 21 A (**10**, 143; **11**, 390; **13**, 326; **17**, 499).

SODIUM PARATUNGSTATE 26-HYDRATE
$Na_{10}[H_2W_{12}O_{42}].26H_2O$

J. Cryst. Spect. Res. **16**, 525-535.

$P\bar{1}$, 11.811, 12.486, 12.206, 82.29, 115.12, 113.76, Z = 1, R = 0.040. The complex anion is as found in related structures (**16**, 231; **37**A, 254; **38**A, 274; **39**A, 245; **42**A, 277; **46**A, 273), with twelve edge- and corner-sharing WO_6 octahedra. One Na ion is disordered, and there is extensive hydrogen bonding.

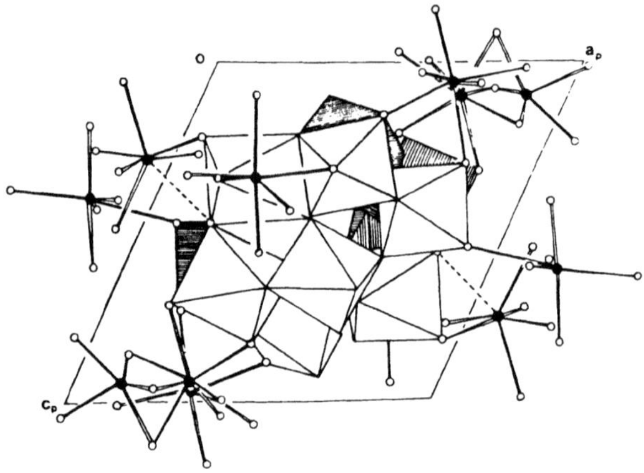

POTASSIUM TUNGSTEN BRONZE (HEXAGONAL)
$K_{0.26}WO_3$

Acta Cryst., C**42**, 641-644.

$P6_3$, 7.389, 7.508, Z = 6, neutron data, R = 0.051 (ℓ = 2n only), and X-ray data, R = 0.046. Structure essentially as previously described in higher-symmetry space groups (**17**, 401; **45**, 258). Satellite reflexions in the neutron data indicate a superlattice with K vacancies and oxygen position modulation.

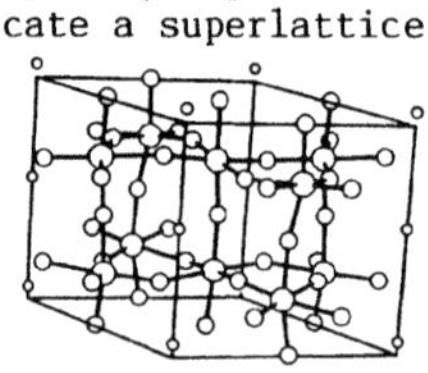

The two lines for each atom are the neutron and X-ray structure parameters, respectively. In the case of the neutron structure, for each atom x, y, z a related atom at $x, y, \frac{1}{2}+z$ was generated, all atoms with occupancies of one-half.

	x	y	z
W	0·477	−0·002	0·25
	0·47305	−0·00006	0·25
O(1*A*)	0·431	0·216	0·279
	0·429	0·221	0·280
O(1*B*)	0·204	0·411	0·224
	0·212	0·413	0·228
O(2)	0·526	0·001	0·0
	0·510	−0·008	0·0
K	0·0	0·0	0·053
	0·0	0·0	0·059

Selected interatomic distances (Å) *and angles* (°)

	Neutron	X-ray		Neutron	X-ray
W–O(1*A*)	1·816 (14)	1·835 (10)	W–O(1*B*^{i})	2·067 (13)	2·050 (11)
W–O(1*A*^{i})	2·029 (16)	2·009 (10)	W–O(2)	1·910 (2)	1·903 (9)
W–O(1*B*)	1·810 (16)	1·835 (11)	W–O(2^{ii})	1·877 (1)	1·880 (3)
K–O(1*A*)	3·24 (2)	3·21 (2)	K–O(1*B*^{iii})	3·61 (3)	3·62 (3)
K–O(1*A*^{iii})	3·45 (3)	3·45 (3)	K–O(2)	3·53 (7)	3·62 (5)
K–O(1*B*)	2·93 (2)	2·93 (2)	K–K^{iv}	3·754 (1)	3·754 (1)

	Neutron	X-ray
O(2)–W–O(2^{ii})	170·1 (3)	167·94 (5)
O(2)–W–O(1*A*)	103·0 (5)	104 (1)
O(2)–W–O(1*A*^{i})	87·9 (4)	90 (1)
O(2)–W–O(1*B*)	91·7 (5)	90 (1)
O(2)–W–O(1*B*^{i})	76·9 (4)	77 (1)
O(1*A*)–W–O(1*A*^{i})	85·4 (7)	86·5 (6)
O(1*A*)–W–O(1*B*)	97·4 (4)	99·4 (4)
O(1*A*^{i})–W–O(1*B*^{i})	83·7 (4)	81·3 (4)
O(1*B*)–W–O(1*B*^{i})	93·5 (7)	92·8 (6)
W–O(2)–W^{v}	168·5 (3)	173 (3)

Symmetry operations: (i) $1-x, x-y, z$; (ii) $1-x, -y, \frac{1}{2}+z$; (iii) $-x, -y, -\frac{1}{2}+z$; (iv) $-x, -y, \frac{1}{2}+z$; (v) $1-x, -y, -\frac{1}{2}+z$.

MAGNESIUM TUNGSTATE DIHYDRATE
$MgWO_4.2H_2O$

J. Solid State Chem., **65**, 118-126.

$P2_1/c$, 5.917, 10.243, 8.566, 90.05, Z = 4, R = 0.066. (100) Layers of edge-sharing $MgO_4(H_2O)_2$ octahedra linked by sharing corners with WO_4 tetrahedra. Mg-O = 2.03-2.11, W-O = 1.75-1.83(1) A. Thermal dehydration gives a monohydrate.

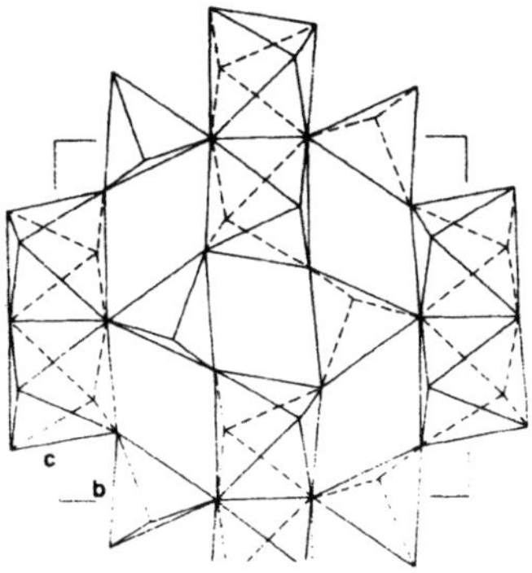

	x	y	z
W	0.29456	0.21469	−0.00091
Mg	0.5248	0.5020	0.1853
O1	0.322	0.3558	0.6855
O2	0.438	0.3835	0.365
O3	−0.005	0.721	0.548
O4	0.414	0.3789	0.009
O5	0.160	0.905	0.314
O6	0.210	0.590	0.216

RUBIDIUM DECATUNGSTOSILICATE HYDRATE
$Rb_8SiW_{10}O_{36}.10H_2O$

Inorg. Chem., **25**, 2114-2119.

$P2_1/c$, 12.426, 19.319, 19.743, 96.33, Z = 4, R = 0.070. γ-Polyanion, derived from Keggin type by 60° rotation of two W_3O_{13} groups and removal of two octahedra.

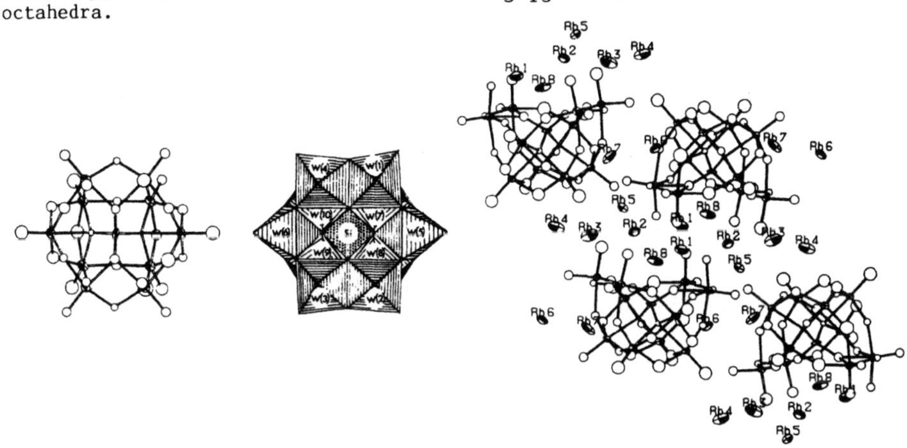

PHOSPHATE TUNGSTEN BRONZES
$P_4W_{12}O_{44}$, $P_4W_{16}O_{56}$

J. Solid State Chem., **61**, 324-331.

$P2_12_12_1$, a = 5.2927, 5.2943, b = 6.5604, 6.5534, c = 23.549, 29.700 A, Z = 1, R = 0.033, 0.043. Structures related to that of $P_4W_8O_{32}$ (**48**A, 232), with zigzag chains of WO_6 octahedra and PO_4 tetrahedra.

Projection onto (100) of (a) $P_4W_{12}O_{44}$ (b) $P_4W_{16}O_{56}$.

PHOSPHOTUNGSTATES

$K_9(H_3O)_3[(Ce_3O_3.2H_2O)(W_9PO_{34})_2].\sim26H_2O$ (I)

$K_{11}\{[Cu_3(NO_3)](W_9PO_{34})_2\}(H_3O)_2.\sim28H_2O$ (II)

Inorg. Chem., 25, 1577-1584.

I, C2/m, 33.219, 15.903, 17.385, 98.41, at 173K, Z = 4, R = 0.060. II, $P2_1/m$, 12.224, 33.502, 12.346, 118.29, Z = 2, R = 0.065. Complex anions with 6- and 7-coordinate Ce, and 5-coordinate Cu. A Ni compound has also been studied, but disorder prevents a detailed structure description.

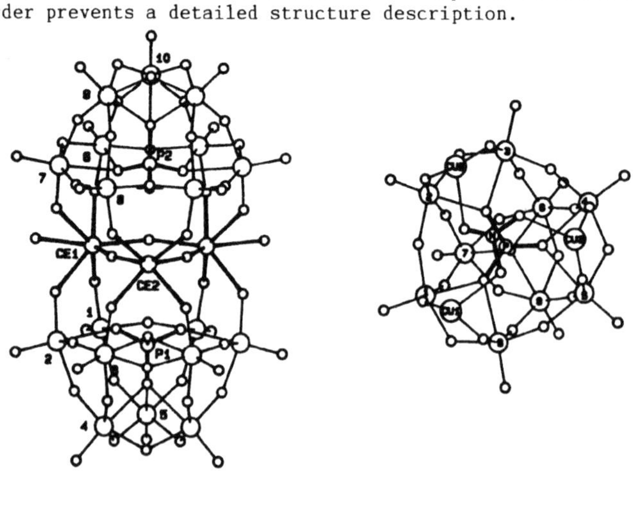

POTASSIUM TUNGSTOPHOSPHATE HYDRATE
$K_4H_2[P_2W_{21}O_{71}(OH_2)_3].28H_2O$

J. Chem. Soc., Dalton, 2237-2242 (1986).

$P6_3/mmc$, 16.295, 19.771, Z = 2, R = 0.037. The anion has C_{2v} symmetry, but is disordered in a D_{3h} site. It contains two $PW_9O_{34}^{9-}$ units linked through three $WO(OH_2)^{4+}$ units, with the H_2O ligand directed towards the interior of the anion.

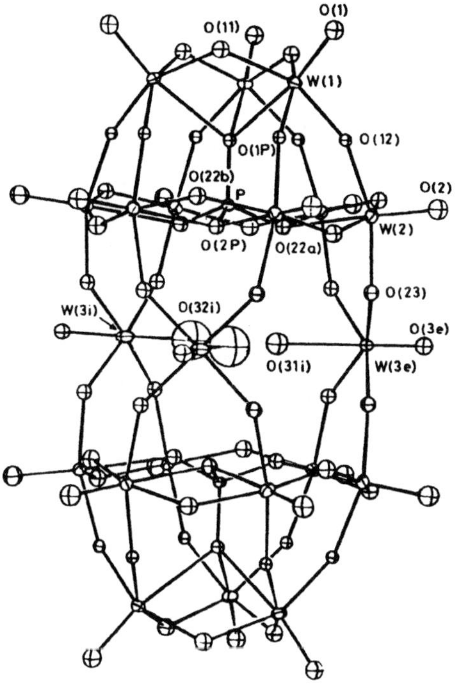

POTASSIUM COBALTOTUNGSTOPHOSPHATE HYDRATE
POTASSIUM ZINCTUNGSTOARSENATE HYDRATE
$K_{10}[M_4(H_2O)_2(XW_9O_{34})_2].nH_2O$ (M = Co, X = P, n = 23; M = Zn, X = As, n = 22)

J. Chem. Soc., Dalton, 2699-2705 (1986).

$P2_1/n$, a = 15.794, 15.842, b = 21.360, 21.327, c = 12.312, 12.308, β = 91.96, 92.42°, Z = 2, R = 0.084, 0.066. Anions have approximate C_{2h} symmetry, and contain two $XW_9O_{34}^{9-}$ groups linked by four MO_6 octahedra (two M have a water ligand each).

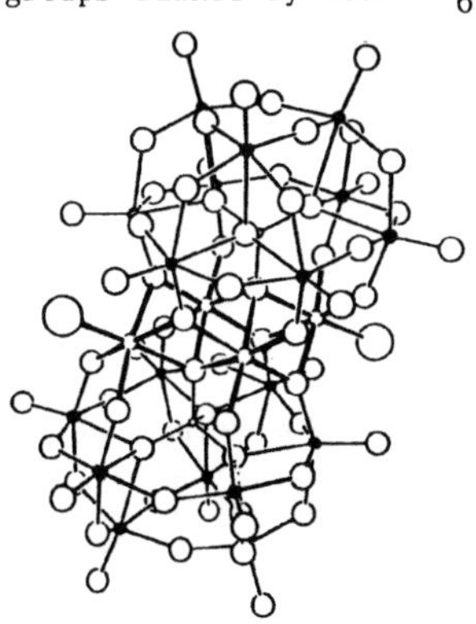

SODIUM HEXATUNGSTOTELLURATE(VI) 22-HYDRATE
$Na_6TeW_6O_{24}.22H_2O$

Acta Cryst., C**42**,1115-1118.

$P\bar{1}$, 10.269, 10.575, 11.104, 90.95, 115.13, 104.96, Z = 1, R = 0.064. Anderson type II polyanion with approximate D_3 symmetry, as for a related molybdate (**18**, 536; **40**A, 201); the central Te has a slightly-flattened octahedral environment, mean Te-O = 1.924(11) A.

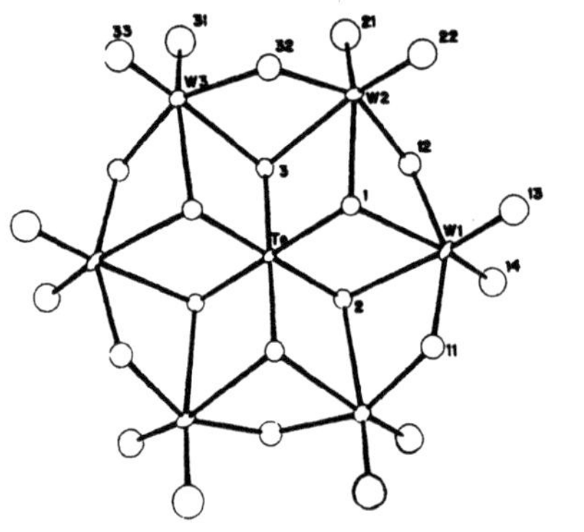

atomic positional ($\times 10^4$; $\times 10^5$ *for* W)

	x	*y*	*z*
W(1)	15458	29573	17698
W(2)	−4787	5841	26785
W(3)	−20159	−23663	9117
Te	0	0	0
Na(1)	3920	−1597	4330
Na(2)	5614	2426	8442
Na(3)	7997	4112	4505
O(1)	1234	755	1848
O(2)	97	1811	−322
O(3)	−1652	−244	444
O(11)	2904	2836	1016
O(12)	−255	2237	1996
O(13)	2886	3338	3443
O(14)	1192	4450	1287
O(21)	−2191	517	2733
O(22)	925	1095	4288
O(31)	−3691	−2355	959
O(32)	−547	−1242	2588
O(33)	−1642	−3816	1442
O(w1)	−2257	2732	−2135
O(w2)	5556	188	8818
O(w3)	−203	4747	3502
O(w4)	3994	6248	4767
O(w5)	6103	−1276	3783
O(w6)	4078	465	3517
O(w7)	2602	−3408	−564
O(w8)	4814	5664	2470
O(w9)	1762	−1958	4817
O(w10)	2150	3681	5799
O(w11)	3908	−2955	7637

POTASSIUM NIOBIUM TUNGSTATES
$KNbWO_6$, $K_{1.88}NbWO_6$, $KNbWO_6.0.69D_2O$

Solid State Ionics, **18-19**, 799-801.

Fd3m, 10.3628, 10.5285, 10.5019, neutron powder data. Similar $NbWO_6$ octahedral frameworks in the three compounds, but with different K sites.

HYDROGEN MOLYBDENUM TUNGSTEN OXIDES
$D_x(Mo,W)O_3$ (x = 0.56, 0.80, 0.81)

J. Solid State Chem., **63**, 154-159.

Cubic, Im3, a = 7.5558, 7.587, 7.574 A, neutron powder data. Structures as for $D_{0.53}WO_3$ (39A, 244), with D in two 24(g) sites, occupancies 0.093, 0.133, 0.135.

SODIUM LITHIUM MANGANATE(II)
$Na_3Li_5Mn_5O_9$

Z. anorg. Chem., **538**, 123-130.

$R\bar{3}m$, 3.356, 26.120 A, Z = 1, R = 0.044. 5Mn, 5Li, 6 O(2) in 6(c): 0,0,z, z = 0.0795, 0.3168, 0.2291; 3Na in 3(b): 0,0,1/2; 3 O(1) in 3(a): 0,0,0. Partially-occupied Mn and Li sites have tetrahedral, and Na has octahedral coordination. Mn-O = 2.04, 2.08, Li-O = 1.98, 2.29, Na-O = 2.53 A.

POTASSIUM MANGANATE
$K_{1.33}Mn_8O_{16}$

Acta Cryst., B**42**, 162-167.

I4/m, 9.866, 2.872, Z = 1, R = 0.019. 1.33 K in 4(e): 0,0,0.3754; Mn, O(1), O(2) in 8(h): x,y,0, x = 0.34891,0.15445,0.54206, y = 0.16621, 0.20295, 0.16496. Hollandite-type structure (**13**, 190), with K shifted 0.36 A along **c** from the 0,0,1/2 site. Diffuse scattering indicates K ordering.

RUBIDIUM MANGANATE
$Rb_{0.2}MnO_2$

Mineral, J., **13**, 130-140.

A2/m, 15.04, 2.886, 14.64, 92.4, R = 0.122, Psilomelane-type structure (**17**, 429), with strings of MnO_6 octahedra forming tunnels along **b**, which contain an almost continuous distribution of electron density for the Rb ions.

BARIUM MANGANATE(V) CHLORIDE
$Ba_5(MnO_4)_3Cl$

Z. anorg. Chem., **542**, 71-88.

$P6_3/m$, 10.469, 7.760, Z = 2, R = 0.042. Apatite structure.

		x	y	z
Ba(1)	4f(3)	1/3	2/3	0,0012
Ba(2)	6h(m)	0,2578	0,0119	1/4
Mn	6h(m)	0,3711	0,4038	1/4
Cl	2b($\bar{3}$)	0	0	0
O(1)	6h(m)	0,4813	0,3281	1/4
O(2)	6h(m)	0,4680	0,5906	1/4
O(3)	12i(1)	0,2584	0,3544	0,0749

HOLLANDITE
$(Ba,K)Mn_8O_{16}.0.4H_2O$

Mineral. J., **13**, 119-129.

I2/m, 10.006, 2.866, 9.746, 91.17, R = 0.050. Hollandite-type structure (**13**, 190), with MnO_6 octahedra, and Ba, K, and H_2O in the tunnel site.

TETRAAMMINECOPPER(II) PERMANGANATE
$Cu(NH_3)_4(MnO_4)_2$

Acta Cryst., C**42**, 942-945.

$P2_1/m$, 5.413, 9.093, 10.749, 96.18, Z = 2, R = 0.024. Isolated units with (4+2)-coordination at Cu.

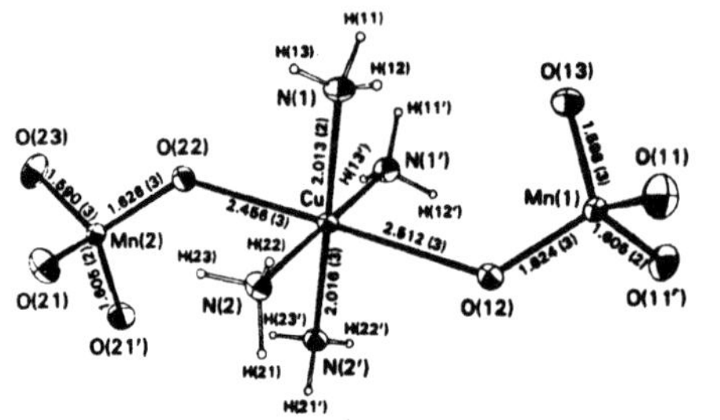

	x	y	z
Cu	0·30013	¼	0·24751
Mn(1)	0·0706	¼	0·56189
Mn(2)	0·5314	¼	0·93229
O(11)	0·9572	0·3948	0·6196
O(12)	−0·0018	¼	0·5889
O(13)	0·6337	¼	0·4085
O(21)	0·6228	0·6050	0·1119
O(22)	0·5843	¼	0·0841
O(23)	0·7924	¼	0·8772
N(1)	0·4981	0·9083	0·6583
N(2)	0·0985	0·0932	0·1506
H(11)	0·364	0·616	0·608
H(12)	0·421	0·035	0·386
H(13)	0·547	0·465	0·304
H(21)	0·050	0·622	0·883
H(22)	0·094	0·021	0·185
H(23)	0·830	0·574	0·908

HOLLANDITE (SILVER)
$Ag_{1.8}Mn_8O_{16}$

Rev. Chim. Minér., **23**, 48–54.

I4/m, 9.725, 2.885, Z = 1, R = 0.061. Hollandite structure (**13**, 190), with Ag having square-planar coordination.

MAGNESIUM PERRHENATE TETRAHYDRATE
$Mg(ReO_4)_2.4H_2O$

Z. anorg. Chem., **532**, 193–196.

P1, 7.692, 7.026, 6.469, 108.279, 92.388, 120.418, Z = 1, R = 0.074. ReO_4 tetrahedra and $MgO_2(H_2O)_4$ octahedra linked by corner-sharing and hydrogen bonding. Re-O = 1.56–1.98, Mg-O = 2.09, 2.10, Mg-OH_2 = 1.82–2.10 A.

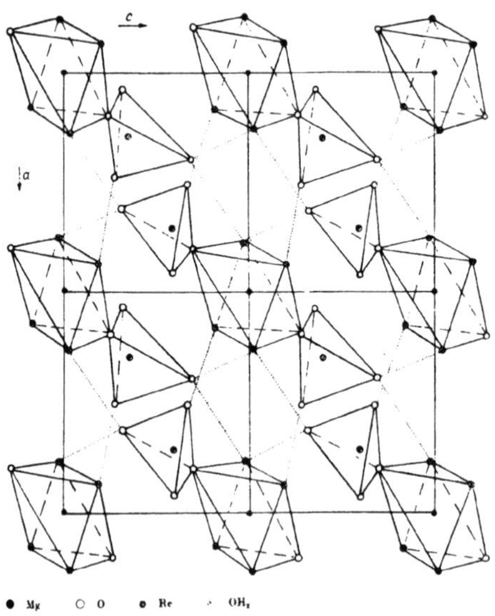

		x	y	z
Mg		0	0	0
Re	(1)	0.6992	0.3752	0.5948
Re	(2)	0.2821	0.6041	0.4218
$O(H_2O)$	(1)	0.234	0.937	0.009
$O(H_2O)$	(2)	0.773	0.067	0.983
$O(H_2O)$	(3)	0.117	0.174	0.809
$O(H_2O)$	(4)	0.839	0.839	0.160
O	(5)	0.070	0.681	0.401
O	(6)	0.920	0.320	0.664
O	(7)	0.615	0.307	0.309
O	(8)	0.357	0.706	0.683
O	(9)	0.493	0.241	0.727
O	(10)	0.488	0.769	0.339
O	(11)	0.519	0.680	0.719
O	(12)	0.199	0.305	0.291

LANTHANUM RHENIUM OXIDE
$La_3Re_2O_{10}$

J. Less-Common Metals, **116**, 293-299.

C2/m, 7.901, 7.866, 7.115, 115.44, Z = 2, R = 0.043. Isolated Re_2O_{10} groups of two edge-sharing ReO_6 octahedra, connected by 8-coordinate La ions. Re-O = 1.87-2.05, Re...Re = 2.484 (short), La-O = 2.42-2.72 A.

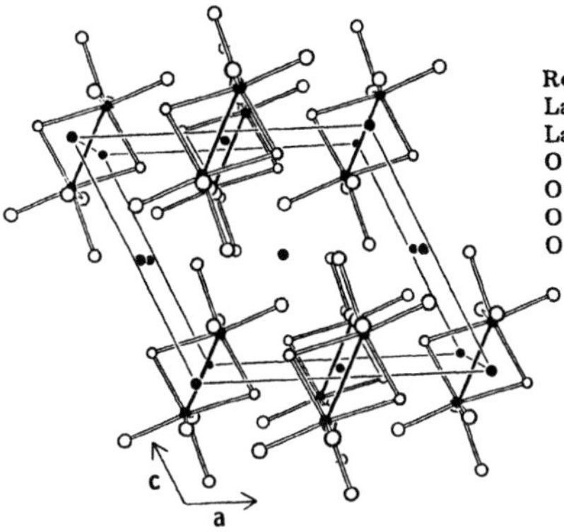

	x	y	z
Re	0.63546	0.0	0.17428
La1	0.0	0.0	0.0
La2	0.0	0.25651	0.50
O1	0.6455	0.0	0.9022
O2	0.9037	0.0	0.2811
O3	0.6365	0.7648	0.2122
O4	0.6754	0.0	0.4733

LITHIUM FERRATES
$LiFe_5O_8$, $Li_{2.8}Fe_5O_8$

Mater. Res. Bull., **21**, 583-592.

$P4_132$, 8.331, 8.389, Z = 8. Powder data. $LiFe_5O_8$ has a spinel structure, 2 x $Fe[Li_{0.5}Fe_{1.5}]O_4$, with long-range Li/Fe order giving a primitive unit cell. Further lithiation displaces Fe from the tetrahedral to octahedral sites, giving a rocksalt-type structure.

CALCIUM FERRITE
$CaFe_4O_7$

Mater. Res. Bull., **21**, 985-994.

C2, 10.409, 6.005, 31.640, 96.30, Z = 14, R = 0.066. Structure corresponds to $Ca_{12}Fe_{60}O_{100}$ per cell, which suggests some Ca/Fe disorder. Two blocks stacked along c: one a sheet of FeO_5 trigonal bipyramids surrounded by two mixed Fe/Ca sheets, and the other a triple layer of Fe atoms with tetrahedral and octahedral coordinations.

STRONTIUM HEXAFERRATE
$Sr_4Fe_6O_{13}$

Mater. Res. Bull., **21**, 175-181.

Iba2, 11.103, 18.924, 5.572, Z = 4, R = 0.11. Perovskite-derived structure with octahedrally- and five-coordinated Fe(III).

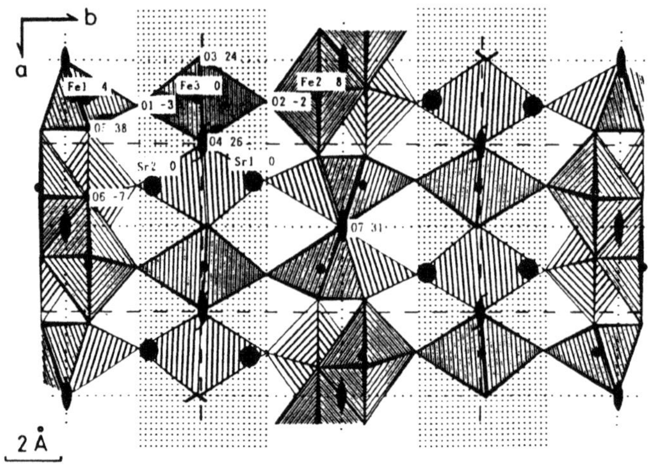

	x	y	z
Sr1	0.3648	0.3432	0.0(
Sr2	0.3805	0.1626	0.0
Fe1	0.1247	0.0401	0.0
Fe2	0.1141	0.4563	0.0
Fe3	0.120	0.2442	-0.0
O1	0.141	0.135	-0.0
O2	0.126	0.365	-0.0
O3	-0.008	0.242	0.2
O4	0.251	0.250	0.2
O5	0.211	0.039	0.3
O6	0.411	0.039	-0.0
O7	0.5	0.5	-0.1

BARIUM MONOFERRITE
$BaFe_2O_4$

Z. anorg. Chem., **538**, 71-77.

Pmcn, 17.347, 9.336, 10.882, Z = 16, R = 0.087. Stuffed tridymite structure. Fe - 4 O = 1.82-1.88(3) A. Another form has been described previously (**37**A, 258 ; **38**A, 276).

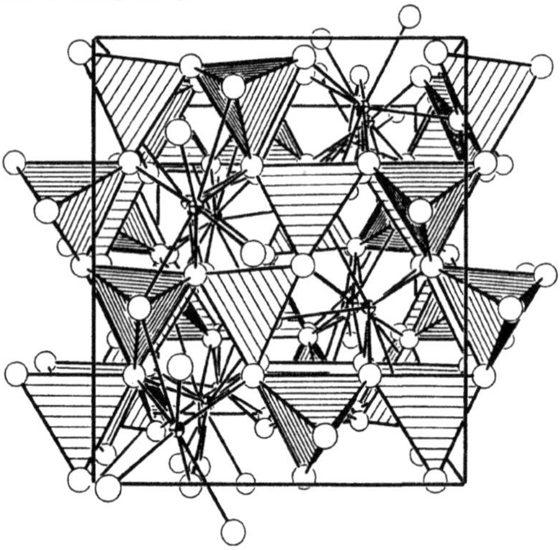

		x	y	z
Ba1	(4c)	0,25	0,2586	0,8826
Ba2	(4c)	0,25	0,7192	0,1426
Ba3	(8d)	0,9892	0,2237	0,8811
Fe1	(8d)	0,3521	0,4102	0,6363
Fe2	(8d)	0,3518	0,9295	0,8838
Fe3	(8d)	0,3861	0,9123	0,3895
Fe4	(8d)	0,0941	0,4117	0,1355
O1	(8d)	0,642	0,081	0,777
O2	(8d)	0,617	0,748	0,272
O3	(8d)	0,643	0,932	0,515
O4	(8d)	0,604	0,075	0,273
O5	(8d)	0,647	0,256	0,536
O6	(8d)	0,600	0,563	0,513
O7	(8d)	0,492	0,622	0,892
O8	(4c)	0,25	0,118	0,107
O9	(4c)	0,25	0,530	0,349

BARIUM MAGNESIUM FERRATE
$BaMg_2Fe_{16}O_{27}$

J. Magn. Magn. Mater., **58**, 247-253.

$P6_3/mmc$, 5.9060, 32.915, Z = 2, R = 0.015. Hexagonal ferrite (**52**A, 235) with sequence RSSR*S*S*, where R and S are ferrite structural blocks. Mg is in the octahedral and tetrahedral sites of the spinel S-blocks.

BARIUM ANTIMONY NICKEL FERRATE
$Ba_2Sb_2Ni_3Fe_8O_{22}$

J. Solid State Chem., **61**, 223-229.

$P\bar{3}m1$, 5.882, 14.237, Z = 1, powder data. 6L ferrite, with a $(hcc)_2$ stacking of two cubic BaO_3 and two hexagonal O_4 layers.

		x	y	z	Occupation (%)
Ba	2(d)	$\frac{1}{3}$	$\frac{2}{3}$	0.417	100
Te_1	2(d)	$\frac{1}{3}$	$\frac{2}{3}$	0.955	100Fe
Te_2	2(c)	0.0	0.0	0.384	100Ni
Oc_1	2(d)	$\frac{1}{3}$	$\frac{2}{3}$	0.680	66Sb; 34Ni
Oc_2	1(a)	0.0	0.0	0.0	67Sb; 33Ni
Oc_3	6(i)	0.174	0.348	0.172	100Fe
O_1	2(c)	0.0	0.0	0.236	100
O_2	2(d)	$\frac{1}{3}$	$\frac{2}{3}$	0.084	100
O_3	6(i)	0.163	0.325	0.909	100
O_4	6(i)	0.496	0.992	0.228	100
O_5	6(i)	0.183	0.366	0.585	100

BARIUM COBALT FERRITE
$BaCo_2Fe_{16}O_{27}$

Phys. Status Solidi, A, **96**, 385-395.

$P6_3/mmc$, 5.9043, 32.910, Z = 2, R = 0.024, and neutron data at 550°C, R = 0.040. W-type hexagonal ferrite (**16**, 248; **21**, 281; **52A**, 235), with possible Co sites determined. [See following report.]

STRONTIUM ZINC FERRITE
$SrZn_2Fe_{16}O_{27}$

Z. Kristallogr., **175**, 139-144.

$P6_3/mmc$, 5.908, 32.745, Z = 2, R = 0.061. W-type ferrite (**16**, 248; **21**, 281; **52A**, 235), with a sequence of 14 cubic and hexagonal close-packed oxygen layers, with Sr replacing one-quarter of the O in every seventh layer. Zn partially occupies tetrahedral sites, and one Fe^{3+} ion, Fe(7), has trigonal bipyramidal coordination (but may be split into two tetrahedral sites).

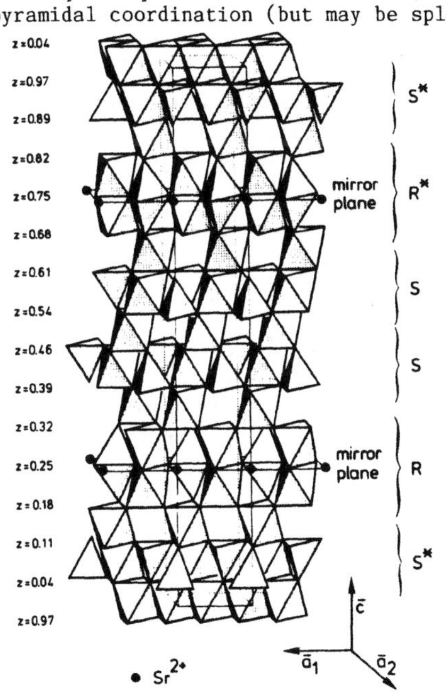

		x	z	s.o.f.*
Sr	2b	0	$\frac{1}{4}$	1
Fe(1)	6g	$\frac{1}{2}$	0	1
Zn(1)	4e	0	0.0566	$\frac{3}{4}$
Fe(2)	4e	0	0.0564	$\frac{1}{4}$
Fe(3)	4f	$\frac{1}{3}$	−0.0753	1
Fe(4)	4f	$\frac{1}{3}$	0.0941	$\frac{3}{4}$
Zn(2)	4f	$\frac{1}{3}$	0.0940	$\frac{1}{4}$
Fe(5)	12k	0.8363	0.1514	1
Fe(6)	4f	$\frac{1}{3}$	0.2086	1
Fe(7)	2d	$\frac{1}{3}$	$-\frac{1}{4}$	1
O(1)	12k	0.8202	0.0353	1
O(2)	4f	$\frac{1}{3}$	0.0344	1
O(3)	12k	0.5115	0.1107	1
O(4)	4e	0	0.1150	1
O(5)	12k	0.1628	0.1805	1
O(6)	4f	$\frac{1}{3}$	−0.1801	1
O(7)	6h	0.4847	$\frac{1}{4}$	1

$y = 2x$; * s.o.f.: site occupation factor

BARIUM LANTHANUM IRON(III) OXIDE
$BaLaFeO_4$

Mater. Res. Bull., **21**, 733-738.

I4/mmm, 3.914, 13.175, Z = 2, powder data. Ba,La in 4(e): 0,0,0.361; Fe in 2(a): 0,0,0; O(1) in 4(e): 0,0,0.165; O(2) in 4(c): 1/2,0,0. K_2NiF_4-type structure, as for the Ca and Sr compounds (**46**A, 282, 283).

NEODYMIUM COPPER RUTHENIUM OXIDES
$(Nd,Cu)_2Ru_2O_{6.625}$

Mater. Res. Bull., **21**, 1489-1493.

Fd3m, 10.25, Z = 8, R = 0.025. Nd,Cu in 16(d): 1/2,1/2,1/2; Ru in 16(c): 0,0,0; O(1) in 48(f): x,1/8,1/8, x = 0.3290; O(2) in 8(b): 3/8,3/8,3/8. Pyrochlore structure (**33**A, 190; x in this reference is probably referred to origin at $\bar{1}$).

$NdCu_3Ru_4O_{12}$

Mater. Res. Bull., **21**, 1131-1136.

Im3, 7.455, Z = 2, R = 0.025. Nd in 2(a): 0,0,0; Cu in 6(b): 0,1/2,1/2; Ru in 8(c): 1/4,1/4,1/4; O in 24(g): 0.1721,0.3050,0. Perovskite-type structure, as for related materials (**41**A, 260; **43**A, 350; **45**A, 235).

LITHIUM COBALTATE(II)
Li_6CoO_4

Z. anorg. Chem., **534**, 61-68.

$P4_2/nmc$, 6.536, 4.654, Z = 2, R = 0.096. Li(1) in 8(f): 0.0341,0.9659,3/4; Li(2) in 4(d): 3/4,3/4,0.2985; Co in 2(a): 3/4,1/4,3/4; O in 8(g): 3/4,0.5143,0.5351. Li and Co ions have tetrahedral coordinations. Co-O = 2.00, Li-O = 1.89-2.11 A.

LANTHANUM COBALTATE
$LaCoO_3$

J. Solid State Chem., **61**, 301-307.

$R\bar{3}c$, 5.3416-5.150 A, 60.990-60.296°, at 4.2-1248K, Z = 1, neutron powder data. La in 2(a): 1/4,1/4,1/4; Co in 2(b): 0,0,0; O in 6(e): 0.1978,0.3021,3/4, at 4.2K. Distorted perovskite. Possibly $R\bar{3}$ at 668K.

LITHIUM RHODATE(III)
$LiRhO_2$

Z. anorg. Chem., **535**, 20-30.

$F4_132$, 8.4127, Z = 16, R = 0.091. Rh in 16(c): 1/8,1/8,1/8; Li in 16(d): 5/8,5/8,5/8; O in 32(e): x,x,x, x = 0.8850. NaCl superstructure.

LITHIUM IRIDATE, PLATINATE, and CERATE(IV)
Li_8MO_6 (M = Ir, Pt, Ce)

Z. anorg. Chem., **536**, 81-91.

$R\bar{3}$, a = 5.4151, 5.4146, 5.6430, c = 15.0584, 15.0143, 16.0271 A, Z = 3, R = 0.046, 0.026, 0.030. Ir, Pt, Ce in 3(a): 0,0,0; Li(1) in 18(f): 0.6227,0.6444,0.8776; Li(2) in 6(c): 0,0,0.3424; O in 18(f): 0.6740,0.6378,0.4102 (parameters for Ir compound, similar values for others). Li_8SnO_6-type (**33**A, 288; **34**A, 263; **43**A, 183; **52**A, 242).

POTASSIUM LITHIUM IRIDATE(V)
KLi_6IrO_6

Z. anorg. Chem., **537**, 106-114.

$R\bar{3}m$, 8.1961, 7.0996 A, Z = 3, R = 0.014. Ir in 3(a): 0,0,0; K in 3(b): 0,0,1/2; Li in 18(f): 0.3395,0,0; O in 18(h): 0.1170,0.8830,0.1562. Stuffed variant of α-Li_6UO_6 type (**52**A, 244); Ir-6 O = 2.00, K-12 O = 2.95,3.07, Li-4 O = 1.93, 1.95 A.

LANTHANUM NICKELATE (ORTHORHOMBIC)
La_2NiO_4

Mater. Res. Bull., **21**, 787-796.

Pncb, 5.468, 5.535, 12.547, Z = 4, X-ray and neutron powder data. La(1) in 4(g): 0,0,0.3774; La(2) in 4(h): 1/2,0,0.1367; Ni(1) in 2(a): 0,0,0; Ni(2) in 2(c): 1/2,0,1/2; O(1) in 4(g): 0,0,0.195; O(2) in 4(h): 1/2,0,0.326; O(3) in 8(m): 0.226,0.250,0.007. Distorted K_2NiF_4-type structure.

LITHIUM PALLADATE(II)
Li_2PdO_2

Z. anorg. Chem., **536**, 77-80.

Immm, 3.7534, 2.9818, 9.3158, Z = 2, R = 0.035. Pd in 2(a): 0,0,0; O in 4(j): 0,1/2,0.6486; Li in 4(i): 0,0,0.7003. Li_2CuO_2-type (**35**A, 216), as previously described (**39**A, 253). Pd-4 O = 2.03, Li-4 O = 1.94, 2.05 A.

BISMUTH SILVER OXIDE
$Bi_4Ag_{18}O_{12}$

C.R. Acad. Sci. Paris, Ser 2, **302**, 631-633.

Cmcm, 10.630, 14.229, 23.91, Z = 4, R = 0.057. The structure contains oxygen-centred Ag_3Bi tetrahedra, which share corners to form $Bi_4Ag_{18}O_{12}$ units, which link to six neighbouring units.

POTASSIUM SODIUM AURATE(I)
KNa_2AuO_2

Z. anorg. Chem., **537**, 115-122.

Pnnm, 10.256, 5.472, 4.000, Z = 2, R = 0.062. Au in 2(a): 0,0,0; K in 2(d): 0,1/2,1/2; Na in 4(g): 0.2648,0.3996,0; O in 4(g): 0.1442,0.7530,0. Isolated linear $[O\text{-}Au\text{-}O]^{3-}$ ions, linked by 3-coordinate Na and 4-coordinate K; Au-O = 2.01, Na-O = 2.29, 2.35, K-O = 2.85 A.

CAESIUM POTASSIUM CADMATE
$Cs_2K_2Cd_3O_5$

Z. Naturforsch., **41**B, 1513-1518.

$P\bar{1}$, 9.832, 9.180, 6.535, 108.3, 113.7, 99.9, Z = 2, R = 0.059. Cd ions have (3+1) tetrahedral coordinations, Cd-O = 2.15-2.21, 2.30-2.44 A; Cs ions have 3 and 5 O neighbours at 2.81-3.31, with further neighbours at 3.56-3.80 A; K ions have fivefold trigonal bipyramidal coordinations, K-O = 2.63-3.06 A.

	x (10^4)	y	z
Cs(1)	6500	4221	7114
Cs(2)	2521	2909	6426
K(1)	3050	9469	2482
K(2)	9807	4306	2225
Cd(1)	929	1410	9682
Cd(2)	8306	1695	3960
Cd(3)	5032	1829	1036
O(1)	351	8780	8022
O(2)	199	7170	2062
O(3)	5976	9961	2748
O(4)	2646	6360	7060
O(5)	3456	2547	2353

SODIUM PRASEODYMATE and TERBATE
$NaPrO_2$, $NaTbO_2$

Z. anorg. Chem., **537**, 97-105.

$I4_1/amd$, a = 4.7619, 4.6311, c = 10.9609, 10.3739 A, Z = 4, R = 0.043, 0.035. Na in 4(b): 0,0,1/2; Pr or Tb in 4(a): 0,0,0; O in 8(e): 0,0,z, z = 0.2213,0.2221. α-$LiFeO_2$ type (**9**, 176; **26**, 370; **27**, 515; **29**, 322; **50**A, 113). Pr-6 O = 2.40, 2.43, Tb-6 O = 2.30, 2.33, Na-6 O = 2.33-3.06 A.

BARIUM NICKEL NEODYMIUM OXIDE
$BaNiNd_2O_5$

Z. anorg. Chem., **532**, 10-16.

Immm, 3.829, 5.932, 11.649, Z = 2, R = 0.073. Ba in 2(a): 0,0,0; Ni in 2(c): 1/2,1/2,0: Nd in 4(j): 1/2,0,0.7023; O(1) in 8(ℓ): 0,0.756,0.356; O(2) in 2(d): 1/2,0,1/2. NiO_6 octahedra (compare the square-planar coordination in the Pt compound, **52**A, 239), linked by 10-coordinate Ba (bicapped cube) and 7-coordinate Nd (monocapped trigonal prism). Ni-O = 1.92, 2.21, Ba-O = 2.96, 2.97, Nd-O = 2.35, 2.36, 2.50(2) A.

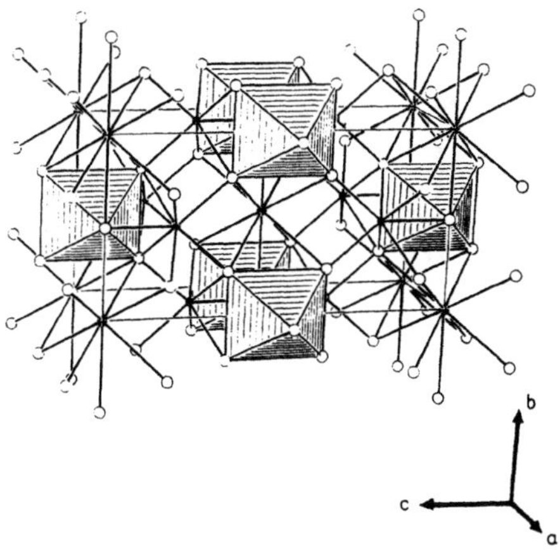

BARIUM NICKEL YTTERBIUM OXIDE
BARIUM COPPER(II) SAMARIUM OXIDE
$BaNiYb_2O_5$, $BaCuSm_2O_5$

Z. anorg. Chem., **540/541**, 243-250.

Pbnm, a = 6.950, 7.283, b = 12.151, 12.405, c = 5.639, 5.744 A, Z = 4, R = 0.058, 0.054. Isostructural with $BaMLn_2O_5$, M = Cu, Zn (**49**A; **50**A), but not with $BaNiNd_2O_5$ (preceding report). Tetragonal forms of $BaMLn_2O_5$ oxides are also known (**48**A, 243; **51**A, 255; **52**A, 239; following report).

		x	y	z
Yb1	(4c)	0.1243	0.2919	0.25
Yb2	(4c)	0,4004	0.0742	0.25
Ba	(4c)	0,9228	0,9035	0.25
Ni	(4c)	0,6954	0,6555	0,25
O1	(8d)	0,171	0.433	0.996
O2	(8d)	0,364	0,226	0,497
O3	(4c)	0.075	0.098	0.25
Sm1	(4c)	0.1146	0,2890	0.25
Sm2	(4c)	0.3946	0.0739	0.25
Ba	(4c)	0.9321	0,9058	0.25
Cu	(4c)	0,7132	0,6592	0,25
O1	(8d)	0.168	0,437	0.989
O2	(8d)	0.351	0.226	0.506
O3	(4c)	0.076	0.095	0.25

BARIUM PALLADIUM NEODYMATE and SAMARATE
BARIUM COPPER NEODYMATE
$BaPdNd_2O_5$, $BaPdSm_2O_5$, $BaCuNd_2O_5$

Mh. Chem., **117**, 465-470.

P4/mbm, a = 6.7387, 6.6667, 6.7027, c = 5.9002, 5.8800, 5.8194 A, Z = 2, R = 0.057, 0.065, 0.052. Essentially isostructural with $BaPtNd_2O_5$ (**52**A, 239), as previously described for $BaCuNd_2O_5$ (**51**A, 255).

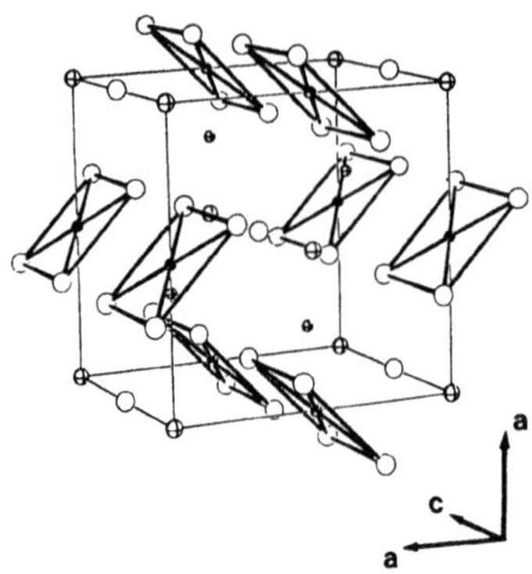

$BaPdNd_2O_5$		x	y	z
Ba	(2a)	0.0	0.0	0.0
Pd	(2d)	0.0	0.5	0.0
Nd	(4h)	0.1738	0.6738	0.5
O_I	(8k)	0.360	0.860	0.749
O_{II}	(2b)	0.0	0.0	0.5
$BaPdSm_2O_5$				
Ba	(2a)	0.0	0.0	0.0
Pd	(2d)	0.0	0.5	0.0
Sm	(4h)	0.1736	0.6736	0.5
O_I	(8k)	0.360	0.860	0.748
O_{II}	(2b)	0.0	0.0	0.5
$BaCuNd_2O_5$				
Ba	(2a)	0.0	0.0	0.0
Cu	(2d)	0.0	0.5	0.0
Nd	(4h)	0.1739	0.6739	0.5
O_I	(8k)	0.362	0.862	0.763
O_{II}	(2b)	0.0	0.0	0.5

RUBIDIUM LITHIUM TERBATE
$Rb_2Li_{14}Tb_3O_{14}$ (I)

LITHIUM TERBATE
$Li_6Tb_2O_7$ (II)

Z. anorg. Chem., **539**, 127-140.

I. Immm, 12.831, 7.909, 7.369, Z = 2, R = 0.047. Isostructural with related Pb compounds (**49**A; **50**A) with $Tb_3O_{14}^{16-}$ ions of three edge-sharing TbO_6 octahedra, linked by 4-coordinate Li^+ and (6+4)-coordinate Rb^+ ions. Tb-O = 2.11-2.33 A.

II. $P2_1/a$, 10.563, 6.135, 5.466, 109.67, Z = 2, R = 0.094. NaCl variant with anion vacancies. Tb-6 O = 2.10-2.24, Li-5 O = 1.86-2.68 A.

$Rb_2Li_{14}[Tb_3O_{14}]$

	x	y	z
Tb1	0,2716	0,0	0,0
Tb2	0,0	0,0	0,0
Rb	0,5	0,0	0,2707
Li1	0,5	0,3474	0,0
Li2	0,2308	0,3586	0,0
Li3	0,1385	0,1600	0,2756
O1	0,2542	0,0	0,7095
O2	0,0	0,0	0,2899
O3	0,1292	0,1815	0,0
O4	0,3699	0,2140	0,0

$Li_6Tb_2O_7$

	x	y	z
Tb	0,1340	0,0109	0,2729
O1	0,0	0,0	0,5
O2	0,0065	0,7630	0,0201
O3	0,2562	0,0048	0,0392
O4	0,2296	0,2932	0,4672
Li1	0,3499	0,7323	0,1976
Li2	0,1351	0,7687	0,7750
Li3	0,1276	0,5299	0,1505

SODIUM DIURANATE(VI)
$Na_2U_2O_7$

J. Less-Common Metals, **119**, 83-90.

$R\bar{3}m$, 3.911, 17.857 A, Z = 3/2, R = 0.082. 3 U in 36(i): -0.0163,-0.0600,0.0054; 3 Na in 3(b): 0,0,1/2; 4.5 O(1) in 6(c): 0,0,0.1048; 6 O(2) in 18(h): 0.0455,-0.0455,0.3514. Isostructural with $CaUO_4$ (**11**, 323; **34**A, 289), with oxygen vacancies, i.e. $Na(UO_{1.5})O_2$ by analogy with $Ca(UO_2)O_2$. U-O = 1.87 (x 2), 2.28 (x 6), Na-O = 2.51 (x 6), 2.61 (x 2) A.

CALCIUM URANATE(VI)
Ca_3UO_6

Acta Cryst., C**42**, 523-525.

P2$_1$/n, 5.7292, 5.9562, 8.2991, 90.56, Z = 2, neutron powder data. Deformed perovskite structure; previous study in **31**A, 160; **34**A, 289.

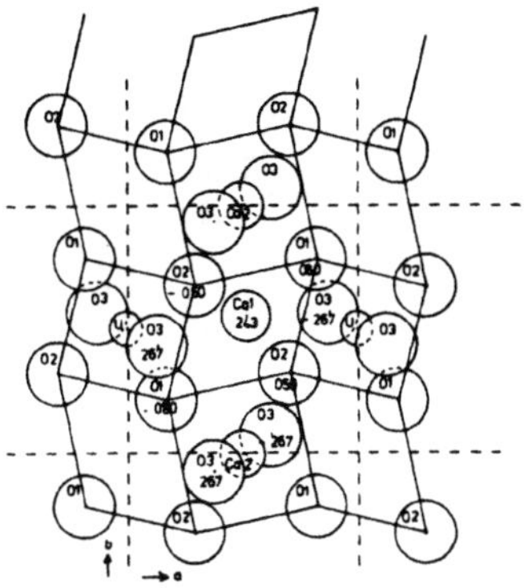

	x	y	z
Ca(1)	0·5151	0·5522	0·2430
Ca(2)	0·5	0·0	0·0
U	0·0	0·5	0·0
O(1)	0·1760	0·2141	−0·0802
O(2)	0·2948	0·6769	−0·0546
O(3)	0·3747	0·9324	0·2669

Atomic distances (Å) *in* Ca_3UO_6

Ca(1)–O(1)	2·344 (4)	Ca(2)–O(1)	2·343 (2)
–O(1)	2·635 (4)	–O(2)	2·298 (3)
–O(1)	3·165 (5)	–O(3)	2·370 (2)
–O(2)	2·861 (5)		
–O(2)	2·818 (4)	U–O(1)	2·091 (2)
–O(2)	2·351 (4)	–O(2)	2·045 (2)
–O(3)	2·412 (4)	–O(3)	2·096 (2)
–O(3)	2·345 (4)		

LEAD (II) URANATE(VI)
$PbUO_4$

Acta Cryst., C**42**, 1684–1685.

Pbcm, 5.536, 7.968, 8.212, Z = 4, R = 0.048. Isostructural with the Ba salt (**11**, 441; **34**A, 289; **42**A, 294), with sheets of corner-sharing UO_6 octahedra, linked by 7-coordinate Pb ions.

	Site symmetry	x	y	z
U	$\bar{1}$	0	0	0
Pb	*m*	0·4954	0·2022	¼
O(1)	1	0·2915	0·9190	0·0765
O(2)	2	0·1587	¼	0
O(3)	*m*	0·8701	0·0374	¼

Interatomic distances (Å) *and angles* (°)

U–O(1)	1·848 (18)	(×2)	U–U	3·984 (1)	(×2)
U–O(2)	2·177 (11)	(×2)	U–U	4·106 (1)	(×2)
U–O(3)	2·195 (9)	(×2)	U–Pb	3·786 (2)	(×2)
Pb–O(1)	2·897 (16)	(×2)	U–Pb	3·822 (2)	(×2)
Pb–O(1)	2·531 (15)	(×2)	Pb–Pb	3·984 (1)	(×2)
Pb–O(2)	2·799 (18)	(×2)	Pb–Pb	4·176 (1)	(×2)
Pb–O(3)	2·455 (26)				

U–O(1)–Pb	103·6 (6)	C(2)–U–O(3)	89·6 (7)
U–O(1)–Pb	146·8 (8)	O(2)–U–O(3)	90·4 (7)
U–O(2)–U	132·4 (13)	O(1)–Pb–O(1)	58·9 (3)
U–O(2)–Pb	113·2 (3)	O(1)–Pb–O(1)	115·9 (5)
U–O(2)–Pb	98·3 (3)	O(1)–Pb–O(1)	171·9 (7)
U–O(3)–U	138·5 (13)	O(1)–Pb–O(2)	59·0 (3)
U–O(3)–Pb	110·5 (6)	O(1)–Pb–O(2)	102·0 (5)
O(1)–U–O(1)	180·0	O(1)–Pb–O(3)	85·0 (6)
O(2)–U–O(2)	180·0	C(1)–Pb–O(1)	68·5 (4)
O(3)–U–O(3)	180·0	O(1)–Pb–O(2)	78·7 (5)
O(1)–U–O(2)	88·1 (8)	O(1)–Pb–O(2)	129·1 (4)
O(1)–U–O(2)	91·9 (8)	O(1)–Pb–O(3)	88·4 (7)
O(1)–U–O(3)	89·1 (8)	O(2)–Pb–O(2)	94·4 (8)
O(1)–U–O(3)	90·9 (8)	O(2)–Pb–O(3)	129·5 (4)

Pb_3UO_6

Acta Cryst., C**42**, 1275–1277.

Pnam, 13.719, 12.351, 8.213, Z = 8, neutron powder data. Previous X-ray results (**32**A, 317) are confirmed.

	x	y	z
U	0·4146	0·2605	0·0016
Pb(1)	0·0311	0·0277	¼
Pb(2)	0·0057	0·5145	¼
Pb(3)	0·3521	0·7310	¼
Pb(4)	0·1832	0·2629	¼
Pb(5)	0·2683	0·0036	0·0300
O(1)	0·3692	0·2915	¼
O(2)	0·0380	0·7237	¼
O(3)	0·2022	0·0854	¼
O(4)	0·3449	0·9151	¼
O(5)	0·2291	0·7265	0·0366
O(6)	0·4321	0·0982	0·0516
O(7)	0·1085	0·9202	0·0537
O(8)	0·0523	0·2050	0·0577

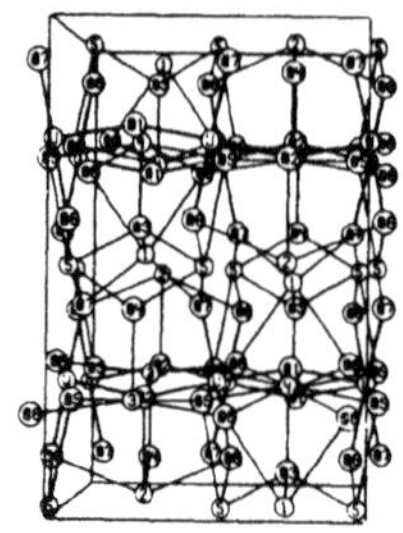

Metal–oxygen distances (Å) for Pb_3UO_6 calculated from (a) neutron data and (b) X-ray data

	(a)	(b)		(a)	(b)
U–O(1)	2·167 (9)	2·15	Pb(3)–O(4)	2·276 (16)	2·31
U–O(2)	2·214 (9)	2·19	Pb(3)–O(5)	2·434 (10) 2×	2·45
U–O(5)	2·040 (11)	2·03	Pb(3)–O(2)	2·611 (15)	2·62
U–O(6)	2·060 (11)	2·05	Pb(3)–O(8)	2·869 (9) 2×	2·88
U–O(7)	2·049 (11)	2·04	Pb(4)–O(3)	2·208 (13)	2·18
U–O(8)	1·991 (11)	1·90	Pb(4)–O(8)	2·496 (10) 2×	2·57
Pb(1)–O(7)	2·343 (11) 2×	2·34	Pb(4)–O(5)	2·682 (9) 2×	2·62
Pb(1)–O(3)	2·453 (14)	2·53	Pb(4)–O(1)	2·576 (14)	2·64
Pb(1)–O(8)	2·716 (10) 2×	2·75	Pb(5)–O(3)	2·260 (8)	2·24
Pb(1)–O(1)	3·149 (15)	3·14	Pb(5)–O(4)	2·359 (10)	2·36
Pb(1)–O(7′)	3·210 (11) 2×	3·18	Pb(5)–O(7)	2·430 (10)	2·38
Pb(2)–O(4)	2·371 (15)	2·30	Pb(5)–O(6)	2·539 (10)	2·54
Pb(2)–O(6)	2·369 (11) 2×	2·40	Pb(5)–O(5)	2·805 (11)	2·91
Pb(2)–O(2)	2·621 (17)	2·72	Pb(5)–O(5′)	3·467 (11)	3·35
Pb(2)–O(6′)	2·817 (10) 2×	2·76			

CAESIUM POTASSIUM NIOBOTITANOURANATE
$(Cs_{0.75}K_{0.25})(Nb,Ti)U_2O_{11}$

Acta Cryst., C**42**, 136-138.

Amam, 7.630, 10.923, 13.609, Z = 4, R = 0.055. Two domains, one with Cs, the other with K. U has pentagonal bipyramidal and Nb,Ti trigonal bipyramidal coordination.

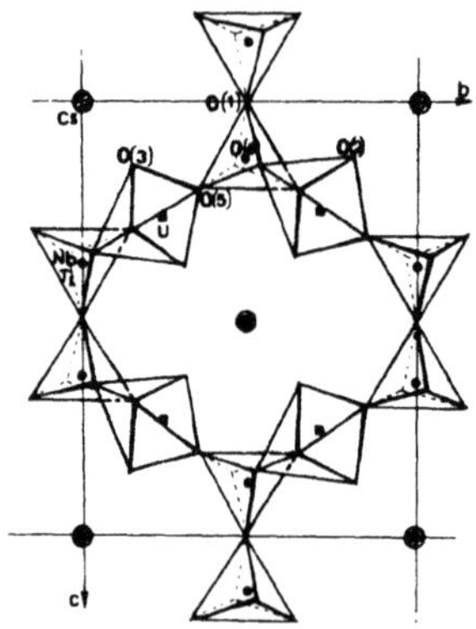

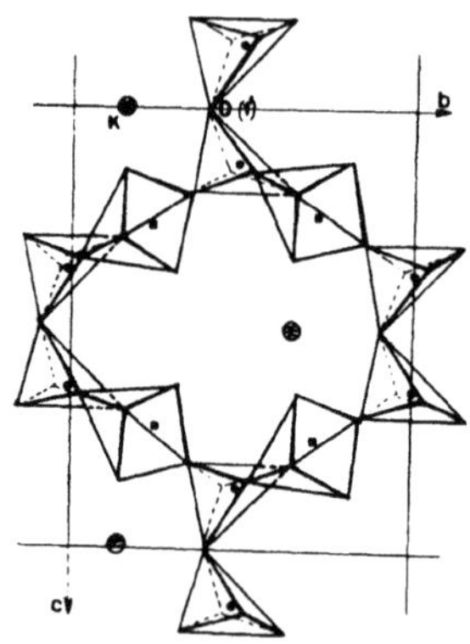

	x	y	z
U	1/4	22691	26336
Nb	0	0	3692
0,75 Cs	0	0	0
0,75 O(1)	0	1/2	0
0,125 O(1′)	488	3994	0
O(2)	1/4	8161	1248
O(3)	1/4	1397	1502
O(4)	1/4	5413	1467
O(5)	347	3532	1994
0,125 K	188	1584	0

Distances cation–anion (Å)

U–O(2)	1,81 (1)	Cs–O(3)	3,185 (5) (×4)
–O(3)	1,81 (1)	–O(2)	3,249 (5) (×4)
–O(5)	2,315 (5) (×2)		
–O(4)	2,368 (5)	K–O(1′)	2,64 (5)
–O(5)	2,396 (5) (×2)	–O(2)	2,68 (3) (×2)
		–O(3)	2,71 (2) (×2)
(Nb,Ti)–O(1)	1,780 (1)		
–O(5)	1,874 (5) (×2)		
–O(4)	1,972 (2) (×2)		

STRONTIUM HYDROXIDE MONOHYDRATE
$Sr(OH)_2.H_2O$

BARIUM HYDROXIDE MONOHYDRATE (ORTHORHOMBIC and MONOCLINIC)
β- and γ-$Ba(OH)_2.H_2O$

BARIUM HYDROXIDE TRIHYDRATE
$Ba(OH)_2.3H_2O$

Z. anorg. Chem., **538**, 131-142.

$Sr(OH)_2.H_2O$ and β-$Ba(OH)_2.H_2O$, $Pmc2_1$, a = 3.6478, 3.8947, b = 6.1981, 6.3657, c = 6.7131, 6.9523 A, Z = 2, R = 0.020, 0.018. Structure as previously described (32A, 245), with Sr and Ba having bicapped trigonal prismatic coordinations to 6 OH and 2 H_2O.

γ-$Ba(OH)_2.H_2O$, $P2_1/m$, 7.049, 4.184, 6.333, 111.45, Z = 2, R = 0.060. Structure related to that of rare-earth hydroxides $M(OH)_3$ ($P6_3/m$). Ba has tricapped trigonal prismatic coordination to 6 OH at mean distance of 2.78 A and 3 H_2O at 2.90 A.

$Ba(OH)_2.3H_2O$, Pnma, 7.640, 11.403, 5.965, Z = 4, R = 0.019. Ba has bicapped trigonal prismatic coordination to 2 OH and 6 H_2O.

In all the structures the water molecules form strong hydrogen bonds to neighbouring OH^- ions, but the hydroxide hydrogen atoms do not participate in hydrogen bonding.

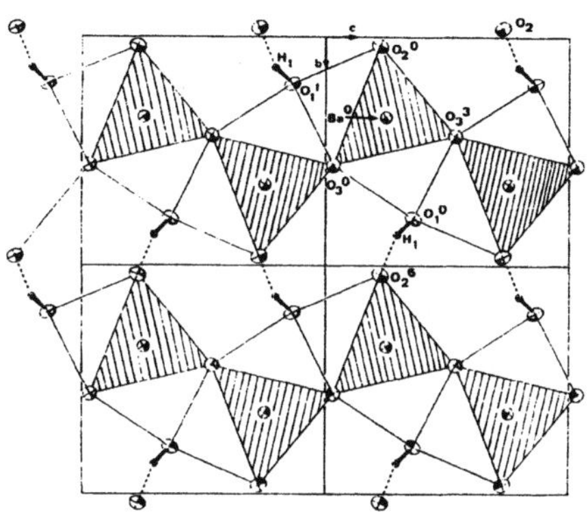

$Ba(OH)_2 \cdot H_2O$ o.-rh.

	x	y	z
$Sr(OH)_2 \cdot H_2O$			
Sr	0.0	0,37190	0,25
		[0,3720]	
O(1)	0,0	0,7845	0,3760
		[0.792]	[0,376]
O(2)	0,5	0,0647	0,2427
		[0,037]	[0,241]
O(3)	0,5	0,5626	0,0350
		[0,563]	[0,039]
H(1)	0,29	0,884	0,330
$Ba(OH)_2 \cdot H_2O$ o.-rh.			
Ba	0,0	0,35547(3)	0,25
O(1)	0,0	0,7923(7)	0,3614
O(2)	0.5	0,0393(6)	0,2266
O(3)	0.5	0,5635(6)	0,0331
H(1)	0,22	0,865(6)	0,293

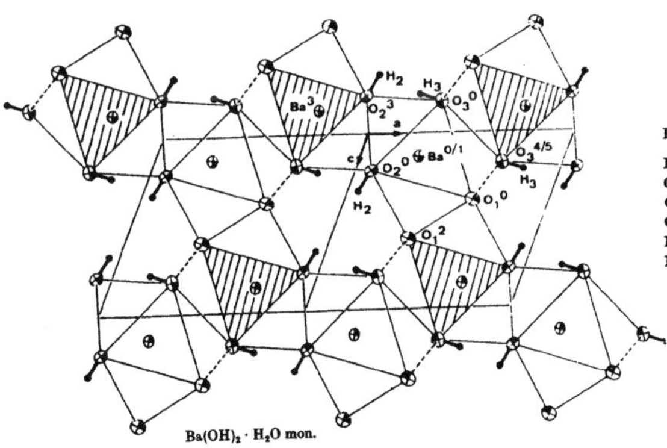

$Ba(OH)_2 \cdot H_2O$ mon.

$Ba(OH)_2 \cdot H_2O$ mon.			
Ba	0,28351	0,25	0,1360
O(1)	0,622	0,75	0,390
O(2)	0,077	0,75	0.213
O(3)	0,298	0,75	−0,178
H(2)	0,05	0,75	0,33
H(3)	0,18	0,75	−0,22

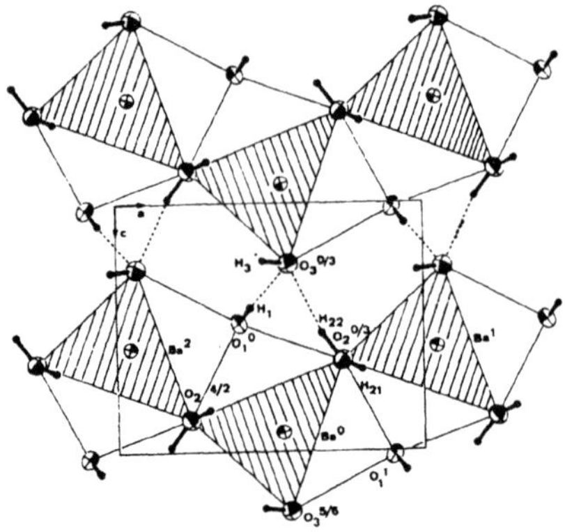

$Ba(OH)_2 \cdot 3\,H_2O$

$Ba(OH)_2 \cdot 3\,H_2O$

Ba	0,54526	0,25	0,92144
O(1)	0,4005	0,25	0,4810
O(2)	0,7371	0,3987	0,6312
O(3)	0,5648	0,4170	0,2422
H(1)	0,439	0,313	0,419
H(21)	0,80	0,46	0,67
H(22)	0,67	0,41	0,52
H(3)	0,48	0,46	0,23

MOTUKOREAITE
$[Mg_6Al_3(OH)_{18}][Na_{0.6}(SO_4)_{1.3}(CO_3)_{0.7}(H_2O)_{12}]$

Neues Jb. Miner., Mh., 263-272.

$R\bar{3}m$, 9.172, 33.51 A, Z = 3, R = 0.114. Double layer structure, with excess positive charge in the ordered brucite-like main layers compensated by additional anions in the also-ordered interlayers.

			x	y	z
3	Al(1)	1	0	0	1/2
18	Mg	1	0.3344	0	1/2
6	Al(2)	1	1/3	2/3	0.5014
3	Na	0.66	0	0	0
6	S	0.66	2/3	1/3	0.0402
18	O(1)	1	0.1061	2x	0.5289
18	O(2)	1	0.4408	2x	0.5305
18	O(3)	1	2y	0.2289	0.5292
36	O(4)	½·0.66	0.1327	0.4301	0.0553
36	O(5)	1/2	0.2162	0.1690	0.0520
36	O(6)	½·0.66	0.5495	0.3923	0.0563
6	O(7)	0.66	1/3	2/3	0.0042

TRICALCIUM ALUMINATE HEXAHYDRATE
$Ca_3[Al(OH)_6]_2$ $3CaO.Al_2O_3.6H_2O$

Z. Anal. Chem., **324**, 124-126.

Ia3d, 12.575, Z = 8, R = 0.03. Ca in 24(c): 0,1/4,1/8; Al in 16(a): 0,0,0; 24(d) site vacant: O in 96(h): -0.0282,0.0531,0.1396; H in 96(h): -0.091,0.054,0.163. Hydrogarnet structure (**52**A, 335).

BARIUM HYDROXYGERMANATE
$BaGe_3O_6(OH)_2$

Acta Cryst., C**42**, 513-515 (1986) and C**43**, 173-174 (1987).

C2/c, 12.703, 4.9047, 9.418, 105.46, Z = 4, R = 0.042.

Eight anions and one Ba cation define a regular h.c.p. layer parallel to the (201) plane. Ge atoms occupy octahedral interstices and a sheet made of three Ge octahedra is formed in the (100) plane by edge sharing. The Ge(1) octahedron shares four edges and the Ge(2) and Ge(3) octahedra share three. Ba is coordinated by twelve anions.

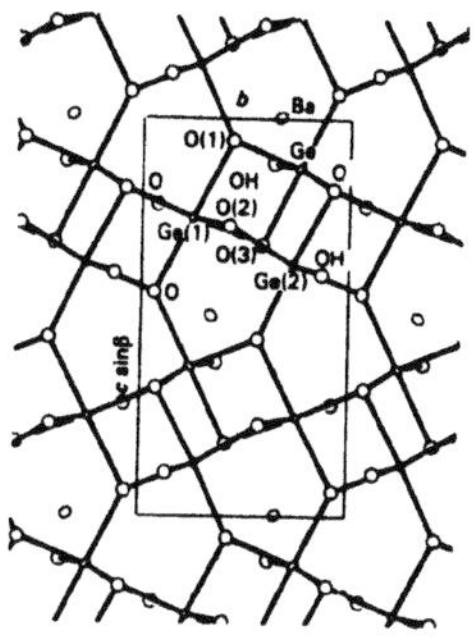

Fractional coordinates ($\times 10^4$)

	x	y	z
Ba	0	6622	−2500
Ge(1)	2500	2500	0
Ge(2)	1922	7341	1215
O(1)	2294	4359	−1915
O(2)	1269	4231	255
O(3)	3193	5819	703
OH	4375	6337	−1392

Interatomic distances (Å) *and interbond angles* (°) *with e.s.d.'s in parentheses for* $BaGe_3O_6(OH)_2$

		Distance	Angle O–M–O
Ge(1) octahedron			
Ge(1)–O(1)	(×2)	1·975 (6)	
–O(2)	(×2)	1·851 (7)	
–O(3)	(×2)	1·885 (6)	
O(1)–O(2)	(×2)	2·701 (10)	89·8 (3)
–O(2^{v})	(×2)	2·712 (9)	90·2 (3)
–O(3)	(×2)	2·529 (8)	81·8 (3)
–O(3^{v})	(×2)	2·918 (9)	98·2 (3)
O(3)–O(2)	(×2)	2·492 (9)	83·7 (3)
–O(2^{v})	(×2)	2·783 (9)	96·3 (3)
Ba polyhedron			
Ba–O(1)	(×2)	3·029 (7)	
–O(2)	(×2)	2·911 (6)	
–O(2^{ii})	(×2)	3·010 (7)	
–O(3^{iii})	(×2)	2·764 (6)	
–OH^{ii}	(×2)	2·740 (8)	
–OH^{i}	(×2)	2·980 (7)	

	Distance	Angle O–M–O
Ge(2) octahedron		
Ge(2)–O(1^{i})	1·891 (6)	
–O(1^{iii})	1·922 (7)	
–O(2)	1·853 (6)	
–O(3)	1·952 (7)	
–O(3^{iii})	1·990 (6)	
–OH^{iii}	1·819 (7)	
O(1^{i})–O(1^{iii})	2·794 (14)	94·3 (3)
–O(2)	2·726 (8)	93·4 (3)
–O(3)	2·776 (10)	92·5 (3)
–OH^{iii}	2·726 (9)	94·6 (3)
O(3^{iii})–O(1^{iii})	2·529 (8)	80·5 (3)
–O(2)	2·740 (9)	90·9 (3)
–O(3)	2·515 (12)	79·3 (3)
–OH^{iii}	2·792 (11)	94·2 (3)
O(1^{iii})–O(3)	2·766 (10)	91·2 (3)
–OH^{iii}	2·734 (9)	93·9 (3)
O(2)–O(3)	2·492 (9)	81·8 (3)
–OH^{iii}	2·647 (10)	92·2 (3)

Symmetry code: (i) $+x, 1-y, \frac{1}{2}+z$; (ii) $+x, 1-y, -\frac{1}{2}+z$; (iii) $\frac{1}{2}-x, 1\frac{1}{2}-y, -z$; (iv) $\frac{1}{2}-x, \frac{1}{2}+y, -\frac{1}{2}-z$; (v) $\frac{1}{2}-x, -\frac{1}{2}+y, -\frac{1}{2}-z$; (vi) $\frac{1}{2}-x, \frac{1}{2}-y, -z$.

NICKEL HYDROXIDE
$Ni(OD)_2$

Acta Cryst., B**42**, 51-55.

P$\bar{3}$m1, 3.126, 4.593, Z = 1, neutron powder data. Ni in 1(a): 0,0,0; O in 2(d): 1/3,2/3,z, z = 0.2221; D in 2(d): z = 0.4275. Structure as previously described (**1**, 163, 195, 775; **4**, 8; **13**, 213; **37**A, 269), with anisotropic thermal motion of D atom similar to that in the Ca and Mg compounds (**21**, 238; **32**A, 242). O-D = 0.973(4) A (corrected for thermal motion).

SODIUM TRIHYDROXOZINCATE
$NaZn(OH)_3$

Z. anorg. Chem., **536**, 137-146.

$P4_2bc$, 10.839, 5.308, at 190K, Z = 8, R = 0.049. Chains of corner-sharing $Zn(OH)_4$ tetrahedra, linked by (5+1)-coordinate Na ions and hydrogen bonds (compare the previous description in terms of planar $Zn(OH)_3$ groups in $P4_2/mbc$ (**26**, 396)). Zn-O = 1.94-2.01, Na-O = 2.35-2.49, 2.83 A.

AMMONIUM NEPTUNIUM HYDROXIDE
$NH_4Np(OH)_5$

J. Less-Common Metals, **121**, 405-414.

$P2_1$, 8.704, 8.704, 7.067 (**c** unique), γ = 127.16, Z = 4, R = 0.030. Layers of 9-coordinate Np ions, linked by ammonium ions. Np-OH = 2.17-2.43(4) A.

HEXAAMMINECOBALT(III) DIHYDROXOTETRAOXONEPTUNATE (VII) DIHYDRATE
$[Co(NH_3)_6][NpO_4(OH)_2].2H_2O$

Radiokhimija, **28**, 690-694.

$P2_1/c$, 6.778, 8.449, 12.448, 119.30, Z = 2, R = 0.032. Octahedral cations, distorted octahedral anions, and water molecules.

GALLIUM THALLIUM OXYSULPHIDE
$Ga_9Tl_3O_2S_{13}$

Acta Cryst., C**42**, 1111-1113.

$P\bar{6}$, 9.598, 7.433, Z = 1, R = 0.050. GaS_3O and GaS_4 tetrahedra, linked by TlS_9 and TlS_{11} polyhedra.

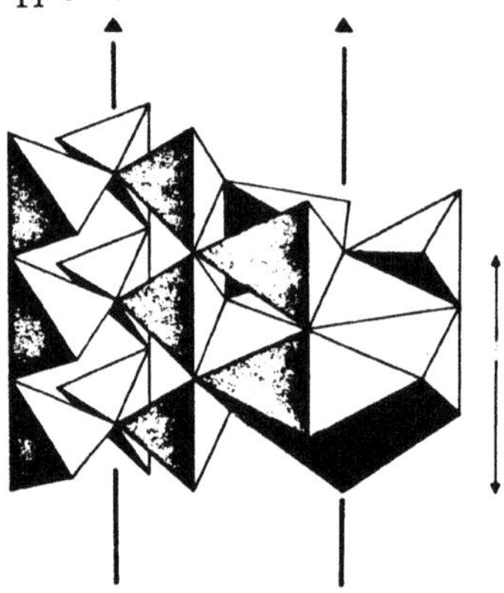

	x	*y*	*z*
Tl(1)	0	0	0
Tl(2)	0	0	$\frac{1}{2}$
Tl(3)	$\frac{1}{3}$	$\frac{2}{3}$	$\frac{1}{2}$
Ga(1)	0,7533	0,1941	0,2649
Ga(2)	0,0736	0,4569	0
S(1)	$\frac{1}{3}$	$\frac{2}{3}$	0
S(2)	0,659	0,042	0
S(3)	0,664	0,025	$\frac{1}{2}$
S(4)	0,0231	0,3165	0,2554
O	$\frac{2}{3}$	$\frac{1}{3}$	0,261

Tl(1)–S(4)	3,494 (6) ×6	Ga(1)–O	1,894 (3) Å
Tl(1)–S(2)	3,497 (9) ×3	Ga(1)–S(3)	2,244 (6)
		Ga(1)–S(4)	2,247 (5)
Tl(2)–S(3)	3,348 (9) ×3	Ga(1)–S(2)	2,346 (5)
Tl(2)–S(4)	3,451 (6) ×6		
		Ga(2)–S(4)	2,237 (6) ×2
Tl(3)–S(3)	3,315 (9) ×3	Ga(2)–S(1)	2,291 (4)
Tl(3)–S(4)	3,668 (6) ×6	Ga(2)–S(2)	2,30 (1)
Tl(3)–S(1)	3,717 (2) ×2		

DJERFISHERITE
$K_6LiFe_{23}S_{26}Cl$

J. Electrochem. Soc., **133**, 2644-2649.

Pm3m, 10.3527, Z = 1, neutron powder data. 23 Fe in 24(m); 0.3681,0.3681,0.1319; S(1) in 12(h): 0.2489,1/2,0; S(2) in 8(g): x,x,x, x = 0.2298; S(3) in 6(f): 0.2514,1/2,1/2; K in 6(e): 0.2991,0,0; Li in 1(b): 1/2,1/2,1/2; Cl in 1(a): 0,0,0. Results in agreement with previous studies (**46**A, 298).

ERBIUM URANIUM OXYSULPHIDE
$(UO)_2ErS_3$

Acta Cryst., C**42**, 1109-1111; Z. anorg. Chem., **540/541**, 59-66.

I4/mmm, 3.785, 20.83, Z = 2, R = 0.037. Alternate UO and ErS_3 sheets, the former containing OU_4 tetrahedra.

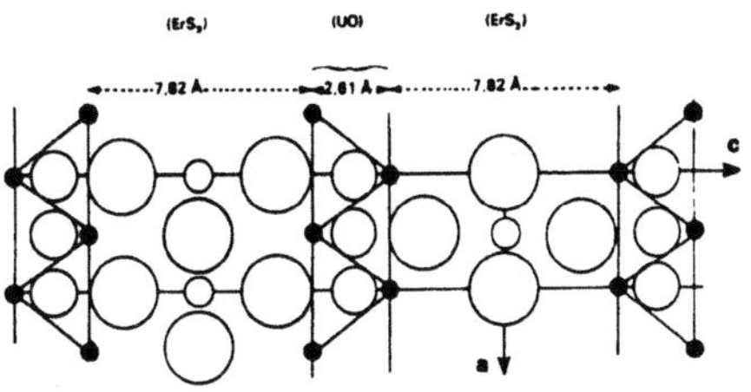

		x	y	z
U	4(*e*)	0	0	0,18755
Er	2(*b*)	0	0	½
S(1)	4(*e*)	0	0	0,3712
S(2)	2(*a*)	0	0	0
O	4(*d*)	0	½	¼

U–O	2,2984 (7)	× 4 Å
U–S(1)	2,945 (2)	× 4
Er–S(2)	2,6785 (7)	× 4
Er–S(1)	2,685 (5)	× 2
O–U	2,2984 (7)	× 4

ALKALI-METAL TRISELENOARSENATE HYDRATES
$KAsSe_3.H_2O$, $MAsSe_3.0.5H_2O$ (M = Rb, Cs)

Z. anorg. Chem., **535**, 179-185.

C2/c, a = 21.891, 23.447, 24.239, b = 5.853, 5.796, 5.909, c = 11.369, 11.446, 11.756 A, β = 101.02, 118.68, 119.53°, Z = 8, R = 0.039, 0.057, 0.068. $AsSe_3$ trigonal pyramids linked into chains by Se-Se bonds, and connected by metal ions. As-Se = 2.314-2.440, Se-Se = 2.347-2.349(2) A.

BARIUM TETRASELENOARSENATE HYDROXIDE DIHYDRATE
$Ba_2AsSe_4(OH).2H_2O$

Z. Naturforsch., **41**B, 436-438.

Cm, 8.992, 9.967, 6.340, 104.51, Z = 2, R = 0.028. Tetrahedral $AsSe_4^{3-}$ anions, OH^- anions, water molecules, and Ba^{2+} ions coordinated to 6 Se and 3 O. As-Se = 2.303-2.329(1), Ba-Se = 3.32-3.59, Ba-O = 2.71-2.88 A.

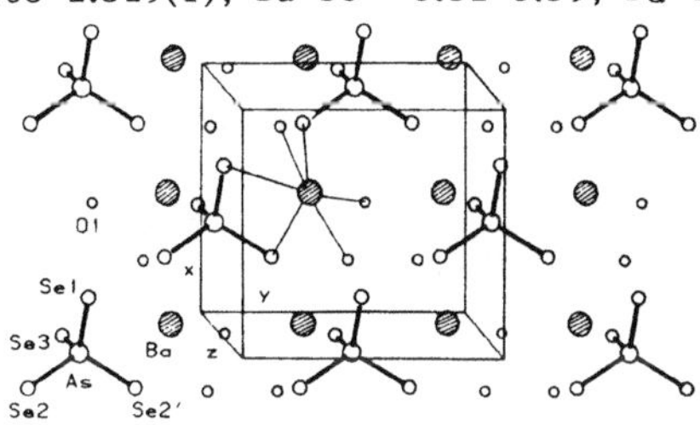

DECASELENIUM BIS(FLUOROSULPHATE)
$Se_{10}(SO_3F)_2$

Acta Cryst., C**42**, 13-16.

$P2_12_12_1$, 8.627, 13.114, 15.135, Z = 4, R = 0.075. Se_{10}^{2+} cations and fluorosulphate anions.

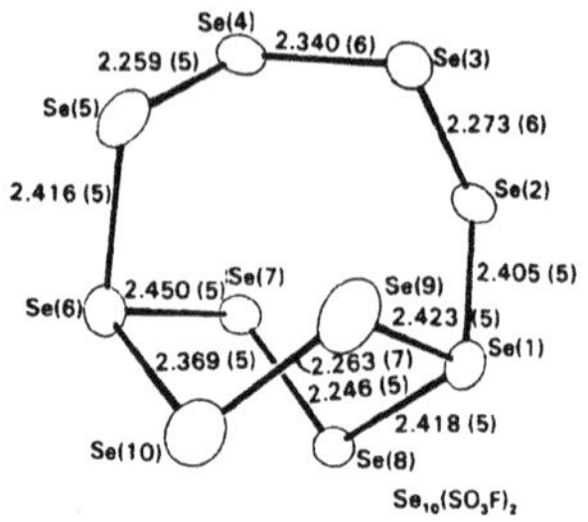

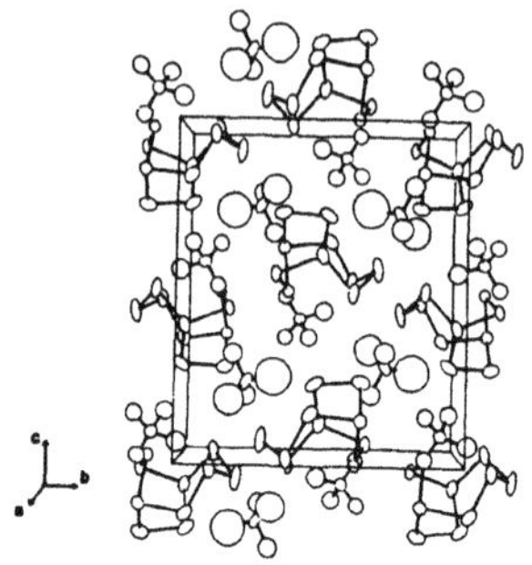

BORIC ACID
H_3BO_3

Acta Cryst., **B42**, 545-552.

P$\bar{1}$, 7.0187, 7.035, 6.3472, 92.49, 101.46, 119.76, at 105K, Z = 4, R = 0.038. Structure as previously described (**3**, 87, 411; **17**, 523; **18**, 394; **31A**, 165), with electron density now determined. B-O = 1.368(1) A.

SODIUM LITHIUM BORATE
$Na_4Li_5(BO_3)_3$

Z. anorg. Chem., **536**, 101-113.

C2/c, 12.388, 7.296, 9.738, 107.29, Z = 4, R = 0.090. Planar borate anions linked by (5+1)- and (4+4)-coordinate Na and 4- and (4+2)-coordinate Li. B-O = 1.37, Na-O = 2.25-2.78, Li-O = 1.87-2.04, 2.43 A.

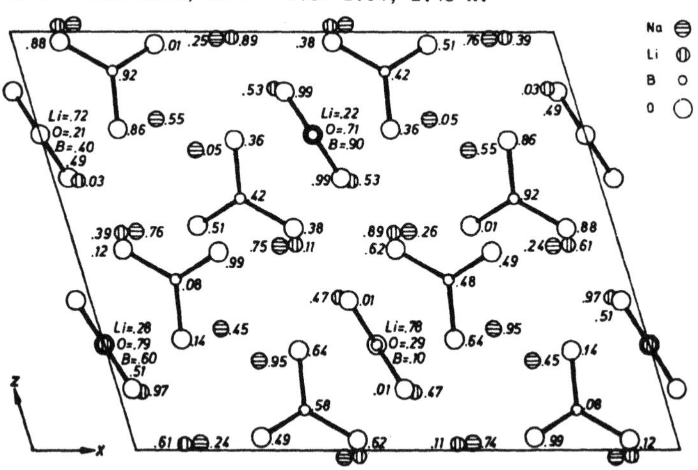

CAESIUM LITHIUM BORATE
$CsLi_5(BO_3)_2$

Z. anorg. Chem., **536**, 92-100.

C2/c, 11.795, 9.433, 8.096, 132.76, Z = 4, R = 0.057. Planar borate anions linked by 12-coordinate Cs and 4-coordinate Li ions. B-O = 1.37-1.40, Cs-O = 3.23-3.84, Li-O = 1.90-2.03 A.

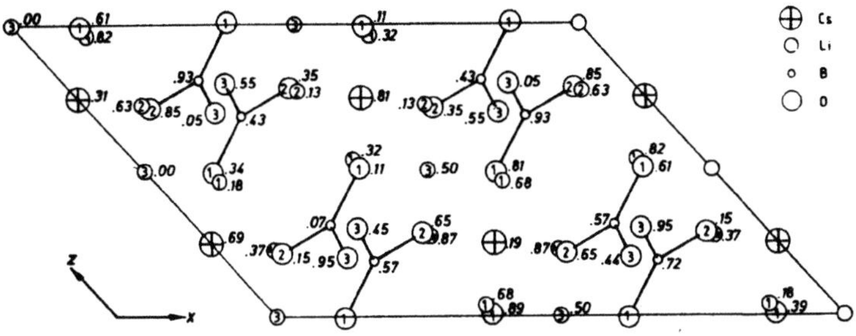

RHODIZITE
$(K,Cs)Al_4Be_4(B,Be)_{12}O_{28}$

Miner. Mag., **50**, 163-172.

$P\bar{4}3m$, 7.318, Z = 1, R = 0.034. K,Cs in 1(a): 0,0,0; Al in 4(e): x,x,x, x = 0.3600; Be in 4(e): x = 0.7429; B,Be in 12(h): x,1/2,0, x = 0.2482; O(1) in 4(e): x = 0.6189; O(2) in 12(i): x,x,z, x = 0.3637, z = 0.0997; O(3) in 12(i): x = 0.1347, z = 0.6013. Structure as previously described (**9**, 186; **32**A, 495). Formula is $(K,Cs)_{0.9}Al_4Be_{4.5}B_{11.35}O_{28}$.

KARLITE
$(Mg,Al)_{6.5}(BO_3)_3(OH)_4Cl_{0.4}$

Neues. Jb. Miner., Mh., 253-262.

$P2_12_12$, 17.637, 17.967, 3.1040, Z = 4, R = 0.042. Framework of six MgO_6 octahedra, one MgO_5 trigonal bipyramid, and three BO_3 triangles, with eight-sided channels along **c** which contain partially-occupied Cl^- sites.

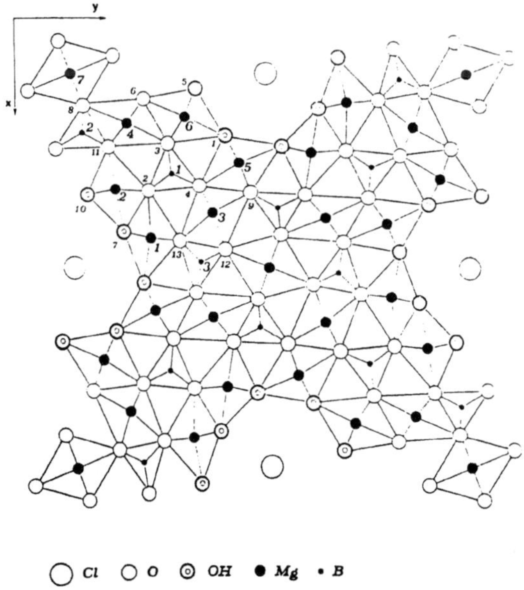

NORDENSKIOLDITE
$CaSn(BO_3)_2$

Neues Jb. Miner., Mh., 111-114.

$R\bar{3}$, 4.858, 16.080, Z = 3, R = 0.039. Ca in 3(a): 0,0,0; Sn in 3(b): 0,0,1/2; B in 6(c): 0,0,0.2395; O in 18(f): 0.2663,-0.0313,0.2417. Dolomite-type structure, as previously described (**3**, 406), with planar BO_3 groups and slightly elongated CaO_6 and SnO_6 octahedra. B-O = 1.377, Ca-O = 2.377, Sn-O = 2.048(1) A.

HENMILITE
$Ca_2Cu(OH)_4[B(OH)_4]_2$

Amer. Min., **71**, 1234-1239.

$P\bar{1}$, 5.7617, 7.9774, 5.6488, 109.61, 91.473, 83.686, Z = 1, R = 0.026. $B(OH)_4$ tetrahedra, linked by Cu^{2+} ions with square-planar coordination, and by CaO_8 polyhedra which share an edge to form dimers. B-O = 1.466-1.482, Cu-O = 1.939,1.949, Ca-O = 2.395-2.726(2) A.

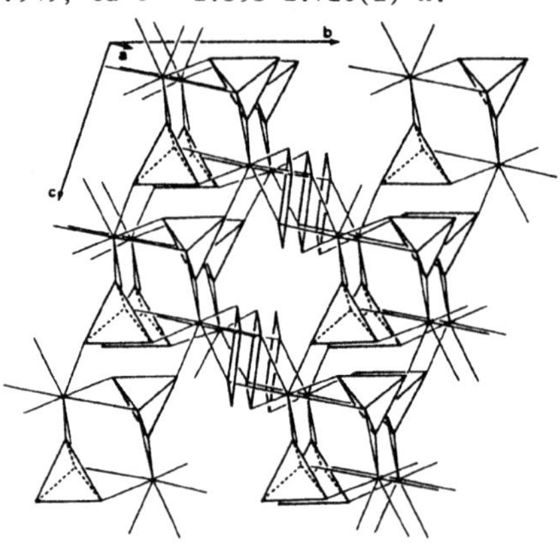

	x	y	z
Ca	0.50791	0.24095	0.22595
Cu	0	0	0
B	0.2790	0.4068	0.7480
O(1)	0.3061	−0.0192	0.1455
O(2)	0.8404	0.0684	0.3239
O(3)	0.2467	0.2907	0.8972
O(4)	0.1523	0.3538	0.5106
O(5)	0.7879	0.4055	0.0978
O(6)	0.5261	0.3983	0.6742
H(1)	0.705	0.040	0.735
H(2)	0.923	0.065	0.427
H(3)	0.226	0.194	0.828
H(4)	0.043	0.418	0.526
H(5)	0.904	0.385	0.058
H(6)	0.619	0.387	0.768

MOYDITE
$(Y,Ln)[B(OH)_4](CO_3)$

Canad. Miner., **24**, 665-673 and 675-678.

Pbca, 9.089, 12.244, 8.926, Z = 8, R = 0.051. Sheets of corner-sharing YO_9 polyhedra, reinforced by carbonate groups, with sheets linked by edge-sharing with $B(OH)_4$ tetrahedra.

	x	y	z
Y	0.0146	0.4991	0.2272
B	0.0032	0.2490	0.2415
C	0.2622	0.5454	0.0268
O1	-0.2792	0.4688	-0.1692
O2	-0.1267	0.4550	0.0164
O3	0.3680	0.5565	-0.0617
O4	0.1019	0.3172	0.1534
O5	-0.0944	0.3256	0.3258
O6	0.0879	0.1796	0.3505
O7	-0.0844	0.1700	0.1538

O4,O5,O6 and O7 are OH-sites.

MOYDITE: BOND LENGTHS (Å) AND ANGLES (°)

Y-O1a	2.491(5)	B-O4	1.456(24)	O4-B-O7	115(2)
O1b	2.364(5)	O5	1.494(23)	O5-B-O7	111(1)
O2a	2.341(6)	O6	1.504(24)	O6-B-O7	104(1)
O2b	2.465(6)	O7	1.477(23)	O4-B-O5	106(1)
O3	2.270(6)	Mean	1.483		109
O4	2.455(8)				
O5	2.505(7)	C-O1	1.292(9)	O1-C-O2	114(1)
O6	2.496(8)	O2	1.290(10)	O1-C-O3	123(1)
O7	2.431(8)	O3	1.252(10)	O2-C-O3	123(1)
Mean	2.424	Mean	1.278		120

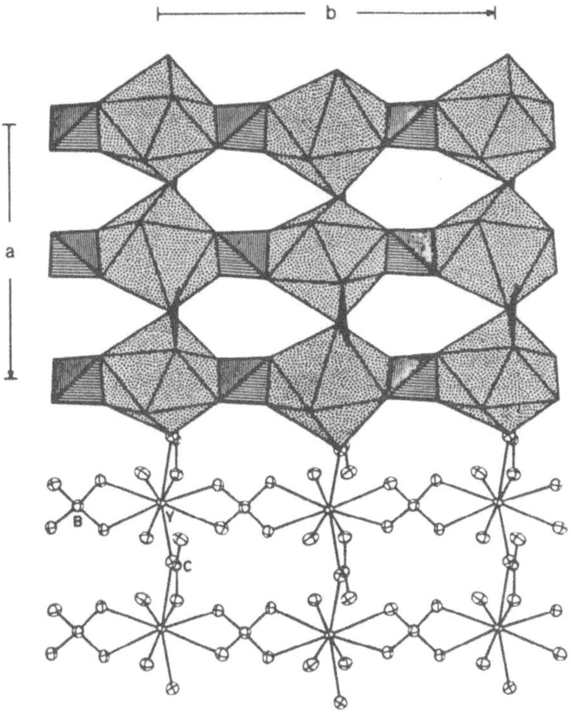

SODIUM SESQUICARBONATE DIHYDRATE
$Na_3H(CO_3)_2.2H_2O$

Mitt. Oesterr. Mineral. Ges., **131**, 7-14.

C2/c, 20.4218, 3.4913, 10.3326, 106.452, Z = 4, neutron data. Results in agreement with **49**A.

DOLOMITE
$CaMg(CO_3)_2$

Amer. Min., **71**, 795-804.

$R\bar{3}$, 4.8069-4.827, 16.002-16.279 A, at 24-700°C, Z = 3, R = 0.021-0.023 at 4 temperatures. Structure as previously described (**1**, 303, 317, 324; **23**, 419; **43**A, 235; **48**A, 265; **50**A).

HUNTITE
$CaMg_3(CO_3)_4$

Amer. Min., **71**, 163-166.

R32, 9.5027, 7.8212, Z = 3, powder data. Structure as previously described (**27**, 561), with Ca in the trigonal prismatic and Mg in the octahedral sites. Ca-O = 2.41, Mg-O = 2.03-2.11, C-O = 1.27-1.29(2) A.

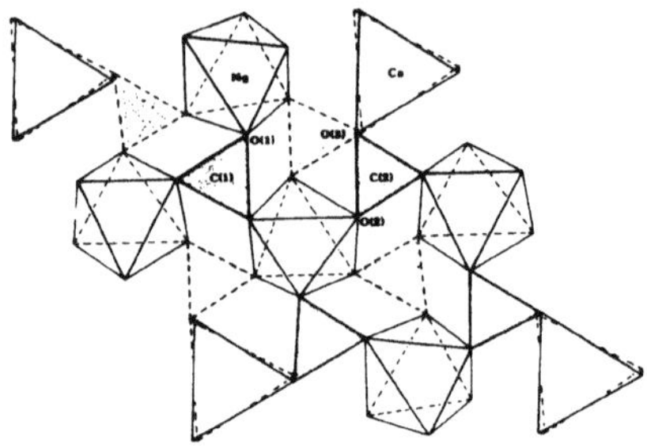

	x	y	z
Ca	0	0	0
Mg	0.5443	0	0
C(1)	0	0	1/2
C(2)	0.4535	0	1/2
O(1)	0.8663	0	1/2
O(2)	0.5895	0	1/2
O(3)	0.4569	0.1355	0.5161

ARAGONITE
$CaCO_3$

Tschermaks Min. Petr. Mitt., **35**, 127-131.

Pmcn, 4.9611, 7.9672, 5.7407, Z = 4, neutron radiation, R = 0.021. Ca: 1/4,0.41508,0.24046; C: 1/4,0.76211,0.08518; O(1): 1/4,0.92224,0.09557; O(2): 0.47347,0.68065,0.08726. Structure as previously described (**1**, 295, 317, 318; **37A**, 280); the carbonate group is slightly non-planar, with C 0.028 A out of the oxygen plane.

DEFERNITE
$Ca_6(CO_3)_2(OH,Cl)_8.2H_2O$ (idealized)

Schweiz. Mineral. Petrogr. Mitt., 65, 153-158 (1985).

Pnam, 17.860, 22.775, 3.658, Z = 4, R = 0.142 (poor quality crystals). Some atoms, particularly in the carbonate groups, have not been determined definitively. The structure contains carbonate and hydroxide anions, linked by Ca ions.

COBALT(II) CARBONATE
$CoCO_3$

NICKEL CARBONATE
$NiCO_3$

Acta Cryst., **C42**, 4-5.

$R\bar{3}c$, a = 4.6618, 4.6117, c = 14.963, 14.735, Z = 6, R = 0.040, 0.057. Calcite-type structures (**1**, 292), x(O) = 0.7234, 0.7201, M-O = 2.110(1), 2.076(2) A.

SODIUM BIS(CARBONATO)CUPRATE(II)
$Na_2Cu(CO_3)_2$

Acta Cryst., **B42**, 430-436.

$P2_1/a$, 6.170, 8.171, 5.648, 116.24, Z = 2, R = 0.026. Structure as previously described (**38A**, 302); electron-density now studied.

	x	y	z
Cu	0	0	0
Na	0·67160	0·09405	0·35690
O(1)	0·32537	−0·08168	0·21023
O(2)	0·36936	−0·28169	−0·03111
O(3)	0·40360	−0·33579	0·37097
C	0·36569	−0·23665	0·18683

BAYLEYITE
$Mg_2[UO_2(CO_3)_3].18H_2O$

Tschermaks Min. Petr. Mitt., **35**, 133-146.

$P2_1/a$, 26.560, 15.256, 6.505, 92.90, Z = 4, R = 0.029. $Mg(H_2O)_6$ octahedra, $UO_2(CO_3)_3$ units (hexagonal bipyramidal coordination at U), and lattice water molecules, all linked by hydrogen bonds only; one Mg octahedron has disordered water sites. Mg-O = 2.04-2.11, U-O = 1.79(uranyl), 2.42-2.46 A.

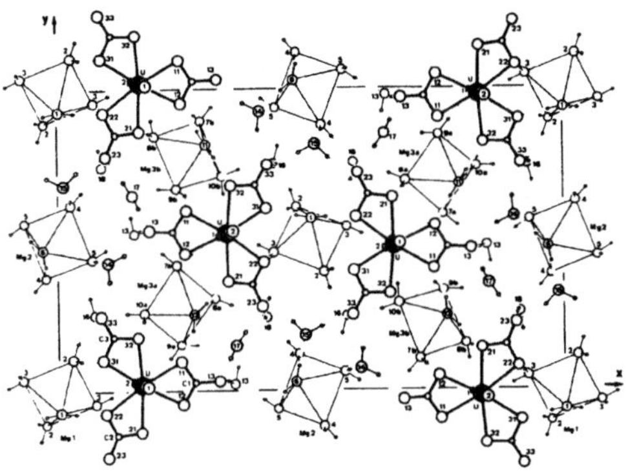

SWARTZITE
STRONTIUM SWARTZITE
$MMg[UO_2(CO_3)_3].12H_2O$ (M = Ca, Sr)

Neues Jb. Miner., Mh., 481-492.

$P2_1/m$, a = 11.080, 11.216, b = 14.634, 14.739, c = 6.439, 6.484 A, β = 99.43, 99.48°, Z = 2, R = 0.023, 0.026. Isolated dinuclear groups, $M(H_2O)_6[UO_2(CO_3)_3]^{2-}$, of $M(H_2O)_6O_2$ square antiprisms sharing edges with UO_2O_6 hexagonal bipyramids, isolated $Mg(H_2O)_6^{2+}$ octahedra, and a system of hydrogen bonds.

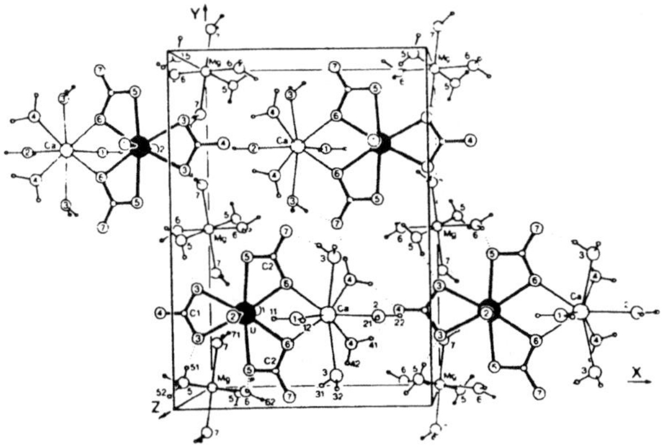

SCHROCKINGERITE
$NaCa_3[UO_2(CO_3)_3](SO_4)F.10H_2O$

Tschermaks. Min. Petr. Mitt., **35**, 1-18.

$P\bar{1}$, 9.634, 9.635, 14.391, 91.41, 92.33, 120.26, Z = 2, R = 0.026. (001) layers of $UO_2(CO_3)_3^{4-}$ anions (hexagonal bipyramidal coordination at U), $NaO_3(H_2O)_3$ octahedra, three types of $CaO_5F(H_2O)_2$ polyhedra, Ca_3F pyramids, and Ca-bonded SO_4 tetrahedra; layers are linked by hydrogen bonds, directly and via interlayer water molecules.

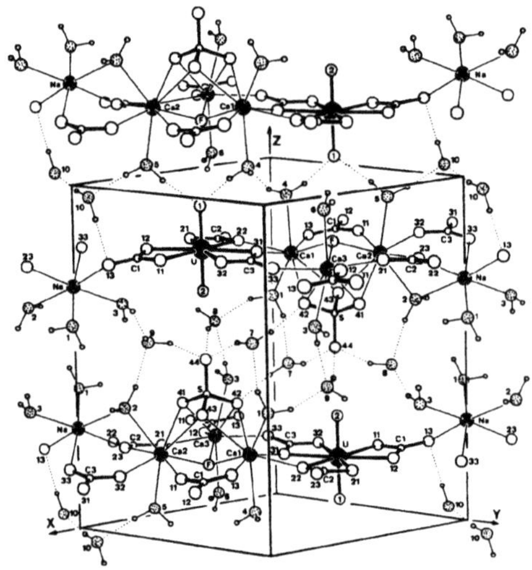

STRONTIUM TRICARBONATODIOXOURANATE(VI) OCTAHYDRATE
$Sr_2[UO_2(CO_3)_3].8H_2O$

Acta Cryst., C**42**, 1678-1681.

$P2_1/c$, 11.379, 11.446, 25.653, 93.40, Z = 8, R = 0.037. Isolated anions in which U has hexagonal bipyramidal coordination, linked by 8- and 9-coordinated Sr ions, with coordinated and lattice water molecules. U-O = 1.77-1.80, 2.39-2.47(1) A.

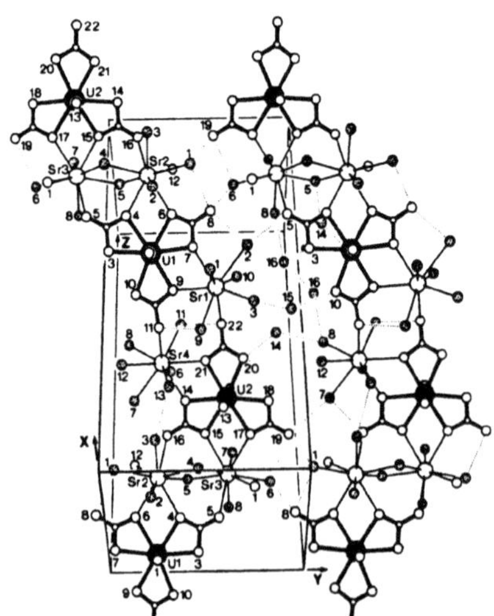

THALLIUM(I) TRICARBONATODIOXOURANATE(VI)
$Tl_4[UO_2(CO_3)_3]$

Acta Cryst., C**42**, 1682-1684.

C2/c, 10.684, 9.309, 12.726, 94.95, Z = 4, R = 0.037. Isostructural with the K and NH_4 salts (**38**A, 302; **46**A, 308), with anions containing U with hexagonal bipyramidal coordination, linked by 9-coordinate Tl ions. U-O = 1.80, 2.41-2.43(1) A.

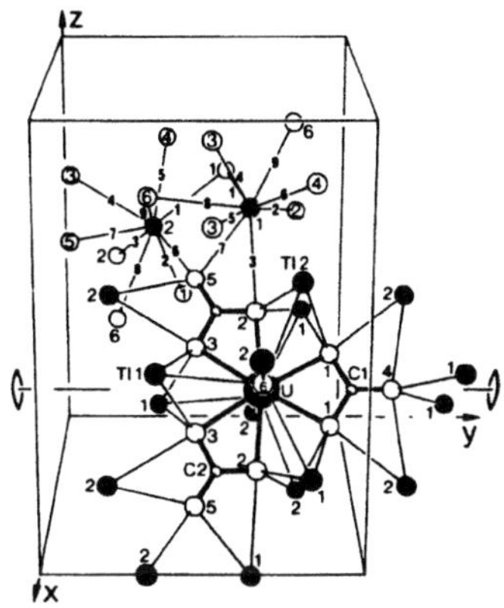

	x	*y*	*z*
Tl(1)	0·6833	0·3441	0·3524
Tl(2)	0·6033	0·8207	0·5468
U	0·5	0·6925	0·25
C(1)	0·5	0·9988	0·25
C(2)	0·6150	0·5412	0·0897
O(1)	0·5422	0·9225	0·1740
O(2)	0·5979	0·6748	0·0851
O(3)	0·5615	0·4735	0·1643
O(4)	0·5	1·1327	0·25
O(5)	0·6776	0·4701	0·0269
O(6)	0·6546	0·6914	0·3187

POTASSIUM CALCIUM NITRITE
$K_2Ca(NO_2)_4$

Soobšč. Akad. Nauk Gruz.SSR, **122**, 81-84.

C2/m, 9.4466, 5.454, 10.908, γ = 88.57°, Z = 2, R = 0.038. Nitrite anions linked by octahedrally-coordinated Ca, and tetrahedrally-coordinated K.

BARIUM NITRITE
$Ba(NO_2)_2$

Z. Kristallogr., **175**, 111-116.

$P2_12_12_1$, 7.6331, 8.3671, 6.7336, Z = 4, R = 0.035. Bent nitrite anions, linked by Ba ions which are coordinated to 9 O and 2 N. N-O = 1.236-1.265(8) A, O-N-O = 113.1, 114.0°, Ba-O = 2.776-3.010(6), Ba-N = 3.059, 3.125(7) A.

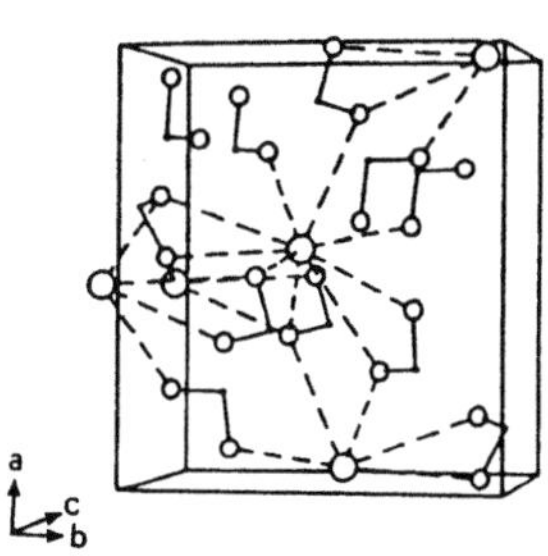

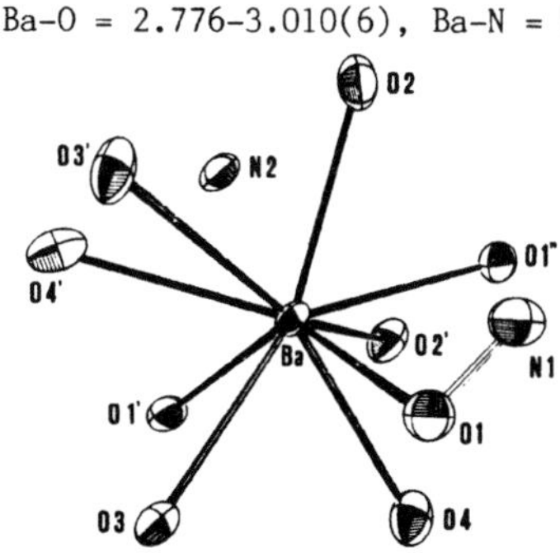

	x	*y*	*z*
Ba	0.4624	0.5396	0.1077
N(1)	0.5342	0.8650	−0.1052
N(2)	0.2793	0.2245	0.0077
O(1)	0.6225	0.8428	0.0513
O(2)	0.4114	0.4786	−0.2932
O(3)	0.7158	0.5930	0.4090
O(4)	0.3309	0.7673	0.3681

POTASSIUM trans-DIAMMINETETRANITROCOBALTATE(III)
$K[Co(NH_3)_2(NO_2)_4]$

Inorg. Chim. Acta, **96**, 99-110 (1985); **115**, 25-29.

$P2_12_12_1$, 11.275, 12.806, 6.696, Z = 4, X-ray and neutron data, R = 0.037, 0.056. trans-Octahedral anions, with asymmetric orientation of the nitro groups and disorder of the H atoms of one of the ammine groups.

POTASSIUM BARIUM HEXANITROCOBALTATE
$K_2Ba[Co(NO_2)_6]$

Acta Cryst., C**42**, 11-13.

Fm3, a = 10.653, Z = 4, R = 0.027. Structure as previously described (**31**A, 181; **37**A, 283), with no thermal anisotropy of the N atom.

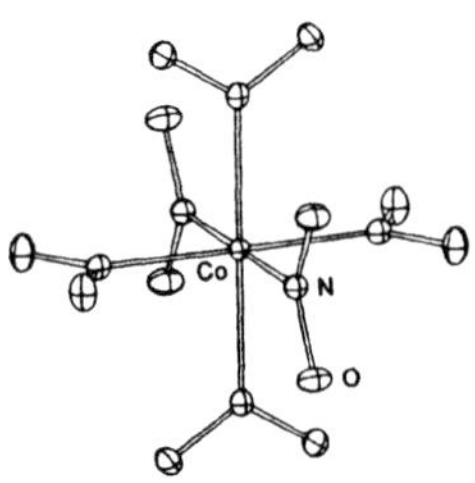

	x	y	z
Co	0	0	0
O	0	2504	998
N	0	1887	0
Ba	0	5000	0
K	2500	2500	2500

Interatomic distances (Å) *and angles* (°) *for complexes of the type* $MM'_2[Co(NO_2)_6]$

	$BaK_2[Co(NO_2)_6]$	$PbK_2[Co(NO_2)_6]$	$PbRb_2[Co(NO_2)_6]$
Co–N	2·010 (3)	2·024 (3)	2·046 (3)
N–O	1·250 (3)	1·236 (4)	1·243 (2)
Ba–O or Pb–O	2·864 (2)	2·782 (3)	2·814 (2)
K–O or Rb–O	3·107 (2)	3·054 (2)	3·096 (1)
O–N–O	116·6 (3)	117·2 (4)	116·7 (1)
Co–N–O	121·7 (2)	121·4 (2)	121·7 (1)

MERCURY (I) NITRITE
$Hg_2(NO_2)_2$ (I)

LITHIUM SILVER NITRITE
$LiAg_2(NO_2)_3$ (II)

Acta Cryst., C**42**, 1-4.

I. $P2_1/c$, 4.435, 10.344, 6.301, 108.74, Z = 2, R = 0.048. Nearly-planar $NO_2...Hg_2...O_2N$ units.

II. C2/c, 10.166, 7.575, 8.479, 91.27, Z = 4, R = 0.033. Nearly-planar $Ag...NO_2...Ag$ units, linked by Li ions. One nitrite is disordered.

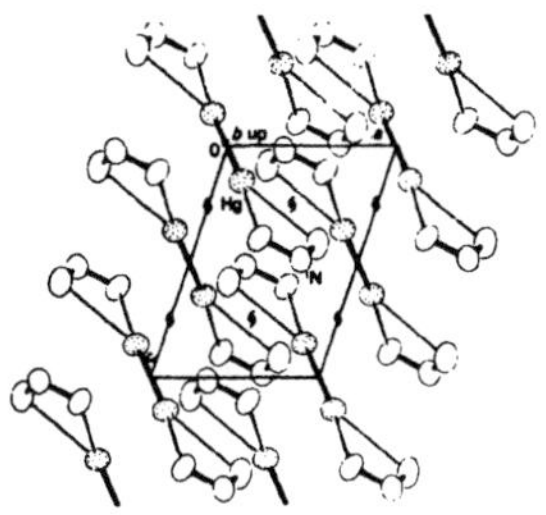

(a)

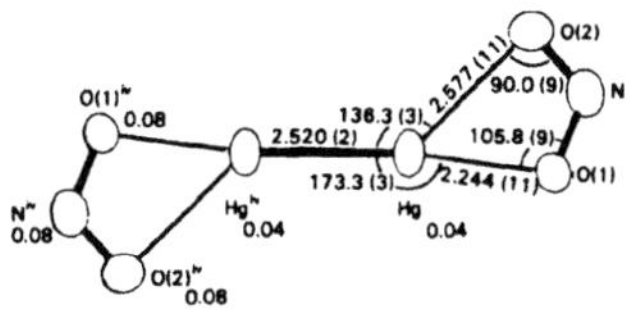

(b)

(a) Projection of the structure of (I) along **b** and (b) the $NO_2\cdots Hg_2\cdots O_2N$ unit. Atoms are also labeled with their deviations (Å) from the plane of one NO_2. Distances in Å, angles in °.

Interatomic distances (Å) *and angles* (°)

Compound (I)			
N–O(1)	1·253 (17)	Hg–O(1^{i})	2·826 (13)
N–O(2)	1·221 (17)	Hg–N	2·852 (13)
O(1)–N–O(2)	113·8 (13)	Hg–O(2^{ii})	2·955 (15)
		Hg–N^{iii}	3·066 (11)
Hg–O(1)	2·244 (11)		
Hg–O(2)	2·577 (11)	Hg–Hg^{iv}	2·520 (2)

Symmetry code: (i) $x, -y+\frac{1}{2}, z-\frac{1}{2}$; (ii) $x-1, y, z$; (iii) $x-1, -y+\frac{1}{2}, z-\frac{1}{2}$; (iv) $-x, -y, -z$.

Compound (II)			
N(1)–O(11)	1·264 (5)	Ag–O(11^{v})	2·925 (3)
N(1)–O(12)	1·236 (5)	Ag–O(12^{iv})	2·925 (3)
O(11)–N(1)–O(12)	116·2 (4)	Ag–N(2^{iii})	2·949 (8)
		Ag–O(21^{i})	2·979 (8)
N(2)–O(21)	1·239 (10)		
N(2)–O(22)	1·290 (11)	Li–O(11^{vi})	2·118 (4)
O(21)–N(2)–O(22)	111·9 (8)	Li–O(11^{ii})	2·118 (4)
		Li–O(22^{vii})	2·136 (9)
Ag–N(2^{i})	2·154 (8)	Li–O(22^{viii})	2·136 (9)
Ag–N(1^{ii})	2·327 (4)	Li–O(12)	2·154 (3)
Ag–O(22^{iii})	2·379 (9)	Li–O(12^{ix})	2·154 (3)
Ag–O(11^{iv})	2·394 (4)	Li–O(21)	2·165 (8)
Ag–O(21^{iii})	2·565 (9)	Li–O(21^{x})	2·165 (8)
Ag–O(12)	2·908 (4)		

Symmetry code: (i) $1-x, y, \frac{1}{2}-z$; (ii) $\frac{1}{2}-x, \frac{1}{2}+y, \frac{1}{2}-z$; (iii) $x, -y, -\frac{1}{2}+z$; (iv) $\frac{1}{2}-x, \frac{1}{2}-y, -z$; (v) $\frac{1}{2}+x, \frac{1}{2}+y, z$; (vi) $x, -y, \frac{1}{2}+z$; (vii) $-\frac{1}{2}+x, \frac{1}{2}+y, z$; (viii) $1-x, -y, 1-z$; (ix) $\frac{1}{2}-x, \frac{1}{2}-y, 1-z$; (x) $\frac{1}{2}-x, -\frac{1}{2}+y, \frac{1}{2}-z$.

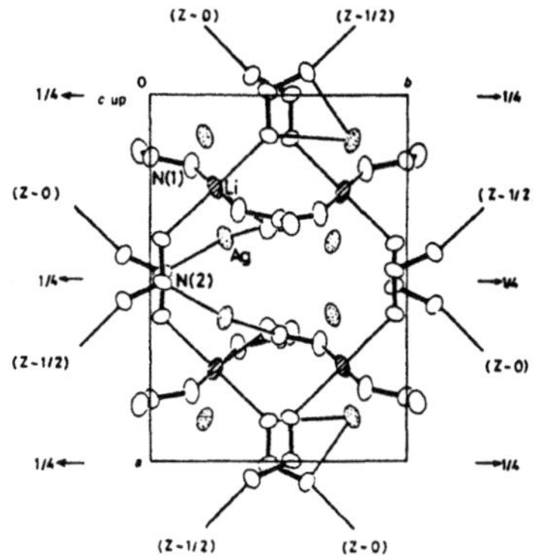

A view of the structure of (II) perpendicular to *ab*. Only one of the two possible orientations for $NO_2^-(2)$ is drawn for simplicity. Hatched and dotted ellipsoids represent Li^+ and Ag^+ ions respectively.

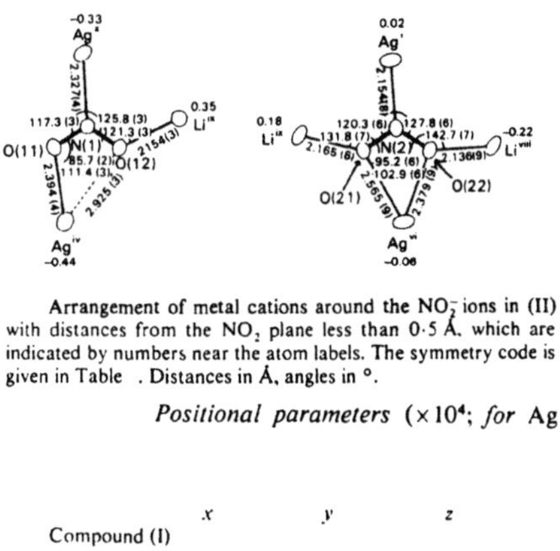

Arrangement of metal cations around the NO_2^- ions in (II) with distances from the NO_2 plane less than 0·5 Å, which are indicated by numbers near the atom labels. The symmetry code is given in Table . Distances in Å, angles in °.

Positional parameters (×10^4; *for* Ag ×10^5)

	x	*y*	*z*
Compound (I)			
Hg	1531	808	1441
N	6559	2045	4855
O(1)	3716	2366	3932
O(2)	7254	1012	4184
Compound (II)			
Ag	38397	28952	6251
Li	2500	2500	5000
N(1)	1600	7	2465
N(2)*	5175	459	4918
O(11)	1417	−431	1039
O(12)	1830	1586	2715
O(21)*	3963	423	5041
O(22)*	5697	−1071	5163

* 50% site-occupation factor.

POTASSIUM NITRATE (PHASE IV)
KNO_3

Physica B+C, **136**, 503-506.

Pnma, 7.4867, 5.5648, 6.7629, Z = 4, neutron powder data. K, N, O(1) in 4(c): x,1/4,z, x = 0.4863, 0.3465, 0.4411, z = 0.6897, 0.1321, 0.2806; O(2) in 8(d): 0.2985,0.4432,0.0556. High-pressure phase.

LEAD(II) NITRATE
$Pb(NO_3)_2$

Acta Cryst., C**42**, 133-135.

Pa3, 7.8586, Z = 4, neutron data, R = 0.019. Pb in 4(a): 0,0,0; N in 8(c): x,x,x, x = 0.34865; O in 24(d): 0.27772,0.28869,0.47730. Structure as previously described (**1**, 304; 313, 787; **2**, 73, 386; **21**, 354); N-O = 1.247(1), Pb-12 O = 2.748, 2.869(1) A.

LIKASITE
$Cu_3(OH)_5(NO_3).2H_2O$

Neues Jb. Miner., Mh., 101-110.

Pcmn, 5.830, 6.775, 21.711, Z = 4, R = 0.108. Previous studies in **43**A, 242; **45**A, 381. (001) sheets of Cu ions coordinated to (4+2), (4+1), and 4 O atoms, with the nitrate groups involved in the longer Cu-O bonds; sheets are linked only by hydrogen bonds.

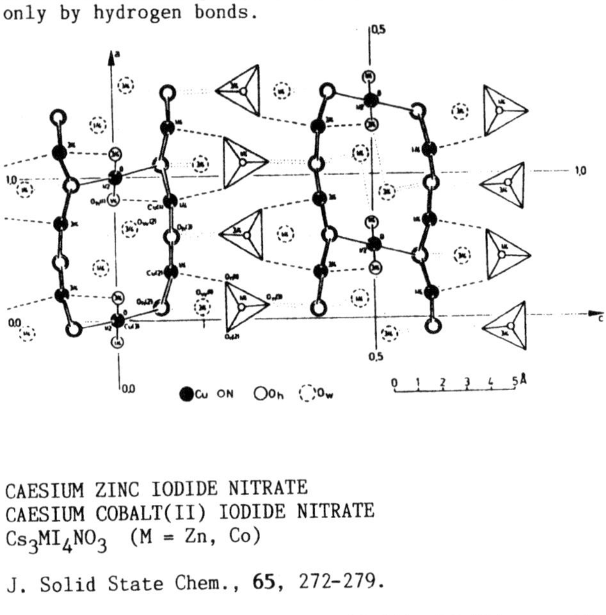

			x	y	z
Cu(1)	m	4c	0.8224	1/4	0.1055
Cu(2)	m	4c	0.3284	1/4	0.1003
Cu(3)	$\bar{1}$	4a	0	0	0
N	m	4c	0.080	1/4	0.242
$O_n(1)$	m	4c	0.261	1/4	0.210
$O_n(2)$	m	4c	−0.103	1/4	0.214
$O_n(3)$	m	4c	0.088	1/4	0.299
$O_h(1)$	m	4c	0.158	3/4	0.002
$O_h(2)$	1	8d	0.076	0.060	0.086
$O_h(3)$	1	8d	0.573	0.055	0.109
$O_w(1)$	m	4c	0.082	3/4	0.170
$O_w(2)$	m	4c	0.628	3/4	0.030

CAESIUM ZINC IODIDE NITRATE
CAESIUM COBALT(II) IODIDE NITRATE
$Cs_3MI_4NO_3$ (M = Zn, Co)

J. Solid State Chem., **65**, 272-279.

Pnma, a = 10.114, 10.078, b = 11.601, 11.621, c = 14.290, 14.336 A, Z = 4, R = 0.051, 0.049. The structures are similar to that of Cs_3CoI_5 (**44**A, 338), with the nitrate anion in one of the iodide sites. They contain MI_4^{2-} tetrahedra and nitrate ions, linked by Cs^+ ions. Zn-I = 2.60-2.64, Co-I = 2.60-2.64 A.

$Cs_3ZnI_4NO_3$

	X	*Y*	*Z*
Cs_1	0.5773	0.250	0.9482
Cs_2	−0.0983	−0.0114	0.18004
Zn	0.1760	0.250	0.0990
I_1	0.3016	0.0613	0.05665
I_2	−0.0427	0.250	−0.0018
I_3	0.1250	0.250	0.2781
O_1	0.791	0.750	0.253
O_2	0.622	0.665	0.214
N	0.676	0.750	0.2322

$Cs_3CoI_4NO_3$

	X	*Y*	*Z*
Cs_1	0.5777	0.250	0.9490
Cs_2	−0.0982	−0.0116	0.18050
Co	0.1762	0.250	0.0985
I_1	0.3005	0.0612	0.0569
I_2	−0.0410	0.250	−0.0017
I_3	0.1267	0.250	0.2271
O_1	0.784	0.750	0.250
O_2	0.620	0.660	0.213
N	0.675	0.750	0.2317

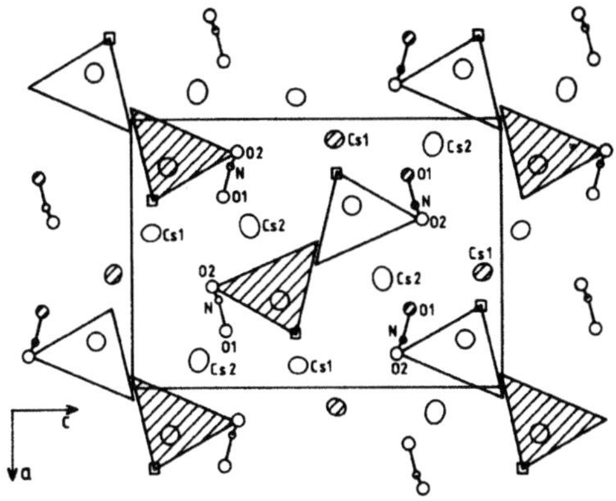

MERCURY(I) NITRATES
$[(Hg_2)_2O(NO_3)]NO_3.HNO_3$ (I), $[(Hg_2)_5(OH)_4(NO_3)_2](NO_3)_4$ (II),
$[Hg_2(OHg)_2](NO_3)_2$ (III)

Acta Cryst., C42, 385-389.

I, Aba2, 11.222, 20.178, 9.354, Z = 8, R = 0.064. II, P$\bar{1}$, 9.990, 9.091, 7.657, 94.05, 101.25, 109.75, Z = 1, R = 0.067. III, $P2_1/c$, 5.532, 6.659, 11.578, 98.84, Z = 2, R = 0.060. I contains mercuriooxonium infinite chains with solvated HNO_3 molecules hydrogen bonded to one of the nitrate ions, II finite four-oxonium-link chains, and III folded infinite oxonium layers containing Hg(I) and Hg(II) ions.

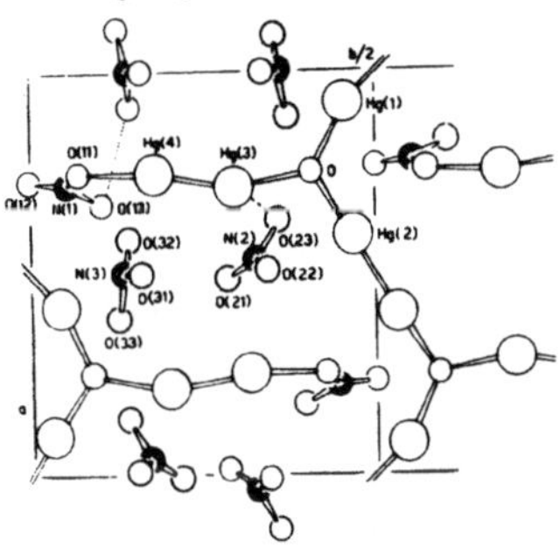

The crystal structure of $[(Hg_2)_2O(NO_3)]NO_3.HNO_3$ projected in the *ab* plane. The hydrogen bond between solvated HNO_3 and the NO_3^- ion is shown by a dotted line, the closest contact between Hg(3) and O(23) by a broken line.

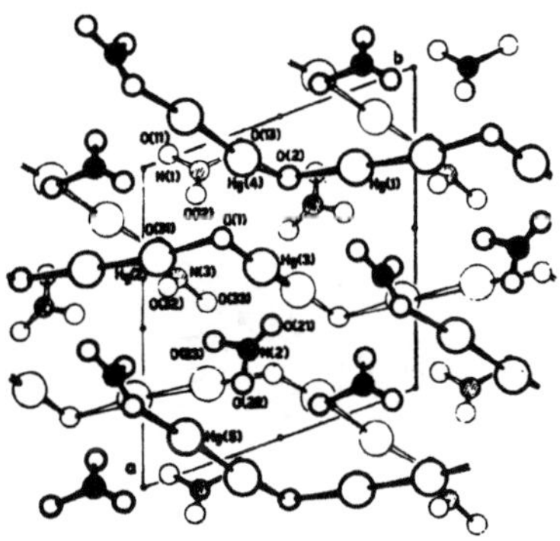

The crystal structure of $[(Hg_2)_5(OH)_4(NO_3)_2](NO_3)_4$ projected in the *ab* plane.

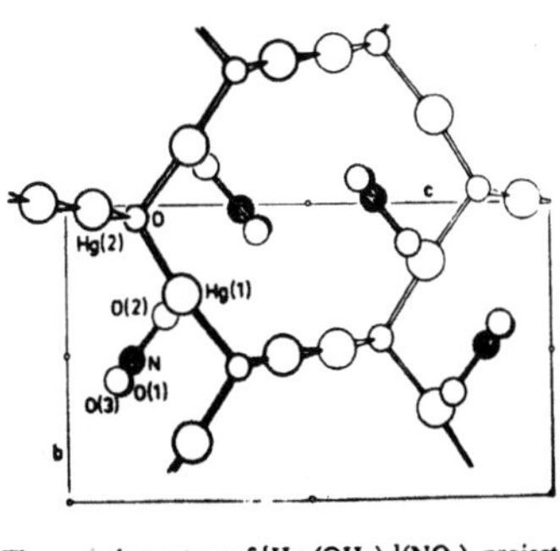

The crystal structure of $[Hg_2(OHg)_2](NO_3)_2$ projected in the *bc* plane. The Hg^{II}–O–Hg^{II}– infinite zigzag chains running along **b** are interconnected in a layer by Hg^I–Hg^I pairs.

NEODYMIUM DIHYDROXIDE NITRATE
GADOLINIUM DIHYDROXIDE NITRATE
$Ln(OH)_2NO_3$ (Ln = Nd, Gd)

J. Solid State Chem., **63**, 231-236.

$P2_1$, a = 6.420, 6.340, b = 3.838, 3.715, c = 7.746, 7.728 A, β = 98.18, 96.95°, Z = 2, R = 0.025, 0.048. Isostructural with the Pr compound (**42**A, 327), with nine-coordinate Ln ions. Nd-O = 2.43-2.68, Gd-O = 2.27-2.64(2) A.

POSITIONAL PARAMETERS ($x, z \times 10^4$ AND $y \times 10^3$)

	x	y	z		x	y	z
Nd	2663	257	955	Gd	2631	259	914
N	2237	646	4363	N	2260	621	4325
O(1)	1216	758	−738	O(1)	1223	722	−757
O(2)	5240	786	1436	O(2)	5194	788	1398
O(3)	1343	768	2891	O(3)	1288	778	2850
O(4)	3420	390	4307	O(4)	3462	377	4176
O(5)	1911	799	5726	O(5)	1914	723	5679

EUROPIUM(III) NITRATE PENTAHYDRATE
$Eu(NO_3)_3.5H_2O$ $[Eu(NO_3)_3(H_2O)_4].H_2O$

Acta Cryst., C**42**, 1450-1452.

$P\bar{1}$, 10.638, 9.568, 6.704, 76.12, 84.68, 63.72, Z = 2, R = 0.079. Isostructural with the Y analogue (**49**A), with layers of 10-coordinate Eu clusters (bicapped square antiprisms) linked by hydrogen bonds.

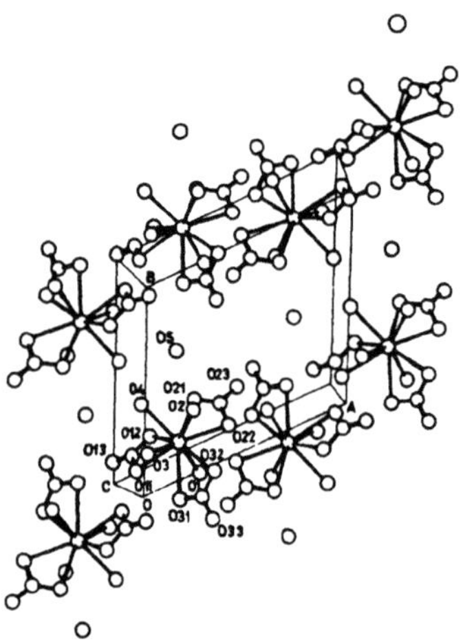

	x	y	z
Eu	0·2053	0·1508	0·2534
N(1)	−0·0558	0·2167	0·0311
N(2)	0·3742	0·2881	−0·0217
N(3)	0·3560	−0·1881	0·4741
O(11)	−0·0113	0·1185	0·2046
O(12)	0·0213	0·2837	−0·0521
O(13)	−0·1638	0·2449	−0·0514
O(21)	0·2430	0·3627	−0·0085
O(22)	0·4316	0·1487	0·0889
O(23)	0·4423	0·3481	−0·1395
O(31)	0·2381	−0·1381	0·3951
O(32)	0·4071	−0·0834	0·4397
O(33)	0·4186	−0·3167	0·5801
O(1)	0·2794	−0·0067	−0·0192
O(2)	0·3056	0·2408	0·4738
O(3)	0·1054	0·1140	0·5934
O(4)	0·0286	0·4076	0·3014
O(5)	0·2193	0·5486	0·5002

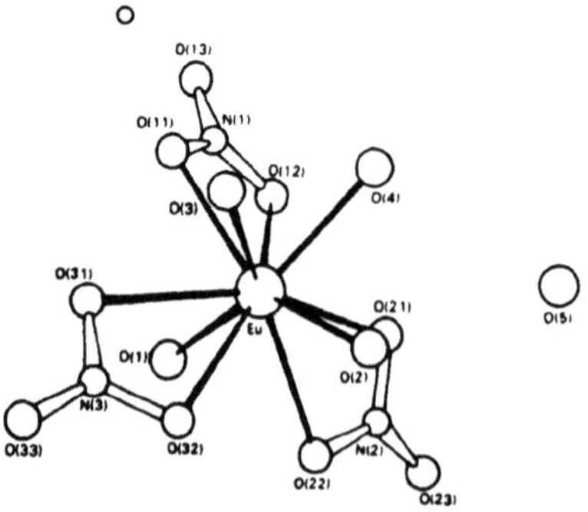

Bond lengths (Å) *and selected bond angles* (°) *with e.s.d.'s in parentheses*

Eu–O(11)	2·52 (1)	N(1)–O(11)	1·28 (2)
Eu–O(12)	2·63 (1)	N(1)–O(12)	1·27 (2)
Eu–O(21)	2·50 (1)	N(1)–O(13)	1·21 (2)
Eu–O(22)	2·55 (1)	N(2)–O(21)	1·26 (2)
Eu–O(31)	2·58 (1)	N(2)–O(22)	1·26 (2)
Eu–O(32)	2·47 (1)	N(2)–O(23)	1·23 (2)
Eu–O(1)	2·50 (1)	N(3)–O(31)	1·25 (2)
Eu–O(2)	2·40 (1)	N(3)–O(32)	1·31 (2)
Eu–O(3)	2·44 (1)	N(3)–O(33)	1·19 (2)
Eu–O(4)	2·42 (1)		

O(11)–Eu–O(12)	49·5 (6)	O(31)–Eu–O(1)	68·1 (6)
O(11)–Eu–O(31)	67·8 (6)	O(31)–Eu–O(3)	67·5 (6)
O(11)–Eu–O(1)	78·0 (7)	O(32)–Eu–O(1)	82·2 (7)
O(11)–Eu–O(3)	73·7 (7)	O(32)–Eu–O(2)	72·0 (7)
O(11)–Eu–O(4)	80·7 (7)	O(32)–Eu–O(3)	79·7 (7)
O(12)–Eu–O(21)	67·0 (7)	O(2)–Eu–O(3)	72·3 (7)
O(12)–Eu–O(1)	68·6 (6)	O(2)–Eu–O(4)	72·5 (7)
O(12)–Eu–O(4)	70·3 (7)	O(3)–Eu–O(4)	71·4 (7)
O(21)–Eu–O(22)	50·4 (7)	O(11)–N(1)–O(12)	116 (2)
O(21)–Eu–O(1)	85·9 (7)	O(11)–N(1)–O(13)	123 (2)
O(21)–Eu–O(2)	81·1 (7)	O(12)–N(1)–O(13)	121 (2)
O(21)–Eu–O(4)	72·1 (7)	O(21)–N(2)–O(22)	117 (2)
O(22)–Eu–O(32)	70·8 (7)	O(21)–N(2)–O(23)	121 (3)
O(22)–Eu–O(1)	69·3 (7)	O(22)–N(2)–O(23)	122 (3)
O(22)–Eu–O(2)	72·8 (7)	O(31)–N(3)–O(32)	115 (2)
O(31)–Eu–O(32)	50·3 (6)	O(31)–N(3)–O(33)	125 (3)
		O(32)–N(3)–O(33)	120 (3)

CAESIUM NEPTUNYL CHLORIDE NITRATE HYDRATE
$Cs_4(NpO_2)_3Cl_6(NO_3).H_2O$

Radiokhimija, **28**, 304-311.

$Cmc2_1$, 19.60, 13.26, 8.54, Z = 4, R = 0.104. (010) layers with Np in pentagonal bipyramidal coordination, linked by Cs ions.

CAESIUM BARIUM POLYPHOSPHATE
$Cs_4Ba(PO_3)_6$

Acta Cryst., C**42**, 928-930.

P31c, 11.549, 9.114, Z = 2, R = 0.054. Chains along c of corner-sharing PO_4 tetrahedra with a four-tetrahedra repeat, linked by 9-coordinate Ba and 9- and 7-coordinate Cs ions. P-O = 1.62-1.64 (chains), 1.41-1.55 A (terminal), P-O-P = 127, Ba-O = 2.76-2.90, Cs-O = 2.99-3.43 A.

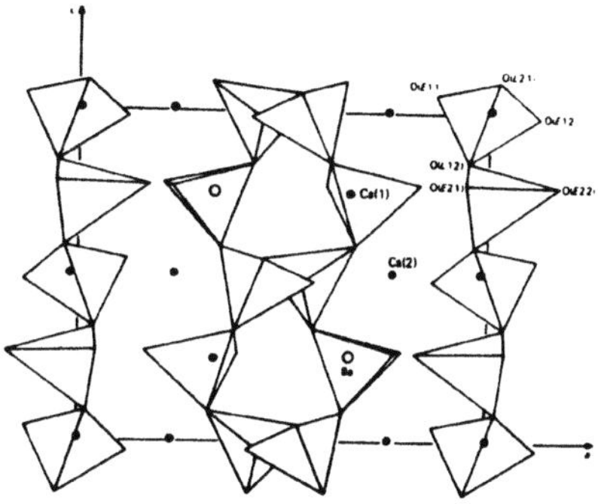

	x	*y*	*z*
Ba	$\frac{1}{3}$	$\frac{2}{3}$	0·7592
Cs(1)	[illegible]	[illegible]	$\frac{1}{4}$*
Cs(2)	0·2354	0·0069	0·5066
P(1)	0·5503	0·0057	0·0006
P(2)	0·3082	0·2847	0·2510
O(*E*11)	0·546	0·120	0·045
O(*E*12)	0·451	0·580	−0·024
O(*L*12)	0·382	0·426	0·337
O(*E*21)	0·162	0·209	0·271
O(*E*22)	0·392	0·221	0·267
O(*L*21)	0·346	0·319	0·586

* Fixed parameter.

LITHIUM LEAD POLYPHOSPHATE
$LiPb_2(PO_3)_5$

Acta Cryst., C**42**, 647-651.

P2/n, 12.289, 9.689, 5.523, 91.01, Z = 2, R = 0.029. Chains of corner-sharing PO_4 tetrahedra with a five-tetrahedra repeat, linked by chains of edge-sharing PbO_7 polyhedra, which are cross-linked by LiO_5 polyhedra. P-O = 1.48-1.49 (terminal), 1.59-1.60 (chain) A, P-O-P = 129-134°, Pb-O = 2.45-2.93, Li-O = 1.97-2.44 A.

	x	*y*	*z*
Pb	0,40772	0,68596	0,40676
P(1)	0,0815	0,1914	0,4208
P(2)	0,1505	0,4587	0,6078
P(3)	$\frac{1}{4}$	0,0019	$\frac{1}{4}$
O(*L*12)	0,1267	0,3461	0,405
O(*L*22)	$\frac{1}{4}$	0,388	$\frac{3}{4}$
O(*L*31)	0,3081	0,1046	0,067
O(*E*11)	0,0264	0,1717	0,657
O(*E*12)	0,0208	0,1656	0,191
O(*E*21)	0,1858	0,5889	0,488
O(*E*22)	0,0584	0,4700	0,778
O(*E*31)	0,1678	−0,0735	0,099
Li	0,025	0,015	0,94

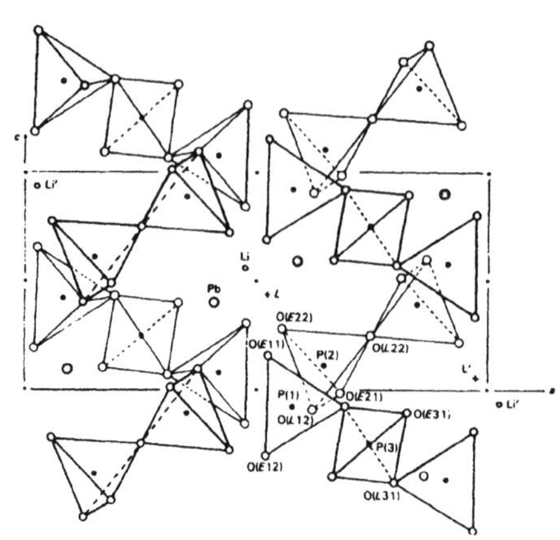

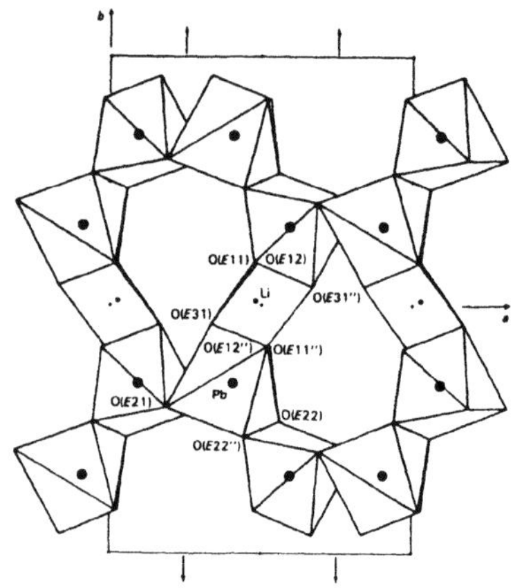

ZIRCONIUM PHOSPHITES
α-$Zr(HPO_3)_2$ (I), $Zr(HPO_4)_{0.85}(HPO_3)_{1.15}\cdot 0.5H_2O$ (II)

J. Solid State Chem., **63**, 455-460.

I. $P\bar{3}$, 5.418, 5.579, Z = 1, powder data. II. C2, 9.33, 5.39, 14.35, 115.7, Z = 4, powder data. Structures proposed from comparison with that of α-$Zr(HPO_4)_2.H_2O$ (**34**A, 328; **43**A, 255).

SODIUM MANGANESE(II) DIHYDROGENPHOSPHITE MONOHYDRATE
$NaMn(H_2PO_3)_3.H_2O$

Acta Cryst., C**42**, 1281-1283.

Pbca, 9.156, 14.960, 14.899, R = 0.042. Isostructural with the Co(II) analogue (**49**A), with $HPO_2(OH)^-$ anions, linked by octahedrally coordinated Mn^{2+} and Na^+ ions.

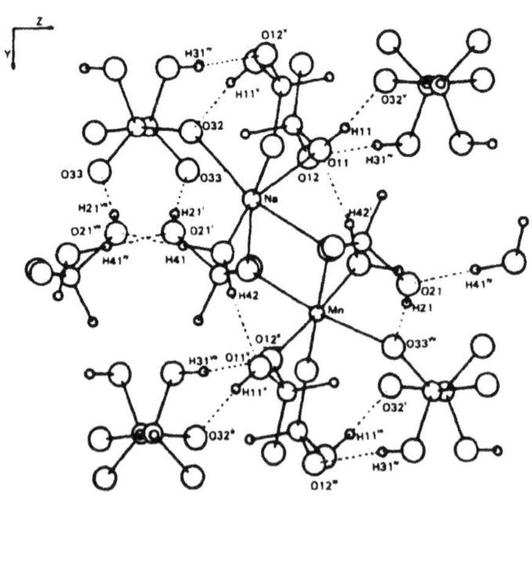

	x	y ($\times 10^4$)	z
Mn	4464·6	8402·6	5453·5
Na	5574	6532·4	4367
P(1)	2343·6	5317·1	5080·7
P(2)	7380·8	7245·2	6218·5
P(3)	4601·2	5312·0	2392·5
O(11)	3741	5728	5535
O(12)	1019	5847	5349
O(13)	2285	4335	5259
O(21)	7354	7943	7017
O(22)	6027	7362	5664
O(23)	8808	7331	5734
O(31)	5011	4356	2046
O(32)	5169	5392	3329
O(33)	5120	6026	1752
O(4)	7731	7400	3817

Bond distances (Å) *and angles* (°) *for* Mn···O *and* Na···O < 3·2 Å

P(1)–O(11)	1·573 (3)	Mn–O(33^{iv})	2·199 (3)
P(1)–O(12)	1·503 (3)	Mn–O(4^{i})	2·268 (3)
P(1)–O(13)	1·494 (2)	Na–O(32)	2·332 (3)
P(2)–O(21)	1·583 (3)	Na–O(22)	2·334 (2)
P(2)–O(22)	1·500 (2)	Na–O(23^{i})	2·351 (2)
P(2)–O(23)	1·498 (2)	Na–O(13^{v})	2·416 (2)
P(3)–O(31)	1·566 (3)	Na–O(4)	2·501 (3)
P(3)–O(32)	1·494 (3)	Na–O(11)	2·701 (3)
P(3)–O(33)	1·509 (3)	O(11)–O(32^{v})	2·582 (4)
Mn–O(22)	2·137 (2)	O(21)–O(33^{iv})	2·592 (4)
Mn–O(13^{iii})	2·144 (2)	O(31)–O(12^{vi})	2·716 (4)
Mn–O(23^{i})	2·167 (2)	O(4)–O(21^{vii})	2·752 (4)
Mn–O(12^{ii})	2·171 (3)	O(4)–O(11^{ii})	3·103 (4)

O(11)–P(1)–O(12)	109·6 (2)	O(4^{i})–Mn–O(33^{iv})	88·6 (1)
O(11)–P(1)–O(13)	109·7 (1)	O(4^{i})–Mn–O(12^{ii})	175·1 (1)
O(12)–P(1)–O(13)	116·3 (1)	O(12^{ii})–Mn–O(13^{iii})	94·5 (1)
O(21)–P(2)–O(22)	108·9 (1)	O(12^{ii})–Mn–O(33^{iv})	96·0 (1)
O(21)–P(2)–O(23)	108·6 (1)	O(13^{iii})–Mn–O(33^{iv})	94·0 (1)
O(22)–P(2)–O(23)	116·4 (1)	O(11)–Na–O(22)	79·27 (9)
O(31)–P(3)–O(32)	107·3 (2)	O(11)–Na–O(32)	90·1 (1)
O(31)–P(3)–O(33)	111·2 (2)	O(11)–Na–O(23^{i})	86·34 (8)
O(32)–P(3)–O(33)	115·1 (2)	O(11)–Na–O(13^{v})	96·7 (1)
O(22)–Mn–O(23^{i})	86·37 (6)	O(11)–Na–O(4)	158·9 (1)
O(22)–Mn–O(4^{i})	90·72 (9)	O(22)–Na–O(4)	81·66 (9)
O(22)–Mn–O(12^{ii})	91·07 (9)	O(22)–Na–O(23^{i})	77·91 (7)
O(22)–Mn–O(33^{iv})	88·36 (9)	O(22)–Na–O(13^{v})	87·16 (9)
O(22)–Mn–O(13^{iii})	173·66 (7)	O(22)–Na–O(32)	165·1 (1)
O(23^{i})–Mn–O(4^{i})	85·93 (9)	O(32)–Na–O(13^{v})	83·6 (1)
O(23^{i})–Mn–O(12^{ii})	89·66 (9)	O(32)–Na–O(4)	106·7 (1)
O(23^{i})–Mn–O(13^{iii})	90·68 (8)	O(32)–Na–O(23^{i})	112·16 (9)
O(23^{i})–Mn–O(33^{iv})	172·33 (9)	O(4)–Na–O(13^{v})	73·36 (8)
O(4^{i})–Mn–O(13^{iii})	83·47 (9)	O(4)–Na–O(23^{i})	98·44 (9)
		O(23^{i})–Na–O(13^{v})	163·98 (9)

Symmetry code: (i) $x - \frac{1}{2}, \frac{3}{2} - y, 1 - z$; (ii) $\frac{1}{2} + x, \frac{3}{2} - y, 1 - z$; (iii) $\frac{1}{2} - x, \frac{1}{2} + y, z$; (iv) $x, \frac{3}{2} - y, \frac{1}{2} + z$; (v) $1 - x, 1 - y, 1 - z$; (vi) $\frac{1}{2} - x, 1 - y, z - \frac{1}{2}$; (vii) $x, \frac{3}{2} - y, z - \frac{1}{2}$.

CAESIUM DIHYDROGEN PHOSPHATE (HIGH-PRESSURE)
CsH_2PO_4

Jpn. J. Appl. Phys., Part 1, **24**, 935-937 (1985).

$P2_1$ (pseudo-$P2_1/a$), 15.625, 6.254, 4.886, 108.08, at 100.7K and 3.6 kbar, Z = 4, neutron radiation, R = 0.072. Large relative displacements (about 1 A) of Cs and PO_4 from the paraelectric and ferroelectric phases. The H bonds along **b** in this phase show antiferroelectric ordering.

BOBIERRITE
MAGNESIUM PHOSPHATE OCTAHYDRATE (SYNTHETIC)
$Mg_3(PO_4)_2.8H_2O$

Amer. Min., **71**, 1229-1233.

Bobierite, C2/c, 4.667, 27.926, 10.067, 105.01, Z = 4, R = 0.041. Synthetic, C2/m, 10.034, 13.407, 4.657, 105.09, Z = 2, R = 0.025. The synthetic material is isostructural with vivianite (**13**, 308; **46A**, 326), and the mineral has a closely related structure with doubled **b** axis. Both structures contain edge-sharing octahedral dimers $Mg_2O_6(H_2O)_4$ and $MgO_2(H_2O)_4$ octahedra linked by PO_4 tetrahedra to form (010) sheets. The arrangement of sheets along **b** is different in the two structures.

	x	*y*	*z*
Bobierrite (polymorph I)			
Mg(1)	0	0.87840	¼
Mg(2)	0	0.06681	¼
Mg(3)	0	0.17283	¼
P	0.3861	0.11913	0.0648
O(1)	0.3715	0.11974	−0.0916
O(2)	0.7142	0.11951	0.1452
O(3)	0.2291	0.16505	0.0958
O(4)	0.2312	0.07345	0.0958
O(w1)	0.3191	0.32540	0.3571
O(w2)	0.3088	0.43350	0.3473
O(w3)	0.2857	0.22389	0.3534
O(w4)	0.2800	0.01297	0.3507
Synthetic $Mg_3(PO_4)_2 \cdot 8H_2O$ (polymorph II)			
Mg(1)	0	0	0
Mg(2)	0	0.39054	0
P	0.31452	0	0.3843
O(1)	0.1575	0	0.3709
O(2)	0.3947	0	0.7144
O(3)	0.3455	0.0956	0.2303
O(w1)	0.4013	0.3866	0.1916
O(w2)	0.1010	0.2787	0.2802

STRUVITE
$MgNH_4PO_4.6H_2O$

Acta Cryst., B**42**, 253-258.

$Pmn2_1$, 6.955, 6.142, 11.218, Z = 2, neutron radiation, R = 0.032. Structure as previously described (**19**, 440; **35**A, 329; **51**A, 276), with a completely ordered ammonium group forming one normal and several bifurcated hydrogen bonds. N-H...O = 2.800-3.498, O-H...O = 2.630-2.649, 2.695, 3.141(5) A.

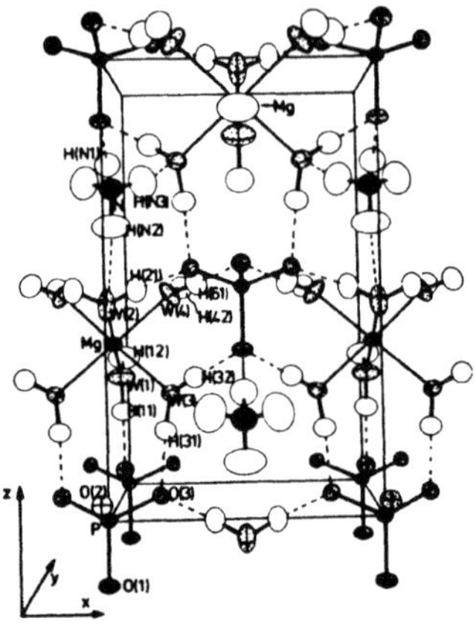

	x	y	z
P	0	−0·0069	0·0019
Mg	0	0·3766	0·3741
O(1)	0	−0·0236	−0·1351
O(2)	0	−0·2382	0·0558
O(3)	0·1823	0·1139	0·0436
W(1)	0	0·6829	0·2878
W(2)	0	0·0768	0·4664
W(3)	0·2179	0·2618	0·2643
W(4)	0·2115	0·4852	0·4874
N	0	0·3657	0·7351
H(11)	0	0·7192	0·2017
H(12)	0	0·8174	0·3299
H(21)	0·1157	0·0070	0·4999
H(31)	0·1989	0·2007	0·1824
H(32)	0·3169	0·1702	0·3020
H(41)	0·3200	0·3904	0·5120
H(42)	0·2511	0·6354	0·5027
H(N1)	0	0·2186	0·7832
H(N2)	0	0·3311	0·6473
H(N3)	0·1148	0·4557	0·7535

HENEUITE
$CaMg_5(CO_3)(PO_4)_3(OH)$

Neues Jb. Miner., Mh., 351-359.

$P\bar{1}$, 6.311, 10.843, 8.676, 95.01, 93.41, 101.04, Z = 2, R = 0.026. PO_4 tetrahedra and CO_3 triangles, linked by CaO_8 polyhedra, MgO_6 octahedra, and one MgO_5 trigonal bipyramid.

BARIUM HYDROXYAPATITE
$Ba_5(PO_4)_3OH$

Kristallografija, **31**, 233-236 [Soviet Physics - Crystallography, **31**, 136-138].

$P6_3$, 10.1904, 7.721, Z = 2, R = 0.039. Apatite structure, with lower-symmetry space group preferred.

	x	y	z
Ba_1	0.7393	−0,0189	0,2552
Ba_2	1/3	2/3	0
Ba_3	1/3	2/3	1/2
P	−0.0322	0.3699	0.256
O_1	0.576	0.118	0.24
O_2	0.260	0.331	0.908
O_3	0.089	0.728	0.593
O_4	0.343	0.859	0.25
OH	0	0	0.148

ALUMINUM and GALLIUM PHOSPHATES and ARSENATES
$AlPO_4$, $AlAsO_4$, $GaPO_4$, $GaAsO_4$

J. Solid State Chem., **61**, 384-396.

$P3_121$, a = 4.927-4.950, 5.022-5.028, 4.874-4.897, 4.986-4.994, c = 10.918-10.948, 11.215-11.207, 11.033-11.021, 11.375-11.361 A, at 173, 293, 373K, Z = 3, R = 0.011-0.024. α-Quartz structure (**3**, 424; **45**A, 303; **50**A).

		$AlPO_4$			$AlAsO_4$		
		173°K	293°K	373°K	173°K	293°K	373°K
M^{III}	x	0,4651	0,466643	0,4679	0,4577	0,4585	0,4587
X^{V}	x	0,4655	0,46685	0,4685	0,45625	0,45704	0,4574
O(1)	x	0,4189	0,4163	0,4263	0,4070	0,4075	0,4097
	y	0,2933	0,2922	0,2887	0,2980	0,2969	0,2957
	z	0,0637	0,06437	0,0654	0,05281	0,0531	0,0534
O(2)	x	0,4146	0,4156	0,4164	0,4043	0,4051	0,4048
	y	0,2593	0,2576	0,2548	0,2860	0,2850	0,2835
	z	0,5491	0,55032	0,5512	0,54695	0,5479	0,5476

$GaPO_4$			$GaAsO_4$		
173°K	293°K	373°K	173°K	293°K	373°K
0,4560	0,4565	0,4570	0,4498	0,4504	0,4509
0,4558	0,4565	0,4569	0,4496	0,4499	0,4502
0,4067	0,4069	0,4091	0,399	0,400	0,397
0,3186	0,3169	0,3163	0,309	0,309	0,309
0,0588	0,0595	0,0598	0,0473	0,0475	0,0491
0,4087	0,4087	0,4086	0,398	0,399	0,400
0,2738	0,2714	0,2704	0,303	0,304	0,298
0,5386	0,5390	0,5392	0,5416	0,5422	0,5399

ALUMINUM PHOSPHATE HYDRATE
$AlPO_4.1.5H_2O$

Acta Cryst., C**42**, 1118-1120.

Pbca, 19.3525, 9.7272, 9.7621, Z = 16, R = 0.033. Framework of PO_4 and AlO_4 tetrahedra and $AlO_4(H_2O)_2$ octahedra; one H_2O of each octahedron lies in a 6-ring, and the other forms a zigzag chain with the third water molecule.

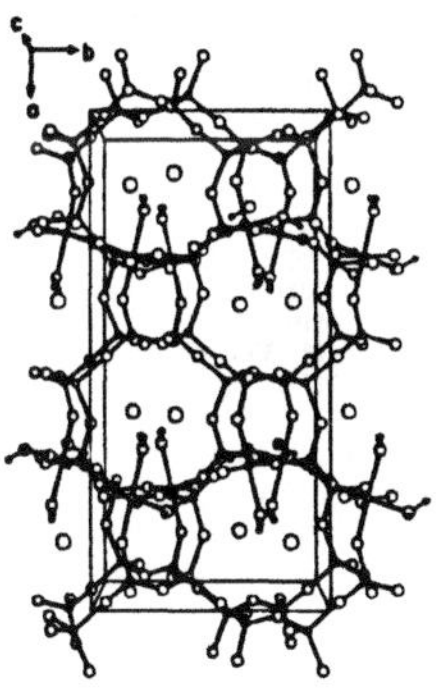

	x	y	z
P(1)	0·44769	0·34071	0·13498
P(2)	0·29050	0·06565	0·33879
Al(1)	0·44808	0·08470	0·32419
Al(2)	0·28207	0·33224	0·14654
O(1)	0·37433	0·3765	0·1701
O(2)	0·46047	0·1662	0·4804
O(3)	0·49520	0·4497	0·1971
O(4)	0·46889	0·2004	0·1942
O(5)	0·27397	0·0582	0·4913
O(6)	0·36522	0·0190	0·3136
O(7)	0·28329	0·2158	0·2953
O(8)	0·25855	0·4758	0·2592
O(9)	0·18446	0·2831	0·1193
O(10)	0·30512	0·1730	0·0340
O(11)	0·11929	0·1440	0·3627
H(1)	0·15752	0·2988	0·0523
H(2)	0·16096	0·2574	0·1883
H(3)	0·28123	0·1025	0·0541
H(4)	0·30108	0·1902	0·9490

(Å)

P(1)–O(1)	1·501 (2)	P(2)–O(5)	1·524 (2)
P(1)–O(2)	1·531 (2)	P(2)–O(6)	1·536 (2)
P(1)–O(3)	1·529 (2)	P(2)–O(7)	1·528 (2)
P(1)–O(4)	1·538 (2)	P(2)–O(8)	1·506 (2)
Mean	1·525	Mean	1·524
Al(1)–O(2)	1·735 (2)	Al(2)–O(1)	1·851 (2)
Al(1)–O(3)	1·724 (2)	Al(2)–O(5)	1·860 (2)
Al(1)–O(4)	1·744 (2)	Al(2)–O(7)	1·842 (2)
Al(1)–O(6)	1·730 (2)	Al(2)–O(8)	1·835 (2)
Mean	1·734	Al(2)–O(9)	1·967 (2)
		Al(2)–O(10)	1·951 (2)
		Mean	1·884

Hydrogen bonds

H(1)···O(11)	2·070 (5)	O(9)–H(1)···O(11)	162·8 (3)	O(9)···O(11)	2·893 (5)
H(2)···O(11)	2·183 (4)	O(9)–H(2)···O(11)	164·5 (3)	O(9)···O(11)	3·011 (4)
H(3)···O(5)	1·990 (3)	O(10)–H(3)···O(5)	148·7 (3)	O(10)···O(5)	2·716 (3)
H(4)···O(7)	1·790 (3)	O(10)–H(4)···O(7)	159·4 (3)	O(10)···O(7)	2·603 (3)
				O(11)···O(4)	3·013 (4)
				O(11)···O(8)	3·047 (4)

ALUMINUM PHOSPHATE MOLECULAR SIEVE
$AlPO_4.(CH_3)_4NOH$

J. Phys. Chem., **90**, 6122-6125.

$P2_12_12_1$, 10.3325, 14.6405, 9.5112, Z = 4, powder data. Framework with 4-, 6-, and 8-rings of AlO_4 and PO_4 tetrahedra, with the organic species in the 8-rings.

GALLIUM PHOSPHATE HYDRATE - ISOPROPYLAMINE
$3GaPO_4.C_3H_9N.H_2O$

Acta Cryst., C**42**, 144-147.

$P2_1/n$, 8.700, 18.146, 9.087, 107.28, Z = 4, R = 0.047. $AlPO_4$-type molecular sieve framework, with enclathrated isopropylamine disordered over two sites.

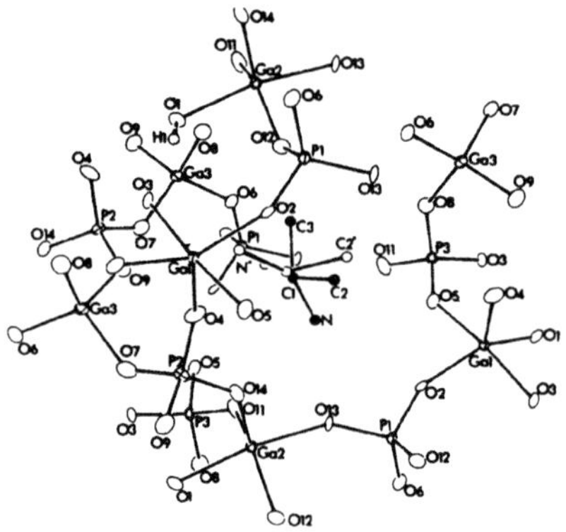

THALLIUM(I) PHOSPHATE
Tl_3PO_4

Acta Cryst., C**42**, 1686-1687.

$P6_3$, 8.369, 5.111, Z = 2, R = 0.017. Previous study in **39**A, 363. Phosphate tetrahedra, linked by Tl^+ ions with three short Tl-O bonds and a fourth O at 3.26 A.

Interatomic distances (Å) *and angles* (°)

$Tl—O(1)^{iv}$	2·553 (5)	$O(1)^{iv}—Tl—O(2)^{iii}$	84·98 (15)
$Tl—O(2)^{vi}$	2·529 (7)	$O(1)^{iv}—Tl—O(2)^{vi}$	84·24 (32)
$Tl—O(2)^{iii}$	2·555 (8)	$O(2)^{iii}—Tl—O(2)^{vi}$	80·28 (18)
$P—O(1)^{i}$	1·536 (18)	$O(1)^{i}—P—O(2)^{i}$	108·7 (6)
$P—3O(2)^{i,ii,iii}$	1·547 (8)	$O(2)^{i}—P—O(2)^{ii}$	110·2 (5)
$Tl^{i}—4Tl^{v,vii,viii,ix}$	3·690 (3)	$Tl^{xi}—O(1)—Tl^{x}$	107·3 (3)
		$Tl^{xi}—O(1)—P^{i}$	111·5 (3)

Symmetry code: (i) x, y, z; (ii) $1 - y, 1 + x - y, z$; (iii) $y - x, 1 - x, z$; (iv) $1 - x, 1 - y, \frac{1}{2} + z$; (v) $y, y - x, \frac{1}{2} + z$; (vi) $1 + x - y, x, z - \frac{1}{2}$; (vii) $y, y - x, z - \frac{1}{2}$; (viii) $x - y, x, z - \frac{1}{2}$; (ix) $x - y, x, \frac{1}{2} + z$; (x) $y, 1 - y + x, z - \frac{1}{2}$; (xi) $1 - x, 1 - y, \frac{1}{2} - z$.

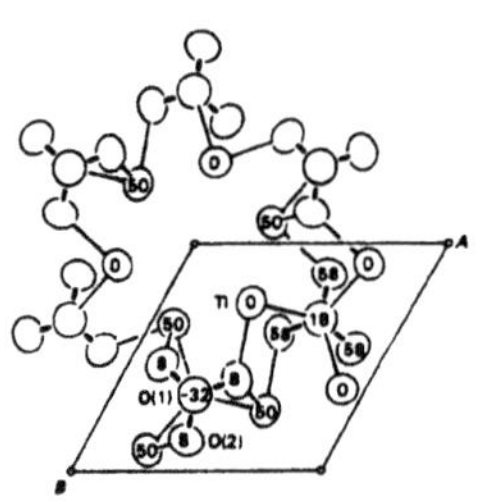

	x	y	z
Tl	0·35393	0·26182	0
P	$\frac{1}{3}$	$\frac{2}{3}$	−0·0161
O(1)	$\frac{1}{3}$	$\frac{2}{3}$	−0·3166
O(2)	0·3966	0·8645	0·0811

THALLIUM(I) DIHYDROGENPHOSPHATE
TlD_2PO_4

Z. Kristallogr., **175**, 145-149.

$P2_1/c$, 6.512, 4.518, 14.319, 91.63, Z = 4, R = 0.038. Structure as previously described (**45**A, 306). Deuterium positions not determined, and the material may not be fully deuterated.

THALLIUM(I,III) HYDROGEN PHOSPHATE HYDRATE
$Tl_3[Tl_{0.5}(H_3O)_{0.5}]H_{14}(PO_4)_8.4H_2O$

Acta Cryst., C**42**, 511-512.

C2/c, 17.367, 9.820, 17.87, 90.17, Z = 4, rather than $P\bar{1}$ (**52**A, 272).

ANTIMONY(V) OXIDE PHOSPHATE
$SbOPO_4$

J. Solid State Chem., **63**, 81-85.

C2/c, 6.791, 8.033, 7.046, 115.90, Z = 4, R = 0.028. Chains along **c** of corner-shared SbO_6 octahedra, cross-linked by corner-sharing with PO_4 tetrahedra. Sb-O = 1.90-1.99, P-O = 1.52, 1.54(1) A.

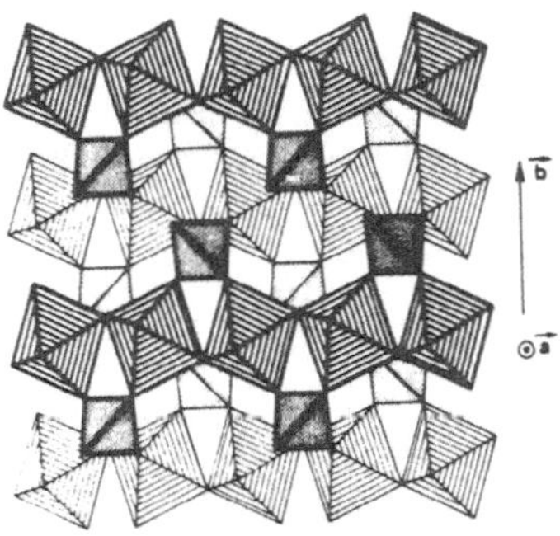

		x	y	z
Sb	4a	0	0	0
P	4e	0	0.3348	¼
O(1)	4e	0	0.9105	¼
O(2)	8f	0.5884	0.2751	0.6254
O(3)	8f	0.1848	0.4440	0.4012

POTASSIUM ANTIMONY PHOSPHATES
K_2SbPO_6

J. Solid State Chem., **63**, 409-413.

Pnma, 9.429, 5.891, 11.030, Z = 4, R = 0.038. Rutile-like chains along **b** of edge-sharing SbO_6 octahedra bridged by phosphate tetrahedra, with chains linked by 11- and 12-coordinate K ions. Sb-O = 1.96-2.02, P-O = 1.50, 1.58, K-O = 2.65-3.97(1) A.

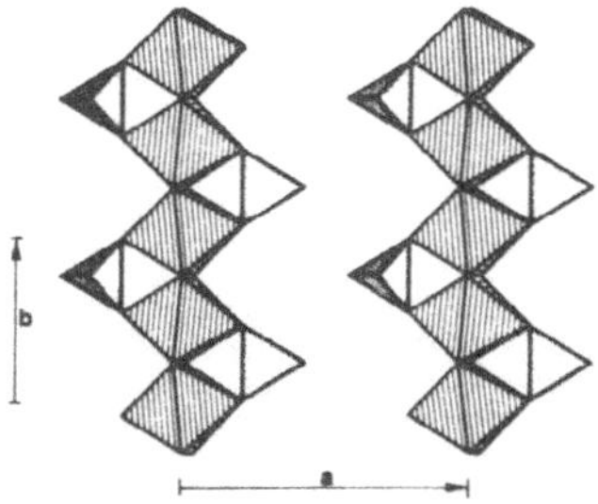

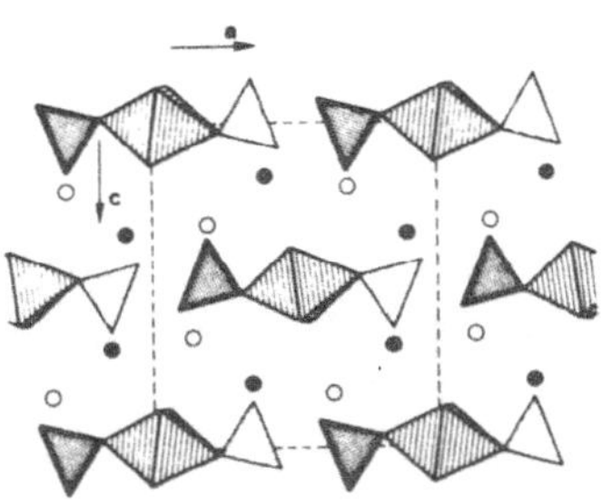

		x	y	z
Sb	4a	0	0	0
K(1)	4c	0.1724	$\frac{1}{4}$	0.3060
K(2)	4c	0.1192	$\frac{1}{4}$	0.6606
P	4c	0.8038	$\frac{1}{4}$	0.5094
O(1)	4c	0.467	$\frac{1}{4}$	0.3817
O(2)	4c	0.501	$\frac{1}{4}$	0.6167
O(3)	4c	0.929	$\frac{1}{4}$	0.426
O(4)	4c	0.830	$\frac{1}{4}$	0.644
O(5)	8d	0.2893	0.538	0.5270

$K_5Sb_5P_2O_{20}$

Mater. Res. Bull., **21**, 1231-1238.

Pnnm, 23.443, 18.452, 7.149, Z = 6, R = 0.044. $Sb_5P_2O_{20}{}^{5-}$ framework of edge- and corner-sharing SbO_6 octahedra and PO_4 tetrahedra, with K ions in large tunnels.

POTASSIUM BISMUTH PHOSPHATE OXIDE
$K_2Bi_3(PO_4)_3O$

J. Solid State Chem., **62**, 351-359.

Pnma, 13.139, 10.413, 9.239, Z = 4, R = 0.048 (at 295K, also 0.058 at 795K). PO_4 tetrahedra linked by BiO_6 octahedra. One PO_4 tetrahedron is disordered at 795K, and is ordered at 295K, with doubling of c and possible space group $P2_1ca$.

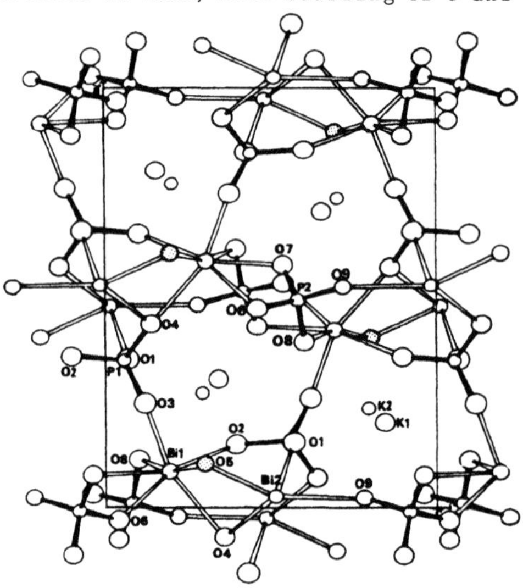

LITHIUM SCANDIUM PHOSPHATE
$Li_3Sc_2(PO_4)_3$

Kristallografija, **31**, 488-494, 592-594, and 595-596 [Soviet Physics - Crystallography, **31**, 287-291, 348-349, and 350-351].

293K, $P2_1/n$ (c unique), 8.853, 12.273, 8.802, 90.00, Z = 4, R = 0.09 and R =

0.044 (twinned crystal). 573K, Pcan, 8.828, 12.399, 8.823, Z = 4 [not 8], R = 0.029. Isostructural with the Fe analogue (**52**A, 282), with a framework of PO_4 tetrahedra and ScO_6 octahedra, and 4- and 5-coordinate Li ions (in the Pcan form, Li(2) and Li(3) have occupancies = 0.25).

293K

	x	y	z
Sc1	0.2473	0.1075	0.4552
Sc2	0.7500	0.3932	0.4636
P1	0.1027	0.1523	0.1000
P2	0.6020	0.3476	0.1098
P3	0.0379	0.4956	0.2526
O1	0.4294	0.3361	0.0991
O2	0.9302	0.1638	0.0964
O3	0.3350	0.2704	0.4792
O4	0.8239	0.2254	0.4879
O5	0.1518	0.0367	0.0667
O6	0.6512	0.4646	0.0822
O7	0.4516	0.0727	0.3575
O8	0.9367	0.4109	0.3191
O9	0.1509	0.4284	0.1607
O10	0.6211	0.0635	0.1366
O11	0.1555	0.1893	0.2587
O12	0.6486	0.3102	0.2697
Li1	0.296	0.325	0.270
Li2	0.582	0.191	0.401
Li3	0.916	0.249	0.291

573K

	x	y	z
Li1	0.292	0.319	0.277
Li2	0.091	0.308	0.419
Li3	0.451	0.230	0.304
Sc	0.2492	0.1096	0.4609
P1	0.1060	0.1502	0.1025
P2	0.0468	0.50	0.25
O1	0.4332	0.3390	0.1011
O2	0.3282	0.2735	0.4820
O3	0.1525	0.0360	0.0645
O4	0.4501	0.0653	0.3592
O5	0.1481	0.4199	0.1664
O6	0.1639	0.1818	0.2608

NASICON
$Na_3Sc_2(PO_4)_3$

J. Phys. Chem. Solids, **47**, 843-854.

α-Phase, Bb, 8.900, 8.900, 22.769, γ = 119.86° (trigonal setting), Z = 6, R = 0.045. β-Phase, R$\bar{3}$c, a = 8.928, c = 22.276 A, Z = 6, R = 0.024. γ-Phase, R$\bar{3}$c, a = 8.949, c = 22.230 A, Z = 6, R = 0.023. Nasicon-type structures (**51**A, 283; **52**A, 276), with increasing amount of PO_4 disorder.

GLASERITE-TYPE PHOSPHATES
$K_3Sc(PO_4)_2$, $K_2CsSc(PO_4)_2$, $K_2CsHo(PO_4)_2$

Rev. Chim. Minér., **22**, 666-675 (1985).

K_3Sc, K_2CsSc, P$\bar{3}$m1, a = 5.430, 5.484, c = 7.629, 7.916 A, Z = 1, R = 0.067 for K_3Sc. Glaserite-type structures (**1**, 378; **46**A, 348). K_2CsHo, monoclinic, 9.803, 5.660, 8.096, 90.0°, Z = 2, R = 0.028. Monoclinically-distorted glaserite-type structure.

NASICON
$NaZr_2P_3O_{12}$

Mater. Res. Bull., **21**, 23-31.

R$\bar{3}$c, 8.813, 22.80, at 295K, Z = 6, R = 0.040, 0.038 at 295, 993K. NASICON structure (**33**A, 397; **42**A, 342; **52**A, 276).

VANADIUM(III) SILICOPHOSPHATE
$V_3P_5SiO_{19}$

J. Solid State Chem., **65**, 168-177.

$P6_3$, 14.4671, 7.4605, Z = 4, R = 0.047. Three-dimensional framework of corner-sharing PO_4 tetrahedra, Si_2O_7 groups, and V_2O_9 pairs of face-sharing octahedra.

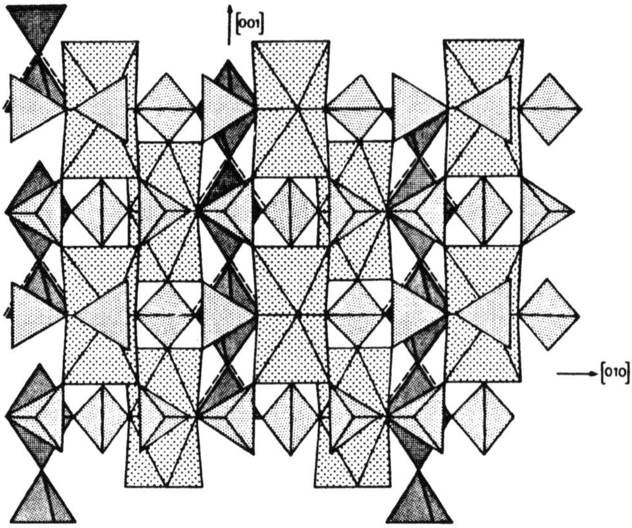

	x	y	z
V(1)	0.3450	0.3210	0.056136
V(2)	0.3393	0.3248	0.4445
P(1)	0.2119	0.4334	0.2492
P(2)	0.5163	0.0961	0.2608
P(3)	0.2334	0.0841	0.2505
P(4)	0	0	0.2911
Si(1)	$-\frac{1}{3}$	$\frac{1}{3}$	0.2177
Si(2)	$\frac{1}{3}$	$-\frac{1}{3}$	0.2975
O(1)	0.3043	0.2070	0.2510
O(2)	0.4599	0.3974	0.2532
O(3)	0.2664	0.3654	0.2452
O(4)	0.2449	0.0415	0.0741
O(5)	0.2475	0.0319	0.4169
O(6)	0.2672	−0.1423	0.0838
O(7)	0.2688	−0.1571	0.4231
O(8)	0.4587	0.0742	0.0866
O(9)	0.4489	0.0504	0.4227
O(10)	0.2449	−0.3063	0.2223
O(11)	−0.4227	0.2194	0.2975
O(12)	0	0	−0.0019
O(13)	0.1185	0.0677	0.2527
O(14)	$-\frac{1}{3}$	$\frac{1}{3}$	0.0080

TANTALUM PHOSPHATE BRONZE
β-$TaPO_5$

J. Solid State Chem., **65**, 331-342.

$P2_1/c$, 13.07, 5.281, 13.24, 120.42, Z = 8 (Pnma subcell with c' = c/2), powder data and electron diffraction patterns. ReO_3-slabs of TaO_6 octahedra, connected by PO_4 tetrahedra, with empty pentagonal tunnels.

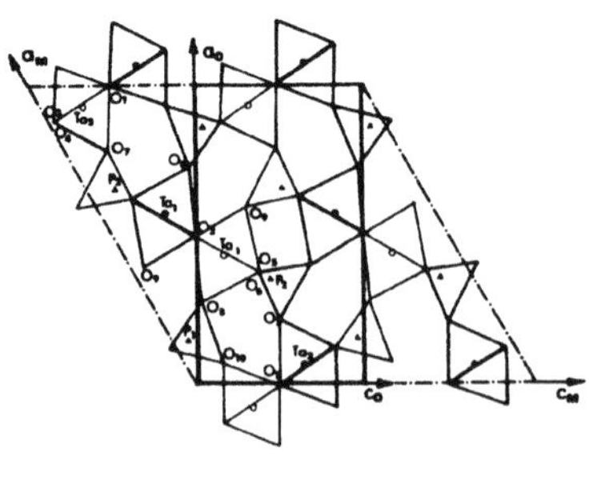

	x	y	z
Ta(1)	0.4323	0.248	0.3007
Ta(2)	0.9389	0.249	0.1327
P(1)	0.150	0.25	0.053
P(2)	0.652	0.25	0.084
O(1)	0.0	0.0	0.25
O(2)	0.50	0.50	0.25
O(3)	0.878	0.0	0.016
O(4)	0.878	0.50	0.018
O(5)	0.378	0.50	0.377
O(6)	0.377	0.0	0.377
O(7)	0.787	0.25	0.133
O(8)	0.279	0.25	0.154
O(9)	0.597	0.25	0.452
O(10)	0.078	0.25	0.117

RUBIDIUM TANTALUM PHOSPHATE
$RbTa(PO_4)_2$

Izv. Akad. Nauk SSSR, Neorg. Mater., **22**, 1369-1373.

C2/m, 8.610, 5.2156, 8.420, 93.18, Z = 2, R = 0.035. Ta coordinated octahedrally to 6 O from six phosphate tetrahedra, and layers of Rb^+ ions.

CHROMIUM(III) PHOSPHATE
α-$CrPO_4$

Z. anorg. Chem., **543**, 111-116.

Imma, 10.403, 12.898, 6.299, Z = 12, R = 0.038. Pairs of edge-sharing $Cr(2)O_6$ octahedra and single $Cr(1)O_6$ octahedra, linked by phosphate tetrahedra, with large channels along **a** and **b**.

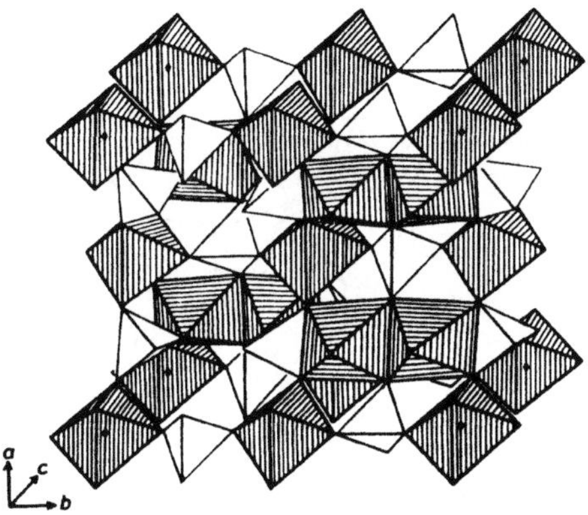

		x	y	z
Cr(1)	4a	0	0	0
Cr(2)	8g	1/4	0,63389	1/4
P(1)	4e	0	1/4	0,9175
P(2)	8g	1/4	0,42642	1/4
O(1)	8h	0	0,6504	0,9568
O(2)	8i	0,8773	1/4	0,7732
O(3)	16j	0,2762	0,3637	0,0552
O(4)	16j	0,1389	0,5098	0,2146

∢ 03-Cr2-03 = 180,0°
∢ 02-Cr2-02 = 83,2°

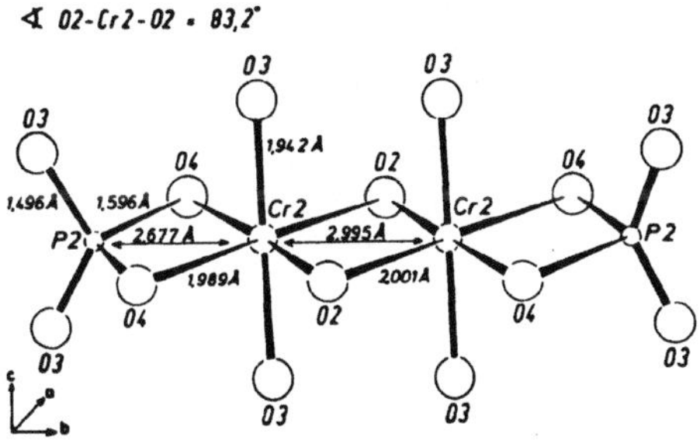

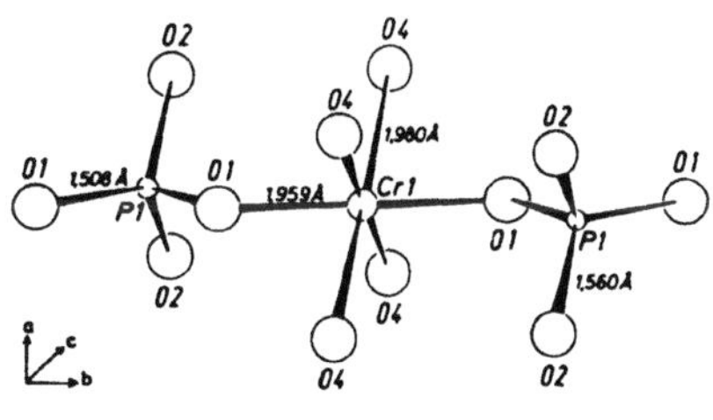

MANGANESE(II) DIHYDROGENPHOSPHATE DIHYDRATE
$Mn(H_2PO_4)_2.2H_2O$

Z. Kristallogr., **173**, 193-198 (1985).

$P2_1/n$, 5.3774, 10.170, 7.315, 94.79, Z = 2, R = 0.035. Isostructural with the Cd analogue (**39**A, 288), with (101) layers of corner-sharing $PO_2(OH)_2^-$ tetrahedra and MnO_6 octahedra; layers are linked by hydrogen bonds. P-O = 1.507, P-OH = 1.568, Mn-O = 2.165-2.227(2) A.

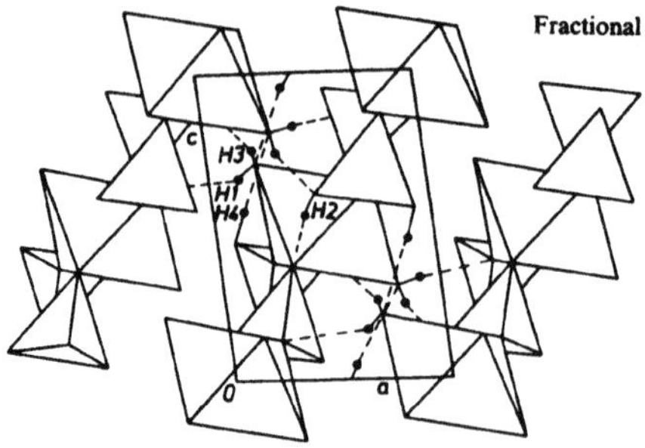

Fractional atomic coordinates for $Mn(H_2PO_4)_2 \cdot 2H_2O$ (multiplied by 10^5 for P atoms, and by 10^4 for O and H atoms).

	x	y	z
Mn	½	½	½
P	24510	21452	33840
O(1)	2341	3614	3628
O(2)	1646	1676	1471
O(3)	5183	1648	3902
O(4)	705	1437	4683
O(W)	2300	5359	7095
H(1)	1060	5740	6730
H(2)	5990	2160	4680
H(3)	1900	4580	7569
H(4)	1320	1390	5490

SWITZERITE
$Mn_3(PO_4)_2.7H_2O$

Amer. Min., **71**, 1221-1223 and 1224-1228.

$P2_1/a$, 8.528, 13.166, 11.812, 110.05, Z = 4, R = 0.089. Sheets of PO_4 tetrahedra and MnO_6 octahedra, linked by hydrogen bonding via water molecules. P-O = 1.51-1.56, Mn-O = 2.10-2.40(2) A. The tetrahydrate, which was previously described as switzerite (**45**A, 308), is now renamed metaswitzerite.

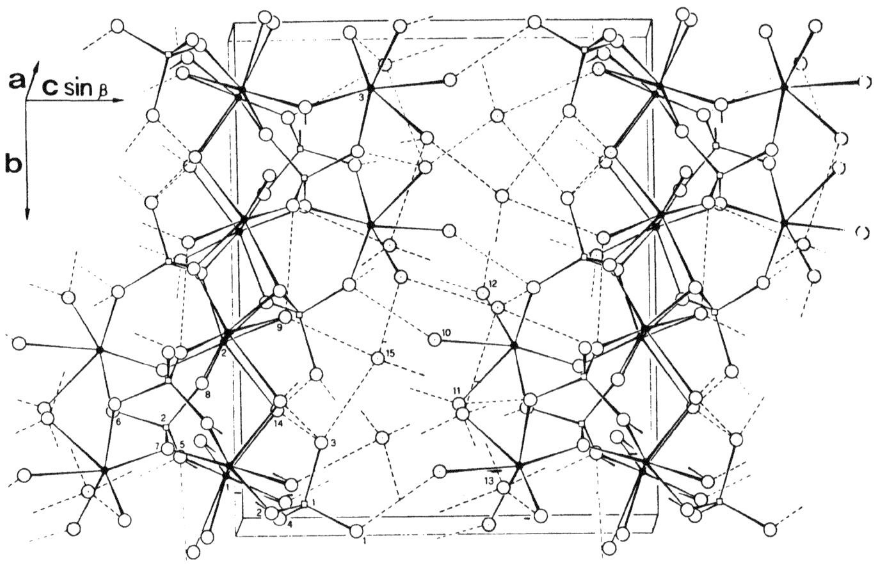

	x	y	z
Mn(1)	−0.0495	0.8831	−0.0218
Mn(2)	0.0561	0.6165	−0.0172
Mn(3)	0.5827	0.1243	0.3363
P(1)	0.3346	0.9442	0.1667
P(2)	0.1859	0.7886	−0.1746
O(1)	0.4013	0.0088	0.2783
O(2)	−0.0635	0.5325	0.0856
O(3)	0.8479	0.6696	0.1964
O(4)	0.6486	0.5262	0.1002
O(5)	0.0895	0.8708	−0.1374
O(6)	0.0832	0.7499	−0.2995
O(7)	0.6485	0.1651	0.1751
O(8)	0.2246	0.6997	−0.0841
O(9)	0.3135	0.5698	0.1174
O(10)	−0.0463	0.6111	−0.5191
O(11)	−0.2715	0.7280	−0.4574
O(12)	−0.2909	0.5156	−0.4036
O(13)	−0.1552	0.9346	0.6287
O(14)	0.0537	0.7477	0.0943
O(15)	0.4303	0.6651	0.3397

SODIUM MANGANESE(II) PHOSPHATE
$NaMnPO_4$

J. Solid State Chem., **61**, 379-383.

Pmnb, 6.9041, 9.0882, 5.1134, Z = 4, R = 0.044. Isostructural with $Na(Fe,Zn)PO_4$ and the mineral maricite (**40**A, 295; **43**A, 255) with PO_4 tetrahedra, linked by pairs of edge-sharing MnO_6 octahedra and by NaO_6 octahedra. P-O = 1.53-1.56, Mn-O = 2.08-2.40, Na-O = 2.33-2.72(1) A.

	10^4x	10^4y	10^4z
Mn	0	0	0
P	¼	8228	5298
Na	¼	8515	5257
O(1)	¼	8787	2436
O(2)	4313	8776	6697
O(3)	¼	6509	5414

POTASSIUM MANGANESE(II) PHOSPHATE
$KMn_4(PO_4)_3$

Kristallografija, **31**, 258-263 [Soviet Physics - Crystallography, **31**, 151-154].

Pmcn, 6.550, 16.028, 9.977, Z = 4, R = 0.034. Isostructural with the Fe analogue (**48**A, 380), with PO_4 tetrahedra linked by MnO_5 polyhedra and 6-coordinate K ions (distorted trigonal prism). P-O = 1.51-1.57, Mn-O = 2.06-2.34, K-O = 2.66-2.83(1) A.

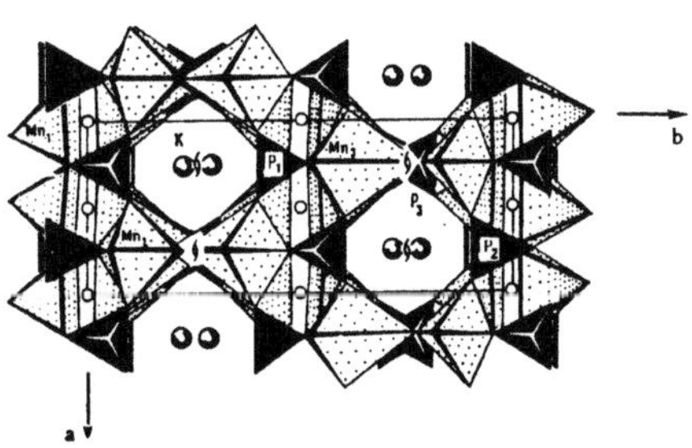

	x	y	z
Mn_1	−0.0012	0.05052	0.24552
Mn_2	0.25	0.87428	0.0728
Mn_3	0.25	0.86770	0.4668
P_1	0.25	0.4395	0.0237
P_2	0.25	0.4465	0.4510
P_3	0.25	0.7226	0.2950
K	0.25	0.2831	0.2401
O_1	0.25	0.5190	0.1154
O_2	0.25	0.0264	0.3767
O_3	0.0605	0.3885	0.0493
O_4	0.25	0.0518	0.1053
O_5	0.25	0.5397	0.4031
O_6	0.0600	0.4071	0.3941
O_7	0.25	0.8152	0.2557
O_8	0.0620	0.0770	0.2423
O_9	0.25	0.7284	0.4492

BOBFERGUSONITE
$Na_2Mn_5FeAl(PO_4)_6$

Canad. Miner., **24**, 599-604 and 605-614.

$P2_1/n$, 12.776, 12.488, 11.035, 97.21, Z = 4, R = 0.038. Structure related to that of wyllieite and alluaudite (**37**A, 296; **40**A, 247), but with different cation ordering giving a doubled axis. Two layers, one with MO_6 octahedral chains, M = Mn, Fe^{3+}, Al, cross-linked by phosphate tetrahedra, the other with two types of MO_6 octahedral chains, M = Na, Mn, not cross-linked within the layer.

	x	y	z
M1	0.13194	0.23856	-0.00190
M2	0.62989	0.23753	-0.00107
M3	0.29562	0.14643	0.72421
M4	0.79574	0.14588	0.72617
M5	0.46125	0.16153	0.27999
M6	0.96067	0.16220	0.27950
X1	0.24944	0.00103	-0.00188
X2	0	0	0
X3	1/2	0	0
X4	0.3734	0.0169	0.4972
X5	0.8738	0.0166	0.5004
P1	0.3816	0.2183	0.0033
P2	0.8826	0.2152	0.0067
P3	0.2013	0.1141	0.2578
P4	0.7041	0.1134	0.2598
P5	0.0597	0.0957	0.7384
P6	0.5588	0.1000	0.7400
O1	0.2980	0.2105	0.5460
O2	0.7979	0.2081	0.5458
O3	0.4655	0.2182	0.4503
O4	0.9637	0.2141	0.4457
O5	0.3290	0.3698	0.4182
O6	0.8267	0.3666	0.4151
O7	0.4426	0.3508	0.6102
O8	0.9444	0.3475	0.6048
O9	0.1125	0.1720	0.3151
O10	0.6180	0.1743	0.3182
O11	0.1343	0.1522	0.6592
O12	0.6357	0.1544	0.6613
O13	0.1098	0.4102	0.3729
O14	0.6116	0.4133	0.3713
O15	0.1714	0.4022	0.6215
O16	0.6709	0.4005	0.6197
O17	0.3011	0.1851	0.2765
O18	0.8056	0.1819	0.2753
O19	0.4591	0.1710	0.7370
O20	0.9596	0.1642	0.7351
O21	0.2776	0.5080	0.1785
O22	0.7773	0.5072	0.1757
O23	0.4651	0.4846	0.8149
O24	0.9717	0.4900	0.8150

SITE-CORRELATION TABLE FOR THE ALLUAUDITE STRUCTURE AND DERIVATIVES

Alluaudite	Wyllieite	Bobfergusonite
M(1)	M(1)	M1, M2
M(2)	M(2a)	M3, M4
	M(2b)	M5, M6
X(1)	X(1b)	X1
	X(1a)	X2, X3
X(2)	X(2)	X4, X5
P(1)	P(1)	P1, P2
P(2)	P(2a)	P3, P4
	P(2b)	P5, P6
O1	O(1a)	O1, O2
	O(1b)	O3, O4
O2	O(2a)	O5, O6
	O(2b)	O7, O8
O3	O(3a)	O9, O10
	O(3b)	O11, O12
O4	O(4a)	O13, O14
	O(4b)	O15, O16
O5	O(5a)	O17, O18
	O(5b)	O19, O20
O6	O(6a)	O21, O22
	O(6b)	O23, O24

BOBFERGUSONITE: INITIAL AND FINAL M- AND X-SITE POPULATIONS

Site	Population
	Initial
M1, M2	1.00 Mn
M3, M4	0.02 Zn + 0.04 Mg + 0.02 Fe^{2+} + 0.09 Fe^{3+} + 0.83 Mn
M5, M6	0.69 Al + 0.31 Fe^{3+}
X1	0.55 Mn + 0.19 Ca + 0.26 Na
X2, X3	1.00 Na
X4, X5	0.40 Na
	Final
M1	0.92(1) Mn
M2	0.91(1) Mn
M3, M4	0.82 Mn + 0.06(1) Fe^{3+} + 0.12(1) Al
M5	0.48(1) Fe^{3+} + 0.34(1) Al + 0.04 Fe^{2+} + 0.09 Mg + 0.05 Zn
M6	0.80(1) Al + 0.20(1) Fe^{3+}
X1	0.81 Mn + 0.19 Ca
X2	0.85(1) Na
X3	0.81(1) Na
X4	0.59(1) Na
X5	0.58(1) Na

IRON(II) HYDROGEN PHOSPHATE HYDRATE
$Fe_5H_2(PO_4)_4.4H_2O$ (I)

SODIUM IRON(II,III) PHOSPHATE
$Na_2Fe(III)Fe(II)_2(PO_4)_3$ (II)

Inorg. Chem., **25**, 2279-2280.

I, C2/c, 17.492, 9.038, 9.349, 96.23, at 173K, Z = 4, R = 0.039. II, C2/c, 11.998, 12.328, 6.500, 114.185, Z = 4, R = 0.024 [positional parameters not given]. Hureaulite and alluaudite type structures (**37**A, 296; **39**A, 287).

IRON PHOSPHATE HYDRATE
$Fe_3(PO_3OH)_4.4H_2O$

Amer. Min., **71**, 222-226.

$P2_1/n$, 5.152, 16.629, 8.749, 90.84, Z = 2, R = 0.049. Phosphate tetrahedra linked by octahedrally-coordinated Fe(II) and Fe(III). P-OH = 1.60, P-O = 1.52, Fe(II)-O = 2.06-2.22, Fe(III)-O = 1.93-2.10(1) A.

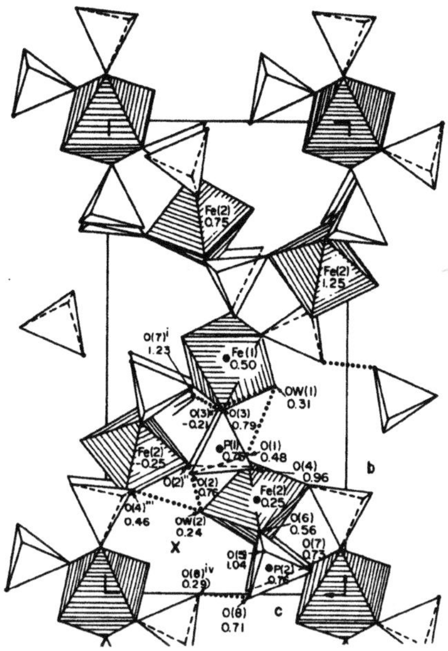

	x	y	z
Fe(1)=Fe^{2+}	0.5000	0.5000	0.5000
Fe(2)=Fe^{3+}	0.2539	0.2029	0.3683
P(1)	0.7454	0.3060	0.4938
P(2)	0.7645	0.0794	0.3054
O(1)	0.4768	0.2988	0.4210
O(2)	0.764	0.2648	0.6493
O(3)	0.7872	0.4008	0.5156
O(4)	0.9579	0.2772	0.3872
O(5)	1.0410	0.1061	0.3417
O(6)	0.563	0.1381	0.3559
O(7)	0.7337	0.0573	0.1362
O(8)	0.709	-0.0006	0.3986
OW(1)	0.310	0.4432	0.3059
OW(2)	0.237	0.1769	0.6010
H1W(1)	0.3641	0.3933	0.3536
H2W(1)	0.1271	0.4381	0.2923
H1W(2)	0.0612	0.2001	0.6442
H2W(2)	0.3762	0.1948	0.6852
H(1)	0.8738	0.4089	0.6177
H(2)	0.6415	0.0019	0.4965

OXONIUM IRON(III) PHOSPHATE HYDRATE
$(H_3O)Fe_3(HPO_4)_2(H_2PO_4)_6.4H_2O$

Acta Cryst., C**42**, 525–528.

C2/c, 16.797, 9.528, 17.609, 90.64, Z = 4, R = 0.091.

In the structure, iron(III) orthophosphate layers perpendicular to the *c* axis alternate with water layers. The iron phosphate layers contain FeO_6 octahedra which share vertices with $O_2P(OH)_2$ and $O_3P(OH)$ tetrahedra. Within this network are hollows that are occupied by oxonium ions. The pseudo symmetry of the layer is well described by the layer symmetry group 3*p*21. Neighbouring layers are connected by hydrogen-bonded water molecules. Fe—O distances vary from 1·94 (2) to 2·06 (2) Å. The present compound is isostructural with analogous thallium and aluminium orthophosphates.

	x	*y*	*z*
Fe(1)	0·29057	0·4728	0·24444
Fe(2)	½	0·8430	¼
P(1)	0·3365	0·7619	0·3434
O(11)	0·3487	0·6130	0·3135
O(12)	0·4037	0·8618	0·3169
O(13)	0·2553	0·8165	0·3231
O(15)	0·3461	0·7587	0·4315
P(2)	0·4427	0·5649	0·1441
O(21)	0·3881	−0·5304	0·1841
O(22)	0·4570	0·7043	0·1816
O(24)	0·5243	0·4846	0·1290
O(25)	0·4070	0·5841	0·0581
P(3)	0·1276	0·5134	0·3535
O(31)	0·1977	0·4789	0·3119
O(32)	0·0497	0·4952	0·3111
O(34)	0·1323	0·6722	0·3803
O(35)	0·1231	0·4357	0·4352
P(4)	0·3187	0·2066	0·3548
O(41)	0·3447	0·3195	0·3047
O(43)	0·2523	0·1149	0·3251
O(44)	0·3957	0·1136	0·3737
O(45)	0·2962	0·2638	0·4310
O(1)	0	0·766	¼
O(2)	0·2218	0·0457	0·0012
O(3)	0·4959	0·8012	−0·0013

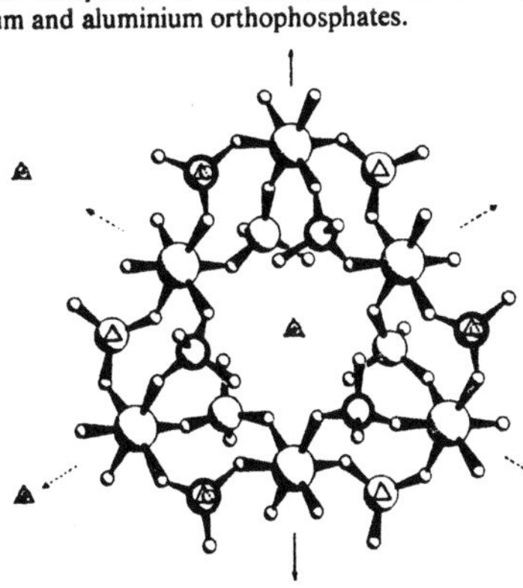

Projection of the oxonium iron phosphate layer along the *c* axis. Fe=large circles (z=0·25); P=medium circles (heavy circles for z = 0·35); O = small circles (+ = oxonium). Triangles [on P(*d*) and oxonium] indicate pseudo threefold axes; arrows indicate twofold axes (solid: parallel **b**; dotted: pseudo).

Selected averaged distances (Å) *and angles* (°)

Type	Range	Average	Number of terms	Standard deviation (r.m.s. deviation of the mean)
Fe—O(P)	1·93–2·05	1·995	9	0·015
P—O(Fe)	1·43–1·55	1·498	9	0·012
P—O(H)	1·51–1·63	1·587	7	0·016
O(Fe)—Fe—O(Fe)	84·3– 97·5	90	15	1·0
O(Fe)—P—O(Fe)	111·5–116·3	113·8	6	0·9
O(Fe)—P(1)—O(H)	104·3–109·0	107·2	3	1·5
O(Fe)—P(*d*)—O(H)	106·8–113·4	109·6	12	0·6
O(H)—P(*d*)—O(H)	100·1–103·2	101·7	3	0·9

POTASSIUM IRON(II,III) PHOSPHATE
$KFe_2(PO_4)_2$

Kristallografija, **31**, 906-912 [Soviet Physics - Crystallography, **31**, 537-541].

$P2_1/a$, 9.127, 10.032, 7.846, 115.87, Z = 4, R = 0.057. Corrugated layers of FeO_5 and FeO_6 polyhedra (probably Fe(II) and Fe(III), respectively), joined by phosphate tetrahedra, with K ions in channels.

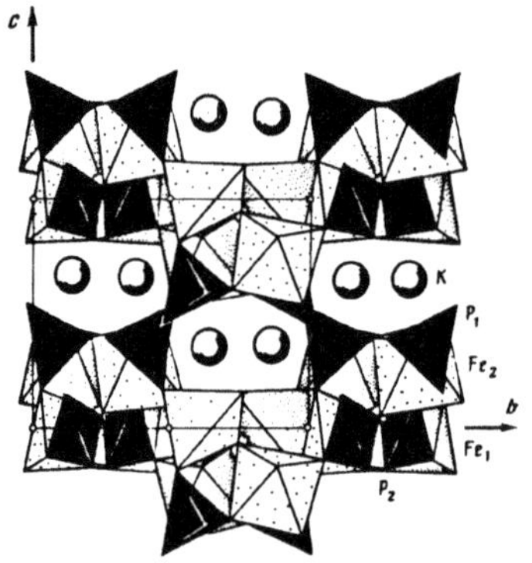

	x	y	z
Fe_1	0.44597	0,88213	0.0671
Fe_2	0.25782	0,12194	0,2709
P_1	0.6479	0.0944	0.3839
P_2	0.1367	0,8381	0.0122
K	0.4251	0.6362	0.3497
O_1	0.2532	0.2833	0.4068
O_2	0.4027	0.2635	0.8421
O_3	0,2415	0.4531	0.8869
O_4	0,4014	0.9684	0.8163
O_5	0.2587	0.2461	0.0508
O_6	0,2538	0.5086	0.5291
O_7	−0.0040	0.3704	0.4067
O_8	0,4861	0.6109	0.8488

POTASSIUM BARIUM IRON(III) PHOSPHATE
$KBaFe_2(PO_4)_3$

J. Solid State Chem., **62**, 16-25.

$P2_13$, 9.8732, 4.2K, Z = 4, neutron powder data. Langbeinite-type structure (**21**, 362; **45**A, 332).

		x	y	z
K/Ba(1)	4*a*	0.0672	0.0672	0.0672
K/Ba(2)	4*a*	0.2959	0.2959	0.2959
Fe(1)	4*a*	0.5855	0.5855	0.5855
Fe(2)	4*a*	0.8526	0.8526	0.8526
P	12*b*	0.6263	0.4606	0.2724
O(1)	12*b*	0.6531	0.5012	0.4185
O(2)	12*b*	0.7607	0.4845	0.1993
O(3)	12*b*	0.5796	0.3160	0.2594
O(4)	12*b*	0.5245	0.5537	0.2031

Bond Lengths (in Å) and Bond Angles (in Degrees) for $KBaFe_2(PO_4)_3$ at 4.2 K

Fe(1)–O(1)	1.964(8)	(×3)	P–O(1)	1.52(1)
Fe(1)–O(2)	2.014(8)	(×3)	P–O(2)	1.53(1)
Fe(2)–O(3)	2.016(8)	(×3)	P–O(3)	1.51(1)
Fe(2)–O(4)	2.012(8)	(×3)	P–O(4)	1.52(1)
O(1)–Fe(1)–O(1)′	94.3		O(1)–P–O(2)	104.9
O(2)–Fe(1)–O(2)′	91.6		O(1)–P–O(3)	112.5
O(1)–Fe(1)–O(2)	80.9		O(1)–P–O(4)	112.5
O(1)–Fe(1)–O(2)′	170.9		O(2)–P–O(3)	111.9
O(1)–Fe(1)–O(2)″	93.8		O(2)–P–O(4)	105.5
			O(3)–P–O(4)	109.4
O(3)–Fe(2)–O(3)′	92.9			
O(4)–Fe(2)–O(4)′	88.1			
O(3)–Fe(2)–O(4)	90.4			
O(3)–Fe(2)–O(4)′	88.5			
O(3)–Fe(2)–O(4)″	176.4			

SODIUM NICKEL PHOSPHATE
$Na_4Ni_7(PO_4)_6$

J. Solid State Chem., **62**, 105-111.

Cm, 10.550, 13.985, 6.398, 104.87, Z = 2, R = 0.041. Layers of PO_4 tetrahedra and NiO_6 octahedra, with Na ions in tunnels. P-O = 1.49-1.59, Ni-O = 1.96-2.35, Na-O = 2.36-3.22(1) A.

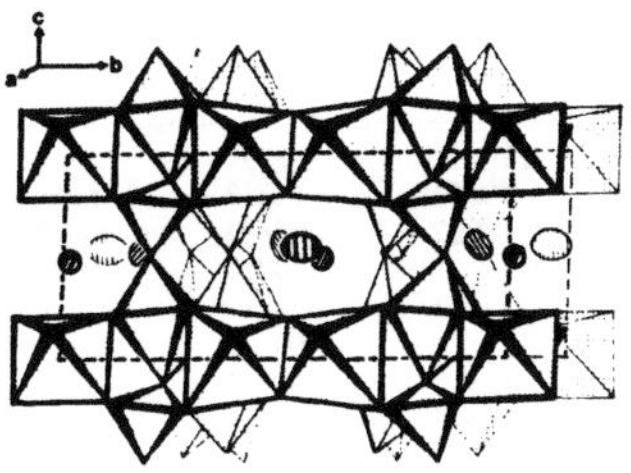

CAESIUM ZINC PHOSPHATE
$CsZnPO_4$

Ferrolectrics, **69**, 283-292.

Form III, room temperature, $P2_1/a$, 18.33, 5.45, 9.25, 90.1, Z = 8, R = 0.043. Form II, 260-310°C, $Pn2_1a$, 9.246, 5.468, 9.359, at 280°C, Z = 4, R = 0.041. Form I, above 310°C, Pnma, 9.191, 5.493, 9.406, at 340°C, Z = 4, R = 0.042. The structures are similar to those of NH_4LiSO_4 at -60°C (**44**A, 349), NH_4LiSO_4 at room temperature (**34**A, 305), and $CsMgPO_4$, respectively. All three structures contain frameworks of corner-sharing PO_4 and ZnO_4 tetrahedra, with Cs ions in cavities.

CADMIUM PHOSPHATE
β'-$Cd_3(PO_4)_2$

J. Chem. Res. (S), 170-171.

$P2_1/c$, 9.170, 10.321, 24.889, 120.8, Z = 12, R = 0.027. Isostructural with the Mn compound (**34**A, 330).

CADMIUM MAGNESIUM PHOSPHATE
β'-$(Cd,Mg)_3(PO_4)_2$

Gazz. Chim. Ital., **116**, 543-547.

$P2_1/c$, a = 9.156, 9.118, b = 10.304, 10.240, c = 8.274, 8.224 A, β = 120.85, 120.84°, for $Cd_{3-x}Mg_x$, x = 0.15, 0.47, Z = 4, R = 0.025, 0.027. Mg substitution disrupts the superstructure (preceding report), and changes **c** to **c**/3, giving a phase isostructural with β-$Zn_3(PO_4)_2$ (**32**A, 357). Mg substitutes randomly at Cd sites, and causes minor disorder of one phosphate group.

CADMIUM PHOSPHATE SILICATE
CADMIUM PHOSPHATE GERMANATE
$Cd_5(PO_4)_2SiO_4$, $Cd_5(PO_4)_2GeO_4$

Z. Kristallogr., **173**, 101-112 (1985).

Pnma, a = 6.692, 6.734, b = 14.99, 15.107, c = 10.140, 10.210 A, Z = 4, R = 0.034, 0.051. Isostructural with silicocarnotite (**37**A, 294), with some P/Si or Ge disorder.

	$Cd_5(PO_4)_2SiO_4$ (× 10^5)			$Cd_5(PO_4)_2GeO_4$		
	x	y	z	x	y	z
Cd(1)	5309	1/4	21397	5588	1/4	21691
Cd(2)	15844	−10867	16312	16009	−11048	16468
Cd(3)	35606	8962	5642	35672	8998	5399
P	734	1/4	58727	723	1/4	58765
A	35271	7160	36721	35416	7069	36679
O(1)	23400	1/4	60776	23444	1/4	61231
O(2)	−4694	1/4	43919	−3703	1/4	44079
O(3)	−8280	16445	64324	−7983	16490	64291
O(4)	40545	−1285	28118	40963	−1563	27776
O(5)	17946	5048	47091	17880	5189	47290
O(6)	29047	14825	26870	29494	14934	26158
O(7)	52989	11037	45478	53245	11321	45947

FLORENCITE
$CeAl_3(PO_4)_2(OH)_6$

INIS Atomindex, 16, 16:062246 (1985).

$R\bar{3}m$, 6.96, 16.33 A, powder data.

NEODYMIUM PHOSPHATE
$NdPO_4$

Kristallografija, **31**, 800-802 [Soviet Physics - Crystallography, **31**, 473-476; in this translation pages 475 and 476 are interchanged].

$P2_1/n$, 6.732, 6.930, 6.383, 103.61, Z = 4, R = 0.040. Monazite structure (**9**, 236). P-O = 1.515-1.555, Nd-10 O = 2.416-2.774, 3.155(8) A.

	x	y	z
Nd	0.28187	0.1583	0.0994
P	0.8020	0.3381	0.1132
O1	0.626	0.288	0.214
O2	0.749	0.496	−0.061
O3	0.976	0.394	0.306
O4	0.885	0.167	0.001

CALCIUM EUROPIUM(III) SULPHOAPATITE
$(Ca,Eu)_{10}(PO_4)_6S_{1+x}$

J. Solid State Chem., **63**, 267-277.

$P6_3$, 9.4560, 6.8409 (for x = 0, i.e. no Eu), Z = 1, powder data for four samples. Apatite structure, with Eu exclusively in the Ca(2) site, and S at 0,0,0.47 (with symmetry reduced to $P6_3$).

GADOLINIUM YTTERBIUM PHOSPHATE
$(Gd,Yb)PO_4$

Inorg. Chim. Acta, **118**, 173-176.

$I4_1/amd$, 6.865, 6.004, Z = 4, R = 0.054. Gd/Yb in 4(a): 0,3/4,1/8; P in 4(b): 0,1/4,3/8; O in 16(h): 0,0.428,0.216. Isostructural with related $LnPO_4$ compounds (**42A**, 352). P-4 O = 1.550, Gd/Yb-8 O = 2.279, 2.381 A.

URANYL DIHYDROGENPHOSPHATE TRIHYDRATE
$UO_2(H_2PO_4)_2.3H_2O$

Solid State Protonic Conduct. Fuel Cells Sens., 191-202 (1984, publ. 1985).

$P2_1/c$, 11.369, 13.899, 7.481, 113.67, Z = 4, R = 0.054. Phosphate groups linked by uranyl ions into ribbons along **a**.

LEAD(II) PYROPHOSPHATE
$Pb_2P_2O_7$

J. Solid State Chem., **62**, 371-376.

$P\bar{1}$, 6.914, 6.966, 12.751, 96.82, 91.14, 89.64, Z = 4, R = 0.069. Nearly-eclipsed pyrophosphate anions linked by 8- and 9-coordinate Pb cations. P-O = 1.44-1.65 A, P-O-P = 131°, Pb-O = 2.43-3.30 A.

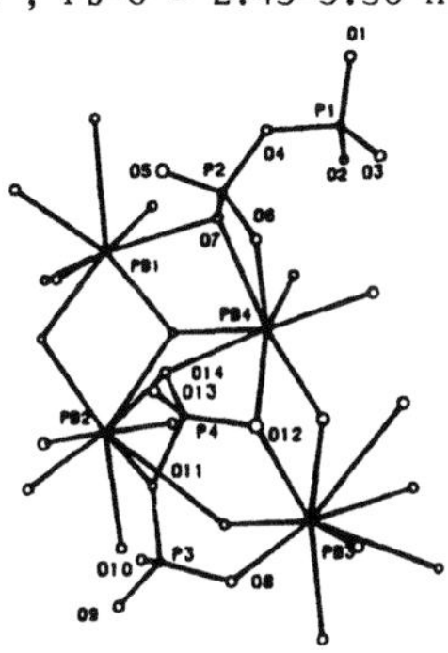

POTASSIUM CHLORO- and BROMOTETRAKIS(DIPHOSPHITO)DIPLATINATE TRIHYDRATE
$K_4[Pt_2(P_2O_5H_2)_4X].3H_2O$ (X = Cl, Br)

POTASSIUM DICHLORO- and DIBROMOTETRAKIS(DIPHOSPHITO)DIPLATINATE DIHYDRATE
$K_4[Pt_2(P_2O_5H_2)_4X_2].2H_2O$ (X = Cl, Br)

J. Amer. Chem. Soc., **105**, 4604-4607 (1983); Inorg. Chem., **25**, 409-412.

Chloro and bromo compounds, P4/mbm, a = 13.249, 13.236, c = 8.146, 8.191 A, Z = 2, R = 0.026, 0.058. Dichloro, $P\bar{1}$, 9.426, 9.534, 8.743, 99.07, 101.66, 118.94, Z = 1, R = 0.038. Dibromo, I4/mmm, 9.483, 15.672, Z = 2, R = 0.029. The chloro and bromo compounds contain -Pt-Pt-X-Pt-Pt-X- chains, with the Pt_2 pairs bridged by four diphosphite ligands; the Cl atom is disordered over two sites. The dichloro and dibromo compounds contain discrete binuclear anions. See also **50A**.

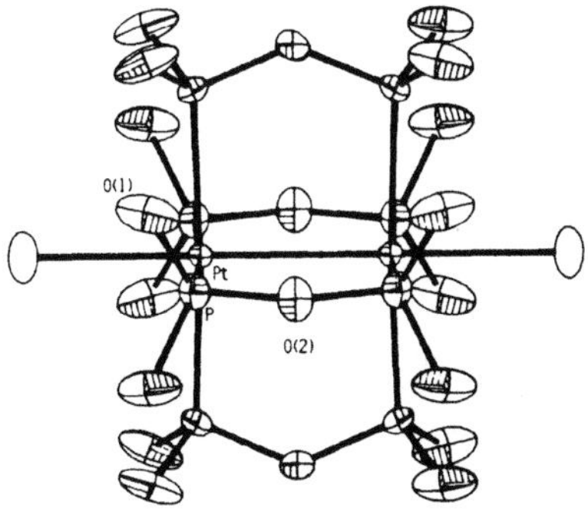

BARIUM TRIMETAPHOSPHATE TETRAHYDRATE
$Ba_3(P_3O_9)_2.4H_2O$

Z. Kristallogr., **174**, 219-224.

C2/m, 16.09, 8.368, 7.717, 95.38, Z = 2, R = 0.038. Cyclic anions, linked by 10- and 7- coordinate Ba ions; Ba(2) has occupancy 0.5. P-O = 1.604-1.611 (ring), 1.475-1.492(6) A (terminal), P-O-P = 123, 130°, Ba-O = 2.72-3.07 A.

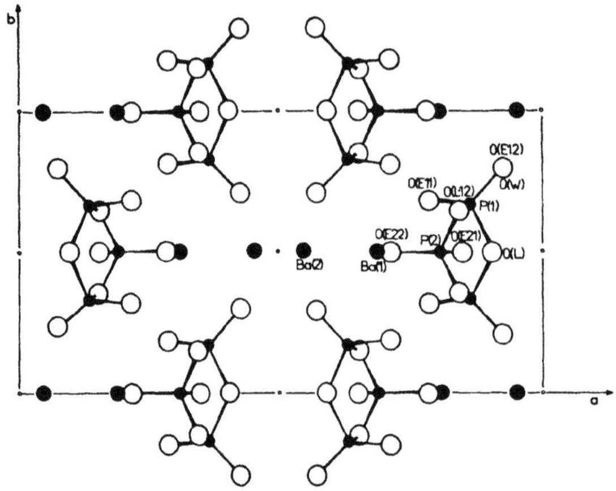

	x	y	z
Ba(1)	0.19029	0	0.95150
Ba(2)	0.04701	0	0.4317
P(1)	0.1357	0.3311	0.7000
P(2)	0.3080	0	0.5976
O(L12)	0.1545	0.3507	0.4997
O(E11)	0.2143	0.3147	0.8157
O(E12)	0.0735	0.2025	0.7069
O(L)	0.0926	$\frac{1}{2}$	0.7327
O(E21)	0.3465	0	0.7812
O(E22)	0.2158	0	0.575
O(W)	0.0788	0.2067	0.1154

SODIUM NICKEL TRIMETAPHOSPHATE HEXAHYDRATE
$Na_4Ni(P_3O_9)_2.6H_2O$

Acta Cryst., C**42**, 268-270.

P$\bar{1}$, 9.186, 8.020, 6.838, 89.17, 102.89, 98.03, Z = 1, R = 0.036. Cyclic anions and $Ni(H_2O)_6$ octahedra, linked by edge-sharing $NaO_5(H_2O)$ and NaO_6 octahedra. P-O = 1.60-1.62 (ring), 1.47-1.48 A (terminal), P-O-P = 125-129°, Ni-O = 2.04-2.08, Na-O = 2.35-2.60 A.

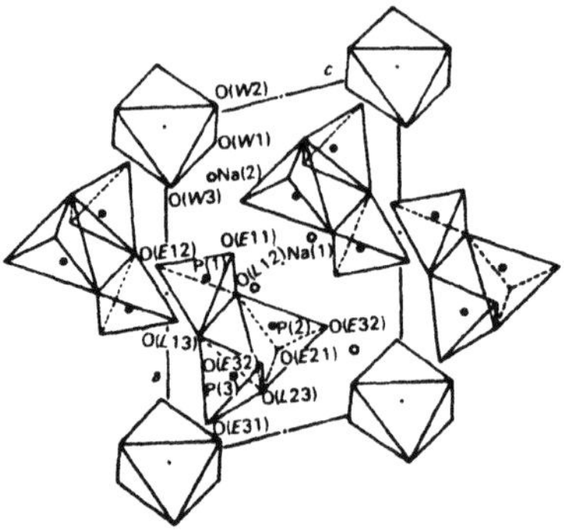

	x	y	z
Ni	0	0	0
P(1)	0,5067	0,7086	0,8454
P(2)	0,7046	0,1375	0,4587
P(3)	0,1987	0,5558	0,7233
Na(1)	0,4328	0,3006	0,6314
Na(2)	0,1904	0,4885	0,2151
O(E11)	0,4311	0,3948	0,2851
O(E12)	0,4015	0,2236	0,9612
O(L12)	0,4429	0,8638	0,7185
O(L13)	0,3539	0,6063	0,8777
O(E21)	0,2528	0,0339	0,5331
O(E22)	0,3207	0,7834	0,3592
O(L23)	0,1723	0,7376	0,6276
O(E31)	0,0811	0,5047	0,8380
O(E32)	0,2195	0,4416	0,5678
O(W1)	0,0849	0,1848	0,2216
O(W2)	0,9615	0,8417	0,2276
O(W3)	0,2110	0,9388	0,0189

AMMONIUM SODIUM MERCURY(II) TRIMETAPHOSPHATE
$(NH_4)_2Na_2Hg(P_3O_9)_2$

Acta Cryst., C**42**, 932-933.

C2/c, 13.524, 8.362, 14.390, 92.58, Z = 4, R = 0.030. Cyclic anions, linked by chains of edge-sharing MO_6 octahedra (M = disordered Hg,Na; (1) = 3.4Hg + 4.6Na, (2) = 0.6Hg + 3.4Na) and by ammonium ions. P-O = 1.60-1.63 (ring), 1.48-1.50 A (terminal), P-O-P = 124-131°, Hg,Na-O = 2.34-2.55, N-O = 2.81-3.10 A.

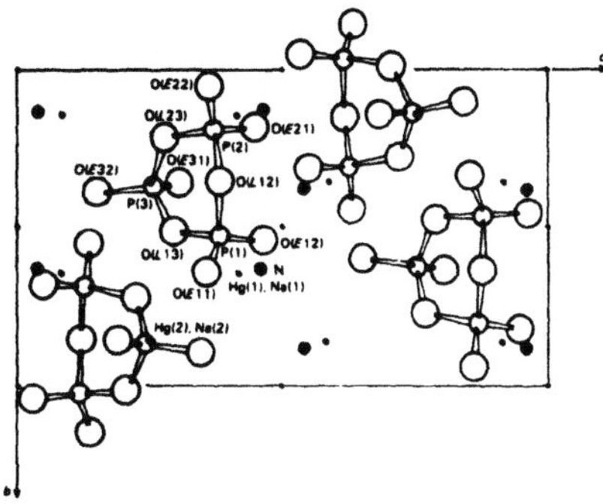

	x	y	z
Hg(1)Na(1)	0·1617 (4)	0·1427	0·4148
Hg(2)Na(2)	0	0·3545	¼
N	0·3489	0·1270	0·0406
P(1)	0·0810	0·0263	0·1183
P(2)	0·4038	0·1878	0·3719
P(3)	0·2720	0·3734	0·2539
O(E11)	−0·0036	0·1446	0·3604
O(E12)	0·1371	0·0472	0·0320
O(L12)	0·4678	0·3512	0·3802
O(E21)	0·3350	0·1856	0·4497
O(E22)	0·4728	0·0528	0·3610
O(L23)	0·3417	0·2194	0·2758
O(E31)	0·1807	0·3600	0·3083
O(E32)	0·2629	0·3960	0·1519
O(L13)	0·3416	0·5186	0·2944

POTASSIUM MERCURY(II) TRIMETAPHOSPHATE
$KHgP_3O_9$

Acta Cryst., C**42**, 930-931.

$P2_12_12_1$, 11.164, 12.46, 5.622, Z = 4, R = 0.042. Isostructural with $NaBaP_3O_9$ (38A, 321), but with Hg(II) in the Na site and K in the Ba site. Cyclic anions, linked by HgO_6 and KO_8 polyhedra. P-O = 1.60-1.64 (ring), 1.46-1.51 A (terminal), P-O-P = 124-129°, Hg-O = 2.26-2.47, K-O = 2.67-2.97 A.

	x	y	z
Hg	0·47903	0·13628	0·09635
K	0·1164	0·0795	0·0975
P(1)	0·4169	0·3719	0·3821
P(2)	0·3085	0·1862	0·5838
P(3)	0·1574	0·3572	0·3946
O(L13)	0·2831	0·4209	0·384
O(L12)	0·4083	0·2801	0·582
O(E11)	0·5014	0·4553	0·476
O(E12)	0·4449	0·3193	0·147
O(L23)	0·1882	0·2561	0·570
O(E21)	0·3176	0·1207	0·361
O(E22)	0·3149	0·1291	0·811
O(E31)	0·1277	0·3160	0·155
O(E32)	0·0687	0·4241	0·526

LITHIUM TETRAMETAPHOSPHATE PENTAHYDRATE
CAESIUM TETRAMETAPHOSPHATE TETRAHYDRATE
$Li_4P_4O_{12}.5H_2O$, $Cs_4P_4O_{12}.4H_2O$

Acta Cryst., C**42**, 129-131 and 131-133.

Li, C2/c, 17.073, 17.029, 13.554, 127.32, Z = 8, R = 0.034. Stacking of cyclic anions and $Li_8O_{16}(H_2O)_6$ groups; two of the five water molecules are zeolitic.

Cs, $P4_1$, 9.466, 21.933, Z = 4, R = 0.035. Alternating cyclic anions and Cs coordination polyhedra along **c**. The Rb salt is isostructural.

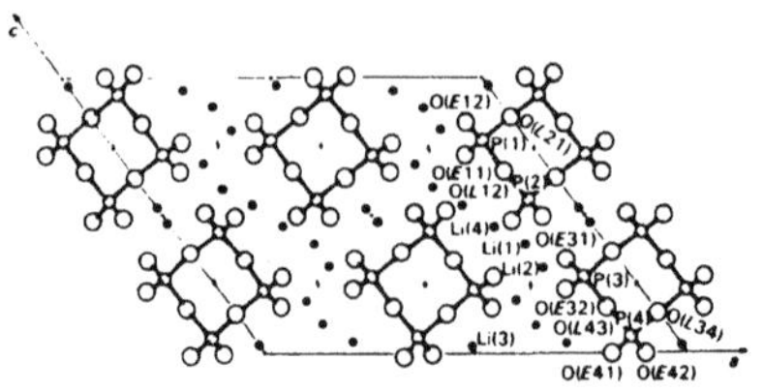

Projection along the *b* axis of the atomic arrangement of $Li_4P_4O_{12}.5H_2O$. The two independent P_4O_{12} anions are superimposed in this projection. Water molecules are not represented.

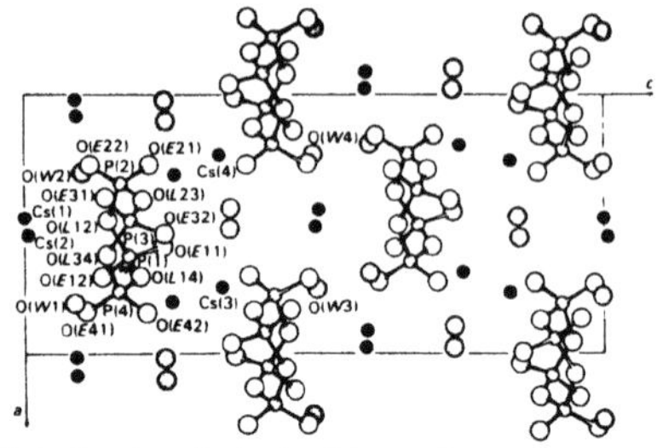

Projection along the *b* axis of the atomic arrangement of $Cs_4P_4O_{12}.4H_2O$.

STRONTIUM TETRAMETAPHOSPHATE HEXAHYDRATE
$Sr_2P_4O_{12}.6H_2O$

Acta Cryst., C**42**, 927-928.

$P\bar{1}$, 6.644, 7.365, 8.618, 101.62, 109.98, 95.65, Z = 2, R = 0.042. Cyclic anions linked by pairs of $SrO_5(H_2O)_3$ polyhedra. P-O = 1.61-1.62 (ring), 1.47-1.49 A (terminal), P-O-P = 129, 132°, Sr-O = 2.54-2.68 A.

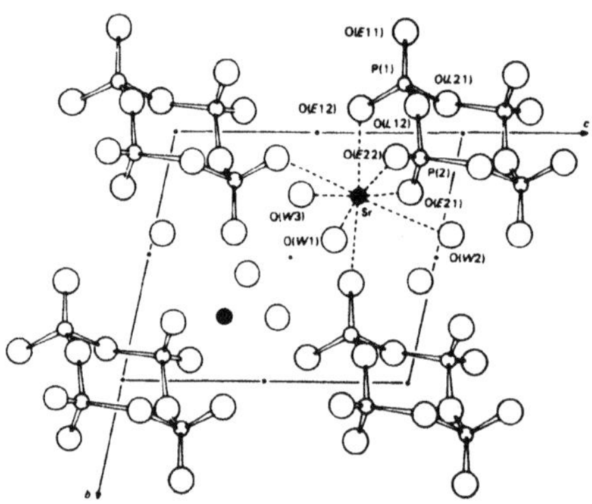

	x	*y*	*z*
Sr	0·59881	0·25201	0·69116
P(1)	0·2363	0·1960	0·2387
P(2)	0·8421	0·9056	0·1247
O(*L*12)	0·0216	0·8825	0·8242
O(*L*21)	0·7048	0·8855	0·9265
O(*E*11)	0·7417	0·5965	0·7254
O(*E*12)	0·6422	0·8946	0·6273
O(*E*21)	0·0062	0·2320	0·8588
O(*E*22)	0·3108	0·0872	0·7835
O(*W*1)	0·7340	0·5717	0·3568
O(*W*2)	0·7368	0·4064	0·0279
O(*W*3)	0·8182	0·2533	0·4895
H(*W*1)	0·74	0·50	0·27
H(*W*2)	0·75	0·52	0·43
H(*W*21)	0·78	0·39	0·12
H(*W*22)	0·83	0·50	0·04
H(*W*31)	0·03	0·73	0·47
H(*W*32)	0·79	0·15	0·39

NIOBYL TETRAPHOSPHATE
$(NbO)_2P_4O_{13}$

Izv. Akad. Nauk SSSR, Neorg. Mater., **22**, 1364-1368.

$P\bar{1}$, 6.586, 8.400, 10.842, 106.37, 90.35, 89.97, Z = 1, R = 0.043. Discrete $P_4O_{13}^{6-}$ anions of four corner-sharing tetrahedra, linked by Nb ions coordinated octahedrally to 5 O from five anions and one niobyl oxygen atom.

HEXAAMMINECOBALT(III) TETRAPOLYPHOSPHATE PENTAHYDRATE
$[Co(NH_3)_6]_2P_4O_{13}.5H_2O$

Z. anorg. Chem., **543**, 152-160.

$P2_1/c$, 15.478, 12.636, 14.511, 112.23, Z = 4, R = 0.052. Tetrapolyphosphate anions, $[O_3P\text{-}O\text{-}PO_2\text{-}O\text{-}PO_2\text{-}O\text{-}PO_3]^{6-}$, octahedral $[Co(NH_3)_6]^{3+}$ cations, and water molecules. P-O = 1.56-1.65 (bridging), 1.44-1.51 A (terminal).

CALCIUM METAARSENATE
$CaAs_2O_6$

Z. Kristallogr., **173**, 313-317 (1985).

P312, 4.824, 5.080, Z = 1, powder data. Ca in 1(a): 0,0,0; As(1) in 1(d): 1/3,2/3,1/2; As(2) in 1(f): 2/3,1/3,1/2; O in 6(ℓ): 0.376,0.008,0.283. Isostructural with $PbSb_2O_6$ (**8**, 156), with octahedral coordinations for Ca and As. As-O = 1.86-1.91, Ca-O = 2.30(4) A.

BARIUM HEXAARSENATE(III)
$Ba_2As_6O_{11}$

Z. anorg. Chem., **538**, 45-52.

$P\bar{1}$, 7.059, 8.951, 5.258, 93.64, 97.43, 71.49, Z = 1, R = 0.030. Polyarsenate anions with double chains, As-O = 1.69-1.89 A, linked into sheets by longer As...O interactions, 2.68 A. Ba-O = 2.71-3.14 A.

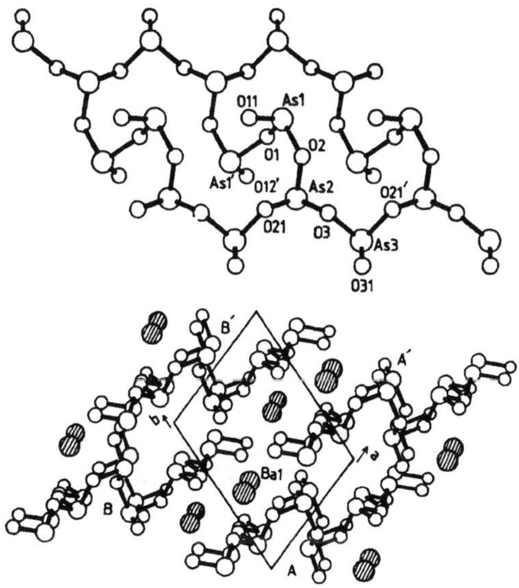

	x	y	z
Ba(1)	0.3061	0.4538	0,7649
As(1)	0,6001	0,0990	0,2277
As(2)	0.1272	0,1729	0.2603
As(3)	−0,1341	0,2595	0,7319
O(1)	0,5362	−0,0779	0,0973
O(2)	0,3617	0,2085	0,3476
O(3)	0,0131	0,3134	0,4966
O(21)	0,0344	0,3095	0,0035
O(31)	−0,3168	0,4356	0.7147
O(11)	0,6031	0,2145	−0,0027
O(12)	0,6523	0,2511	0,1268

SCHULTENITE
$PbHAsO_4$

LEAD(II) HYDROGEN PHOSPHATE
$PbHPO_4$

Tschermaks Min. Petr. Mitt., **35**, 157-166.

P2/c, a = 4.859, 4.6838, b = 6.756, 6.6451, c = 5.843, 5.7817 A, β = 95.40, 97.138°, Z = 2, R = 0.043, 0.032. Previous study in **13**, 309. XO_4 tetrahedra linked across a centre of symmetry by a short O...O hydrogen bond (2.46 A); a transition at 312K from Pc to P2/c corresponds to an ordering of the H atom in the hydrogen bond. P-4 O = 1.53, 1.56, As-4 O = 1.68, 1.70, Pb-6 O = 2.35-2.86(1) A.

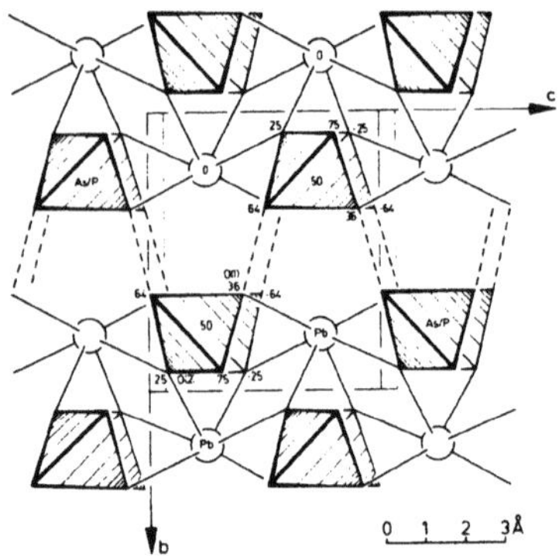

LEAD(II) COPPER(I) ARSENATE
$PbCuAsO_4$

Acta Cryst., C**42**, 774-776.

$P\bar{1}$, 4.832, 5.837, 7.995, 78.68, 74.95, 84.04, Z = 2, R = 0.030. AsO_4 tetrahedra, linked by Cu^+ ions with linear 2-coordination and Pb^{2+} ions with (4+2)-coordination.

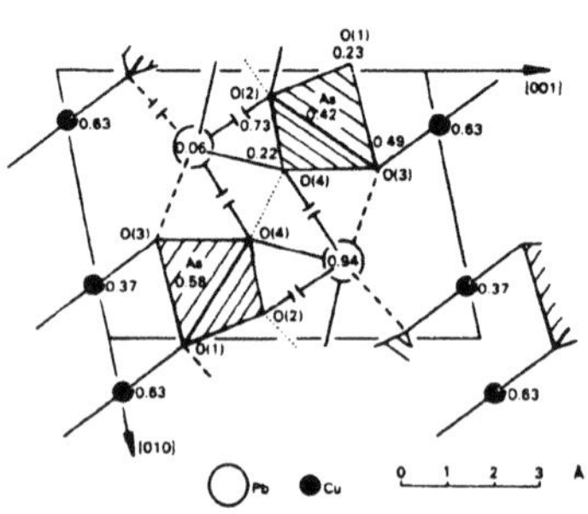

	x	y	z
Pb	0·9387	0·7103	0·6760
Cu	0·6346	0·1969	0·0043
As	0·4218	0·2030	0·6866
O(1)	0·7672	0·0289	0·1919
O(2)	0·7313	0·0953	0·5642
O(3)	0·5091	0·6306	0·1787
O(4)	0·7762	0·6262	0·4391

Some selected bond lengths (Å) *with e.s.d.'s in parentheses for* $PbCuAsO_4$

Pb–O(2)	2·334 (7)	Cu–O(1)	1·848 (8)
Pb–O(4)	2·380 (7)	Cu–O(3)	1·855 (8)
Pb–O(2')	2·462 (6)	As–O(1)	1·699 (7)
Pb–O(4')	2·468 (6)	As–O(2)	1·696 (7)
Pb–O(1)	2·704 (7)	As–O(3)	1·697 (7)
Pb–O(3)	2·945 (7)	As–O(4)	1·681 (6)

LEAD(II) COPPER(II) DIARSENATE
$PbCuAs_2O_7$

Mh. Chem., **117**, 1343-1348.

$P2_1/n$, 5.553, 8.404, 13.011, 91.61, Z = 4, R = 0.036. Diarsenate anions with two corner-sharing AsO_4 tetrahedra, linked into ($\bar{1}$01) sheets by Cu ions with (4+1) tetragonal pyramidal coordination; sheets are linked by 8-coordinate Pb ions. As-O = 1.72, 1.77 (bridging), 1.67-1.69(1), Cu-O = 1.94-2.02, 2.24(1), Pb-O = 2.48-3.00(1) A.

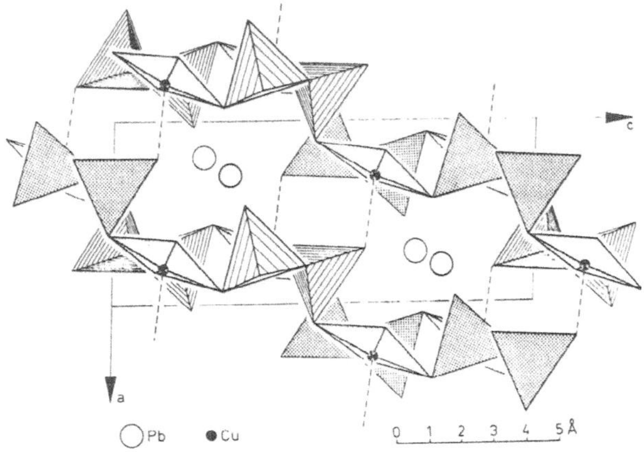

	x	y	z
Pb	0.70796	0.31962	0.72291
As1	0.2490	0.5270	0.8418
As2	0.3273	0.8093	0.9913
Cu	0.1991	0.1358	0.8832
O1	0.2085	0.1036	0.5368
O2	0.3233	0.3999	0.6049
O3	0.8015	0.1759	0.8950
O4	0.8749	0.3487	0.5161
O5	0.0638	0.1203	0.7378
O6	0.5395	0.0493	0.6936
O7	0.1540	0.8375	0.6526

LEAD COPPER ARSENITE CHLORIDE
$Pb_6Cu(AsO_3)_2Cl_7$

Mh. Chem., **117**, 1257-1261.

$R\bar{3}$, 9.8614, 17.089, Z = 3, R = 0.050. Complex $Pb_6(AsO_3)_2Cl_6$ layers linked via Cu^+ and Cl^- ions. A novel element within the structure is a Cl_3Cu-AsO_3 group, Cu-As = 2.34, Cu-Cl = 2.44, As-O = 1.76 A.

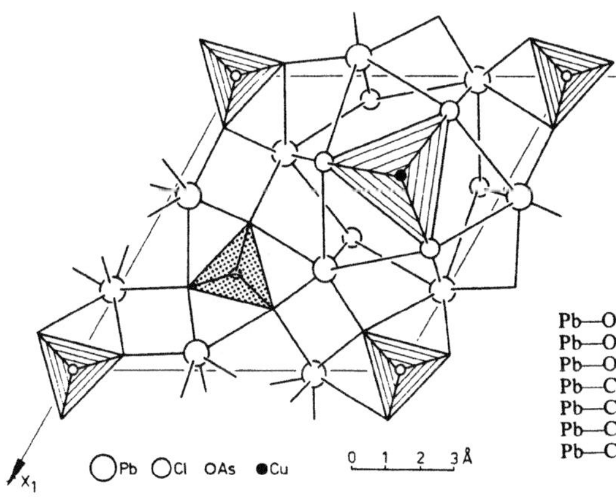

		x	y	z
Pb	18f	0.6535	0.5944	0.2280
As	6c	0	0	0.2352
Cu	6c	1/3	2/3	0.0390
Cl1	3a	0	0	0
Cl2	18f	0.2936	0.4123	0.2543
O	18f	0.5076	0.7141	0.8520

Pb—O	2.34(1) Å	As—O	1.76(1)	3 ×
Pb—O′	2.39(1)	As—Cu	2.34(1)	
Pb—O″	2.97(1)			
Pb—Cl1	3.197(4)	Cu—Cl2	2.44(1)	3 ×
Pb—Cl2	2.828(4)	Cu—As	2.34(1)	
Pb—Cl2′	3.106(4)			
Pb—Cl2″	3.235(5)			

IRON(III) ARSENATE
$FeAsO_4$

Nature, **320**, 46-48.

$P2_1/n$, 7.5636, 8.0795, 5.0117, 104.47, Z = 4, neutron powder data. Isostructural with $LiAlH_4$ (**32**A, 149). As-4 O = 1.66-1.71, Fe-5 O = 1.88-2.06(3) A.

LAMMERITE
$Cu_3(AsO_4)_2$

Amer. Min., **71**, 206-209.

$P2_1/a$, 5.079, 11.611, 5.394, 111.72, Z = 2, R = 0.031. α-PbO_2-type chains along [100] of edge-sharing distorted CuO_6 octahedra, forming (001) layers which are linked by arsenate tetrahedra and additional Cu octahedra. As-O = 1.67-1.71, Cu-O = 1.93-2.03, 2.28-2.92 A. Another synthetic form has been described (**33**A, 419).

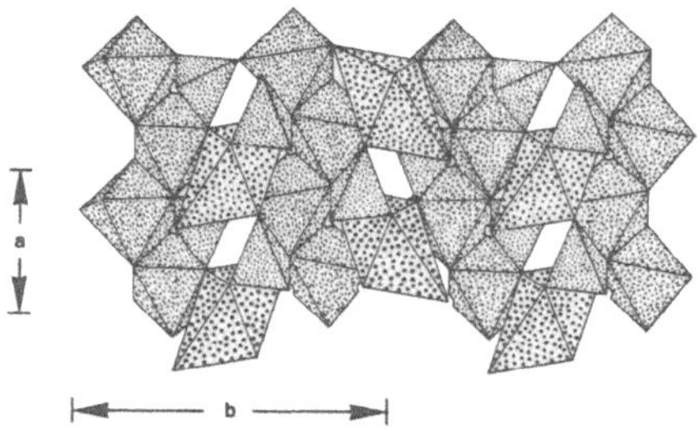

	x	y	z
Cu(1)	0	1/2	1/2
Cu(2)	0.2462	0.8367	0.8563
As	-0.1111	0.8851	0.2625
O(1)	0.1921	0.9173	0.2203
O(2)	-0.0946	0.8157	0.5390
O(3)	0.1717	0.6871	0.9904
O(4)	0.2893	0.9902	0.7332

COPPER(II) DECAHYDROGENTETRAARSENATE
$CuH_{10}(AsO_4)_4$

Acta Cryst., C**42**, 391-393.

$P\bar{1}$, 5.392, 7.632, 8.298, 105.87, 97.63, 93.6, Z = 1, R = 0.039. Chains along **c** of AsO_4 tetrahedra and CuO_6 distorted octahedra.

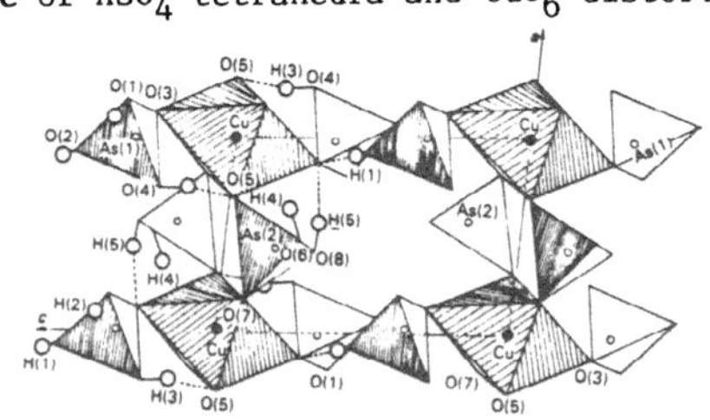

	x	y	z
As(1)	0,04005	0,29141	0,71334
As(2)	0,42009	0,81620	0,79219
Cu	0	0	0
O(1)	0,8742	0,4774	0,7376
O(2)	0,1083	0,2459	0,5116
O(3)	0,8860	0,1132	0,7419
O(4)	0,3125	0,3711	0,8505
O(5)	0,6687	0,8467	0,9436
O(6)	0,4231	0,6126	0,6415
O(7)	0,1439	0,7967	0,8593
O(8)	0,4192	0,9785	0,6857
H(1)	0,106	0,124	0,450
H(2)	0,983	0,570	0,725
H(3)	0,296	0,310	0,932
H(4)	0,569	0,563	0,674
H(5)	0,570	0,038	0,694

	Å		°
		O(1)–As(1)–O(2)	104,9 (2)
As(1)–O(1)	1,705 (4)	O(1)–As(1)–O(3)	113,4 (2)
As(1)–O(2)	1,709 (5)	O(1)–As(1)–O(4)	104,0 (2)
As(1)–O(3)	1,638 (5)	O(2)–As(1)–O(3)	111,2 (2)
As(1)–O(4)	1,699 (4)	O(2)–As(1)–O(4)	108,0 (2)
		O(3)–As(1)–O(4)	114,5 (2)
		O(5)–As(2)–O(6)	109,4 (2)
As(2)–O(5)	1,668 (3)	O(5)–As(2)–O(7)	114,0 (2)
As(2)–O(6)	1,710 (4)	O(5)–As(2)–O(8)	113,7 (2)
As(2)–O(7)	1,669 (4)	O(6)–As(2)–O(7)	104,2 (2)
As(2)–O(8)	1,707 (5)	O(6)–As(2)–O(8)	105,4 (2)
		O(7)–As(2)–O(8)	109,5 (2)
Cu–O(3)	2,538 (5)	O(3)–Cu–O(5)	89,8 (2)
Cu–O(5)	1,998 (3)	O(3)–Cu–O(7)	90,9 (2)
Cu–O(7)	1,943 (3)	O(5)–Cu–O(7)	90,8 (2)

X–H···*Y*	*X*–H	*X*···*Y*	*X*–H···*Y*
O(1)–H(2)···O(7)	0,93 (11)	2,624 (5)	137 (9)
O(2)–H(1)···O(3)	0,93 (11)	2,968 (6)	164 (10)
O(4)–H(3)···O(5)	0,93 (12)	2,687 (6)	163 (10)
O(6)–H(4)···O(1)	0,93 (13)	2,805 (6)	176 (12)
O(8)–H(5)···O(3)	0,89 (08)	2,590 (5)	165 (8)

POTASSIUM SULPHITE
K_2SO_3

Acta Chem. Scand., A**40**, 479-480.

$P\bar{3}m1$, 5.915, 6.968, Z = 2, R = 0.028. K(1) in 1(a): 0,0,0; K(2) in 1(b): 0,0,1/2; K(3) in 2(d): 1/3,2/3,0.6746; S in 2(d): 1/3,2/3,0.1835; O in 6(i): 0.1977,0.3954,0.2700. Trigonal pyramidal sulphite anions, linked by 6- and 9-coordinate K. S-O = 1.515(1) A, O-S-O = 105.2°, K-O = 2.583-3.143(1) A.

MAGNESIUM SULPHITE HEXAHYDRATE
$MgSO_3.6H_2O$

Acta Cryst., B**42**, 552-557.

R3, a = 5.911 A, α = 96.25°, at 120K, Z = 1, Mo and neutron radiations, R = 0.018, 0.038. Structure as previously described (**39**A, 301; **51**A, 299), with deformation density now examined.

SODIUM HYDROXOSULPHITOMETALLATE MONOHYDRATES
$NaM_2OH(SO_3)_2.H_2O$ (M = Mg, Mn, Fe, Co, Zn)

Z. Naturforsch., **41**B, 852-858.

$P\bar{1}$, 8.905, 7.845, 6.315, 113.28, 110.41, 98.00, for Mg compound (similar values for others), Z = 2, R = 0.025-0.068. MO_6 octahedra which share edges and corners to form (010) layers, which are linked by Na ions and hydrogen bonds.

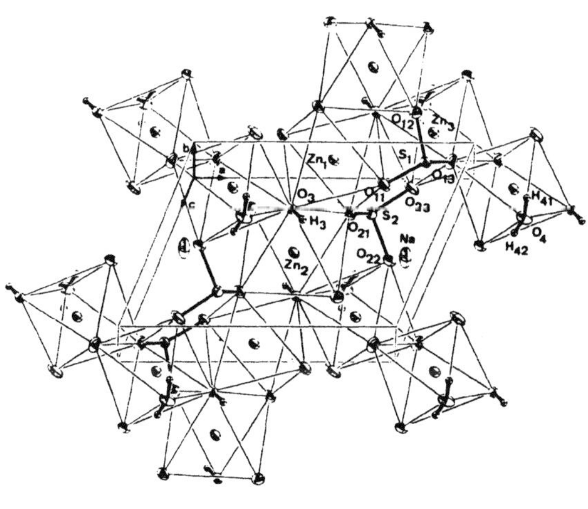

AMMONIUM HYDROGENDISULPHATE (DEUTERATED)
$(ND_4)_3D(SO_4)_2$

Acta Cryst., C**42**, 776-780.

Phase III', P2, 10.087, 5.835, 15.542, 101.71, Z = 4, R = 0.028, at 189K. Phase VI, P1, 10.123, 5.846, 15.476, 89.98, 101.78, 90.01, Z = 4, R = 0.035, 233K. Both phases have pseudo-P2/c symmetry, with ammonium and sulphate tetrahedra linked by normal hydrogen bonds.

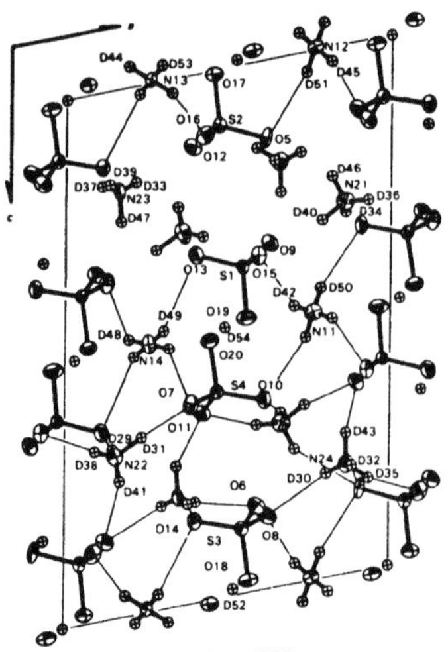
phase (III')

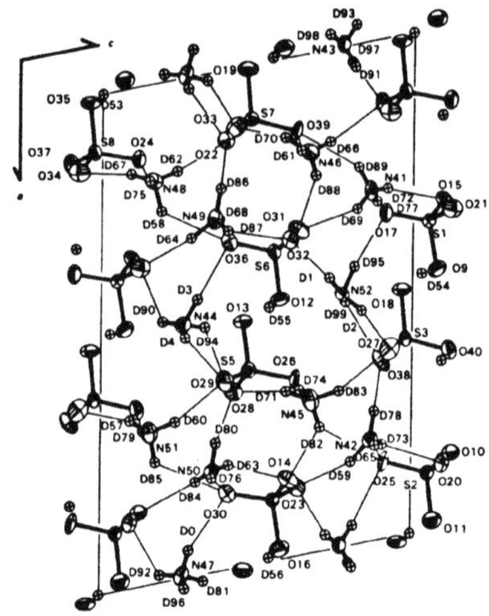
phase (VI)

LITHIUM SULPHATE MONOHYDRATE
$Li_2SO_4.H_2O$

J. Chem. Phys., **85**, 5221-5227.

$P2_1$, 5.4553, 4.8690, 8.1761, 107.337, at 298K, Z = 2, R = 0.017, 0.016, at 80, 298K. Structure as previously described (**51**A, 301), with electron-density distribution now studied.

RUBIDIUM LITHIUM SULPHATE (PHASE IV)
$RbLiSO_4$

Acta Cryst., B**42**, 11-16.

Pn (**c** unique), 9.157, 5.316, 43.654, $\gamma = 89.97^o$, at 446K, Z = 20, R = 0.091. Fivefold commensurate superstructure resulting mainly from rotation of sulphate tetrahedra along **c** (see **45**A, 329; **46**A, 346).

CAESIUM HYDROGEN SULPHATE
$CsDSO_4$

Kristallografija, **31**, 450-454 [Soviet Physics - Crystallography, **31**, 264-267].

$P2_1/c$, 7.789, 8.146, 7.726, 110.87, Z = 4, R = 0.066. Sulphate tetrahedra linked by columns of Cs coordination polyhedra; deuterium atoms were not located, but probably link O(1) and O(2) in a hydrogen bond. A model for the phase transition in $CsHSO_4$ is proposed.

($\times 10^4$, $\times 10^5$ for Cs)

	x	y	z
Cs	78433	12914	79359
S	2478	1273	7205
O_1	3340	711	5923
O_2	4102	2207	8685
O_3	8072	47	1869
O_4	1039	2454	6409

MAGNESIUM BISULPHATE
$Mg(HSO_4)_2$

Kristallografija, **31**, 1220-1221 [Soviet Physics - Crystallography, **31**, 721-722].

$P2_1/n$, 7.299, 8.273, 4.940, 99.90, Z = 2, R = 0.029. Tetrahedral $SO_3(OH)^-$ anions linked in chains along **b** by hydrogen bonds, with further linking via MgO_6 octahedra, S-OH = 1.556, S-O = 1.441-1.453, Mg-O = 2.056-2.075(1) A.

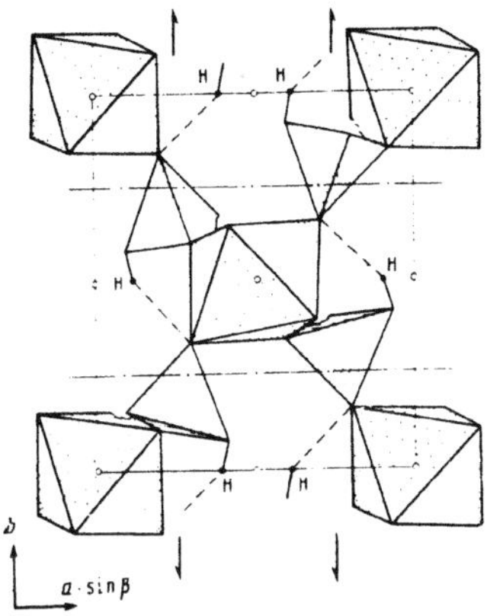

	x	y	z
Mg	0	0	0
S	0.23456	0.17272	0.56458
O_1	0.2045	0.8401	0.9251
O_2	0.0809	0.1509	0.7114
O_3	0.0934	0.5808	0.7787
O_4	0.1974	0.1098	0.2885
H	0.110	0.500	0.753

LEONITE
$K_2Mg(SO_4)_2.4H_2O$

Z. Kristallogr., **173**, 75-79 (1985).

C2/m, 11.769, 9.539, 9.889, 95.31, Z = 4, R = 0.031. Isostructural with the Mn analogue (**7**, 32, 140; **24**, 383; **26**, 447; **33**A, 369), with sulphate tetrahedra connected by 6-coordinate Mg, 9- and 10-coordinate K, and hydrogen bonds; one of the sulphate groups shows slight disorder.

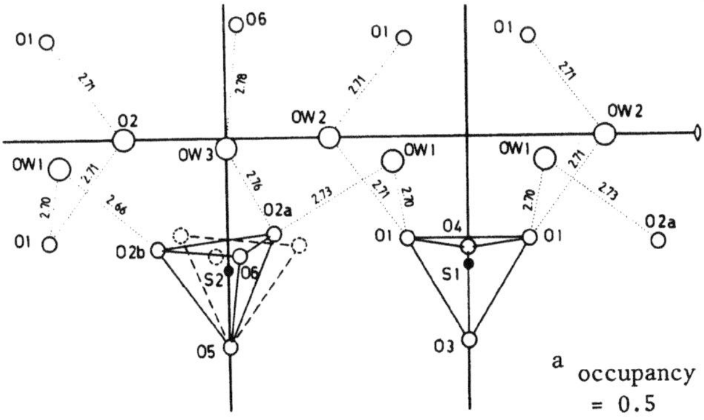

a occupancy = 0.5

		x	y	z
K	8j	0,16770	0,26352	0,24977
Mg1	2a	0,0	0,0	0,0
Mg2	2d	0,0	0,5	0,0
S1	4i	0,20766	0,5	−0,0433
S2	4i	0,21118	0,0	−0,4808
O1	8j	0,1591	0,3736	−0,0433
O2a*	8j	0,1519	0,0975	−0,4004
O2b*	8j	0.1737	0,1467	−0,4515
O3	4i	0,1687	0,0	0,0543
O4	4i	0,1821	0,5	0,0978
O5	4i	0,1648	0,5	0,4387
O6*	8j	0,1891	0,0236	0,3732
OW1	8j	0,0400	0,3429	−0,3585
OW2	4g	0,0	0,2184	0,0
OW3	4i	0,0149	0,0	−0,2090
H1	8j	0,073	0,273	−0,387
H2	8j	0,074	0,359	−0,275
H3	8j	0,049	0,271	−0,024
H4	4i	0,075	0,0	−0,268
H5	4i	0,042	0,0	0,275

LANGBEINITE-TYPE SULPHATES
$K_2M_2(SO_4)_3$

Phys. Chem. Miner., **13**, 17-24.

M = Mg, Ni, Co, Zn, Ca, $P2_13$ and M = Ca, $P2_12_12_1$. SO_4 tetrahedra, distorted MO_6 octahedra (**21**, 362; **45A**, 332).

BARIUM SULPHATE
$BaSO_4$

Zairyo Gijutsu, **4**, 409-414.

Pbnm, 7.144, 8.865, 5.445, Z = 4, R = 0.072, for synthetic material. Structure similar to that of natural barite (**1**, 343).

JURBANITE
$Al(SO_4)(OH).5H_2O$

Z. Kristallogr., **173**, 33-39 (1985).

$P2_1/n$, 8.3965, 12.479, 8.1549, 101.917, Z = 4, R = 0.047. Structure as previously described for synthetic material (**27**, 625), with pairs of edge-sharing $Al(OH)_2(H_2O)_4$ octahedra and isolated SO_4 tetrahedra, connected by hydrogen bonds. Al-O = 1.85-1.91, S-O = 1.45-1.49, O-H...O = 2.62-2.83 A.

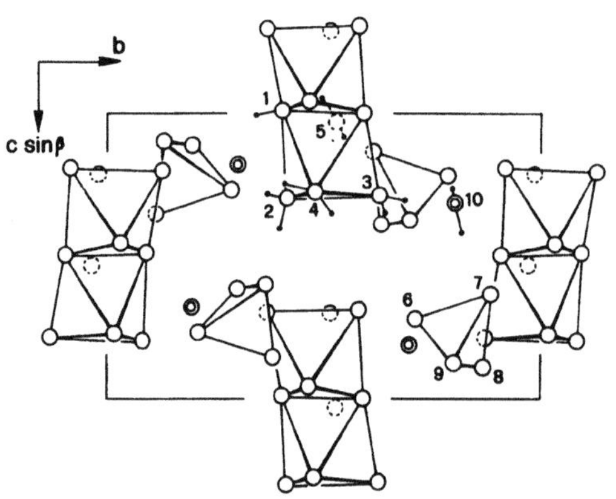

	x	y	z
S	0.1358	0.8151	0.786[illegible]
Al	0.1158	0.5106	0.153[illegible]
O(1) OH	0.0325	0.4075	−0.004[illegible]
O(2) Ow	0.2569	0.4175	0.300[illegible]
O(3) Ow	0.1932	0.6288	0.291[illegible]
O(4) Ow	−0.0346	0.4814	0.292[illegible]
O(5) Ow	0.2890	0.5316	0.036[illegible]
O(6)	0.1899	0.7123	0.724[illegible]
O(7)	0.0789	0.8842	0.639[illegible]
O(8)	0.2773	0.8675	0.896
O(9)	0.0069	0.8000	0.878
O(10) Ow	0.0223	0.8027	0.314
H(1)	0.001	0.345	0.013
H(21)	0.238	0.400	0.410
H(22)	0.354	0.383	0.290
H(31)	0.136	0.688	0.310
H(32)	0.299	0.643	0.354
H(41)	−0.100	0.520	0.355
H(42)	−0.090	0.410	0.285
H(51)	0.395	0.557	0.080
H(52)	0.320	0.478	−0.040
H(101)	−0.097	0.800	0.290
H(102)	0.045	0.818	0.424

VANADIUM(II) SULPHATE HEXAHYDRATE
$VSO_4.6H_2O$

Inorg. Chem., **25**, 3423-3428.

C2/c, 10.081, 7.286, 24.445, 98.78, Z = 8, R = 0.039. Isostructural with the Mg and Co compounds (**27**, 616; **29**, 347), with two independent octahedral $V(H_2O)_6^{2+}$ cations and tetrahedral sulphate anions. V-O = 2.123-2.150(2) A.

POTASSIUM BIS(SULPHATO)VANADATE(III)
$KV(SO_4)_2$

Inorg. Chem., **25**, 1571-1577.

$R\bar{3}$, a = 4.781, c = 23.545 A, at 143K, Z = 3, R = 0.049. (001) triple layers of edge- and corner-sharing VO_6 octahedra and SO_4 tetrahedra, linked by 12-coordinate K ions.

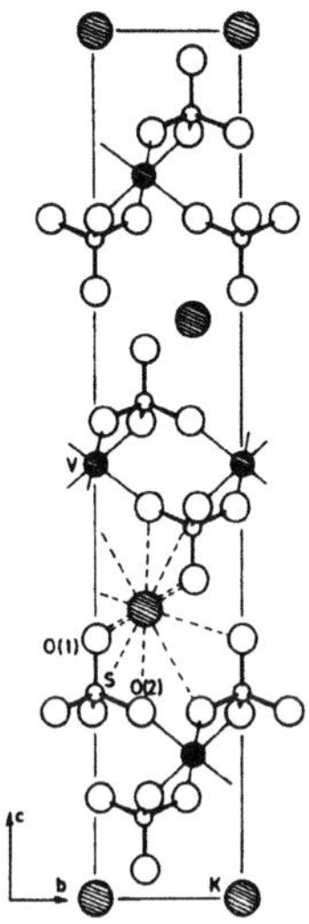

		x	y	z
K	3a	0	0	0
V	3b	0	0	0.5
S	6c	0	0	0.23858
O(1)	6c	0	0	0.30035
O(2)	18f	0.2797	0.3066	0.21688

Interatomic Distances (Å) and Bond Angles (deg) with Standard Deviations in Parentheses for $KV(SO_4)_2$

K–O(1)	2.867 (1) (6×)		
K–O(2)	3.275 (1) (6×)		
V–O(2)	1.996 (4) (6×)	O(2)–V–O(2′)	88.5 (2) (6×)
		O(2)–V–O(2″)	91.5 (2) (6×)
S–O(1)	1.455 (5)	O(1)–S–O(2)	110.0 (2) (3×)
S–O(2)	1.496 (5) (3×)	O(2)–S–O(2′)	109.0 (2) (3×)

KRAUSITE
$KFe(SO_4)_2.H_2O$

Amer. Min., **71**, 202-205.

$P2_1/m$, 7.920, 5.146, 9.014, 102.76, Z = 2, R = 0.037. Structure as previously described (**30**A, 372).

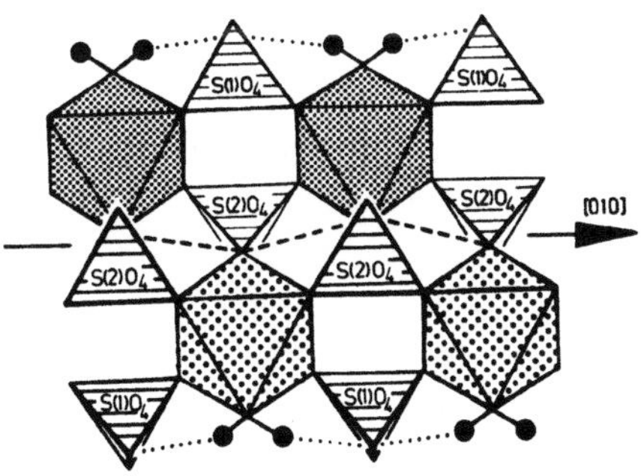

		x	y	z
K	2(e)	0.5746	1/4	0.2150
Fe	2(e)	0.0896	1/4	0.2278
S(1)	2(e)	0.6653	1/4	0.6531
O(1)	2(e)	0.7399	1/4	0.5191
O(2)	2(e)	0.4794	1/4	0.6210
O(3)	4(f)	0.7287	0.0179	0.7499
S(2)	2(e)	0.1554	1/4	0.8817
O(4)	2(e)	0.0619	1/4	0.0062
O(5)	2(e)	0.3414	1/4	0.9354
O(6)	4(f)	0.1014	0.0167	0.7861
O(7)=Ow	2(e)	0.1035	1/4	0.4554
H	4(f)	0.145	0.368	0.488

UNGEMACHITE
$K_3Na_8Fe(SO_4)_6(NO_3)_2.6H_2O$

Amer. Min., **71**, 826-829.

$R\bar{3}$, 10.898, 24.989 A, Z = 3, R = 0.039. $Fe(SO_4)_6$ clusters linked via Na (trigonal-prismatic coordination) to nitrate groups, to form (001) sheets, with further intrasheet linkage provided by 10-coordinate K ions; sheets are connected only by hydrogen bonds.

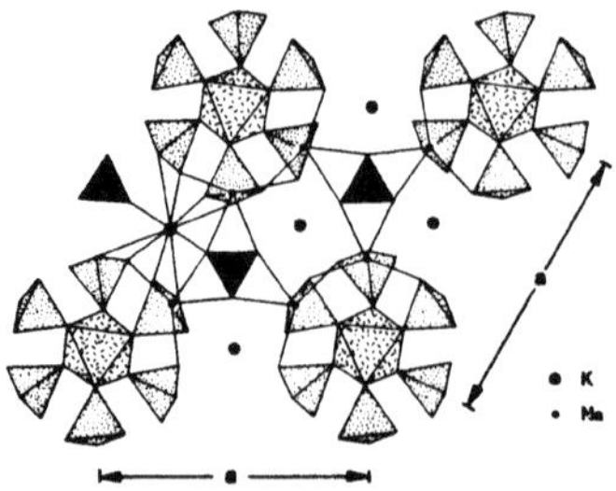

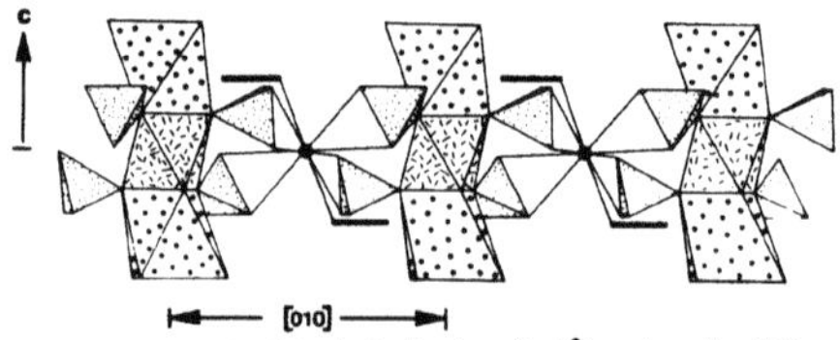

	x	y	z
Fe	0	0	0
S	0.34609	0.17748	0.27813
N	1/3	2/3	0.0951
K	1/2	0	0
Na(1)	0	0	0.1285
Na(2)	0.6440	0.8170	0.09069
O(1)	0.5046	0.2737	0.28304
O(2)	0.2824	0.1394	0.33119
O(3)	0.2953	0.2600	0.24891
O(4)	0.3212	0.0515	0.2485
O(5)	0.1200	-0.1006	0.1700
O(6)	0.4016	0.7990	0.0951
H(5)A	0.774	-0.181	0.192
H(5)B	0.772	-0.066	0.190

Selected polyhedral edge-lengths (Å) and angles (°) in ungemachite

O(1)b-O(1)l	2.680(4) x6	O(1)b-Fe-O(1)l	84.5(1) x6
O(1)b-O(1)c	2.952(4) x6	O(1)b-Fe-O(1)c	95.5(1) x6
<O-O>Fe	2.816	<O-Fe-O>	90.0
O(1)-O(2)	2.431(3)	O(1)-S-O(2)	109.9(1)
O(1)-O(3)	2.369(4)	O(1)-S-O(3)	105.4(1)
O(1)-O(4)	2.400(3)	O(1)-S-O(4)	107.7(2)
O(2)-O(3)	2.406(4)	O(2)-S-O(3)	110.8(2)
O(2)-O(4)	2.402(4)	O(2)-S-O(4)	110.8(2)
O(3)-O(4)	2.425(4)	O(3)-S-O(4)	112.0(2)
<O-O>S	2.406	<O-S-O>	109.4
O(6)-O(6)k	2.164(3) x3	O(6)-N-O(6)k	120.0(1) x3
O(1)b-O(1)n	2.680(4) x3	O(1)b-Na(1)-O(1)n	65.1(1) x3
O(1)b-O(5)	3.378(5) x3	O(1)b-Na(1)-O(5)	88.9(1) x3
O(1)b-O(5)m	3.695(5) x3	O(1)b-Na(1)-O(5)m	100.0(1) x3
O(5)-O(5)m	3.610(4) x3	O(5)-Na(1)-O(5)m	101.6(1) x3
<O-O>Na(1)	3.341	<O-Na(1)-O>	88.9
O(2)g-O(3)h	3.334(4)	O(2)g-Na(2)-O(3)h	88.6(1)
O(2)g-O(4)i	3.273(5)	O(2)g-Na(2)-O(4)i	87.6(1)
O(2)g-O(6)	3.344(5)	O(2)g-Na(2)-O(6)	86.2(1)
O(2)g-O(6)k	3.361(4)	O(2)g-Na(2)-O(6)k	86.8(1)
O(3)h-O(5)j	3.230(4)	O(3)h-Na(2)-O(5)j	84.9(1)
O(3)h-O(6)k	3.177(4)	O(3)h-Na(2)-O(6)k	79.1(1)
O(4)i-O(5)j	3.219(4)	O(4)i-Na(2)-O(5)j	85.1(1)
O(4)i-O(6)	3.216(4)	O(4)i-Na(2)-O(6)	81.1(1)
O(5)j-O(6)	4.195(5)	O(5)j-Na(2)-O(6)	117.3(1)
O(5)j-O(6)k	4.180(5)	O(5)j-Na(2)-O(6)k	116.7(1)
O(6)-O(6)k	2.164(3)	O(6)-Na(2)-O(6)k	50.1(1)
<O-O>Na(2)	3.336	<O-Na(2)-O>	87.6

AMMONIUM NICKEL SULPHATE HEXAHYDRATE
$(ND_4)_2Ni(SO_4)_2.6D_2O$

Aust. J. Chem., **39**, 1023-1028.

$P2_1/a$, 9.073, 12.305, 6.255, 106.9, at 2K, neutron radiation, R = 0.035 (also R = 0.022 for limited data set for the H compound). Tutton's salt structure (**28**, 213; **29**, 354; **44**A, 274); Ni-O = 2.04, 2.07, S-O = 1.46-1.48 A.

TETRAAMMINEPLATINUM(II) DIIODOTETRAAMMINEPLATINUM(IV) HYDROGENSULPHATE DIHYDRATE
$[Pt(NH_3)_4][PtI_2(NH_3)_4](HSO_4)_4.2H_2O$

Acta Cryst., C**42**, 1105-1109.

C2/m, 18.609, 7.563, 11.862, 126.22, Z = 2, R = 0.041. Linear ...I-Pt(IV)-I...Pt(II)... chains along **c** of alternate octahedral and square-planar units; Pt(IV)-I = 2.699(1), Pt(II)...I = 3.237(1) A.

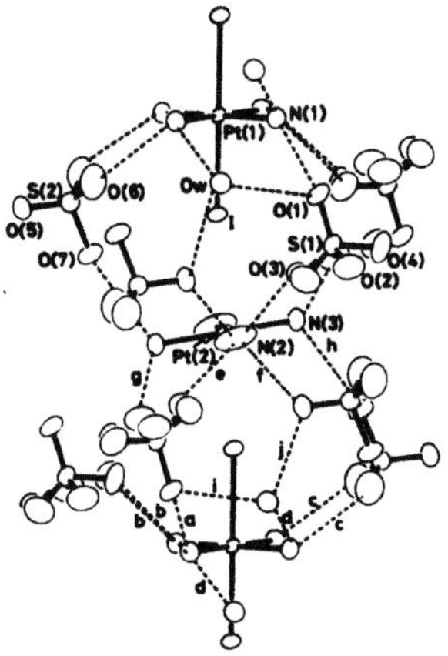

Fractional coordinates (×10⁴, *for* Pt *and* I × 10⁵)

	x	*y*	*z*
Pt(1)	0	0	0
Pt(2)	0	0	50000
I	857	0	23524
N(1)	971	1941	878
N(2)	0	2713	5000
N(3)	1373	0	6482
S(1)	3231	0	1185
S(2)	2961	0	4565
O(1)	3480	0	234
O(2)*	3053	1919	1446
O(3)*	3836	725	2472
O(4)*	2333	784	424
O(5)	3906	0	5527
O(6)	2607	1573	3708
O(7)	2596	0	5450
Ow	5355	0	1605

* Occupancy factor = 0·5

ZINC HYDROXIDE SULPHATE HYDRATES
$3Zn(OH)_2.ZnSO_4.nH_2O$ (n = 5, 3)

Acta Cryst., B**42**, 32-39.

n = 5, $P\bar{1}$, 8.354, 8.350, 11.001, 94.41, 82.95, 119.93, Z = 2, R = 0.042 (some disorder). n = 3, $I\bar{1}$, 8.367, 8.393, 18.569, 90.29, 89.71, 120.53, Z = 4, R = 0.13 (some disorder). Both structures contain modified $Zn(OH)_2$ layers of edge-shared octahedra, with one-seventh of the octahedral sites vacant, and with $Zn(OH)_3(H_2O)$ tetrahedra above and below the vacant sites. Sulphate tetrahedra are connected to both sides of the layers by corner-sharing, and layers are held together by strong hydrogen bonds between coordinated and free water molecules and sulphate oxygens.

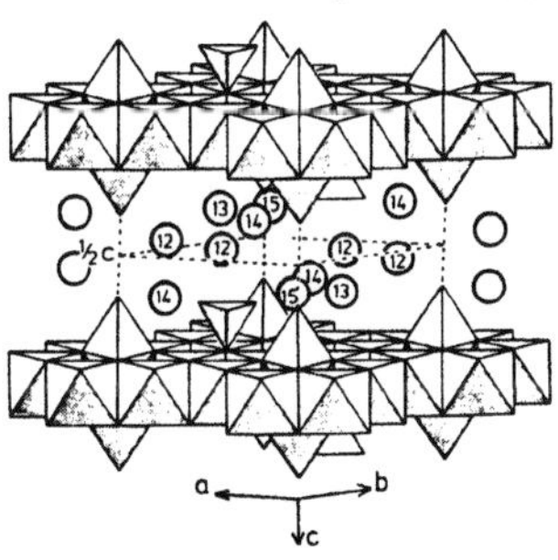

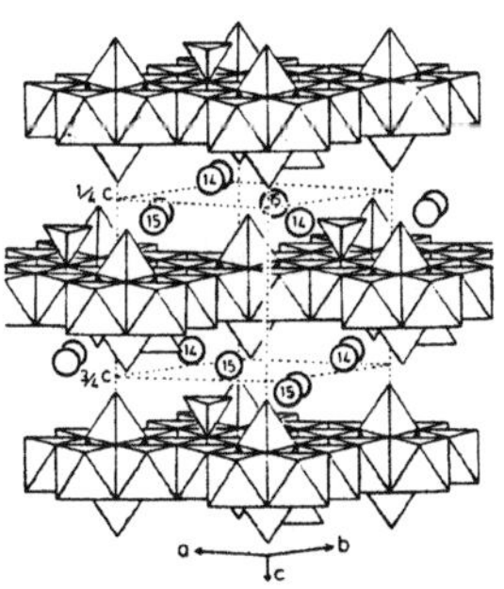

$3Zn(OH)_2.ZnSO_4.5H_2O$

	Population parameter	*x*	*y*	*z*
Zn(1)		0·1307	0·4173	0·9994
Zn(2)		0·4146	0·2858	0·0077
Zn(3)		0·7137	0·1309	0·0084
Zn(4)		0·9778	0·0021	0·1610
S		0·3000	0·6774	0·2603
$O_H(1)$		0·7390	0·9446	0·1091
$O_H(2)$		0·0409	0·8122	0·1078
$O_H(3)$		0·1723	0·2503	0·1151
$O_H(4)$		0·6153	0·5280	0·0843
$O_H(5)$		0·8821	0·3775	0·0870
$O_H(6)$		0·4634	0·1049	0·0856
O(7)		0·3176	0·6672	0·1229
$O_w(8)$		0·9362	0·9810	0·3419
O(9)		0·2858	0·5103	0·3075
O(10)		0·1297	0·6901	0·3017
O(11)		0·4663	0·8423	0·3016
$O_w(12)$		0·9720	0·2859	0·4577
$O_w(13)$		0·7220	0·3381	0·3445
$O_w(14A)$	0·72	0·3461	0·1170	0·3531
$O_w(14B)$	0·28	0·4276	0·1085	0·4757
$O_w(15A)$	0·55	0·6700	0·6209	0·3481
$O_w(15B)$	0·45	0·3455	0·3326	0·5741

$3Zn(OH)_2.ZnSO_4.3H_2O$

	x	*y*	*z*
Zn(1)	0·1330	0·4184	0·0000
Zn(2)	0·4167	0·2866	0·0041
Zn(3)	0·7146	0·1285	0·0088
Zn(4)	0·9977	0·9936	0·0944
S	0·3189	0·6639	0·1516
$O_H(1)$	0·750	0·942	0·067
$O_H(2)$	0·062	0·810	0·063
$O_H(3)$	0·189	0·243	0·066
$O_H(4)$	0·630	0·528	0·051
$O_H(5)$	0·896	0·376	0·055
$O_H(6)$	0·479	0·106	0·052
O(7)	0·327	0·658	0·072
$O_w(8)$	0·991	0·986	0·192
O(9)	0·289	0·488	0·181
O(10)	0·169	0·693	0·171
O(11)	0·500	0·818	0·178
$O_w(14)$	0·502	0·138	0·209
$O_w(15)$	0·747	0·515	0·199

SODIUM trans-DIAQUOTETRASULPHATOZINCATE
$Na_6[Zn(SO_4)_4(H_2O)_2]$

Acta Cryst., C**42**, 949-952.

$P\bar{1}$, 6.052, 6.815, 10.141, 70.07, 77.29, 74.72, Z = 1, R = 0.029. Octahedral complex anions, linked by 6-coordinate Na ions. Zn-OH_2 = 2.032, Zn-OS = 2.135(2) A.

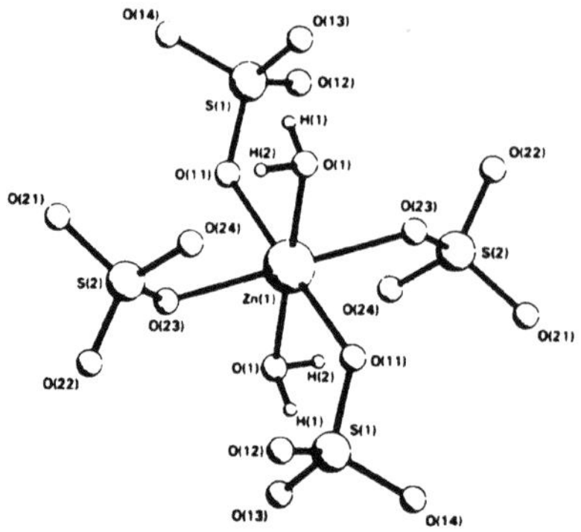

	x	y	z
Zn(1)	1·0	0·0	0·0
S(1)	1·12357	−0·26617	0·32546
S(2)	0·55463	0·20474	0·19979
O(11)	1·1750	−0·1156	0·1815
O(12)	0·9882	−0·1442	0·4207
O(13)	0·9874	−0·4122	0·3179
O(14)	1·3465	−0·3860	0·3727
O(21)	0·3656	0·3519	0·1268
O(22)	0·4630	0·0976	0·3459
O(23)	0·6667	0·0426	0·1238
O(24)	0·7289	0·3231	0·1984
O(1)	0·9656	−0·3041	0·0376
Na(1)	0·5617	0·7081	0·1072
Na(2)	1·0409	0·2150	0·3332
Na(3)	0·6081	−0·2621	0·4636
H(1)	1·012	−0·356	0·119
H(2)	1·083	−0·356	−0·019

POTASSIUM CERIUM(IV) SULPHATE MONOHYDRATE
$K_2Ce(SO_4)_3.H_2O$

Ž. Strukt. Khim., **27**, No. 1, 176-180 [J. Struct. Chem., **27**, 165-168].

C2, 20.600, 7.0744, 18.583, 126.083, Z = 8, R = 0.071. Framework of nine-coordinate Ce ions, with 9-, 10-, and 12-coordinate K ions. Ce-O = 2.27-2.51 A.

HYDRAZINIUM NEODYMIUM SULPHATE MONOHYDRATE
$N_2H_5Nd(SO_4)_2.H_2O$

J. Chem. Soc., Dalton, 119-123 (1986).

$Pca2_1$, 10.951, 9.554, 15.762, Z = 8, R = 0.059. Two independent Nd, each 9-coordinate to 7 O, 1 H_2O, and 1 N (distorted tricapped trigonal prism); bridging sulphate groups result in (001) sheets, which are linked by hydrogen bonds.

TRIS[DI-μ-HYDROXO-TETRAAMMINECHROMIUM(III)]CHROMIUM(III) DITHIONATE TETRAHYDRATE
$[Cr\{(OH)_2Cr(NH_3)_4\}_3](S_2O_6)_3.4H_2O$

Acta Chem. Scand., A**40**, 476-478.

$P2_1/n$, 18.533, 9.736, 20.067, 98.69, Z = 4, R = 0.05. The cation contains a central $Cr(OH)_6$ octahedron sharing edges with three $Cr(OH)_2(NH_3)_4$ octahedra, and has approximate D_3 symmetry. Cations, dithionate anions, and water molecules are linked by hydrogen bonds.

POTASSIUM TRITHIONATE
$K_2S_3O_6$

Z. Kristallogr., **173**, 59-74 (1985).

Pnam, 9.794, 13.690, 5.742, Z = 4, R = 0.028. Structure as previously described (**3**, 138, 502; **45**A, 342). Electron density distribution now studied.

SODIUM HYDROGEN SELENITE
$NaHSeO_3$, $NaDSeO_3$

Kristallografija, **31**, 264-269 [Soviet Physics - Crystallography, **31**, 155-159].

B2/b (**c** unique), 21.98, 10.28, 5.79, γ = 105.1°, Z = 16, neutron radiation, R = 0.036, 0.049. Structure as previously described (**28**, 221), with $O_2Se(OH)^-$ ions linked into dimers by O-H...O hydrogen bonds, probably with dynamic disorder of the H (and D) atoms.

POTASSIUM HYDROGENSELENATE
$KHSeO_4$

Acta Cryst., C**42**, 270-272.

Pbca, 8.690, 10.053, 19.47, Z = 16, R = 0.054. Isostructural with the sulphate (**24**, 394; **42**A, 367).

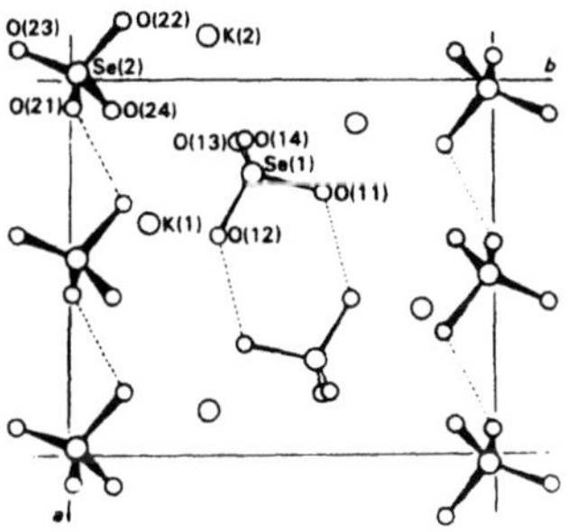

	x	y	z
Se(1)	0·25433	0·42428	0·00284
Se(2)	−0·01870	0·01642	0·20493
K(1)	0·3851	0·1760	0·1255
K(2)	−0·1247	0·3199	0·1189
O(11)	0·2963	0·5931	0·0004
O(12)	0·4176	0·3432	0·0067
O(13)	0·1667	0·3929	−0·0672
O(14)	0·1537	0·4023	0·0700
O(21)	0·0723	0·0030	0·2792
O(22)	−0·1715	0·1182	0·2229
O(23)	−0·0832	−0·1220	0·1778
O(24)	0·0785	0·0985	0·1497

Principal interatomic distances (Å) *and bond angles* (°) *for* $KHSeO_4$

Se(1)–O(11)	1·737 (6)	Se(2)–O(21)	1·654 (7)
Se(1)–O(12)	1·638 (6)	Se(2)–O(22)	1·712 (6)
Se(1)–O(13)	1·594 (8)	Se(2)–O(23)	1·591 (7)
Se(1)–O(14)	1·589 (8)	Se(2)–O(24)	1·598 (7)
O(11)···O(12^{i})	2·57 (1)	O(22)···O(21^{ii})	2·51 (1)
O(11)–Se(1)–O(12)	107·8 (4)	O(21)–Se(2)–O(22)	103·9 (4)
O(11)–Se(1)–O(13)	105·7 (4)	O(21)–Se(2)–O(23)	112·8 (4)
O(11)–Se(1)–O(14)	105·9 (4)	O(21)–Se(2)–O(24)	112·3 (4)
O(12)–Se(1)–O(13)	110·8 (4)	O(22)–Se(2)–O(23)	108·5 (4)
O(12)–Se(1)–O(14)	111·7 (4)	O(22)–Se(2)–O(24)	103·8 (4)
O(13)–Se(1)–O(14)	114·5 (4)	O(23)–Se(2)–O(24)	114·5 (4)

Symmetry code: (i) $1-x$, $1-y$, $-z$; (ii) $x-0·5$, y, $0·5-z$.

K(1)–O(11^{i})	3·020 (8)	K(2)–O(11^{v})	2·895 (8)
–O(12)	2·873 (7)	–O(12^{vi})	2·967 (7)
–O(13^{ii})	2·785 (8)	–O(13^{v})	3·079 (8)
–O(14)	3·223 (8)	–O(13^{vi})	2·979 (8)
–O(14^{i})	2·976 (8)	–O(14)	2·728 (8)
–O(21^{iii})	3·018 (7)	–O(21^{vii})	2·744 (7)
–O(22^{iii})	3·047 (7)	–O(22)	2·896 (7)
–O(23^{iv})	2·849 (8)	–O(23^{viii})	2·846 (8)
–O(24)	2·816 (8)	–O(24)	2·903 (8)

Symmetry code: (i) $0·5-x$, $-0·5+y$, z; (ii) $0·5+x$, $0·5-y$, $-z$; (iii) $0·5+x$, y, $0·5-z$; (iv) $0·5-x$, $0·5+y$, z; (v) $-x$, $1-y$, $-z$; (vi) $-0·5+x$, $0·5-y$, $-z$; (vii) $-x$, $0·5+y$, $0·5-z$; (viii) $-0·5-x$, $0·5+y$, z.

RUBIDIUM SELENATE
Rb_2SeO_4

RUBIDIUM TETRAFLUOROBERYLLATE
Rb_2BeF_4

Jpn. J. Appl. Phys., Part 1, **24**, 591-593 (1985).

Rb_2SeO_4, Pnam, 7.961, 10.794, 6.161, Z = 4, R = 0.039. Rb_2BeF_4, Pna2_1, 7.649, 10.184, 5.870, Z = 4, R = 0.037. Tetrahedral anions, linked by Rb cations. Previous study in **30**A, 270.

RUBIDIUM HYDROGEN SELENATE
$Rb_3H(SeO_4)_2$

Kristallografija, **31**, 178–180 [Soviet Physics – Crystallography, **31**, 105–106].

A2/a, 10.475, 6.090, 15.418, 102.91, Z = 4, R = 0.037. Isostructural with $(NH_4)_3H(SO_4)_2$ (**44**A, 268), with a formally symmetric O...H...O hydrogen bond, in which the hydrogen atom is probably statistically distributed in two sites.

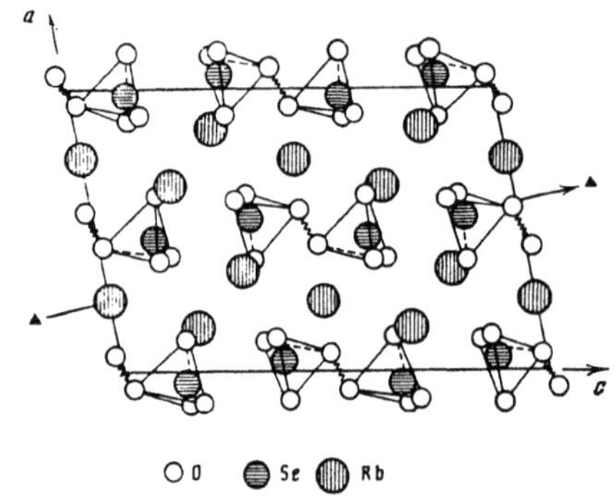

	x	y	z
Rb(1)	1/4	0.7521	0
Rb(2)	0.65176	0.7343	0.19486
Se	0.46199	0.2267	0.11638
O(1)	0.3986	0.0172	0.1567
O(2)	0.4406	0.1767	0.0061
O(3)	0.6190	0.2413	0.1547
O(4)	0.3879	0.4535	0.1286
H	0	0	0

CALCIUM HYDROGENSELENITE MONOHYDRATE
$Ca(HSeO_3)_2.H_2O$ (I)

CALCIUM HYDROGENSELENITE DISELENITE
$Ca_2(HSeO_3)_2(Se_2O_5)$ (II)

J. Solid State Chem., **65**, 363–369.

I, $P\bar{1}$, 6.911, 7.369, 6.739, 90.51, 90.93, 107.46, Z = 2, R = 0.036. II, C2/c, 14.719, 7.059, 11.793, 117.96, Z = 4, R = 0.047. Both structures contain trigonal pyramidal Se-containing anions, linked by Ca ions (8-coordinate in I, 7-coordinate in II).

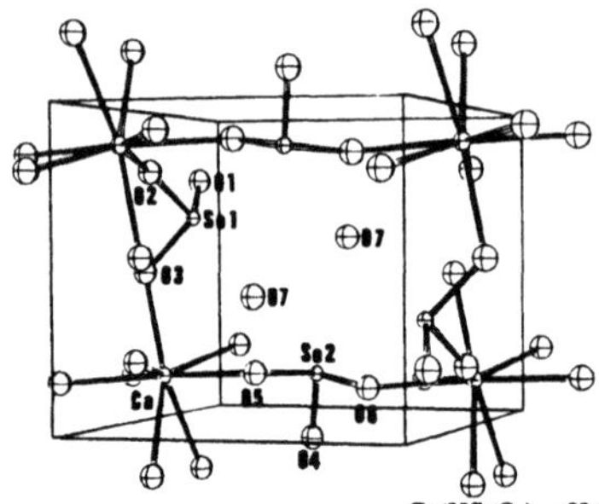

$Ca(HSeO_3)_2 \cdot H_2O$

	x	y	z
Se(1)	.74841	−.03727	.34243
Se(2)	.81965	.59264	.86831
Ca(1)	.77174	.07785	.87837
O(1)	.95759	.04504	.21104
O(2)	.58540	.01975	.19172
O(3)	.78435	.14391	.52888
O(4)	.91911	.65041	.09572
O(5)	.64318	.71534	.85624
O(6)	.65651	.36270	.90111
O(7)	.29716	.55745	.59145

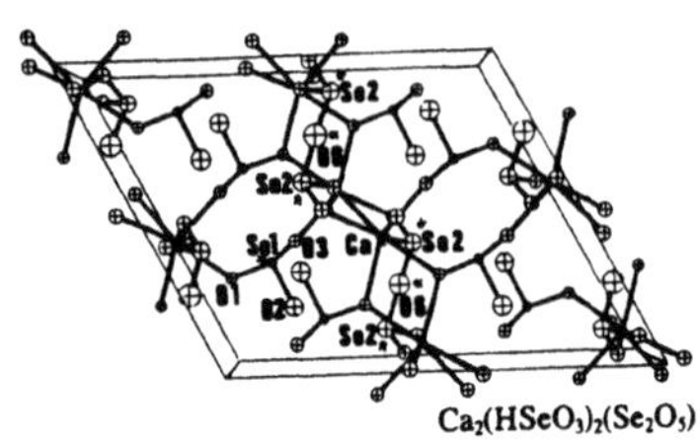

$Ca_2(HSeO_3)_2(Se_2O_5)$

	x	y	z
Se(1)	.2823	.3629	.1606
Se(2)	.0901	.2643	.3937
Ca(1)	.9795	.2488	.5749
O(1)	.3950	.2501	.2061
O(2)	.2846	.3647	.3122
O(3)	.1926	.1969	.0877
O(4)	.0778	.4317	.4830
O(5)	.0253	.0899	.4126
O(6)[a]	.0419	.3609	.2572

[a] Population parameter 0.5

INTERATOMIC DISTANCES (Å) AND ANGLES (°) AROUND SELENIUM ATOMS

	$Ca(HSeO_3)_2 \cdot H_2O$	$Ca_2(HSeO_3)_2(Se_2O_5)$
Se1-O1	1.663(1)	1.686(7)
Se1-O2	1.653(1)	1.773(10)
Se1-O3	1.784(1)	1.671(6)
Se2-O4	1.668(1)	1.649(8)
Se2-O5	1.724(1)	1.636(8)
Se2-O6	1.749(1)	1.744(8)
O1-Se1-O2	99.11 (6)	97.5 (4)
O1-Se1-O3	101.47 (6)	104.8 (3)
O2-Se1-O3	100.56 (6)	99.3 (4)
O4-Se2-O5	101.84 (6)	101.9 (5)
O4-Se2-O6	102.20 (6)	93.7 (5)
O5-Se2-O6	99.34 (6)	102.5 (5)
Se2-O6-Se2	—	131.3 (1.2)

POTASSIUM ANTIMONY(III) SELENATE FLUORIDE HYDRATE
$K_2SeO_4(SbF_3)_2.H_2O$

J. Solid State Chem., **63**, 52-61.

$P2_12_12_1$, 12.374, 14.796, 5.929, Z = 4, R = 0.033. SeO_4 tetrahedra and SbF_3 trigonal pyramids (longer Sb...O and Sb...F contacts and a lone-pair give 3.3.1 geometry at each Sb); K ions have 9- and 6-coordinations to F and O. Se-O = 1.63-1.65, Sb-F = 1.92-1.97, Sb...O,F = 2.44-2.97(1) A.

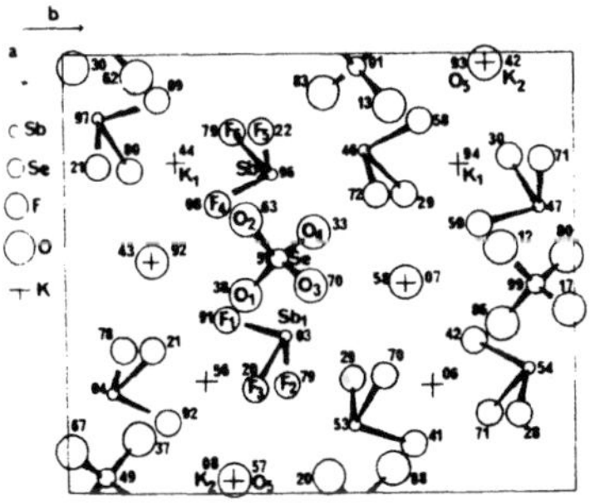

	x	y	z
Sb(1)	0,14704	0,06810	0,9694
Sb(2)	0,77551	0,09170	0,0386
Se	0,46831	0,42232	0,5090
K(1)	0,2467	0,2277	0,4394
K(2)	0,9710	0,3317	0,0825
F(1)	0,1064	0,1830	0,094
F(2)	0,2574	0,0647	0,208
F(3)	0,7638	0,3715	0,197
F(4)	0,3480	0,3024	0,078
F(5)	0,1772	0,3870	0,215
F(6)	0,6780	0,1681	0,207
O(1)	0,5524	0,3563	0,377
O(2)	0,8783	0,1423	0,365
O(3)	0,0317	0,0171	0,300
O(4)	0,5920	0,9892	0,172
O(5)	0,4740	0,1719	0,435

COPPER(II) SELENITE
$CuSeO_3$

Z. Kristallogr., **175**, 61-72.

Form II, Pcab, 5.962, 7.010, 12.240, Z = 8, R = 0.025. Form III, $P2_1/n$, 7.725, 8.241, 8.502, 99.16, Z = 8, R = 0.050. Form IV, $P\bar{1}$, 4.736, 8.579, 10.618, 66.69, 88.53, 88.40, Z = 6, R = 0.051. SeO_3 trigonal pyramids linked by Cu ions which have square-planar 4-coordinations, with one or two more-distant oxygen neighbours. Se-O = 1.667-1.752, Cu-O = 1.908-2.021, 2.194-2.947 A. Form I (**42**A, 383) has 12% closer packing. [See also this volume, p. 239.]

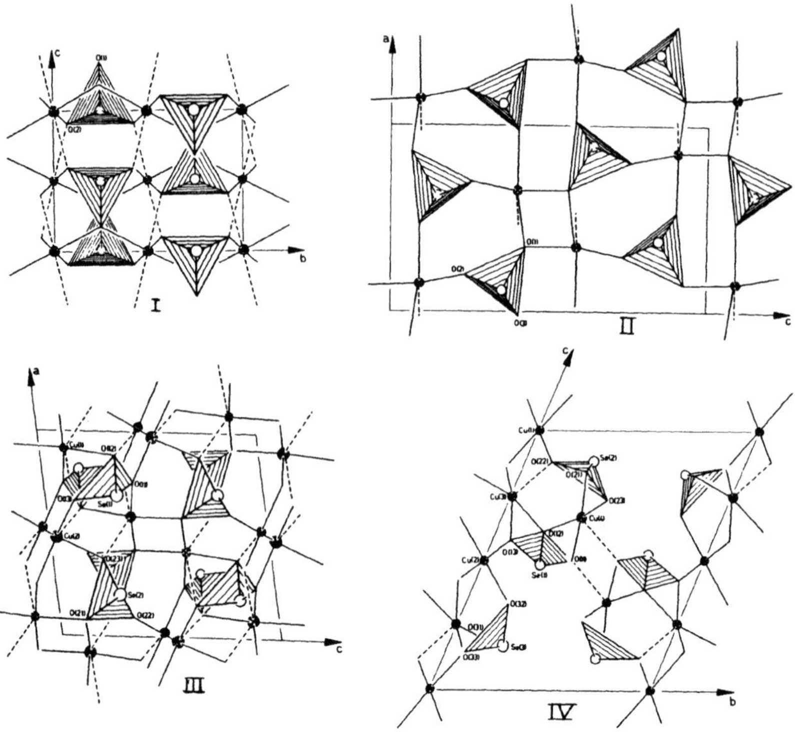

	Cu(SeO$_3$)-II x	y	z
Cu	0,65003	−0,07836	0,40676
Se	0,11183	0,05125	0,35443
O(1)	0,3550	0,0421	0,4307
O(2)	0,2044	−0,0411	0,2361
O(3)	−0,0255	−0,1436	0,4088

	Cu(SeO$_3$)-III x	y	z
Cu(1)	0,91592	0,10237	0,11595
Cu(2)	0,47605	0,79292	0,02795
Se(1)	0,81397	0,47214	0,16966
O(11)	0.7513	0,2837	0,0901
O(12)	0,6163	0,5642	0,1557
O(13)	0,8408	0,4385	0,3694
Se(2)	0,29906	0,36534	0,20897
O(21)	0,4238	0,4200	0,3872
O(22)	0,4074	0,2014	0,1608
O(23)	0,1202	0,2809	0,2688

		Cu(SeO$_3$)-IV x	y	z
Cu(1)	1a	0	0	0
Cu(2)	1b	0	0	½
Cu(3)	2i	0,5302	0,0067	0,7455
Cu(4)	2i	0,9677	0,3570	0,6647
Se(1)	2i	0,4580	0,2689	0,4786
O(11)	2i	0,2142	0,4023	0,4974
O(12)	2i	0,6972	0,2226	0,6122
O(13)	2i	0,3003	0,0751	0,5773
Se(2)	2i	0,4194	0,3071	0,8788
O(21)	2i	0,7692	0,3007	0,8426
O(22)	2i	0,3107	0,1184	0,8753
O(23)	2i	0,2806	0,4466	0,7306
Se(3)	2i	0,9744	0,2530	0,1688
O(31)	2i	0,6293	0,2149	0,1708
O(32)	2i	1,0675	0,1962	0,3331
O(33)	2i	1,1170	0,0780	0,1440

COPPER(II) HYDROGEN SELENITE
$Cu(SeO_2OH)_2$

Z. Kristallogr., **173**, 267-272 (1985).

$P2_1/n$, 5.7662, 7.3521, 6.4474, 93.28, Z = 2, R = 0.023. Hydrogen-bonded anion dimers are linked into (101) sheets by CuO_4 square planes; sheets are linked by longer Cu-O bonds.

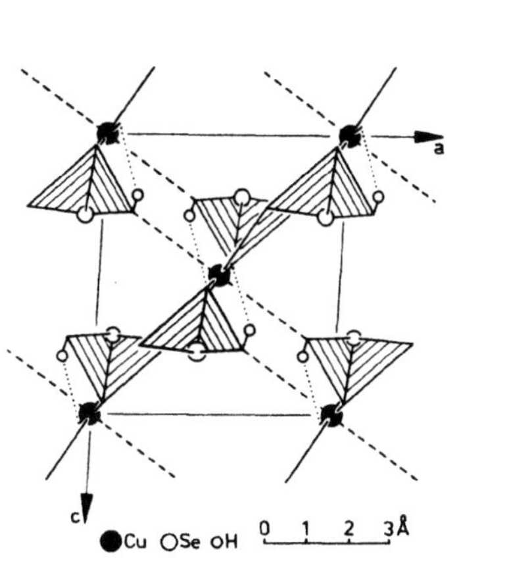

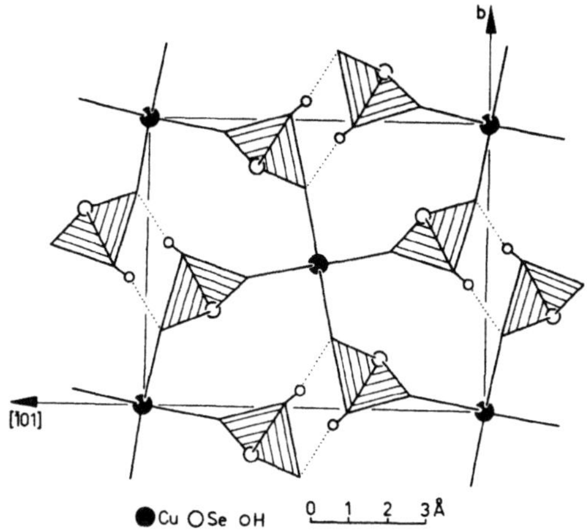

	x	y	z
Se	0.92318	0.66677	0.28436
Cu	0	0	0
O(1)	0.9522	0.7380	0.0384
O(2)	0.6749	0.5516	0.2555
O(3)	0.1243	0.4842	0.2899
H	0.140	0.443	0.211

Cu – O(1) = 1.963(2) 2 ×
Cu – O(2) = 1.957(2) 2 ×
Cu – O(3) = 2.491(3) 2 ×

Se – O(1) = 1.688(2)
Se – O(2) = 1.665(2)
Se – O(3) = 1.773(2)

H – O(3) = 0.60(9)
H. . .O(1) = 2.13(8)

COPPER OXIDE SELENATES
$Cu_2O(SeO_3)$, $Cu_4O(SeO_3)_3$

Mh. Chem., **117**, 887-896.

$Cu_2O(SeO_3)$, form I, $P2_13$, 8.925, Z = 8, R = 0.033; form II, $P2_1/n$, 6.987, 5.953, 8.429, 92.17, Z = 4, R = 0.043.

$Cu_4O(SeO_3)_3$, form I, $P2_1/a$, 15.990, 13.518, 17.745, 90.49, Z = 16, R = 0.062; form II, $P\bar{1}$, 7.992, 8.141, 8.391, 77.34, 65.56, 81.36, Z = 2, R = 0.044.

Cu(1) in $Cu_2O(SeO_3)$ has trigonal bipyramidal coordination; all other Cu atoms in the four structures have square-planar coordinations, some having one or two more-distant O neighbours. The SeO_3 groups are trigonal pyramidal, and the oxide ions are tetrahedrally coordinated by four Cu.

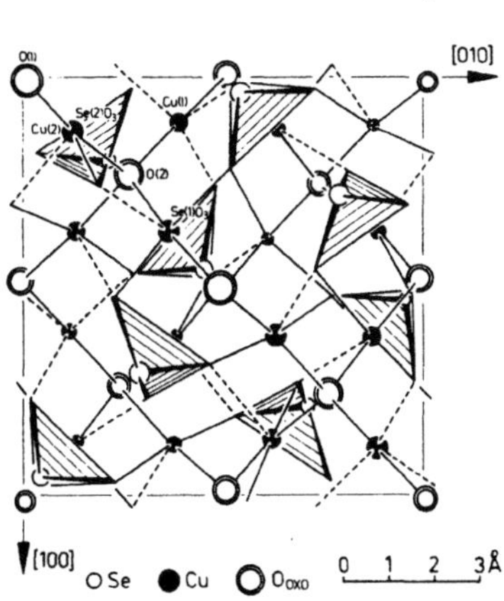

$Cu_2O(SeO_3)$-I

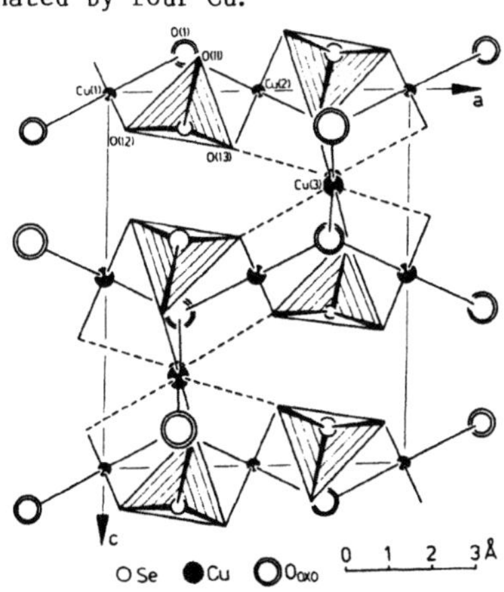

$Cu_2O(SeO_3)$-II

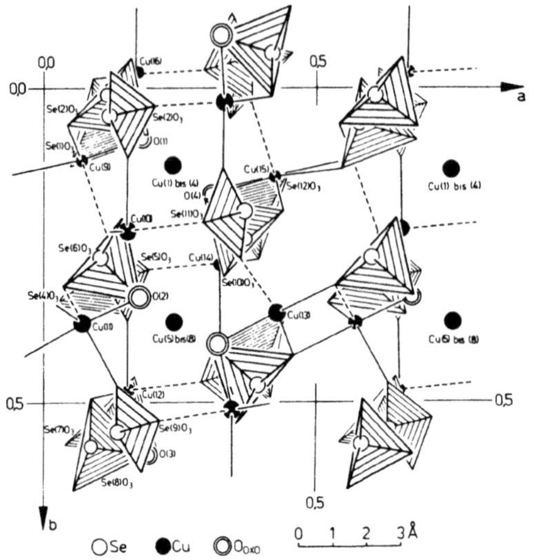

$Cu_4O(SeO_3)_3$-I

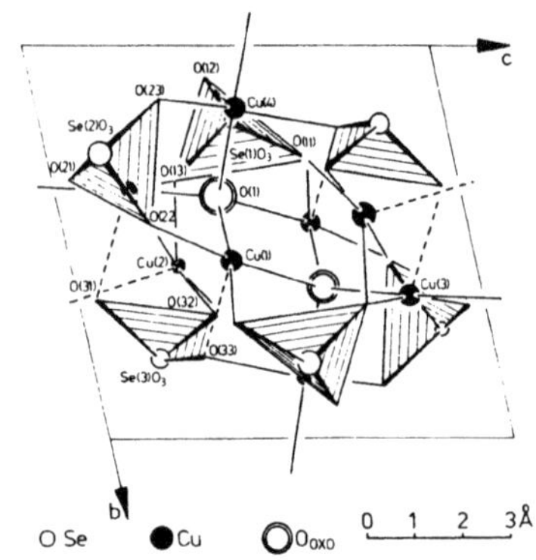

$Cu_4O(SeO_3)_3$-II

LEAD COPPER NITRATE SELENATES
$PbCu_3(OH)(NO_3)(SeO_3)_3.0.5H_2O$ (I), $Pb_2Cu_3O_2(NO_3)_2(SeO_3)_2$ (II)

Mh. Chem. **117**, 1099-1106.

I, $P\bar{1}$, 7.761, 9.478, 9.514, 66.94, 69.83, 81.83, Z = 2, R = 0.042. II, $Cmc2_1$, 5.884, 12.186, 19.371, Z = 4, R = 0.070. I contains a three-dimensional structure of (4+1)- and (4+2)-coordinated Cu ions and SeO_3 trigonal pyramids, with 9-coordinate Pb, and nitrate groups bonded to two Cu. II contains $[Pb_2Cu_3O_2(SeO_3)_2]$ sheets connected only by Pb-O bonds.

$PbCu_3(OH)(NO_3)(SeO_3)_3 \cdot \frac{1}{2}H_2O$

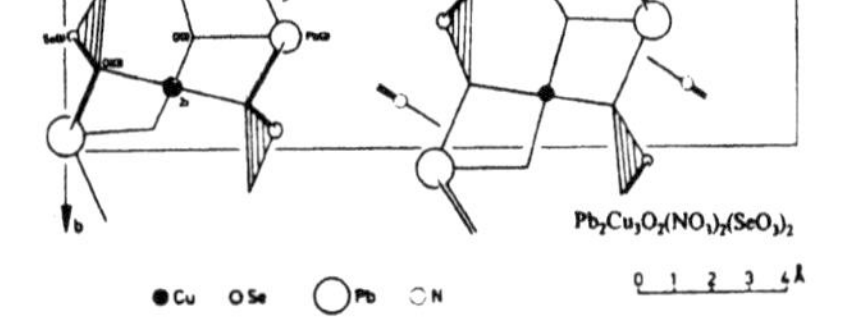

$Pb_2Cu_3O_2(NO_3)_2(SeO_3)_2$

$PbCu_3(OH)(NO_3)(SeO_3)_3.1/2H_2O$

	x	y	z
Pb	0,46864	0,44251	0,23745
Cu(1)	0	0	0
Cu(2)	0	1/2	1/2
Cu(3)	0,3655	-0,1167	0,0958
Cu(4)	0,4403	0,0257	0,3547
O_h	0,4941	0,0507	0,8145
N	0,8235	0,2574	0,3831
$O_n(1)$	0,1964	0,8395	0,4856
$O_n(2)$	0,8706	0,3891	0,3487
$O_n(3)$	0,7873	0,2296	0,2776
Se(1)	0,1230	0,6444	0,1091
O(11)	0,3027	0,5411	0,0370
O(12)	0,2004	-0,1709	0,0059
O(13)	0,8351	0,3845	0,7166
Se(2)	0,2376	0,2313	0,0895
O(21)	0,3066	0,2037	0,2501
O(22)	0,4397	0,2734	-0,0658
O(23)	0,8123	0,9487	0,8823
Se(3)	0,1595	0,1541	0,6087
O(31)	0,3865	0,1025	0,5313
O(32)	0,1535	0,1381	0,7938
O(33)	0,1911	0,3450	0,5107
O_w*	0,495	0,570	0,435

Ow occupancy = 0.5

$Pb_2Cu_3O_2(NO_3)_2(SeO_3)_2$

Pb(1)	0	0,4634	0,0000
Pb(2)	0	0,7660	0,2978
Cu(1)	0	0,6141	0,1465
Cu(2)	0,2491	0,3630	0,1462
O(1)	0	0,770	0,173
O(2)	0	0,457	0,118
Se(1)	0	0,766	0,0097
O(11)	0	0,643	0,052
O(12)	0,222	0,828	0,046
Se(2)	0	0,4607	0,2835
O(21)	0	0,586	0,245
O(22)	0,232	0,403	0,244
N(1)	0	0,133	0,348
$O_n(11)$	0	0,191	0,295
$O_n(12)$	0,183	0,109	0,377
N(2)	0	0,101	-0,048
$O_n(21)$	0	0,047	0,007
$O_n(22)$	0,183	0,130	-0,075

ZINC SELENITE
$ZnSeO_3$

COPPER(II) SELENITE
$CuSeO_3$

Acta Cryst., C**42**, 1285-1287; Z. Kristallogr., **174**, 291-295.

Pcab, Zn, 6.233, 7.200, 11.987, Z = 8, R = 0.035; Cu, 5.954, 7.006, 12.232, Z = 8, R = 0.037. SeO_3 trigonal pyramids, linked by ZnO_5 or CuO_5 trigonal bipyramids. Se-O = 1.68-1.72 A, O-Se-O = 99-101°, Zn-O = 1.97-2.20, Cu-O = 1.96-2.22 A. High-pressure forms in **42**A, 383. [See also this volume, p. 235.]

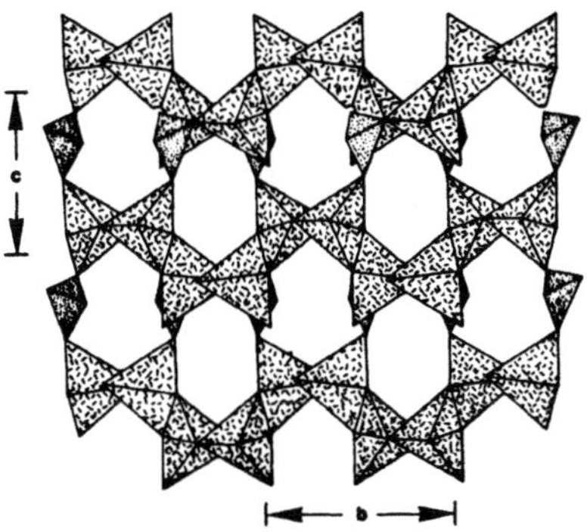

	x	y	z
$ZnSeO_3$			
Zn	0·1353	0·6076	0·4045
Se	0·1119	0·0445	0·3575
O(1)	0·2007	−0·0343	0·2328
O(2)	0·3394	0·0311	0·4342
O(3)	−0·0162	−0·1517	0·4071
$CuSeO_3$			
Cu	0·1500	0·5783	0·4067
Se	0·1118	0·0514	0·3544
O(1)	0·2050	−0·0419	0·2359
O(2)	0·3534	0·0425	0·4305
O(3)	−0·0263	−0·1432	0·4085

LANTHANUM SELENATE DODECAHYDRATE
$La_2(SeO_4)_3.12H_2O$

Lanthanide Actinide Res., **1**, 169-177.

C2/c, 10.669, 20.393, 10.744, 110.12°, Z = 4, R = 0.11. Layers of SeO_4 tetrahedra and LaO_9 polyhedra (distorted square antiprism plus one capping atom). Half of the water molecules are not coordinate to La, but are involved in hydrogen bonding. La-O = 2.42-2.65, 3.06 A.

SODIUM LANTHANUM DISELENATE DIHYDRATE
$NaLa(SeO_4)_2.2H_2O$

Kristallografija, **31**, 1081-1086 [Soviet Physics - Crystallography **31**, 640-643].

$P2_1/c$, 11.421, 7.135, 11.178, 107.51°, Z = 4, R = 0.060. Framework of edge- and corner-sharing LaO_9 polyhedra and SeO_4 tetrahedra, with 8-coordinate Na.

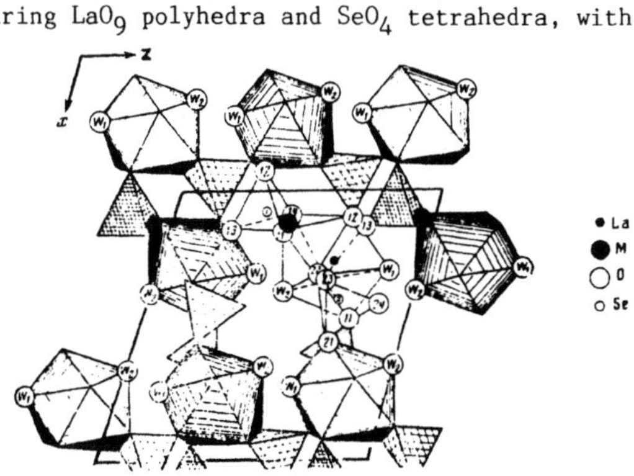

CERIUM(IV) SELENITE
$Ce(SeO_3)_2$

Acta Cryst., C**42**, 1475-1477.

$P2_1/n$, 7.008, 10.587, 7.262, 107.00, Z = 4, R = 0.032. CeO_6 strings along **a** of edge-sharing CeO_8 Archimedean antiprisms, linked by SeO_3 trigonal prisms (with the lone-pair completing a tetrahedron at Se). Se-O = 1.68-1.73 A, O-Se-O = 92-105°, Ce-O = 2.25-2.53(1) A.

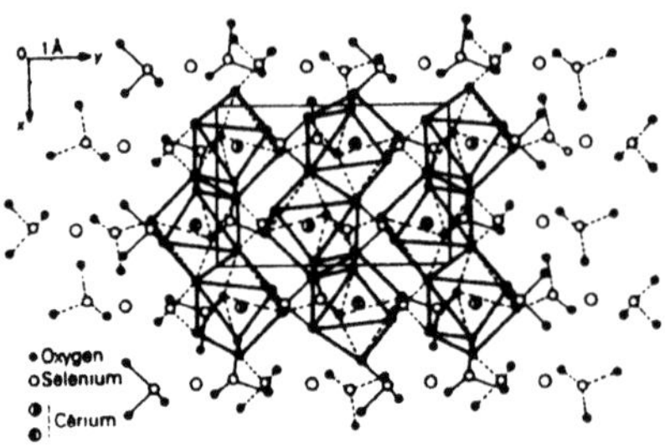

(× 10^4)	x	y	z
Ce(1)	2602	960	141
Se(2)	2391	2836	4308
Se(3)	2786	−691	−4355
O(4)	1417	2404	2002
O(5)	4057	3995	4086
O(6)	704	3953	4564
O(7)	1861	206	−2885
O(8)	5256	−889	−3152
O(9)	1947	−2138	−3947

POTASSIUM LITHIUM TELLURATE(VI)
$K_3Li_3TeO_6$

J. Less-Common Metals, **124**, 251-262.

Pnma, 11.695, 5.987, 9.689, Z = 4, R = 0.075. Isolated TeO_6 octahedra, linked by Li ions which have tetrahedral and square-planar coordinations, and by 6- and 7-coordinate K ions. Te-O = 1.93-1.95 A.

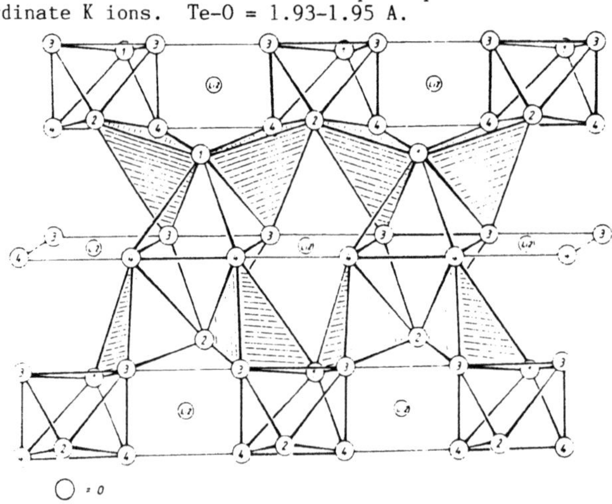

POTASSIUM SODIUM TELLURATE(VI)
$K_4Na_2TeO_6$

J. Less-Common Metals, **125**, 223-231.

$P2_1/a$, 7.255, 9.833, 6.667, 122.74, Z = 2, R = 0.094. Isolated TeO_6 octahedra, linked by 6-coordinate K and (5+1)-coordinate Na ions. Te-O = 1.94-1.98, K-O = 2.64-3.11, Na-O = 2.22-2.65, 3.06 A.

		x	y	z
Te	(2a)	0.0000	0.0000	0.0000
K(1)	(4e)	0.5140	0.8231	0.4988
K(2)	(4e)	0.2462	0.4867	0.3169
Na	(4e)	0.2315	0.7427	0.0093
O(1)	(4e)	0.8100	0.4957	0.1094
O(2)	(4e)	0.5604	0.3673	0.2521
O(3)	(4e)	0.0603	0.8535	0.2207

CAESIUM TELLURATES
Cs_2TeO_3 (1), $Cs_2Te_2O_5$ (2), $Cs_2Te_4O_9$ (3), $Cs_2Te_4O_{12}$ (4)

Acta Cryst., C**42**, 520-523.

1, P321, 6.790, 7.972, Z = 2, R = 0.045. 2, Pbca, 12.102, 12.092, 10.760, Z = 8, R = 0.060. **3**, I$\bar{4}$2d, 10.783, 20.599, Z = 8, R = 0.056. **4**, R$\bar{3}$m, 7.2921, 18.332, Z = 3, R = 0.043.

Structures (1) and (2) are related to the perovskite structure, (3) and (4) are of the inverse pyrochlore type. In structures (1)–(3) Te^{IV} atoms are octahedrally coordinated by oxygens and holes, the number of holes varying from 1 to 3. In $Cs_2Te_4O_{12}$ the Te^{IV} atom is at the centre of a very nearly regular octahedron of oxygens, an exceptional coordination for tetravalent Te. In this octahedron the length of the Te—O bonds is 2·104 (7) Å.

Bond lengths (Å) *and angles* (°) *around* Te *atoms*

(1) Cs_2TeO_3			
Te—O (3×)	1·84 (2)	O—Te—O$^{i}_{1}$	101 (1)
(2) $Cs_2Te_2O_5$			
Te(1)—O(2)	1·83 (1)	O(2)—Te(1)—O(3)	100·9 (5)
—O(3)	1·84 (1)	O(2)—Te(1)—O(5)	98·2 (4)
—O(5)	1·98 (1)	O(3)—Te(1)—O(5)	99·2 (4)
Te(2)—O(4)	1·83 (1)	O(1)—Te(2)—O(4)	101·2 (4)
—O(1)	1·84 (1)	O(1)—Te(2)—O(5)	98·8 (4)
—O(5)	1·97 (1)	O(4)—Te(2)—O(5)	98·3 (4)
(3) $Cs_2Te_4O_9$			
Te(1) O(3)	1·86 (1)	O(3)—Te(1)—O(2)	97·6 (6)
—O(2)	1·87 (1)	O(3)—Te(1)—O(4)	98·1 (6)
—O(4)	2·02 (1)	O(3)—Te(1)—O(2_1)	89·6 (5)
—O(2^{ii}_1)	2·31 (1)	O(2)—Te(1)—O(4)	89·9 (6)
		O(2)—Te(1)—O(2_1)	88·7 (5)
		O(4)—Te(1)—O(2_1)	172·3 (5)
Te(2)—O(1)	1·84 (2)	O(1)—Te(2)—O(4_1)	98·1 (6)
—O(4^{iii}_1)	1·96 (1)	O(1)—Te(2)—O(5)	95·1 (6)
—O(5)	1·98 (1)	O(1)—Te(2)—O(3)	82·5 (6)
—O(3)	2·40 (1)	O(1)—Te(2)—O(1_1)	92·2 (5)
—O(1^{iv}_1)	2·51 (1)	O(4_1)—Te(2)—O(5)	82·3 (6)
		O(4_1)—Te(2)—O(3)	84·2 (5)
		O(4_1)—Te(2)—O(1_1)	169·3 (5)
		O(5)—Te(2)—O(3)	165·8 (4)
		O(5)—Te(2)—O(1_1)	94·2 (5)
		O(3)—Te(2)—O(1_1)	99·9 (5)
(4) $Cs_2Te_4O_{12}$			
Te(1)—O(2)	(6×) 2·104 (7)	O(2)—Te(1)—O(2^{vi}_2)	89·7 (2)
Te(2)—O(2^{v}_1)	(2×) 1·886 (6)	O(1)—Te(2)—O(1^{vii}_1)	86·0 (2)
—O(1)	(4×) 1·941 (6)	O(1)—Te(2)—O(2_1)	88·6 (3)

Symmetry code: (i) $1-y, x-y, z$; (ii) $-y, x, -z$; (iii) $x, \frac{1}{2}-y, \frac{1}{4}-z$; (iv) $1-y, x, -z$; (v) $\frac{2}{3}-x, \frac{1}{3}-2x, \frac{1}{3}-z$; (vi) $2x, x, 1-z$; (vii) $2x, x, -z$.

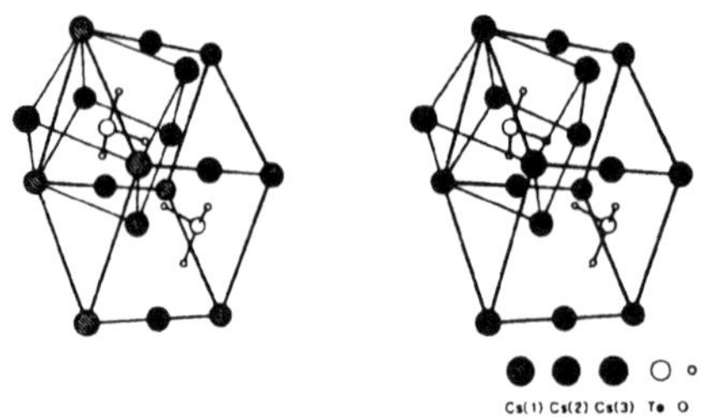

Cs_2TeO_3. Outlines of unit cell (heavy) and subcell (light) are drawn.

	x	y	z
Cs(1)	0·0	0·0	0·0
Cs(2)	0·0	0·0	0·5
Cs(3)	0·3333	0·6667	0·3589
Te	0·6667	0·3333	0·1644
O	0·837	0·611	0·269

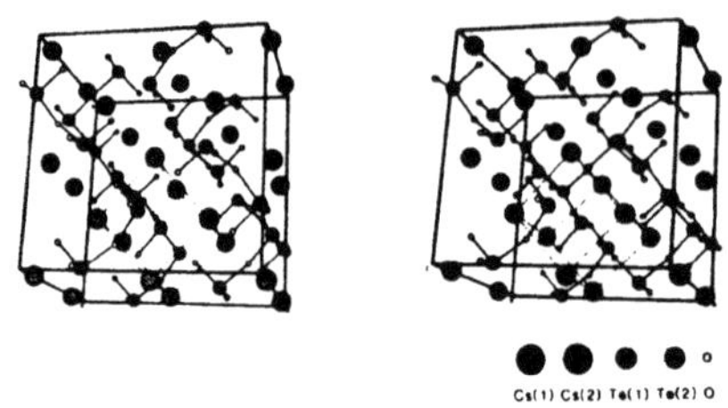

$Cs_2Te_2O_5$. Outlines of unit cell (heavy) and subcell (light) are drawn.

	x	y	z
Cs(1)	0·24742	0·24601	0·24745
Cs(2)	0·49552	0·49488	0·21370
Te(1)	0·26329	0·00935	−0·00048
Te(2)	0·48853	0·23672	0·00289
O(1)	0·4980	0·2267	0·1730
O(2)	0·1534	0·1130	−0·0106
O(3)	0·2744	−0·0078	0·1690
O(4)	0·3857	0·3464	−0·0148
O(5)	0·3913	0·1089	−0·0280

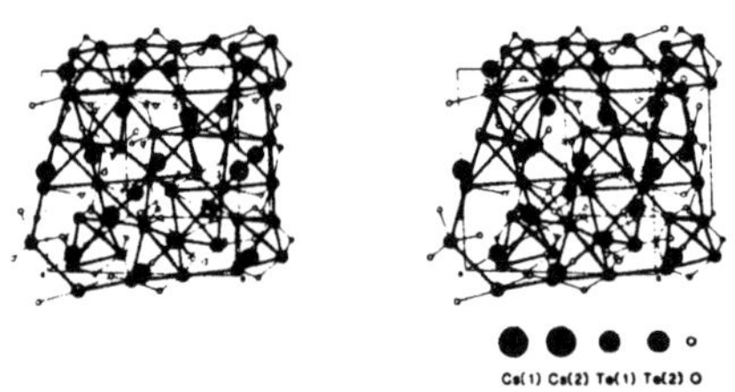

$Cs_2Te_4O_9$. For clarity only the part from $z = -0{\cdot}05$ to $z = 0{\cdot}65$ is drawn. To show the correspondence with $Cs_2Te_4O_{12}$ the Te–Te tetrahedra are indicated by a heavy outline.

	x	y	z
Cs(1)	0·7205 (2)	0·25	0·125
Cs(2)	0·0	0·5	0·00586
Te(1)	0·1250	0·1214	0·06183
Te(2)	0·3655	0·3905	0·07226
O(1)	0·297	0·506	0·0170
O(2)	0·171	0·069	−0·0213
O(3)	0·172	0·286	0·0533
O(4)	0·286	0·054	0·0975
O(5)	0·5	0·5	0·1035

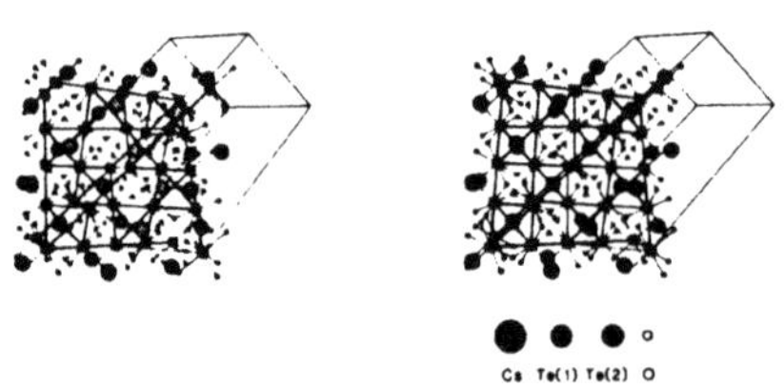

$Cs_2Te_4O_{12}$. The contents of the pseudo-cubic *F* cell and the edges of the hexagonal cell are shown. Te–Te tetrahedra are indicated by a heavy outline.

	x	y	z
Cs	0·0	0·0	0·12200
Te(1)	0·0	0·0	0·5
Te(2)	0·5	0·0	0·0
O(1)	0·2036	−0·2036	−0·0172
O(2)	0·1364	−0·1364	0·4341

CAESIUM POTASSIUM TELLURATE(VI)
$Cs_2K_2TeO_5$

J. Less-Common Metals, **124**, 29-40.

$P4_2/mnm$, 9.162, 9.592, Z = 4, R = 0.096. Isolated trigonal bipyramidal TeO_5^{4-} ions, linked by 6- and 8-coordinate Cs and 6-coordinate K. Te-O = 1.88-1.94(1) A.

LEAD(II) TELLURATE(IV) (TETRAGONAL)
$PbTeO_3$

Acta Cryst., C**42**, 1688-1690.

$P4_1$, 5.304, 11.900, Z = 4, R = 0.048. TeO_3^{2-} trigonal pyramids, linked by 6-coordinate Pb^{2+} to form channels along **c**.

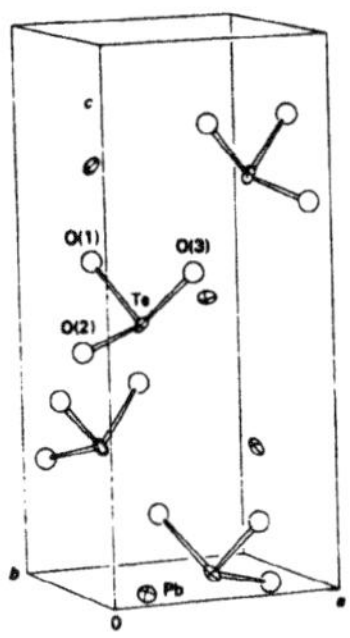

	x	y	z
Pb	0,2449	0,2491	0,0
Te	0,2506	0,2527	0,4892
O(1)	0,127	0,479	0,594
O(2)	−0,062	0,124	0,453
O(3)	0,396	0,017	0,589

Te–O(1)	1,85 (2)	Te–O(1^{i})	2,92 (2)
Te–O(2)	1,84 (2)	Te–O(3^{ii})	2,87 (2)
Te–O(3)	1,89 (2)		
Pb–O(1^{iii})	2,45 (2)	Pb–O(1^{iv})	2,69 (2)
Pb–O(2^{v})	2,27 (2)	Pb–O(2^{i})	2,69 (2)
Pb–O(3^{vii})	2,49 (2)	Pb–O(3^{vi})	2,60 (2)
O(1^{iii})–O(2^{v})	3,07 (2)		
O(1^{iii})–O(3^{vii})	3,19 (3)		
O(2^{v})–O(3^{vii})	3,10 (2)		
O(1)–Te–O(2)	94,5 (6)	O(1)–Te–O(3)	98,8 (8)
O(2)–Te–O(3)	105,5 (6)		
O(1^{iii})–Pb–O(2^{v})	80,8 (5)	O(2^{v})–Pb–O(3^{vi})	86,6 (5)
O(1^{iii})–Pb–O(3^{vi})	74,3 (6)	O(2^{v})–Pb–O(3^{vii})	81,1 (5)
O(1^{iii})–Pb–O(3^{vii})	80,3 (6)	O(3^{vi})–Pb–O(3^{vii})	153,2 (7)

(i) $y, -x, z-\frac{1}{4}$; (ii) $y, 1-x, z-\frac{1}{4}$; (iii) $1-y, x, z-\frac{3}{4}$; (iv) $-x, 1-y, z-\frac{1}{2}$; (v) $-x, -y, z-\frac{1}{2}$; (vi) $1-x, -y, z-\frac{1}{2}$; (vii) $-y, x, z-\frac{3}{4}$.

LEAD TELLURATES
$Pb_2[M_{1.5}Te_{0.5}]O_{6.5}$ (M = Ti, Zr, Sn, Hf)

Z. anorg. Chem., **537**, 213-218.

Fd3m, 10.3529, 10.7406, 10.5962, 10.7072, Z = 8, R = 0.014-0.047. Pb in 16(c); M/Te in 16(d); O in 48(f): x = 0.436, 0.421, 0.428, 0.424 (origin at $\bar{3}$m) and 8(a). Pyrochlore structures, with six-coordination for M/Te and 7-coordination (plus stereochemically-active lone-pair) for Pb.

TELLURIUM(IV) SELENIUM(IV) OXIDE
Te_3SeO_8

J. Solid State Chem., **63**, 172-178.

$P\bar{1}$, 4.854, 6.902, 11.515, 102.02, 100.80, 69.84, Z = 2, R = 0.027. 12-Membered Te-O ring, with Se connected to the ends of the ring. Te-O = 1.87-2.86, Se-O = 1.68-1.72(1) A.

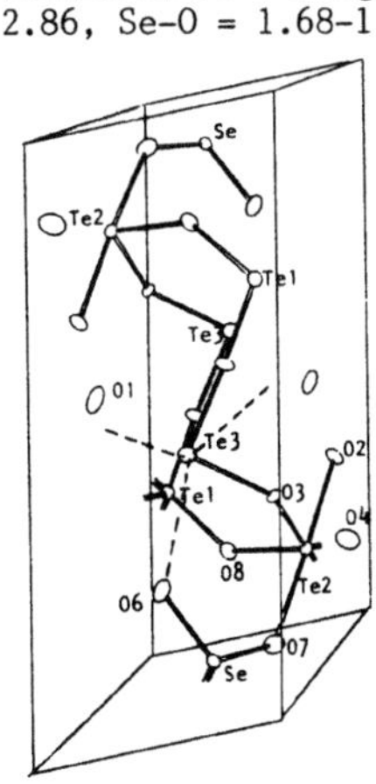

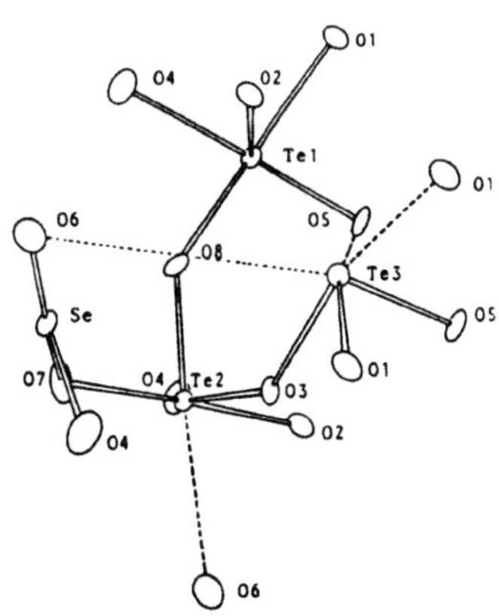

	X	Y	Z
Te_1	0.85619	0.13327	0.30347
Te_2	0.10732	0.46100	0.19217
Te_3	0.48783	0.74175	0.42561
Se	0.35118	0.82542	0.09127
O1	0.25564	0.02322	0.43257
O2	0.01969	0.35212	0.32735
O3	0.15705	0.67711	0.31960
O4	0.08057	0.04930	0.14056
O5	0.65098	0.30900	0.44532
O6	0.58480	0.77677	0.21661
O7	0.17890	0.64196	0.07014
O8	0.51464	0.30389	0.19715

MERCURY(II) TELLURATE(IV)
$HgTeO_3$

Acta Cryst., C**42**, 917-918.

$P\bar{1}$, 6.139, 7.361, 7.459, 84.76, 65.72, 87.11, Z = 4, R = 0.065. TeO_3 trigonal pyramids (mean Te-O = 1.89 A), connected by longer Te...O contacts (2.54-3.12(2) A) to form a framework of distorted TeO_6 octahedra. Hg ions have octahedral and monocapped trigonal prismatic coordinations, Hg-O = 2.07-2.94(2) A.

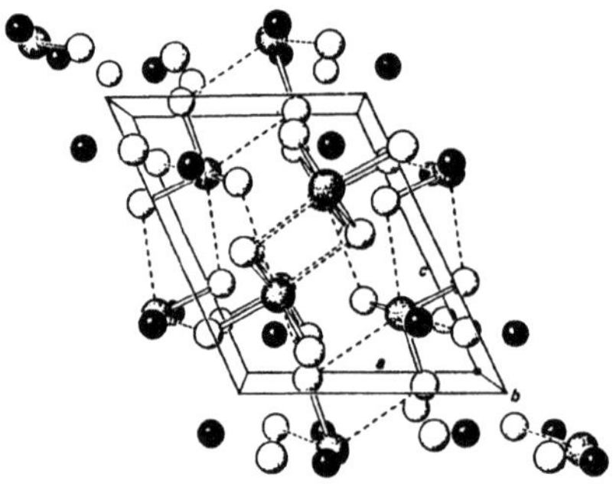

	x	y	z
Hg(1)	0·7926	0·0557	0·7664
Hg(2)	0·2066	0·2794	0·8416
Te(1)	0·6863	0·1596	0·3192
Te(2)	0·2346	0·4808	0·2433
O(1)	0·661	0·016	0·138
O(2)	0·268	0·043	0·528
O(3)	0·029	0·186	0·188
O(4)	0·644	0·302	0·715
O(5)	0·089	0·439	0·642
O(6)	0·730	0·449	0·017

TITANIUM(IV) PERCHLORATE
$Ti(ClO_4)_4$

Inorg. Chem., **25**, 1386-1390.

C2/c, 12.451, 7.814, 12.826, 108.13, at 173K, Z = 4, R = 0.051. Molecular units, with flattened dodecahedral geometry at Ti.

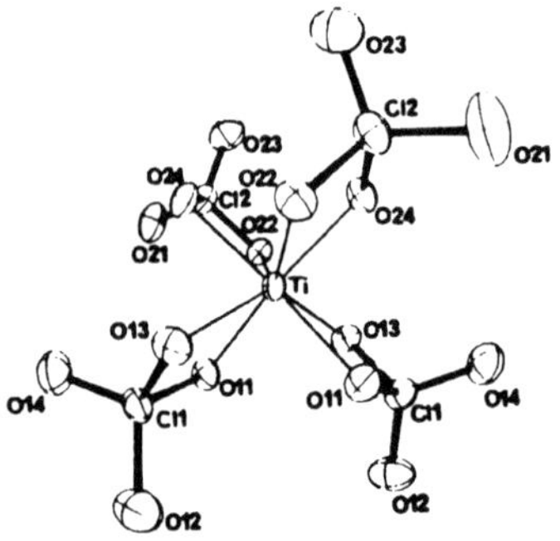

	x	y	z
Ti	0.0	0.0345	0.250
Cl1	0.8249	−0.1867	0.1640
Cl2	0.9410	0.2693	0.0849
O11	0.4296	0.3540	0.1316
O12	0.3239	0.1428	0.1917
O13	0.3523	0.4260	0.2642
O14	0.2274	0.3728	0.0833
O21	0.8687	0.2349	−0.0206
O22	0.0375	0.1448	0.1166
O23	0.9804	0.4370	0.1022
O24	0.8868	0.2172	0.1705

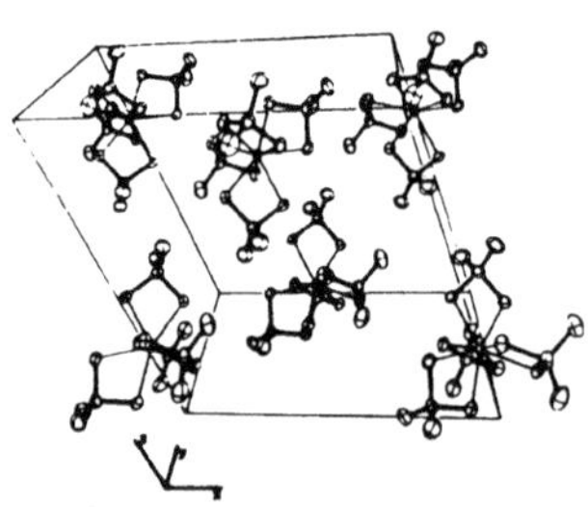

Bond Distances (Å) and Angles (deg) in $Ti(ClO_4)_4$

Ti–O11	2.058 (3)	Cl1–O13	1.507 (3)
Ti–O13	2.084 (3)	Cl1–O11	1.520 (3)
Ti–O22	2.094 (3)	Cl2–O23	1.392 (4)
Ti–O24	2.042 (3)	Cl2–O21	1.399 (4)
Cl1–O12	1.380 (3)	Cl2–O22	1.500 (3)
Cl1–O14	1.408 (3)	Cl2–O24	1.512 (4)
O24–Ti–O22	66.3 (1)	O13–Cl1–O11	97.2 (2)
O11–Ti–O13	66.4 (1)	O23–Cl2–O21	115.1 (2)
O12–Cl1–O14	115.7 (2)	O23–Cl2–O22	110.8 (2)
O12–Cl1–O13	111.2 (2)	O23–Cl2–O24	110.6 (2)
O12–Cl1–O11	110.5 (2)	O21–Cl2–O22	110.9 (2)
O14–Cl1–O11	110.3 (2)	O21–Cl2–O24	110.6 (2)
O14–Cl1–O13	110.4 (2)	O22–Cl2–O24	97.4 (2)

SILVER PERCHLORATE (CUBIC)
$AgClO_4$

Z. Kristallogr., **175**, 283-297.

Fm3m, 7.03 A (at 170°C), Z = 4. 4 Ag in 4(b): 1/2,1/2,1/2; 4 Cl in 4(a): 0,0,0; 16 O in 96(j): 0,0.120,0.160. Orientationally disordered perchlorate groups.

STRONTIUM BROMATE MONOHYDRATE
BARIUM BROMATE MONOHYDRATE
BARIUM IODATE MONOHYDRATE
LEAD(II) CHLORATE MONOHYDRATE
LEAD(II) BROMATE MONOHYDRATE
$M(XO_3)_2.H_2O$

Z. anorg. Chem., **535**, 31-38.

C2/c, Z = 4, R = 0.019-0.057. Trigonal pyramidal anions, linked by 11-coordinate metal ions and hydrogen bonding via the water molecule.

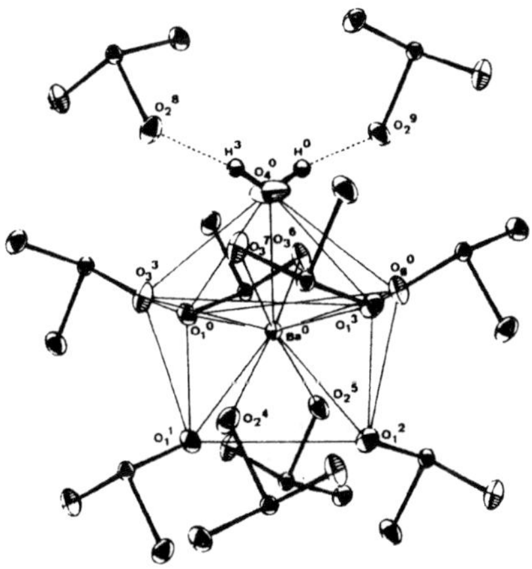

	a (pm)	b	c	β/°
$Sr(BrO_3)_2 \cdot H_2O$	887,0	761,1	936,9	91,82
$Ba(BrO_3)_2 \cdot H_2O$	907,1	789,6	963,4	93,25
$Ba(IO_3)_2 \cdot H_2O$	904,8	798,7	991,1	92,13
$Pb(ClO_3)_2 \cdot H_2O$	881,1	774,4	913,1	93,13
$Pb(BrO_3)_2 \cdot H_2O$	895,2	770,9	937,9	92,20

	x	y	z
$Sr(BrO_3)_2 \cdot H_2O$			
Sr	0,0	0,40078	0,25
Br	0,25461	0,27408	0,55595
O(1)	0,0752	0,3369	0,5601
O(2)	0,2607	0,0913	0,6538
O(3)	0,2704	0,2050	0,3892
O(4)	0,0	0,0696	0,25
$Ba(BrO_3)_2 \cdot H_2O$			
Ba	0,0	0,39254	0,25
Br	0,25100	0,27695	0,55733
O(1)	0,0770	0,3401	0,5606
O(2)	0,2565	0,0991	0,6483
O(3)	0,2641	0,2188	0,3931
O(4)	0,0	0,047	0,25
$Ba(IO_3)_2 \cdot H_2O$			
Ba	0,0	0,39129	0,25
I	0,25927	0,28374	0,5618
O(1)	0,0660	0,3406	0,5633
O(2)	0,2580	0,0936	0,6617
O(3)	0,2620	0,1997	0,3947
O(4)	0,0	0,0530	0,25
H	0,073	−0,003	0,224
$Pb(ClO_3)_2 \cdot H_2O$			
Pb	0,0	0,37054	0,25
Cl	0,2432	0,2692	0,5525
O(1)	0,0837	0,335	0,5538
O(2)	0,2482	0,105	0,6353
O(3)	0,2663	0,228	0,3944
O(4)	0,0	0,060	0,25
$Pb(BrO_3)_2 \cdot H_2O$			
Pb	0,0	0,3837	0,25
Br	0,2491	0,2713	0,5554
O(1)	0,073	0,338	0,554
O(2)	0,253	0,090	0,652
O(3)	0,267	0,204	0,389
O(4)	0,0	0,061	0,25

SODIUM HYDROGEN PERIODATE HYDRATE
$Na_5HI_2O_{10}.14H_2O$

Z. anorg. Chem., **538**, 159-165.

P$\bar{1}$, 8.450, 8.533, 9.066, 76.96, 62.94, 86.98, Z = 1, R = 0.063. Dimeric $H_2I_2O_{10}^{4-}$ anion containing two edge-sharing octahedra, as in a related K compound (**30A**, 350; **43A**, 296) [in the present structure, this requires an additional H atom; the atomic positional parameter table lists one of the water H atoms at 1/2,0,0]. Na ions have 6-coordinations, and there is an extensive hydrogen bond system.

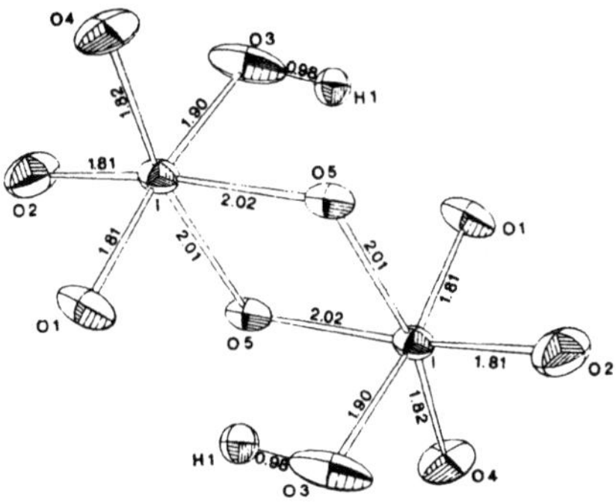

POTASSIUM HYDROGEN IODATE CHLORIDE
$K_2H(IO_3)_2Cl$

Wuli Xuebao [Acta Physica Sinica], **35**, 1102-1107.

$Pca2_1$, 15.032, 6.574, 8.571, Z = 4, neutron radiation, R = 0.033. Pyramidal IO_3 groups, linked in pairs by an O-H...O hydrogen bond (2.603 A), with Cl^- and K^+ ions.

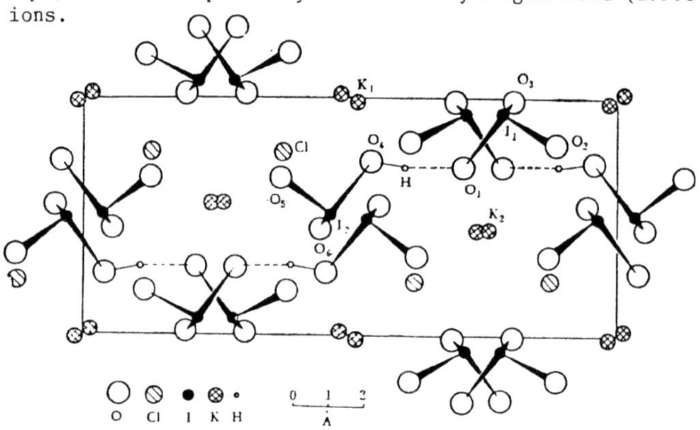

DECA-DODECASIL 3R
$120SiO_2.6M(10).9M(12).6M(19)$, M(10) = empty, M(12) = N_2,
M(19) = 1-aminoadamantane

Z. Kristallogr., **175**, 93-104.

$R\bar{3}m$, 13.860, 40.891, Z = 1, R = 0.197 (twinned crystal).

Its crystal structure is built by corner-sharing $[SiO_4]$ tetrahedra. The tetrahedra are connected to pseudohexagonal layers of face-sharing pentagonal dodecahedra ($[5^{12}]$ cages). These layers are stacked in ABCABC sequence and interconnected by additional $[SiO_4]$ tetrahedra which form six-membered rings between the layers. Thus two new types of cage arise, another small decahedron, $[4^3 5^6 6^1]$ cage and a large 19-hedron, $[4^3 5^{12} 6^1 8^3]$ cage, the latter housing the 1-aminoadamantane guest molecule.

The 19-hedra are interconnected through 8-membered rings (pore width 4.5 Å), forming a two-dimensional pore system between the dodecahedral layers. Untreated deca-dodecasil 3R is a member of the clathrasil family because the guest molecules trapped during synthesis are too large to pass through the 8-membered rings of the 19-hedra. However, on thermal treatment of deca-dodecasil 3R the guest molecules are decomposed and the fragments are driven out of the cages, transforming the clathrasil to a phase possessing zeolitic properties.

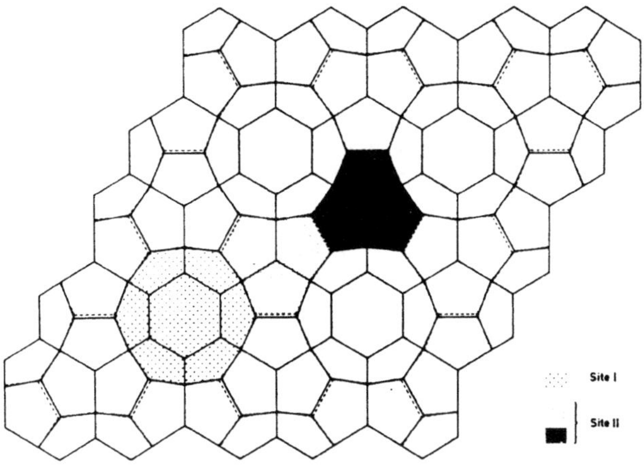

DODECASIL 1H
$34SiO_2.3M.2M'.1M''$

J. Inclusion Phenom., **4**, 85-91.

P6/mmm, 13.825, 11.189, R = 0.054. Framework of corner-sharing SiO_4 tetrahedra, with 5^{12}, $4^3 5^6 6^3$, and $5^{12} 6^8$ cages, the latter containing disordered 1-aminoadamantane molecules (M''). M and M' = N_2

SODIUM ORTHOSILICATE and ORTHOGERMANATE
Na_4SiO_4, Na_4GeO_4

Mh. Chem., **117**, 793-797.

$P\bar{1}$, a = 5.576, 5.688, b = 5.576, 5.701, c = 8.393, 8.583 A, α = 80.92, 81.32, β = 71.84, 71.50, γ = 67.44, 67.95°, Z = 2. Structures as previously described (**48**A, 336; **52**A, 183), with the silicate now described in terms of the reduced triclinic cell. The materials are isotypic, although an Na(3)-O(3) distance of 3.07 A in the germanate indicates only 4-coordination for Na(3), compared to 5-coordination (Na(3)-O(3) = 2.82 A) in the silicate.

RUBIDIUM MAGNESIUM PENTASILICATE
$Rb_2MgSi_5O_{12}$

Z. Kristallogr., **175**, 1-7.

Ia3d, 13.530, Z = 8, powder data. Rb in 16(b): 1/8,1/8,1/8; Si,Mg in 48(g): 1/8,0.665,0.585; O in 96(h): 0.100,0.127,0.717. Pollucite-type structure (**32**A, 480; **34**A, 370; **51**A, 326), with a three-dimensional framework of corner-sharing $(Mg,Si)O_4$ tetrahedra, with 12-coordinate Rb in large cavities. Mg,Si-O = 1.61-1.63, Rb-O = 3.29, 3.60(2) A.

SODIUM CALCIUM HEXASILICATE

$Na_4Ca_4[Si_6O_{18}]$

Acta Cryst., C**42**, 934-937.

$P3_121$, 10.464, 13.168, Z = 3, R = 0.031. Distortion of the $Na_6Ca_3[Si_6O_{18}]$ structure (**52**A, 322), with vacant sites caused by replacement of 2Na by Ca.

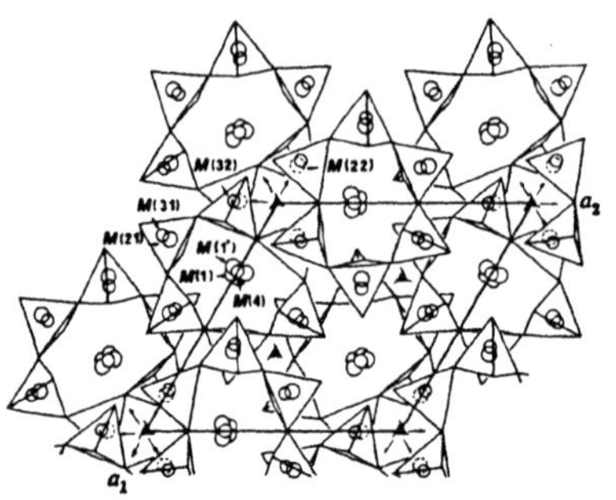

	Position	Occupancy	*x*	*y*	*z*
O(11)	3(*b*)		0·1671	0	5/6
O(12)	3(*b*)		0·5555	0	5/6
O(13)	6(*c*)		0·3442	0·2842	0·8382
O(14)	6(*c*)		0·5849	0·2643	0·8199
O(21)	6(*c*)		0·2468	0·1535	0·6622
O(22)	6(*c*)		0·4782	0·2472	0·0032
O(23)	6(*c*)		0·5513	0·1104	0·6540
O(31)	6(*c*)		0·0652	0·1853	0·7949
O(32)	6(*c*)		0·5965	0·4991	0·8880
O(33)	6(*c*)		0·7990	0·2184	0·7763
Si(1)	6(*c*)		0·1986	0·1526	0·7780
Si(2)	6(*c*)		0·4996	0·3234	0·8958
Si(3)	6(*c*)		0·6243	0·1485	0·7635
M(1)	6(*c*)	{ 0·59 Na	0·3086	0·9830	0·5892
M(1')†	6(*c*)	{ 0·41 Ca	0·2821	0·9704	0·6151
M(21)	6(*c*)	1·00 Na	0·5041	0·3380	0·6642
M(22)‡	3(*a*)		0·165	0·165	0
M(31)	6(*c*)	{ 0·41 Na, 0·59 Ca	0·5262	0·3704	0·1558
M(32)	3(*b*)	1·00 Ca	0·8203	0	5/6
M(4)	6(*c*)	1·00 Ca	0·3039	0	1/3

† The ratio *M*(1')/[*M*(1) + *M*(1')] is 0·286

‡ *M*(22) is vacant and the coordinates are only approximate.

BISMUTH(III) SILICATE

Bi_2SiO_5

Neues Jb. Miner.,Mh.,13-18.

$Cmc2_1$, 15.173, 5.4729, 5.3132, Z = 4, R = 0.052. Bi, 0.1648,0.2174,0.2500; Si, 0,0.6750,0.2000; O(1), 0,0.5220,0.4780; O(2), 0.0920,0.1710,0.6890; O(3), 0.2510,0.4910,0.4730. Isostructural with Bi_2GeO_5 (this volume, p.134), with SiO_3^{2-} tetrahedral chains separated by $Bi_2O_2^{2+}$ layers.

HAFNIUM SILICATE (HAFNON)

$HfSiO_4$

Z. anorg. Chem., **532**, 171-174.

$I4_1/amd$, 6.580, 5.980, Z = 4, R = 0.037. Hf in 4(a): 0,3/4,1/8; Si in 4(b): 0,1/4,3/8; O in 16(h): 0,0.0657,0.1930. Zircon-type structure (**1**, 345).

MANGANESE(II) SILICATE
$MnSiO_3$

Amer. Min., **71**, 781-785.

Tetragonal, $I4_1/a$, 11.774, 11.636, Z = 32, R = 0.052, material synthesized from rhodonite at 1500°C, 12GPa. Tetragonal garnet structure (**34**A, 260), with tetrahedral Si, ordered octahedral Mn and Si, and dodecahedral Mn sites. Si-O = 1.61-1.69 (tetrahedral), 1.76-1.82 (octahedral), Mn-O = 2.13-2.17 (octahedral), 2.16-2.69 A (dodecahedral).

Site	Equi-point	Sym-metry	Atom	X	Y	Z
D1	16f	1	Mn	0.1258	0.0079	0.2590
D2	8e	2	Mn	0.0	0.25	0.6235
Oc1	8c	$\bar{1}$	0.98 Mn +0.02 Si	0.0	0.0	0.5
Oc2	8d	$\bar{1}$	0.98 Si +0.02 Mn	0.0	0.0	0.0
T1	4a	$\bar{4}$	Si	0.0	0.25	0.375
T2	4b	$\bar{4}$	Si	0.0	0.25	0.875
T3	16f	1	Si	0.1262	0.0143	0.7597
O(1)	16f	1	O	0.0302	0.0617	0.6730
O(2)	16f	1	O	0.0465	-0.0411	0.8627
O(3)	16f	1	O	0.2224	0.1099	0.8064
O(4)	16f	1	O	0.2098	-0.0796	0.7039
O(5)	16f	1	O	-0.0649	0.1665	0.4654
O(6)	16f	1	O	-0.1034	0.2152	0.7858

CALCIUM IRON(III) ALUMINOSILICATE
$CaFeAlSiO_6$

Contrib. Mineral. Petrol., **92**, 530-535.

C2/c, 9.7797, 8.7819, 5.3685, 105.78, Z = 4, R = 0.025. M(1) = 0.18Al + 0.82Fe in 4(e): 0,y,1/4, y = 0.90571; M(2) = Ca in 4(e): y = 0.31126; T = 0.50Si + 0.41Al + $0.09Fe^{3+}$ in 8(f): 0.28841,0.09501,0.22004; O(1,2,3) in 8(f): 0.1085,0.0869,0.1300; 0.3646,0.2626,0.3203; 0.3549,0.0210,0.9839. Pyroxene structure (**2**, 130; **40**A, 288).

NICKEL SILICATE
Ni_2SiO_4

IRON(II) SILICATE
Fe_2SiO_4

Phys. Chem. Miner., **13**, 227-232.

Fd3m, a = 8.045-8.092, 8.237-8.340 A for Ni, Fe at 20-700°C, Z = 8, R = 0.013-0.033, 0.021-0.042. Spinels, x(O) = 0.3689-0.3685, 0.3658-0.3654.

ALBITE (LOW)
$NaAlSi_3O_8$

Amer. Min., **71**, 727-733.

$C\bar{1}$, 8.1151, 12.7621, 7.1576, 94.218, 116.803, 87.707, at 13K, Z = 4, neutron radiation, R = 0.022. Structure as previously described (**3**, 164, 546; **22**, 494; **27**, 676; **29**, 398; **34**A, 368; **37**A, 328; **43**A, 305; **46**A, 381), with Al in site T_1O.

ANORTHITE
$CaAl_2Si_2O_8$

J. Geophys. Res., B, **91** (B13), E91-97.

P$\bar{1}$, 8.173, 12.858, 14.170, 93.17, 115.95, 91.22, Z = 8. R = 0.042, lunar sample. Structure as previously described (**27**, 679; **37**A, 328), with some Na ordered in two cation sites.

BERYL
$Be_3Al_2Si_6O_{18}$

EUCLASE
$BeAlSiO_4OH$

Amer. Min., **71**, 977-984.

Beryl, P6/mcc, 9.214-9.127, 9.194-9.064, at 0.001-57 kbar, Z = 2, R = 0.031-0.041 at 1, 19, 36, 57 kbar. Structure as previously described (**1**, 305, 329; **13**, 370; **30**A, 427; **31**A, 221; **33**A, 480; **38**A, 362). All polyhedra are compressed at high pressure.

Euclase, $P2_1/a$, 4.780-4.730, 14.322-14.136, 4.634-4.580, 100.31-100.16, at 0.001-62 kbar, Z = 4, R = 0.040-0.048 at 1, 21, 42, 62 kbar. Structure as previously described (**3**, 148, 516; **27**, 579). Interatomic distances and angles change with increased pressure.

BERYL (ALKALI-RICH)
$[H_2O,Cs,K,Na](Be,Li)_3Al_2(SiO_3)_6$

Amer. Min., **71**, 547-556.

P6/mcc, 9.236-9.259, 9.246-9.264, at 24-800°C, Z = 2, R = 0.050, 0.093, 0.070 at 24, 500, 800°C. Structure as previously described (**1**, 305, 329; **33**A, 480; **38**A, 362; **43**A, 306), with Cs, K, and H_2O at 0,0,1/4 and Na at 0,0,0 in channels parallel to **c**.

BETALOMONOSOVITE
$Na_6Ti_4O_4(Si_2O_7)_2(H_2PO_4)_2$ (idealized)

Kristallografija, **31**, 1070-1075 [Soviet Physics - Crystallography, **31**, 633-636].

P$\bar{1}$, 5.326, 14.184, 14.47, 102.2, 95.5, 90.17 (**c** = **a** + **c** of **41**A, 387), Z = 2, R = 0.087. Structure essentially as previously described (**41**A, 387), with (001) three-layer packets, linked by Na ions and hydrogen bonds.

BIKITAITE (TRICLINIC)
$LiAlSi_2O_6.H_2O$

Neues Jb. Miner., Mh., 241-252.

P1, 8.606, 4.953, 7.599, 89.89, 114.42, 89.96°, Z = 2, R = 0.033. Slight distortion of the previous monoclinic description (**40**A, 282), but with an ordered Al/Si distribution; mean Si-O = 1.608-1.618, mean Al-O = 1.743-1.747 A.

CHABAZITE (CAESIUM EXCHANGED, HYDRATED and DEHYDRATED)
$Cs_{3.0}Ca_{0.4}Al_{3.8}Si_{8.3}O_{24}.9.5H_2O$

Zeolites, **6**, 137-141.

$R\bar{3}m$, a = 9.441, 9.397 A, α = 94.23, 93.74^o, for hydrated and dehydrated forms, R = 0.049, 0.046. Chabazite structure, with Cs mainly in site IV. Only minor changes on dehydration.

CHLORITE-Ia
$(Mg,Cr)_6(Si_3Al)O_{10}(OH)_8$

Canad. Miner., **24**, 649-654.

$C\bar{1}$, 5.329, 9.246, 14.442, 90.00, 97.08, 90.04, Z = 2, R = 0.044. Structure as previously described (**28**, 270), but with disordered tetrahedral sites (mean T-O = 1.650 and 1.652 A).

	x	y	z
T(1)	0.3968	-0.0001	0.1906
T(2)	0.8970	0.1667	0.1908
M(1)	0.0	0.0	0.0
M(2)	-0.0009	0.3334	0.00006
M(3)	0.0012	0.1666	0.50314
M(4)	0.5	0.0	0.5
O(1)	0.3590	0.0000	0.0766
O(2)	0.8588	0.1667	0.0767
OH(1)	0.3578	0.3331	0.0734
O(3)	0.1467	0.0691	0.2319
O(4)	0.6463	0.0967	0.2318
O(5)	0.9383	0.3333	0.2320
OH(2)	0.1517	0.0018	0.4291
OH(3)	0.1359	0.3368	0.4292
OH(4)	0.6415	0.1616	0.4292
H(1)	0.3670	0.3314	0.1405
H(2)	0.0968	0.0189	0.3714
H(3)	0.0816	0.3356	0.3681
H(4)	0.6181	0.1525	0.3730

CLINOPTILOLITE
$Na_3(Mg,Ca)_2(Al,Si)_{36}O_{72}.24H_2O$

HEULANDITE
$(Rb,K)_2(Ca,Sr)_4(Al,Si)_{36}O_{72}.21H_2O$

Stud. Surf. Sci. Catal., **28**, 449-456.

C2/m, a = 17.668, 17.699, b = 17.986, 17.927, c = 7.417, 7.422 A, β = 116.33, 116.47^o, Z = 1, R = 0.059, 0.060. Heulandite-type structures (**33**A, 484; **38**A, 368; **41**A, 380; **43**A, 358; **46**A, 387), with Mg in the M(4) site, and Rb in the M(3) site (35% occupancy).

CLINTONITE (XANTHOPHYLLITE)
$Ca(Mg,Al)_3(Si,Al)_4O_{10}(OH)_2$

Amer. Min., **71**, 1194-1197.

C2/m, 5.2037, 9.0126, 9.8145, 100.26, Z = 2, neutron radiation, R = 0.020. 1M mica structure (**23**, 484; **31**A, 238).

	x	y	z
Ca	0.50	0.50	0.00
M(1)	0.00	0.50	0.50
M(2)	0.50	0.32982	0.50
T	0.57049	0.16705	0.20976
O(1)	0.42771	0.00	0.15161
O(2)	0.86158	0.18853	0.15029
O(3)	0.63088	0.16856	0.38782
O(4)	0.62827	0.50	0.39774
H	0.5966	0.50	0.3001

CORDIERITE (LOW)
$Na_{0.27}(Li,Mg,Fe,Mn)_2(Al,Be)_4Si_5O_{18}.nH_2O$

Amer. Min., **71**, 746-757.

Cccm, 17.054-17.060, 9.759-9.797, 9.293-9.335, for 4 specimens, Z = 4, R = 0.019-0.021. Structure as previously described (**43**A, 310), with Al/Si ordering, and Na in the centre of a six-membered ring of O atoms.

Z. Kristallogr., **174**, 205-217.

Cccm, 17.049, 9.766, 9.324, at 100K, Z = 4, R = 0.019, 0.020, 0.020, at 100, 300, 500K. Structure as described above. (Mg,Fe)-O bond lengths increase with increasing temperature, and the oblate six-membered ring of O atoms about Na becomes slightly more circular.

DANBURITE
$CaB_2Si_2O_8$

Z. Kristallogr., **173**, 293-304 (1985).

Pnam, 8.037-8.101, 8.757-8.801, 7.722-7.786, at 25-910^oC, Z = 4. Structure as previously described (**2**, 153, 558; **23**, 467; **40**A, 283), but Ca is considered as 9-coordinate. With increasing temperature one B-O bond length increases from 1.498(1) to 1.512(3) A.

DUMORTIERITE
$(Al,Fe)_7BSi_3O_{18}$ (idealized)

Amer. Min., **71**, 786-794.

Pmcn, 11.79, 20.21, 4.69, Z = 4, R = 0.029-0.078 for 4 samples. Structure as previously described (**44**A, 305), with about 0.4 OH per formula unit.

EDINGTONITE
$BaAl_2Si_3O_{10}.3.7H_2O$

Neues Jb. Miner., Mh., 541-551.

$P2_12_12$, a = 9.552, 9.536, b = 9.676, 9.649, c = 6.515, 6.496 A, at 293, 100K, Z = 2, neutron radiation, R = 0.031, 0.046. Structure as previously described for orthorhombic edingtonite (**42**A, 404), with ordered Si/Al sites (disorder produces a tetragonal form, **51**A, 337).

First row T = 293 K, second row T = 100 K

	x	y	z
Si(1)	.00	.00	.0137
			.0133
Si(2)	−.1761	.0943	.3879
	−.1768	.0937	.3888
Al	.0924	.1714	.6267
	.0917	.1721	.6263
O(1)	.1739	.3317	.6328
	.1726	.3325	.6340
O(2)	−.0535	.1965	.4670
	−.0535	.1967	.4641
O(3)	.1985	.0375	.5377
	.1986	.0373	.5407
O(4)	.0360	.1338	.8776
	.0330	.1352	.8777
O(5)	−.1351	.0353	.1583
	−.1368	.0326	.1577

	x	y	z
Ba	.50	.00	.6305
			.6369
OW(1)	.1765	.3232	.1487
	.1771	.3220	.1461
H(1)	.1307	.2601	.0558
	.1279	.2615	.0523
H(2)	.2354	.3684	.0617
	.2287	.3750	.0496
OW(2)	.3783	.1227	−.0241
	.3769	.1240	−.0197
H(3)	.3020	.0884	.0485
	.3003	.0928	.0535
H(4)	.4159	.2041	.0341
	.4090	.2066	.0390

ENSTATITE
$MgSiO_3$

Z. Kristallogr., **176**, 159-175.

Pbca, 18.235, 8.818, 5.179, Z = 16, neutron radiation, R = 0.032. Structure as previously described (**2**, 134, 529; **9**, 254; **13**, 361; **26**, 510; **30**A, 425; **40**A, 288; **45**A, 407). Minor variations in bond distances and angles are interpreted in terms of differences in coordination.

	$\times 10^5$		
	x	*y*	*z*
Mg(1)	**37584**	**65393**	**86580**
Mg(2)	**37681**	**48693**	**35882**
Si(A)	**27167**	**34166**	**5031**
Si(B)	**47357**	**13732**	**79827**
O(1A)	**18346**	**34009**	**3471**
O(1B)	**56231**	**34034**	**80017**
O(2A)	**31091**	**50255**	**4319**
O(2B)	**43277**	**48289**	**68909**
O(3A)	**30318**	**22259**	**−16799**
O(3B)	**44762**	**19506**	**60361**

EPISTILBITE
$(Na,Ca)Al_2Si_6O_{16}.5H_2O$ (idealized)

Z. Kristallogr., **173**, 257-265 (1985).

C2, 9.101, 17.741, 10.226, 124.66, Z = 3, R = 0.06. Structure essentially as previously described (**32A**, 488; **41A**, 432), but in C2 rather than C2/m, and with 82:12 disorder of two tetrahedra, as in dachiardite (51A, 335). Al occupancies of the tetrahedral sites vary from 6 to 50%. Two extraframework cation and six water sites have been located; both cations have square antiprismatic coordination, with clusters of four antiprisms.

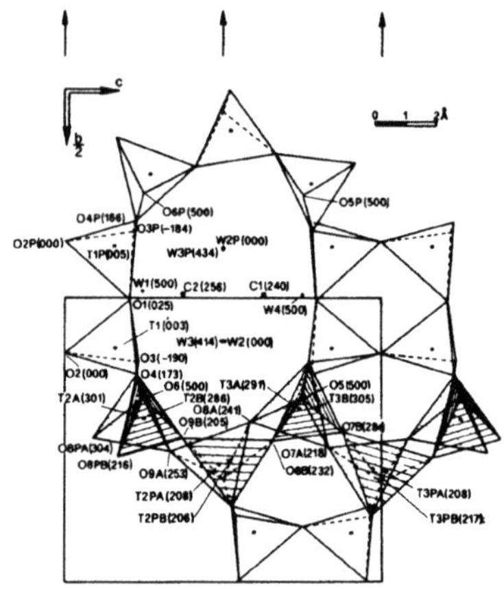

FAYALITE
Fe_2SiO_4

RUTILE
TiO_2

Physica B+C, **139-140**, 333-336.

Fayalite, Pbnm, a = 4.822-4.739, b = 10.488-9.899, c = 6.094-5.979 A, at 0.001-140 kbar, Z = 4, R = 0.074-0.094. Olivine-type structure (**28**, 265; **33**A, 468; **41**A, 382; **43**A, 312).

Rutile, P4/mnm, a = 4.593-4.517, c = 2.959-2.940 A, at 0.001-86 kbar, Z = 2, R = 0.047-0.118, x(O) = 0.3051-0.3076. Structure as previously described (**1**, 155; **45**A, 218).

FRANCISCANITE
$Mn_3V_{0.5}(SiO_4)(O,OH)_3$

Neues Jb. Miner., Mh., 493-499.

P3, 8.1518, 4.8091, Z = 2, R = 0.06. Framework of edge- and corner-sharing MnO_6 octahedra, with Si in tetrahedral holes, and V (partial occupancy) in distorted octahedra arranged in channels along **c**. Mn-O = 2.05-2.34, Si-O = 1.58-1.68, V-O = 1.77-2.30 A.

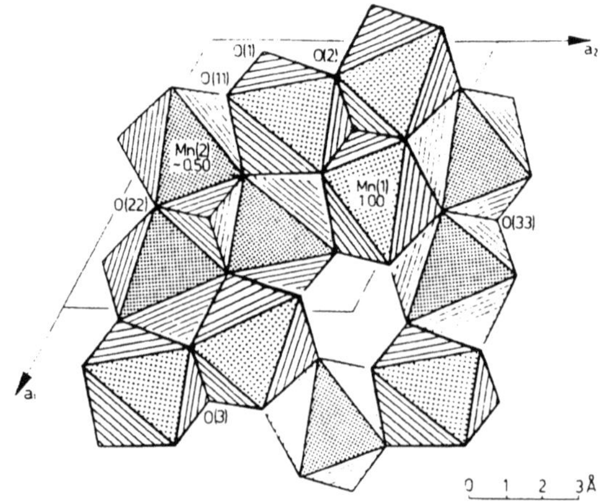

		x	y	z
Mn(1)	3d	0.5848	0.8916	0
Mn(2)	3d	0.4090	0.1151	0.4994
Si(1)	1b	1/3	2/3	0.4342
Si(2)	1c	2/3	1/3	0.9398
V(1)	1a	0	0	0.9609
V(2)	1a	0	0	0.4324
O(1)	3d	0.7889	0.8358	0.7814
O(11)	3d	0.2135	0.1630	0.2647
O(2)	3d	0.3666	0.8677	0.2945
O(22)	3d	0.6277	0.1317	0.7973
O(3)	1b	1/3	2/3	0.7844
O(33)	1c	2/3	1/3	0.2670

GANOPHYLLITE
$K_6Mn_{24}(Si,Al)_{40}O_{96}(OH)_{16}.21H_2O$

Miner. Mag., **50A**, 307-315.

A2/a, 16.60, 27.13, 50.18, 90.14, Z = 8 (subcell, a/3, b/2, c/2). Models proposed for the subcell, differing slightly from a previous model (Mineral. J., **10**, 1), and for the supercell structure.

GISMONDITE
$CaAl_2Si_2O_8.4H_2O$

Zeolites, **6**, 361-366.

$P2_1/c$, 10.011, 10.614, 9.853, 93.11, at 15K, neutron radiation, R = 0.081. Structure as previously described (**28**, 279; **37A**, 348), with ordered Si/Al distribution, and excess Si in the Al(1) site. Fully occupied Ca site, with two disordered coordinations.

Stud. Surf. Sci. Catal., **24**, 481-492.

$P2_1/c$, 10.020, 10.360, 9.830, 92.54, Z = 4, R = 0.039, 0.033 for two natural samples. Gismondite structure (**28**, 279; **37A**, 348), with ordered Si/Al distributions, and an additional partially-occupied water site.

GMELINITE (BARIUM-SUBSTITUTED)
$Ba_4Al_8Si_{16}O_{48}.nH_2O$

Kristallografija, **31**, 879-882 [Soviet Physics - Crystallography, **31**, 519-521].

$P6_3/mmc$, 13.793, 10.021, Z = 1, R = 0.071. Gmelinite structure (**31A**, 227), with a three-dimensional aluminosilicate framework. Ba ions and water molecules are statistically distributed in gaps in the framework.

		Filling coefficient of position	x	y	z
Ba1	4(f)	0.53	2/3	1/3	−0.1060
Ba2	12(k)	0.11	0.1871	0.0935	−0.0436
T(Al, Si)	24(l)	1	0.4381	0.1044	0.0928
O1	12(k)	1	0.4312	0.2156	0.0547
O2	12(k)	1	0.5662	−0.5662	0.0601
O3	12(j)	1	0.4079	0.0667	1/4
O4	12(i)	1	0.3450	0	0
W1	6(h)	0.22	0.390	0.195	3/4
W2	12(j)	0.33	0.141	0.547	3/4
W3	4(e)	0.40	0	0	0.400
W4	6(h)	0.42	0.089	−0.089	3/4
W5	12(k)	0.37	0.227	0.1135	−0.115
W6	12(j)	0.30	0.472	0.190	3/4
W7	12(j)	0,14	0 319	0 186	3/4

GONNARDITE
$(Na,Ca)_2(Si,Al)_5O_{10}.3H_2O$ (idealized)

Neues Jb. Miner., Mh., 219-228.

$I\bar{4}2d$, 13.21, 6.622, Z = 4, R = 0.022. Natrolite-type framework, with disordered Si/Al distribution and one cation site.

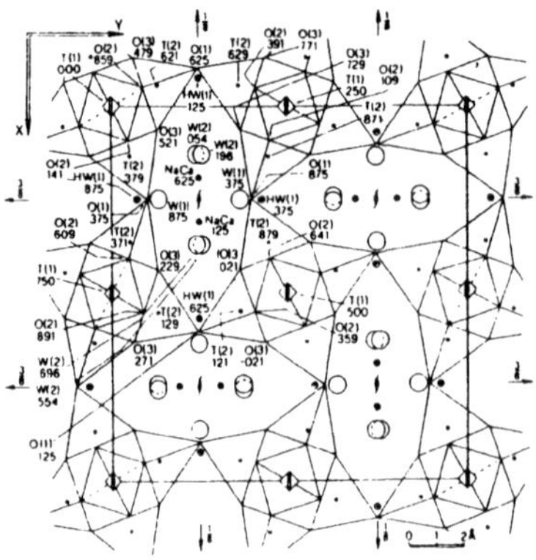

	Occ.	x	y	z
T(1)	100	0.	0.	0.
T(2)	100	0.1332	0.0536	0.3789
O(1)	100	0.2500	0.1037	0.3750
O(2)	100	0.0917	0.4497	0.1094
O(3)	100	0.4369	0.3674	0.0210
NaCa	100	0.3084	0.2500	0.1250
W(1)	100	0.2500	0.3652	0.3750
W(2)	25	0.1280	0.2434	0.0545
HW(1)	100	0.2500	0.4272	0.3750

GOOSECREEKITE
$CaAl_2Si_6O_{16}.5H_2O$

Amer. Min., **71**, 1494-1501.

$P2_1$, 7.401, 17.439, 7.293, 105.44, Z = 2, R = 0.061. Zeolitic framework of 4-, 6-, and 8-membered rings of almost completely ordered corner- and edge-sharing SiO_4 and AlO_4 tetrahedra. Three intersecting channels contain 7-coordinate Ca.

HELVITE
$Mn_8S_2[Be_6Si_6O_{24}]$

Z. Kristallogr., **173**, 305-312 (1985).

$P\bar{4}3n$, 8.294-8.190, at 1 atm-45 kbar, Z = 1, R = 0.025-0.040. Mn in 8(e): x,x,x, x = 0.1693; Si in 6(d): 0,1/2,1/4; Be in 6(c): 1/2,0,1/4; O in 24(i): 0.1418,0.1404,0.4171; S in 2(a): 1/2,1/2,1/2 (parameters at 1 atm). Structure as previously described (**38**A, 368). The BeO_4 tetrahedron shrinks abruptly at 30 kbar.

HYDROSODALITE
$Na_6(AlSiO_4)_6.8H_2O$

SODALITE
$Na_6(AlSiO_4)_6$

Zeolites, **6**, 367-372.

$P\bar{4}3n$, a = 8.848, 9.122, Z = 1, powder data. Structures as previously described (**2**, 150, 562; **32**A, 490; **44**A, 308). The hydrogen bonding of hydrosodalite is destroyed in a two-step dehydration reaction at 400-500K, with a relaxation of the aluminosilicate framework. The 9% volume expansion is reflected in the increase in Si-O-Al from 136 to 156°.

SODIUM GALLOSILICATE
$Na_8[Ga_6Si_6O_{24}](OH)_2.6H_2O$

Zeolites, **6**, 388-391.

$P\bar{4}3n$, 8.856, Z = 1, powder data. Sodalite structure (**2**, 150, 562; **32A**, 490; preceding report); Si-O-Ga = 133°.

LAIHUNITE-3M
$Fe_{1.6}SiO_4$

Amer. Min., **71**, 1455-1460.

$P2_1/b$, 4.805 (unique axis), 10.189, 17.403, α = 91.0, Z = 12, R = 0.116. An olivine-like substructure with c' = c/3 is essentially isomorphous with $M_3(PO_4)_2$ structures (**38A**, 313; **41A**, 323). In the superstructure each octahedral Fe site in the average structure splits into two or three non-equivalent sites. The formula is $Fe(II)_{0.8}Fe(III)_{0.8}\square_{0.4}SiO_4$.

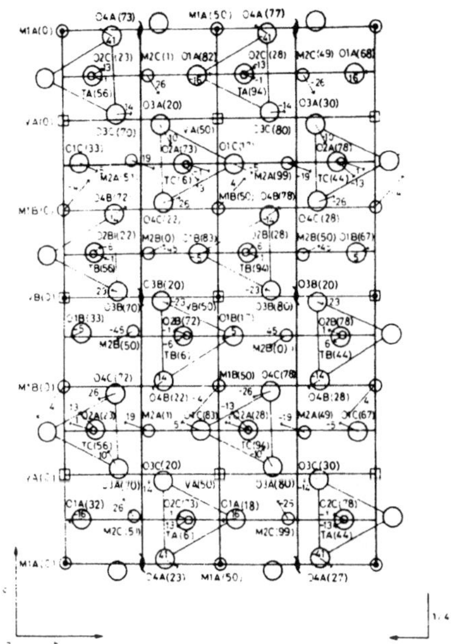

		Occupancy	x	y	z
M1A	2	1.00	0.0	0.0	0.0
M1B	4	0.907	0.0008	0.0089	0.3278
M2A	4	1.00	0.0088	0.2660	0.2488
M2B	4	1.00	−0.0045	0.2788	0.5823
M2C	4	1.00	0.0102	0.2770	0.9109
TA	4	1.00	0.0597	0.4008	0.0837
TB	4	1.00	0.0594	0.4000	0.4174
TC	4	1.00	0.0593	0.4118	0.7483
O1A	4	1.00	0.323	0.0501	0.0851
O1B	4	1.00	0.326	0.0508	0.4181
O1C	4	1.00	0.328	0.0603	0.7506
O2A	4	1.00	0.227	0.0993	0.2545
O2B	4	1.00	0.223	0.1065	0.5857
O2C	4	1.00	0.227	0.1079	0.9193
O3A	4	1.00	0.699	0.1738	0.1815
O3B	4	1.00	0.701	0.1752	0.5116
O3C	4	1.00	0.699	0.1820	0.8448
O4A	4	1.00	0.225	0.3313	0.0094
O4B	4	1.00	0.220	0.3236	0.3429
O4C	4	1.00	0.222	0.3329	0.6785
VA	4	0.00	0.5	0.5	0.1667
VB	4	0.666	0.5	0.5	0.5

LISETITE
$CaNa_2Al_4Si_4O_{16}$

Amer. Min., **71**, 1372-1377 and 1378-1383.

$Pbc2_1$, 8.260, 17.086, 9.654, Z = 4, R = 0.026. Framework of four-membered rings of tetrahedra in the (100) plane, with partial Al/Si ordering, with 7-coordinate Ca and 6-coordinate Na.

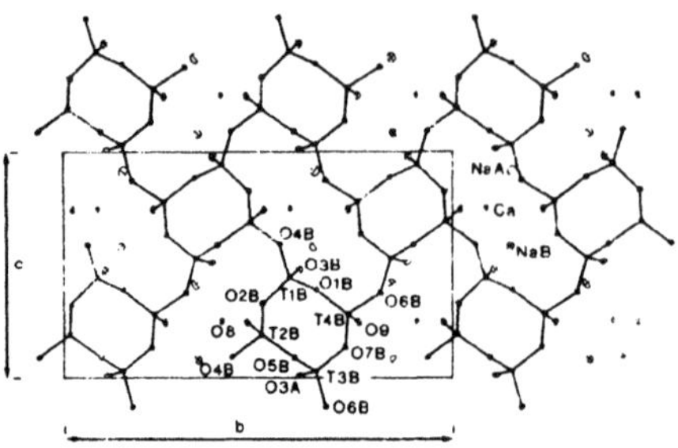

					x	y	z
Ca	Ca	0.92	Na	0.08	42	4115	2500
NaA	Na	0.94	Ca	0.06	5352	3461	4147
NaB	Na	0.95			4671	3508	811
T1A	Al	0.84	Si	0.16	6688	907	604
T2A	Al	0.11	Si	0.89	8174	76	3074
T3A	Al	0.88	Si	0.12	6969	1525	4557
T4A	Al	0.10	Si	0.90	8040	2432	2063
T1B	Al	0.08	Si	0.92	3263	866	4416
T2B	Al	0.91	Si	0.09	1722	146	1905
T3B	Al	0.12	Si	0.88	2994	1554	306
T4B	Al	0.92	Si	0.08	1915	2365	2815
O1A					7908	1672	1091
O2A					6807	180	1864
O3A					4727	1126	121
O4A					7438	645	-1032
O5A					8402	852	3997
O6A					7624	1805	6223
O7A					6934	2326	3434
O1B					2092	1564	3878
O2B					3178	179	3259
O3B					5088	1108	4791
O4B					2583	615	5921
O5B					1611	980	899
O6B					2441	1797	-1257
O7B					3125	2305	1347
O8					-115	-221	2457
O9					-103	2641	2464

MESOLITE

$Na_2Ca_2Al_6Si_9O_{30} \cdot 8H_2O$

Acta Cryst., C**42**, 937-942.

Fdd2, 18.4049, 56.655, 6.5443, Z = 8, R = 0.032. Complete Si/Al order with the same framework topology as natrolite and scolecite; $NaO_4(H_2O)_2$ and $CaO_4(H_2O)_3$ polyhedra.

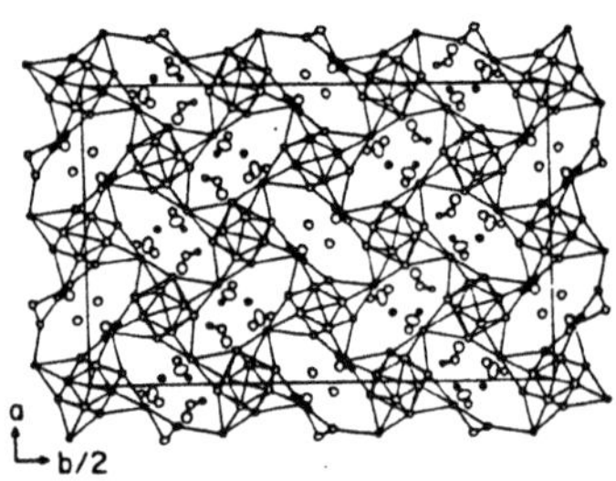

MORDENITE

$(Na,K,Ca)_{6.5}(Al_9Si_{39})O_{96} \cdot 29H_2O$

Z. Kristallogr., **175**, 249-256.

$Cmc2_1$, 18.094, 20.516, 7.524, Z = 1, R = 0.060. Structure as previously described (**26**, 515), but with asymmetry of the extraframework ions giving lower symmetry, so that a previously linear T-O-T angle is probably about 170°. Si/Al distribution is partly ordered, Ca has 8-coordination, and K/H_2O statistically occupy a 6-coordinate site.

MULLITE

$Al_2(Al_{2.8}Si_{1.2})O_{9.6}$

Amer. Min., **71**, 1476-1482.

Pbam, 7.5785, 7.6817, 2.8864, Z = 1, R = 0.025. Average structure as previously described (**16**, 344; **24**, 473; **27**, 708; **28**, 233; **34**A, 366).

	Al	T	T*	Oab	Oc	Oc*	Od
x	0.0	0.14901	0.26247	0.3590	0.5	0.4498	0.1273
y	0.0	0.34026	0.20529	0.4218	0.0	0.0505	0.2186
z	0.0	0.5	0.5	0.5	0.5	0.5	0.0
occ.	1.0	0.56 Al 0.25 Si	0.13 Al 0.06 Si	1.0	0.39	0.19	1.0

MURMANITE
$Na_2Ti_2(Si_2O_7)O_2.2H_2O$ (idealized)

Kristallografija, **31**, 82-87 [Soviet Physics - Crystallography, **31**, 44-48].

P1, 8.700, 8.728, 11.688, 94.31, 98.62, 105.62, Z = 4, R = 0.091. The structure differs somewhat from a previous description (**30**A, 463). It contains three-layer stacks parallel to (001) of Si_2O_7 tetrahedral units, TiO_6 octahedra, and 5-, 6-, and 8-coordinate Na.

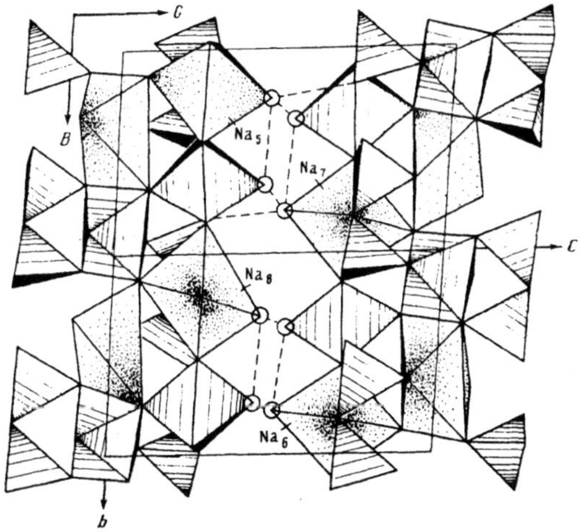

OMPHACITE
(Ca,Na)(Mg,Fe,Al)Si_2O_6

Amer. Min., **71**, 1434-1440.

C2/c, 9.501, 8.654, 5.238, 107.23, Z = 4, R = 0.023. Structure as previously described (**41**A, 394), with some vacancies in the M2 site.

	M1	M2	Si	O1	O2	O3
x	0.00000	0.00000	0.28947	0.11043	0.3608	0.35291
y	0.90540	0.30090	0.09297	0.0779	0.2601	0.0108
z	0.25000	0.25000	0.22831	0.1309	0.3005	0.0034

ORIENTITE
$Ca_8Mn_{10}(SiO_4)_3(Si_3O_{10})_3(OH)_{10}.4H_2O$

Amer. Min., **71**, 176-187.

P2mm, 9.044, 6.091, 19.031, Z = 1, R = 0.069. Structure quite similar to a previous description in Ccmm (**52**A, 340).

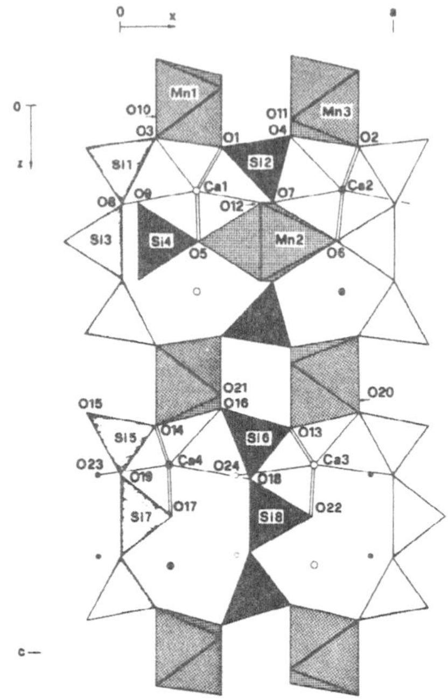

PLAGIOCLASE (An_{48})
$(Na,Ca)(Al,Si)_4O_8$

Amer. Min., **71**, 1399-1408.

$C\bar{1}$, 8.179, 12.880, 7.112, 93.44, 116.21, 90.23, Z = 4, R = 0.076, 0.056 for neutron and X-ray data. Average structure as described for albite (**45**A, 366; **46**A, 381), with Na and Ca distributed over two sites.

	X	Y	Z
	X-ray parameters		
M	0.26775	-0.01756	0.16714
M'	0.27167	0.02751	0.10131
T$_1$o	0.00676	0.16413	0.21481
T$_1$m	0.00323	0.81648	0.23087
T$_2$o	0.68625	0.10900	0.31833
T$_2$m	0.68191	0.87882	0.35629
OA1	0.00424	0.13009	0.98124
OA2	0.58250	0.99185	0.27843
OBo	0.81416	0.10547	0.19152
OBm	0.81622	0.85266	0.24473
OCo	0.01478	0.29124	0.27976
OCm	0.01469	0.68743	0.21517
ODo	0.19740	0.10866	0.38381
ODm	0.18965	0.86665	0.42927

	Neutron parameters		
M	0.26809	-0.01884	0.16802
M'	0.27180	0.02701	0.10219
T$_1$o	0.00694	0.16420	0.21481
T$_1$m	0.00346	0.81651	0.23115
T$_2$o	0.68601	0.10893	0.31813
T$_2$m	0.68193	0.87892	0.35638
OA1	0.00387	0.13025	0.98101
OA2	0.58279	0.99186	0.27866
OBo	0.81405	0.10547	0.19149
OBm	0.81627	0.85251	0.24428
OCo	0.01490	0.29097	0.27964
OCm	0.01466	0.68761	0.21573
ODo	0.19785	0.10857	0.38416
ODm	0.18964	0.86653	0.42901

PROTOPYROXENE
$(Mg,Co,Li,Sc)SiO_3$

Z. Kristallogr., **173**, 87-96 (1985).

Pbcn, a = 9.253, 9.320, b = 8.7716, 8.898, c = 5.3350, 5.346 A, at 25, 1140°C, Z = 8, R = 0.024, 0.080. Isostructural with protoenstatite (**43**A, 320).

Temperature	25°C	1140°C		25°C	1140°C
Atom			Atom		
M1 *x*	0.0	0.0	O1 *x*	0.1187	0.1197
y	0.09846	0.1018	*y*	0.0921	0.0911
z	0.75	0.75	*z*	0.0794	0.078
M2 *x*	0.0	0.0	O2 *x*	0.3752	0.3765
y	0.26355	0.2643	*y*	0.2498	0.2436
z	0.25	0.25	*z*	0.0696	0.073
Si *x*	0.29334	0.2932	O3 *x*	0.3502	0.3507
y	0.09035	0.0888	*y*	0.9822	0.9898
z	0.07291	0.0770	*z*	0.3042	0.317

SILLIMANITE
Al_2SiO_5

Amer. Min., **71**, 742-745.

Pbnm, 7.479, 7.670, 5.769, Z = 4, neutron radiation, R = 0.037. Structure as previously described (**2**, 112, 512; **27**, 707; **28**, 237; **45**A, 367), with ordered Al/Si distribution. Refinements of the kyanite and andalusite structures are also in agreement with previous results (**45**A, 367).

STAUROLITE (COBALT)
$(Fe,Mg,Co)_2Al_9Si_4O_{22}(OH)_2$ (idealized)

Amer. Min., **71**, 1461-1465 and 1466-1472.

C2/m, 7.782, 16.608, 5.648, 90.04, Z = 2, R = 0.034. Structure as previously described (**2**, 113, 514; **22**, 519, **31**A, 234; **33**A, 486; **38**A, 375; **50**A), with some Fe in Al sites.

	x	y	z
Fe	0.3909	0.0	0.2495
Si	0.1342	0.1663	0.2496
Al(1A)	0.5	0.1747	0.0
Al(1B)	0.5	0.1747	0.5
Al(2)	0.2633	0.4103	0.2504
Al(3A)	0.0	0.0	0.0
Al(3B)	0.0	0.0	0.5
U(1)	0.5	0.0	0.0
U(2)	0.5	0.0	0.5
O(1A)	0.2345	0.0	0.9642
O(1B)	0.2351	0.0	0.5356
O(2A)	0.2550	0.1614	0.0152
O(2B)	0.2547	0.1614	0.4839
O(3)	0.0015	0.0891	0.2488
O(4)	0.0211	0.2493	0.2498
O(5)	0.5271	0.0993	0.2499

Cation site	Postulated occupancy
Fe	1.63Fe, 1.44Al, 0.55Co, 0.13Ti, 0.04Zn
Si	7.46Si, 0.44Al
Al(1A)	3.55Al, 0.32Mg
Al(1B)	3.55Al, 0.31Mg
Al(2)	8.00Al
Al(3A)	0.30Mg, 0.56Al, 0.10Fe
Al(3B)	0.30Mg, 0.40Al, 0.10Fe
U(1)	0.12Fe, 0.02Mn
U(2)	0.10Fe, 0.02Mn

STELLERITE
$CaAl_2Si_7O_{18}\cdot 7H_2O$

Cryst. Res. Technol., **21**, 1029-1034.

Fmmm, 13.625, 18.284, 17.850, R = 0.043. Structure as in **41**A, 401.

TETRANATROLITE
$(Na,K)_2(Al,Si)_5O_{10}.2H_2O$ (idealized)

Kristallografija, **31**, 434-439 [Soviet Physics - Crystallography, **31**, 254-257].

$I\bar{4}2d$, 13.141, 6.638, Z = 4, R = 0.076. Natrolite-type structure (**3**, 168), but with Al/Si disorder and higher symmetry. Si,Al-O = 1.66-1.68, Na-O = 2.40-2.66(1) A.

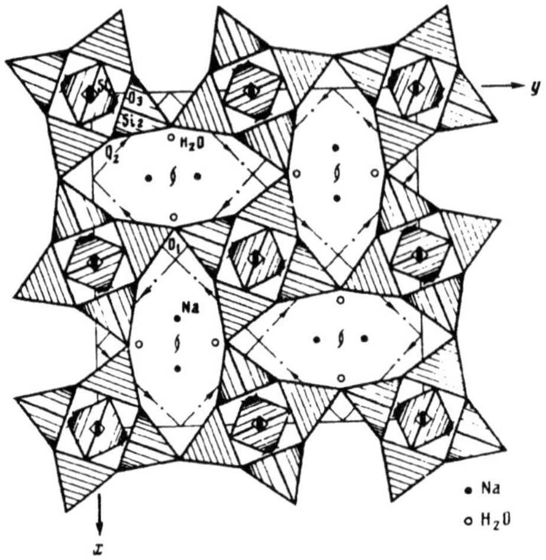

	x	y	z
$(Si, Al)_1$	0	0	0
$(Si, Al)_2$	0.0543	0.1336	0.6210
Na	0.6897	0.2500	0.1250
O_1	0.3967	0.2500	0.1250
O_2	0.1345	0.0598	0.487
O_3	0.0475	0.0931	0.862
O_4	0.131	0.2500	0.1250

TREMOLITE
$Ca_2Mg_5Si_8O_{22}(OH)_2$

Neues Jb. Miner., Mh., 360-366.

C2/m, 9.822, 18.063, 5.290, 104.66, Z = 2, R = 0.078 (film data). Structure as previously described (**2**, 131, 535; **23**, 478; **39**A, 357; **42**A, 417), with vacant A-site and statistical cation distribution in the M sites.

	x	y	z			
O(1)	0,1105	0,0864	0,2181			
O(2)	0,1185	0,1715	0,7248			
O(3)(OH)	0,1100	0	0,7163			
O(4)	0,3666	0,2483	0,7923			
O(5)	0,3469	0,1336	0,0983			
O(6)	0,3438	0,1190	0,5884			
O(7)	0,3380	0	0,2906			
T(1)	0,2804	0,0840	0,2962	0,955 Si	0,045 Al	
T(2)	0,2894	0,1709	0,8045	1 Si		
M(1)	0	0,0881	1/2			
M(2)	0	0,1769	0	0,880 Mg	0,022 Al	
M(3)	0	0	0	0,060 Fe^{2+}	0,038 Fe^{3+}	
M(4)	0	0,2772	1/2	0,100 Mg	0,845 Ca	0,055 Na

URANOPHANE (BETA)
$Ca(UO_2)_2(SiO_3OH)_2.5H_2O$

Amer. Min., **71**, 1489-1493.

$P2_1/a$, 13.966, 15.443, 6.632, 91.38, Z = 4, R = 0.047. Structure essentially as previously described (**39**A, 338), but with a revision of the structural formula on the basis of bond-strength sums.

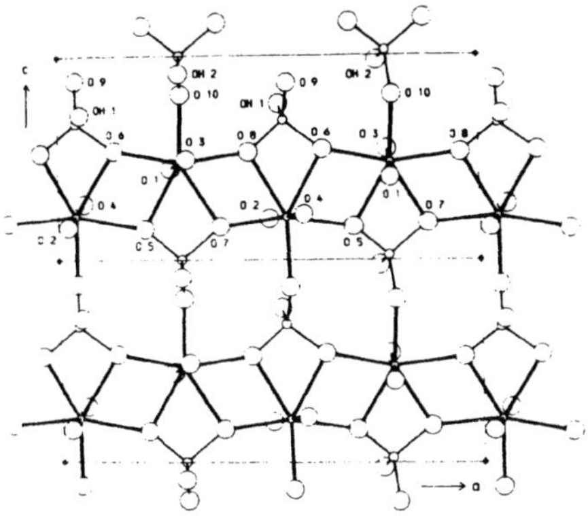

	x	y	z
U1	0.0194	0.2600	0.2631
U2	0.2313	0.7397	0.4707
Ca	0.1899	0.4848	0.2821
Si1	0.0181	0.2224	0.7293
Si2	0.2667	0.1944	0.0638
O1	0.258	0.354	0.479
O2	0.989	0.146	0.231
O3	0.727	0.877	0.429
O4	0.051	0.373	0.300
O5	0.182	0.220	0.208
O6	0.889	0.765	0.409
O7	0.358	0.200	0.223
O8	0.567	0.747	0.420
O9	0.977	0.716	0.075
O10	0.725	0.744	0.138
OH1	0.982	0.880	0.211
OH2	0.745	0.907	0.004
OW1	0.544	0.949	0.167
OW2	0.360	0.491	0.325
OW3	0.804	0.591	0.032
OW4	0.875	0.502	0.370
OW5	0.131	0.005	0.231

VESUVIANITE
$Ca_{19}Al_4Al(Al,Mg)_8Si_{18}O_{69}(OH)_9$

Amer. Min., **71**, 1483-1488.

P4/n, 15.504, 11.808, Z = 2, R = 0.055. Structure essentially as previously described (**2**, 126, 525; **41**A, 396), but in a lower symmetry space group (**50**A).

VESUVIANITE (COPPER)
$Ca_{19}Al_4(Fe,Cu)(Al,Mg,Zn)_8Si_{18}O_{68}(OH,F)_{10}$

Amer. Min., **71**, 1011-1014.

P4/nnc, 15.532, 11.776, Z = 2, R = 0.042. Structure as previously described (**2**, 126, 525; **41**A, 396), with Cu in a Jahn-Teller distorted 5-coordinate site.

YEATMANITE
$Mn_9Sb_2Zn_6Si_4O_{28}$ $[Mn_5Sb_2][Zn_6Mn_4Si_4]O_{28}$

Mineral. J., **13**, 53-64.

$P\bar{1}$, 5.554 [in text, 5.443 in abstract], 11.683, 9.134, 92.06, 101.19, 76.34, Z = 1, R = 0.10. Structure related to that of catoptrite (**42**A, 401), with a Mn_5Sb_2 octahedral sheet sandwiched by two $Zn_3Mn_2Si_2$ tetrahedral sheets; there is a small amount of Fe in one Si site.

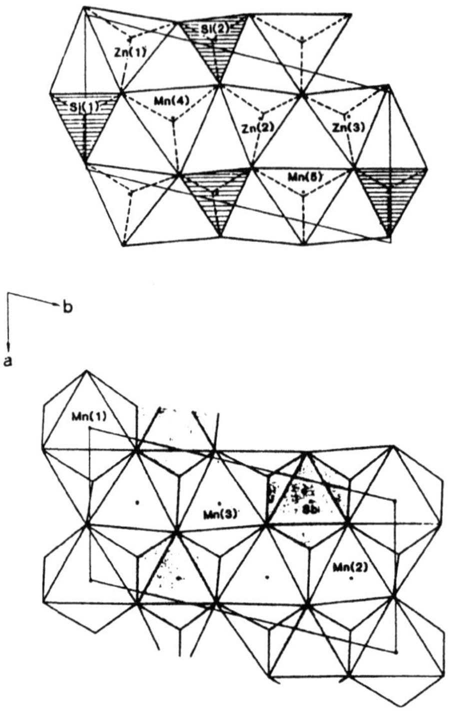

	x	y	z
Mn(1)	.0	.0	.0
Mn(2)	.5710	.8565	-.0060
Mn(3)	.2952	.4249	.0007
Sb	.1415	.7199	-.0021
Mn(4)	.6403	.2845	.3590
Mn(5)	.9337	.7081	.3465
Zn(1)	.2068	.1392	.3489
Zn(2)	.4847	.5716	.3297
Zn(3)	.3436	.8600	.3382
Si(1)	.7651	-.0060	.3086
Si(2)*	.0561	.4275	.3172
O(1)	.6906	-.0039	.1280
O(2)	.1186	.1540	.1190
O(3)	.5959	.2884	.1244
O(4)	.9953	.4108	.1293
O(5)	.3832	.5835	.1080
O(6)	.8775	.7112	.1135
O(7)	.2718	.8447	.1135
O(8)	.9368	.0027	.6445
O(9)	.2936	.1218	.6185
O(10)	.6959	.2880	.5915
O(11)	.1472	.4557	.6479
O(12)	.6486	.5615	.6367
O(13)	.9761	.6960	.6084
O(14)	.4160	.8855	.6266

*Fe was assigned to this site.

YUGAWARALITE

$CaAl_2Si_6O_{16}.4H_2O$

Z. Kristallogr., **174**, 265-281.

Pc, 6.700, 13.972, 10.039, 111.07, at 13K, Z = 2, neutron radiation, R = 0.032. Structure as previously described (**34**A, 374), with Si/Al order and one Ca site coordinated to four framework O (2.48-2.56 A) and four water molecules (2.32-2.54 A) (compare **41**A, 434). Three of the four water molecules show disorder, with multiple O or H sites, or both.

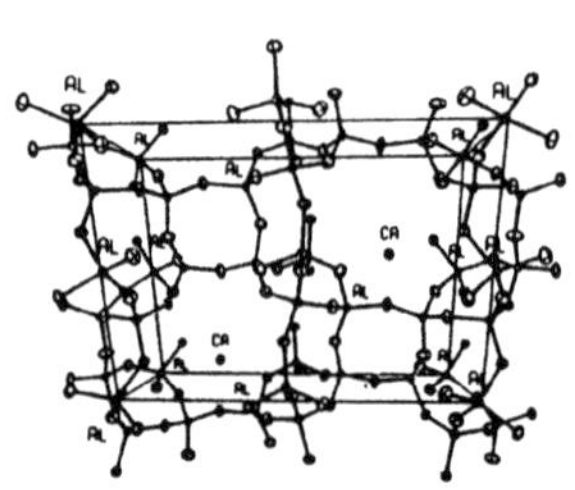

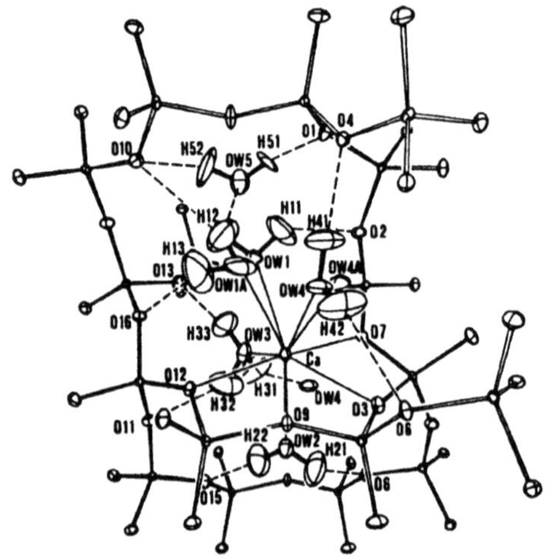

ZEOLITE A (SODIUM MAGNESIUM, DEHYDRATED)
$Na_8Mg_2Al_{12}Si_{12}O_{48}$

J. Solid State Chem., **62**, 184-190.

Fm3c, 24.4244, Z = 8, neutron powder data. Typical zeolite A structure, with two exchangeable-cation sites and a small amount of residual water.

		Population	*x*	*y*	*z*
Si(1)	96(*i*)	96	0	0.0946	0.1837
Al(1)	96(*i*)	96	0	0.1873	0.0879
O(1)	96(*i*)	96	0	0.1085	0.2470
O(2)	96(*i*)	96	0	0.1467	0.1481
O(3)	192(j)	192	0.0538	0.0577	0.1676
Mg(1)	64(*g*)	14.4	0.0952	0.0952	0.0952
Na(1)	64(*g*)	45.2	0.0996	0.0996	0.0996
Na(2)	96(*i*)	21.3	0	0.2144	0.2255
O(4)	64(*g*)	5.9	0.1433	0.1433	0.1433
H(1)	192(*j*)		0.1385	0.1385	0.1865
H(2)	192(*j*)		0.1700	0.1700	0.1215
H(3)	192(*j*)		0.1811	0.1232	0.1522
H(4)	192(*j*)		0.1232	0.1811	0.1522

ZEOLITE 4A (NICKEL-EXCHANGED)
$Ni_{3.5}Na_3Al_{12.5}Si_{12}O_{48}\cdot xH_2O$

J. Phys. Chem., **90**, 3931-3935.

Pm3m, 12.124, Z = 1, R = 0.077. Typical zeolite structure (**44**A, 320; **46**A, 396), with 3 Ni in an 8(g) site on a threefold axis (x = 0.2631) and 0.5Ni + 0.5Al in 1(a).

ZEOLITE 4A (SILVER-EXCHANGED, DEHYDRATED)
$Ag_{7.6}Na_{4.4}Al_{12}Si_{12}O_{48}$

Bull. Korean Chem. Soc., **7**, 190-193.

Pm3m, 12.311 A, R = 0.064. Typical zeolite structure, with metal sites given (including a Ag atom).

ZEOLITE D-RHO
$(D,H)_{2.1}Cs_{1.1}Al_{3.2}Si_{44.8}O_{96}$

J. Phys. Chem., **90**, 4414-4423.

Im3m, 14.850, Z = 1, neutron powder data. Si/Al in 48(i): 1/4,0.1035,1/2-y; O(1) in 48(j): 0,0.2175,0.3830; O(2) in 48(k): 0.1654,x,0.3728; 1.1 Cs in 6(b): 0,1/2,1/2; 1.6 Al in 12(e): 0.35,0,0.

ZEOLITE DZK500 and DZK650
$D_{8.5}Cs_{4.5}Al_{13}Si_{83}O_{192}$, $D_{3.5}Cs_{4.5}Al_8Si_{88}O_{192}$

Zeolites, **6**, 378-387.

Im3m, 18.633, 18.440, Z = 1, neutron powder data. Cs in 8 rings, and some non-framework Al and O.

GALLOSILICATE ZEOLITES
$Na_xGa_xSi_{192-x}O_{384}$ (x = 85, 72, 57)

J. Phys. Chem., **90**, 6858-6864.

Fd3m, a = 25.239, 25.152, 24.999 A, Z = 1, neutron powder data. Faujasite framework (**29**, 404; **32**A, 484), with disordered Si/Ga distribution, and four Na sites.

ZEOLITE L (GALLIUM, DEHYDRATED)
$K_9Ga_9Si_{27}O_{72}$

Mater. Res. Bull., **21**, 661-672 and 1539.

P6/mmm, 18.580, 7.4894, Z = 1, neutron powder data. Structure as previously described (**34**A, 371), with Ga in both tetrahedral sites.

	Site	x	y	z
Si1(T4)	12q	0.0925	0.3556	0.5
Si2(T6)	24r	0.1673	0.4991	0.2132
O1	6k	0.0	0.2753	0.5
O2	6m	0.1654	2x	0.5
O3	12o	0.2658	2x	0.2593
O4	24r	0.1014	0.4135	0.3201
O5	12o	0.4266	2x	0.2743
O6	12p	0.1446	0.4767	0.0
K1	2c	0.3333	2x	0.0
K2	2d	0.3333	2x	0.5
K3	3g	0.0	0.5	0.5
K4	6j	0.0	0.3234	0.0
K5	3f	0.0	0.5	0.0

ZEOLITE (LITHIUM A(BW))
$LiAlSiO_4.H_2O$

Z. Kristallogr., **176**, 67-73; Acta Chem. Scand., A**40**, 500-506; Stud. Surf. Sci. Catal., **28**, 443-447.

$Pna2_1$, 10.313, 8.194, 4.993, Z = 4, R = 0.051, and neutron powder data. Aluminosilicate tetrahedral framework as previously described (**40**A, 290), with Li coordinated tetrahedrally to three framework oxygens and one water molecule. Si-O = 1.610-1.648(1), Al-O = 1.723-1.765(5), Li-O = 1.91-2.00(2) A.

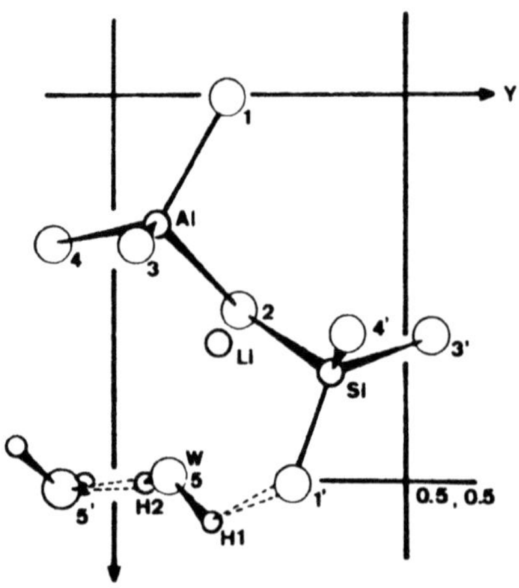

	x	(×10⁴) y	z
Li	1862	6849	2520
Al	1593	810	2500
Si	3544	3757	2492
O1	61	1584	1970
O2	2736	2198	1391
O3	1912	399	5907
O4	1804	−1008	689
O5	4891	903	−2395

ZEOLITE NaZK-4
$Na_9Al_9Si_{15}O_{48}$

Zeolites, **6**, 449-454.

Pm3m, 12.1907, neutron powder data. Zeolite structure, with Si/Al disorder, and Na in the centre of the 6-ring window.

ZEOLITE X (Cd-EXCHANGED, HYDRATED and PARTIALLY DEHYDRATED)
$Cd_{44}Al_{88}Si_{104}O_{384}.xH_2O$ (x = 233, 138)

Zeolites, **6**, 439-444.

Fd3, a = 24.797, 24.818 A, Z = 1, R = 0.082, 0.088. Zeolite structure, with two main Cd sites. Partial dehydration causes movement of some cations.

ZEOLITE Y (SODIUM, BENZENE)
$Na_{56}Al_{56}Si_{136}O_{384}.xC_6H_6$

J. Phys. Chem., **90**, 1311-1318.

Fd3m, a = 24.81-24.85, Z = 1, neutron powder data. Zeolite Y structure (**32**A, 484; **33**A, 497), with three Na sites, and two benzene sites in cavities in the zeolite.

ZEOLITE H^+-ZSM5
$H_xAlSi_{95}O_{192}$

Z. Kristallogr., **176**, 183-192.

Pnma, 20.108, 19.918, 13.392, Z = 1, powder data. Refinement of the structure (**48**A, 369).

ZEOLITE ZSM-5 (TETRAPROPYLAMMONIUM)

Zeolites, **6**, 35-38.

Pnma, 20.100, 19.959, 13.409, R = 0.067. Zeolite structure, with the organic cation at the intersection of two channels.

ZEOLITE ZSM-5 (Ni-EXCHANGED)

Stud. Surf. Sci. Catal., **28**, 415-422.

Pnma, 20.105, 19.96, 13.423. Two possible Ni sites.

TABLE I

Some structural information has also been given for the following materials (listed with abbreviated 1986 references).

Compound	Structure	Reference
Si	Study of electron distribution	Acta Cryst., A42, 271
Ferrifayalite, $Fe_{1.5}SiO_4$	Electron microscope studies of superstructures	Ibid., B42, 17
$Na_2S_2O_3$ $MgS_2O_3.6H_2O$	Charge density studies (see also 50A; 51A, 313)	Ibid., B42, 26
$CaUO_{3.5}$	Defect structure proposed	Ibid., B42, 141
$(Ca,Sr)_2SiO_4$	Relationship between β and γ-forms (37A, 323; 43A, 299; 51A, 328)	Ibid., B42, 146
Cuprite, Cu_2O	Charge-density studies [Structure in 1, 153, 222; 2, 279]	Ibid., B42, 201
MgO	Charge-density study	Ibid., B42, 247
CdI_2	Structural thermal transformations of polytypes	Ibid., B42, 350
Strontium silicate, Sr_2SiO_4	Study of the modulated structure in the β to α' transition	Ibid., B42, 417 and 423
$(Fe,Mn)_3(PO_4)_2$	Sarcopside, with Fe in M2 site	Amer. Min., 71, 136
Staurolite	Review of the structure	Ibid., 71, 1142
$(Fe,Zn)_2SiO_4$	Fayalite and willemite phases	Ibid., 71, 1502
Lithium heptagermanate, $Li_2Ge_7O_{15}$	Pbcn, neutron data, R = 0.067. Structure as previously described (35A, 318), with some large thermal parameters confirmed, and ascribed to large rotational motion and/or disorder	Ann. Rep. Res. React. Inst., Kyoto Univ., 19, 11-22
Orthoenstatite	Two domains. Results in agreement with previous studies	Bull. Acad. Serbe Sci. Arts, Cl. Sci. Nat. Math., Sci. Nat., 26, 79
Gaidonnayite, $Na_2ZrSi_3O_9.2H_2O$	Hydrogen bonding is described for the structure (52A, 329, where y of O(5) should be 0.0149)	Canad. Miner., 24, 417

Farringtonite Graftonite Sarcopside	Review of neutron study of cation distributions	Chem. Scripta, 26A, 115
Tungsten oxides	Review of structures	Ibid., 26, 535
Niobium and tungsten oxides	Review of structures	Ibid., 26, 547
LiF	NaCl	Chungnam Kwahak Yonguchi, 12, 59
$A(Cu,Mn)Mn_4O_{12}$, A = Na, Ca, Y, La	Perovskites	Cryst. Res. Technol., 21, 705
Kaersutite, $(Na,K,Ca)_3(Mg,Fe,Al,Ti)_5$-$(Si,Al)_8(O,OH)_{24}$	Neutron powder data	Ibid., 21, 1517
Spinel, $MgAl_2O_4$	Spinel, but space group considered to be $F\bar{4}3m$	Geokhimija, No. 7 1035
$BkCl_3$ $CfCl_3$	UCl_3 and $PuBr_3$ phases	Inorg. Chem., 25, 3779
SbSBr	Study of 22.8K phase transition	Jpn. J. Appl. Phys., Part 1, 24, 594-596 (1985)
Sr_2LnF_7	Tetragonal fluorite-related superstructure	J. Less-Common Metals, 118, 21
$LiMnRuO_4$	Spinel, a = 8.297 A, Li in tetrahedral site, Mn/Ru in octahedral site	Ibid., 118, 123
$LiFeRuO_4$	Spinel, a = 8.365 A, Li and Fe in tetrahedral and octahedral sites, Ru in octahedral site	
$Bi_{1.25}Na_{0.75}Ru_2O_7$ $Bi_{1.65}Ag_{0.35}Ru_2O_7$	Pyrochlores	Ibid., 119, 151
$Ba_8M_{0.5}Ti_2Nb_{4.5}O_{24}$, M = In, Lu, Yb $Ba_8Ti_3Nb_4O_{24}$	8H hexagonal perovskite, $P\bar{3}m1$ 8H hexagonal perovskite, $P6_3/mmc$	Ibid., 120, 203
$Ba_{11}Re_{1.75}Nb_{1.75}W_{3.5}O_{32}$	11H perovskite, $P\bar{3}m1$	Ibid., 120, 267
$Ba_7Nb_4Ti_2O_{21}$	21R perovskite, $R\bar{3}m$	Ibid., 120, 287
UBr_3	$P6_3/m$, 7.901, 4.428 A at 10K, Z = 2, neutron powder data. Structure as in 41A, 181; x(Br) = 0.2983, y(Br) = 0.3847	Ibid., 121, 151
UCl_4	Structure in 12, 166; 39A, 167. Spin-density study	Ibid., 121, 301

NpO_2	Fluorite, with high pressure phase (tentatively, Cmcm)	J. Less-Common Metals, 121, 461
NpO_2CO_3	UO_2CO_3	Ibid., 121, 477
$LiMn(Ru,Ti)O_4$	Spinel	Ibid., 124, 111
$Li_{1.67}MnRuO_4$	Tetragonal spinel	Ibid., 124, 141
$(Li,Cu)MnRuO_4$	Spinel	Ibid., 124, 155
$Pb_2(MSb)O_{6.5}$, M = Ti, Zr, Sn, Hf	Pyrochlores	J. Mater. Sci. Lett., 5, 675
$KAlF_4$	Study of antiphase domains	J. Phys. C: Solid State Phys., 19, 4623
Graphite (CrO_3, CH_3COOH intercalated)	Incommensurate intercalation of $(CrO_3)_3$ trimer, and acetic acid which links the trimers into a zigzag chain	J. Phys. Chem., 90, 2231
$(Na_2O)_{0.33}NaAlSiO_4$	Cristobalite	J. Solid State Chem., 61, 40
$RbNiCrF_6$	Pyrochlore	Ibid., 61, 188
Pb apatites	Apatite	Ibid., 61, 230
Uranyl selenites	Structural relationships are discussed	Ibid., 61, 308
$InVO_4$ (low-temperature)	α-$MnMoO_4$	Ibid., 62, 11
Zeolite A (lithium exchanged)	Carnegieite, nepheline, β- and γ-eucryptite phases	Ibid., 63, 46
$KNbO_3$-Nb_2O_5 $KNb_{13}O_{33}$	Tetragonal tungsten bronze superstructures $NaNb_{13}O_{33}$	Ibid., 63, 216
$SrFeO_x$, x = 2.5-3.0	Distorted perovskites	Ibid., 63, 237
$(Cu,Zn)Al_2O_4$ $(Cu,Mg)Al_2O_4$	Spinels	Ibid., 63, 336
$CrNbO_4.xVO_2$ $CrNbO_4.xVO_2.yMoO_2$ $FeNbO_4.2VO_2.yMoO_2$	Rutile	Ibid., 64, 81
$Cs_{0.9}Lu_3F_{9.9}$	The structure is described in $P\bar{6}m2$, a = 7.943, c = 4.227 A, Z = 1, R = 0.024, rather than Cm (51A, 139). Lu-F distances are more regular, 2.13-2.25(3) A, and the composition is revised slightly	Ibid., 64, 119

$La(OH)_2Br$ $La(OH)_2Br.1.5H_2O$	$Y(OH)_2Cl$ Hexagonal	J. Solid State Chem., 64, 200
$ThOF_2$	Tysonite-type structure proposed	Ibid., 64, 206
LiX, X = F, Cl, Br, I	NaCl	Ibid., 64, 217
$Li_{2.5}Fe_{2.5}O_4$ $Li_2Zn_{0.5}Mn_{1.5}O_4$ $Li_2Fe_{0.5}Mn_{1.5}O_4$	Spinels, with Li insertion	Ibid., 64, 240
$Na_2WO_2F_4$ LiW_3O_9F	Ordered F/O distributions deduced in these structures (49A; 51A, 161)	Ibid., 65, 27
$LaCO_3OH$	Two polymorphs	Ibid., 65, 94
$Li_2MgSn_3O_8$ $Li_2(Mg,Fe)Sn_3O_8$	Ramsdellite	Ibid., 65, 265
$(Zn,Ge)Fe_2O_4$	Spinel, x(O) = 0.38, some Fe in tetrahedral site	Ibid., 65, 309
$Ca_{17}Yb_{10}F_{64}$ Ca_2ErF_7	Fluorite-type superstructures proposed	Kristallografija, 31, 478 [Soviet Physics - Crystallography, 31, 281]
$(Bi,Mg)_2RuIrO_{6.75}$	Pyrochlore, Fd3m, a = 10.22 A, x(O) = 0.325	Mater. Res. Bull., 21, 307
$LiZnMn_2O_4$ $LiCdMn_2O_4$	Spinels, with Li in 16(c) site	Ibid., 21, 609
$(CdBi)(M,Sb)_2O_7$, M = Cr, Ga, V etc.	Pyrochlores	Ibid., 21, 727
$Bi_{38}Mo_7O_{78}$	Defect fluorite	Ibid., 21, 739
$KTaWO_6.H_2O$	Pyrochlore	Ibid., 21, 773
$[Na(H_2O)_2]_{0.25}MoO_3$	Molybdenum bronze type structure proposed	Ibid., 21, 945
Ln_3RuO_7, Ln = Nd, Sm, Eu	Fluorite-related super-structure, as for La_3NbO_7 (45A, 246)	Ibid., 21, 1103
NASICONS	Further comments on the the structure (52A, 277)	Ibid., 21, 1137
$Sr_2(Mn,Ti)_2O_{5+x}$	Oxygen-deficient perovskite	Ibid., 21, 1147
$Cu_2Cd_3S_2I_4$	Wurtzite	Mh. Chem., 117, 305
$M_x(H_2O)_yCrS_2$ M = Na, K etc.	3R, 2H, and 2O polytypes	Ibid., 117, 1387

$(Mg,Al,Fe,Cr)_3O_4$	Spinels, Fd3m, a = 8.123-8.427 A, x(O) = 0.262	Neues Jb. Miner., Abh., 155, 319
Dickite	O-H(1) dipole is parallel to (001). Structure as previously described (20, 432; 26, 504; 37A, 331)	Neues Jb. Miner., Mh., 19
$Ln(SCN)_3.nH_2O$, Ln = La-Nd, n = 7 Ln = Sm-Gd, n = 6	Ln ions have distorted square antiprismatic coordinations (capped for n = 7)	New Front. Rare Earth Sci., 1, 304
γ-$LiIO_3$	Neutron powder data give results in agreement with previous studies (49A; 52A, 319)	Physica B+C, 136, 493
$MgAl_2O_4$ Fe_3O_4	High-pressure study of these spinels	Phys. Chem. Miner., 13, 215
Ni_2SiO_4 Fe_2SiO_4	High-temperature study of these spinels	Ibid., 13, 227
Magnetite, Fe_3O_4	Inverse spinel, x(O) = 0.2548-0.2540 at 0.0001-4.44GPa	Ibid., 13, 238
Si(high-pressure)	Diamond, β-Sn, simple hexagonal, h.c.p., and b.c.c. phases	Phys. Rev., B, 34, 4679
$BaSn_2Fe_4O_{11}$	Cmcm; structure proposed from comparison with hexagonal $BaTi_2Fe_4O_{11}$ (40A, 210)	Phys. Status Solidi, A, 97, 231
Hollandites	Study of commensurate and incommensurate super-structures	Proc. Roy. Soc., A, 408, 295
Uranyl selenate dihydrate, $UO_2SeO_4.2H_2O$	$UO_2SO_4.2H_2O$-type	Radiokhimija, 28, 279
$FePO_4$	α-Quartz	Rev. Chim. Minér., 23, 99
$K_6[(H_2O)NiTiW_{11}O_{39}]$. xH_2O, x = 13, 11.5	Cubic, 21.3364, 21.9218, Z = 8. Modified Keggin-type anion	Rev. Roum. Chim., 31, 345
β''-Alumina (europium)	Eu in 9(d) sites	Solid State Ionics, 18-19, 645
NASICONS	Neutron powder study of three nonstoichiometric monoclinic materials	Ibid., 21, 213

Andalusite, Al_2SiO_5	Structure essentially as previously described (2, 110, 512; 6, 22; 26, 502; 27, 710; 45A, 367)	Yanshi Kuangwuxue Zashi, 5, 128
Zeolite NaY	Powder diffraction study of the effect of dehydration	Yingyong Kexue Xuebao, 4, 213
YOX, X = Cl, Br, I	PbFCl	Z. anorg. Chem., 532, 31
$Bi_{12}(B,Bi)O_{20}$	Unit cell corresponds to that previously given for γ-Bi_2O_3 (44A, 342)	Ibid., 533, 93
NbO_x	β-$ZrNb_{24}O_{62}$	Ibid., 535, 186
AB_2X_5, A = K, In, Tl B = Sr, Sn, Pb X = Cl, Br, I	$NH_4Pb_2Cl_5$ and $NH_4Pb_2Br_5$ types	Ibid., 536, 35
$Ln_2Mo_2O_7$	Pyrochlores	Ibid., 540/541, 177
$Bi_{12}MO_{20}$, M = Mn etc.	Sillenite-type (5, 70)	Ibid., 542, 153
Wollastonite-2M (parawollastonite), $Ca_3Si_3O_9$	Comparisons of previous studies (33A, 453; 45A, 380; 51A, 354)	Z. Kristallogr., 174, 309
Magnesium chloride, β-$MgCl_2$	Study of stacking faults	Ibid., 176, 1
$CaCuGe_2O_6$ $CuMgSi_2O_6$	Clinopyroxenes	Ibid., 176, 219
BiI_3	Study of stacking faults	Ibid., 176, 263
$NaGaH_4$ $MGaH_4$, M = K, Rb, Cs	$CaSO_4$ $BaSO_4$	Ž. Neorg. Khim., 31, 260
$H_5[PMo_{10}V_2O_{40}].36H_2O$ $H_6[PMo_9V_3O_{40}].30H_2O$	Keggin anions with orientational disorder are confirmed (46A, 266; 51A, 234)	Ž. Strukt. Khim., 27, No. 1, 187 [J. Struct. Chem., 27, 176]

ELECTRON DIFFRACTION

The following compounds have been studied by electron diffraction of the vapours (listed with abbreviated 1986 reference). Bond lengths are in A, angles in degrees.

Chlorofluorophosphoranes,	Trigonal bipyramids		Inorg. Chem., 25, 2828
$PClF_4$	P-F(ax)	1.581 - 1.597	
PCl_2F_3	P-F(eq)	1.535, 1.538	
PCl_3F_2	P-Cl(eq)	2.000 - 2.001	
PCl_4F	P-Cl(ax)	(in PCl_4F) 2.107	

$ZnCl_2$ $ZnBr_2$ ZnI_2	Linear, Zn–X = 2.072, 2.204, 2.401 A	Inorg. Chem., 25, 3163
$OsOCl_4$	C_{4v} square pyramid Os–O 1.663 Os–Cl 2.258 O–Os–Cl 108.3 Cl–Os–Cl 84.4	Ibid., 25, 3659
N,N-Difluoroamino-sulphonyl fluoride FSO_2NF_2	S–F 1.55 S–O 1.41 S–N 1.76 N–F 1.38 Angles at S 106 - 123 Angles at N 102, 104	Z. anorg. Chem., 539, 191
AlF_3 GaF_3	D_{3h} symmetry Al–F 1.626 Ga–F 1.713	Ž. Strukt. Khim., 27, No. 2, 43 [J. Struct. Chem., 27, 212]
$ZrCl_4$	Tetrahedral Zr–Cl 2.319	Ibid., 27, No. 4, 177 [Ibid., 27, 660]
InI_3	Planar In–I 2.641	Ibid., 27, No. 5, 48 [Ibid., 27, 714]

PAPERS REFERRED TO LATER YEARS

Many preliminary notes have not been reported, since fuller accounts will appear at a later date. The compounds studied, and abbreviated 1986 references, are listed below.

$In_{11}Mo_{40}O_{62}$ $In_3Mo_{11}O_{17}$	Angew. Chem., 98, 831
β-MoO_3	Chem. Comm., 336
$Mo_{36}O_{110}(NO)_4(H_2O)_{14}.52H_2O$	Ibid., 835
Zeolite CSZ-1	Ibid., 1211
Gallosilicate zeolite	Ibid., 1295 [see this volume, p. 266]
$(NH_4)_2SbF_5$	Dokl. Akad. Nauk SSSR, 283, 126 (1985)
Karoite (lithium, titanium) $Li_{0.5}Mg_{0.25}Ti_{2.25}O_5$	Ibid., 286, 1134
Elpasolite (iron) $K_2NaH_{1.5}Fe_{0.5}F_6$	Ibid., 287, 126
$K_2Tc_2Cl_6$	Ibid., 288, 389
$Li_3Fe_2(PO_4)_3$ (monoclinic)	Ibid., 288, 634

$LnVO_3$	Dokl. Akad. Nauk SSSR, 288, 881
$Ba(HSO_4)_2$	Ibid., 288, 1376
Kuliokite, $Y_4Al(SiO_4)_2(OH)_2F_5$	Ibid., 289, 1378
Cs_2PbCl_6	Ibid., 289, 1395
$Ga_2(SeO_3)_3.3H_2O$	Ibid., 291, 352
Keivyite, $Yb_2Si_2O_7$	Ibid., 291, 863
Lithosite, $K_3HAl_2Si_4O_{13}$	Ibid., 291, 1370
α-$CrPO_4$	Nature, 322, 620
β-$Zn_7Sb_2O_{12}$	Naturwissenschaften, 73, 562
$Cs_3Cu(NO_2)_5$	Z. Kristallogr., 174, 4
Vanthoffite, $Na_6Mg(SO_4)_4$	Ibid., 174, 5
TeI_4	Ibid., 174, 12
$Na_3Li(MoO_4)_2.6H_2O$	Ibid., 174, 21
Deca-dodecasil 3R	Ibid., 174, 64 [This volume, p. 246]
$Pb_6Re_4O_{17}(OH)_6$	Ibid., 174, 71
$Cs_3As_7.NH_3$	Ibid., 174, 82
$Cd_2Nb_2O_7$	Ibid., 174, 85
M_4TeO_5 (M = Li, Na, K) $Cs_2K_2TeO_5$ $K_3Li_3TeO_6$ KNa_5TeO_6	Ibid., 174, 89
KNa_2AuO_2	Ibid., 174, 90
$Li_6Tb_2O_7$	Ibid., 174, 90
$CsNO_2$	Ibid., 174, 93
$KLi_2(OH)_3$ $CsLi_2(OH)_3$ $CsLi_2(NH_2)_3$	Ibid., 174, 100
Ag_2O_3 Ag_3O_4 KO_3 RbO_3	Ibid., 174, 103
$Al_2Ge_2O_7$ $Al_4Ge_3O_{12}$ Al_2GeO_5	Ibid., 174, 111
$KBGe_2O_6$ $LiBSi_2O_6$ $LiBSiO_4$ $LiBGeO_4$	Ibid., 174, 112

$RbFe(CO)_3NO$	Z. Kristallogr., 174, 114
$KI(CN)_2$ $KI(CN)_2.KI(ICN)_2$	Ibid., 174, 124
Na_3NO_3, $(ONa_3)NO_2$	Ibid., 174, 127
$BaBiO_3$ $BaPb_{0.75}Bi_{0.25}O_3$	Ibid., 174, 140
Zeolite ZSM-22 $24SiO_2.(C_2H_5)_2NH$	Ibid., 174, 141
$H_2SO_4.xH_2O$ (x = 6.5, 8) SO_2X_2 (X_2 = F_2, FCl, Cl_2)	Ibid., 174, 143
Na_3GdCl_6	Ibid., 174, 145
$Ca_5Re_2O_{12}$	Ibid., 174, 149
$Cs_6Zn_5(MoO_4)_8$	Ibid., 174, 152
$KZrF_5.H_2O$	Ibid., 174, 157
$H_3AlF_6.6H_2O$ $H_2TiF_6.3H_2O$ $H_2TiF_6.6H_2O$	Ibid., 174, 162
$KGaGe_2O_6$ $KAlGe_2O_6$ $CsAlGe_2O_6$	Ibid., 174, 163
$In_{11}Mo_{40}O_{62}$	Ibid., 174, 188
$(Mg,Fe)_2SiO_4$	Ibid., 174, 198

SUBJECT INDEX

This index contains the names of the substances printed at the heads of the reports, and some additional entries.

METALS FORMULA INDEX

The entries are in alphabetical order by formula.

GPSR Compliance

The European Union's (EU) General Product Safety Regulation (GPSR) is a set of rules that requires consumer products to be safe and our obligations to ensure this.

If you have any concerns about our products, you can contact us on ProductSafety@springernature.com

In case Publisher is established outside the EU, the EU authorized representative is:

Springer Nature Customer Service Center GmbH
Europaplatz 3
69115 Heidelberg, Germany

Batch number: 10614744

Printed by Printforce, the Netherlands